Major Issues in International Construction Law

Major Issues in International Construction Law

Mauro Rubino-Sammartano
Troy Harris

WILEY

Published by John Wiley & Sons, Inc., Hoboken, New Jersey.

For general information on our other products and services or for technical support, please contact our Customer Care Department within the United States at (800) 762-2974, outside the United States at (317) 572-3993 or fax (317) 572-4002.

Wiley also publishes its books in a variety of electronic formats. Some content that appears in print may not be available in electronic formats. For more information about Wiley products, visit our web site at www.wiley.com.

Library of Congress Cataloging-in-Publication Data

Names: Rubino-Sammartano, Mauro, author. | Harris, Troy, author.
Title: Major issues in international construction law / Mauro Rubino Sammartano, Troy Harris,
Description: Hoboken: John Wiley & Sons Inc, 2026. | Includes bibliographical references and index.
Identifiers: LCCN 2025025562 | ISBN 9781394279159 (hardback) | ISBN 9781394279173 (adobe pdf) | ISBN 9781394279166 (epub)
Subjects: LCSH: Construction contracts. | Construction industry—Law and legislation. | Building laws.
Classification: LCC K891.B8 H36 2026 | DDC 343.07/8624—dc23/eng/20250530
LC record available at https://lccn.loc.gov/2025025562

Cover Design: Wiley
Cover Image: © Stock Studio 4477/Shutterstock

SKY10136893_010926

Table of Contents

List of Contributors

Aarta Alkarimi
Chrysalis International LLP
Dubai, United Arab Emirates
Email: alkarimi@chrysalis-llp.com

Vitor Parreiras Almeida
Mattos Filho
São Paulo, Brazil
Email: vitor.parreiras@mattosfilho.com.br

Albert Bates, Jr.
Troutman Pepper Locke
Pittsburgh, Pennsylvania, USA
Email: albert.bates@troutman.com

Rohit Bhat
Freshfields Singapore
Email: Rohit.Bhat@freshfields.com

Michele Bonomini
Studio Legale Padovan
Milan, Italy
Email: mbonomini@studiopadovan.com

Garrett J. Brown
Long International
Lenexa, Kansas, USA
Email: ProgressiveConsulting07@gmail.com

Scott Chalmers
McCarthy Tetrault
Toronto, Canada
Email: schalmers@mccarthy.ca

Virginie Colaiuta
LMS Legal LLP
London, United Kingdom
Email: virginie.colaiuta@lmslex.com

Jamey B. Collidge
Troutman Pepper Locke
Philadelphia, Pennsylvania, USA
Email: jamey.collidge@troutman.com

Paolo Daino
BonelliErede
Milan, Italy
Email: Paolo.Daino@belex.com

Lauren Davies
Vinson & Elkins LLP
London, United Kingdom
Email: ldavies@velaw.com

Camille Dupuy
Holman Fenwick Willan (HFW)
Paris, France
Email: camille.dupuy@hfw.com

Julien Fouret
Holman Fenwick Willan (HFW)
Paris, France
Email: julien.fouret@hfw.com

Shona Frame
CMS
Glasgow, United Kingdom
Email: Shona.Frame@cms-cmno.com

Giuseppe Giancarlo Franco
Fenwick Elliott
London, United Kingdom
Email: gfranco@fenwickelliott.com

Jeremy Glover
Fenwick Elliott
London, United Kingdom
Email: jglover@fenwickelliott.com

Joy Harb
Holman Fenwick Willan (HFW)
Paris, France
Email: joy.harb@hfw.com

Marlena Harutyunyan
Archipel
Geneva, Switzerland
Email: mharutyunyan@archipel.law

Roberto Hernandez-Garcia
Comad, S.C.
Mexico City, Mexico
Email: rhernandez@comad.com.mx

Kathryn Irwin
Singleton Urquhart Reynolds
Vogel LLP
Toronto, Canada
Email: kirwin@singleton.com

Amy Lacey
Howden (ex-Fenchurch Law)
London, United Kingdom
Email: amy.lacey@
howdenspecialty.com

Maude Lebois
GBS Disputes (Gaillard Banifatemi
Shelbaya Disputes)
Paris, France
Email: mlebois@gbsdisputes.com

Yoonji Lee
Freshfields
Singapore
Email: Yoonji.lee@freshfields.com

Nicholas Lingard
Freshfields
Singapore
Email: Nicholas.lingard@
freshfields.com

Alexandra Michalopoulou
Ioannis Vassardanis & Partners
Athens, Greece
Email: a.michalopoulou@ivlawfirm.gr

Giovanni Minuto
BonelliErede
Milan, Italy
Email: Giovanni.Minuto@belex.com

Thiago Fernandes Moreira
Mattos Filho
Rio de Janeiro, Brazil
Email: thiago.moreira@mattosfilho.com.br

Costanza Mosca
LMS Studio Legale
Milan, Italy
Email: costanza.mosca@lmslex.com

Marco Padovan
Studio Legale Padovan
Milan, Italy
Email: mpadovan@studiopadovan.com

Réka Àgnes Papp
Archipel
Geneva, Switzerland
Email: rpapp@archipel.law

Aline Piteres Porto
Mattos Filho
São Paulo, Brazil
Email: aline.piteres@mattosfilho.com.br

Nicolas Pralica
GBS Disputes (Gaillard Banifatemi Shelbaya Disputes)
London, United Kingdom
Email: npralica@gbsdisputes.com

Nicholas Reynolds
Singleton Urquhart Reynolds Vogel LLP
Toronto, Canada
Email: nreynolds@singleton.com

R. Bruce Reynolds
Singleton Urquhart Reynolds Vogel LLP
Toronto, Canada
Email: BReynolds@singleton.com

Keith Robertson
CMS
Glasgow, United Kingdom
Email: keith.robertson@cms-cmno.com

Mauro Rubino-Sammartano
Member of the Milan Bar, Ch.ArbFCIArb (Lifetime), former member of the Paris Bar
Milan, Italy
Email: mauro.rubino@lawfed.com

Basil Thévignot
LMS Legal LLP
London, United Kingdom
Email: basil.thevignot@lmslex.com

R. Zachary Torres-Fowler
Troutman Pepper Locke
Philadelphia, Pennsylvania, USA
Email: zach.torres-fowler@troutman.com

Eleni Tsakri
Ioannis Vassardanis & Partners
Athens, Greece
Email: e.tsakri@ivlawfirm.gr

Jennifer Varley
Bryan Cave Leighton Paisner LLP
London, United Kingdom
Email: Jennifer.Varley@bclplaw.com

Fabiola Viccei
Studio Legale Padovan
Milan, Italy
Email: fviccei@studiopadovan.com

Laura West
CMS
Edinburgh, United Kingdom
Email: Laura.West@cms-cmno.com

Geoff White
Greenwood Roche
Auckland, New Zealand
Email: gwhite@greenwoodroche.com

Nonkululeko Zondo
Vinson & Elkins LLP
London, United Kingdom
Email: nzondo@velaw.com

Mubarik Khan
McCarthy Tétrault
Toronto, Canada
Email: mubarikkhan@mccarthy.ca

1

The International Construction Contract

*Mauro Rubino-Sammartano**

Introduction

A study of the major issues in international construction inevitably starts with an analysis of the construction contract, which is the center of the construction project and is dominated by contract law.

This study is even more relevant when a contract contains a foreign element, what in large projects may derive from various sources such as the nationality of the parties and the operational center of the contractor's activities vis-a-vis the site of the works. International Construction Contract might be by itself a monograph. Being just one chapter of the book, it must focus on highlighting its key aspects.

Notion

The general definition of a construction contract is a contract by which one party (the Contractor) agrees to carry out building or engineering works for another party (the Owner or Employer) for a consideration.[1] This definition is enriched with respect to international construction contracts[2] due to their foreign element – *i.e.*, when the parties and/or the performance of the works are not all located in the same country. As set out by Art. 21 of the Hague Principles of Choice of Law in International Commercial Contracts

* Member of the Milan Bar, Ch.ArbFCIArb (Lifetime), former member of the Paris Bar.

1 Ian Duncan Wallace, *Hudson's Building and Engineering Contracts*, Sweet and Maxwell, 970.

2 Philip Lane Bruner & Patrick J. O'Connor, *Bruner & O'Connor on Construction Law* (Clark Boardman Callaghan, 2002–2023).

Major Issues in International Construction Law, First Edition. Mauro Rubino-Sammartano and Troy Harris.
Published 2026 by John Wiley & Sons, Inc.

> a contract is international unless each party has its establishment in the same State and the relationships of the parties and all other relevant elements, regardless of the chosen laws, are connected only with that State.[3]

In various legal systems, this definition is enriched by the express addition of an additional element: the contractor has the duty and right to organize what he deems necessary to carry out the works and he is entitled to operate at his own risk.

Construction, particularly international construction, has been the subject of an extensive number of textbooks and articles; only some of them may be mentioned here.[4]

---

3 Approved on March 19, 2015.

4 In the United Kingdom: Hudsons's *Building and Engineering Contracts*, Sweet and Maxwell, 10th ed. 1970; Ian Duncan Wallace, Construction Contracts, *Principles and Policies in Tort and in Contract*, Sweet and Maxwell, 1980; Ian Duncan Wallace, *The International Civil Engineering Contract*, Sweet and Maxwell, 1974; Donald Keating, *Building Contracts*, Sweet and Maxwell, 4th ed. 1978; John Uff, *Construction Law*, Sweet and Maxwell, 4th ed. 1985; Keith Pickavance, *Delay and Disruption Construction Contracts*, 2nd ed. LLP 2000. In the United States: Business Laws Inc editorial staff, *Construction Contracts Deskbook*, 2024 ed. (Business Laws Inc., 2024); Renato Nazzini, *Construction Arbitration and Alternative Dispute Resolution: Theory and practice around the world*, Informa Law 2024; Annette Davis Perrochet, *Construction Industry Formbook* 2023 ed.; Donald Charrett, *Contracts for Construction and Engineering Projects*, 2d ed. (Informa Law from Routledge, Sept. 25, 2023); James Acret & Annette Davis Perrochet, *Construction Arbitration Handbook*, (Clark Boardman Callaghan, 2023); David K. De Wolf, Keller W. Allen & Darlene Barrier Caruso, *Contract law & Practice*, 3rd ed. Thomson West, 2023; Jonathan J. Sweet, *Sweet on Construction Industry Contracts: Major AIA Documents*, 7th ed. (Wolters Kluwer Legal & Regulatory U.S., last updated Dec. 8, 2023); Philip Lane Bruner & Patrick J. O'Connor, *Bruner & O'Connor on Construction Law* (Clark Boardman Callaghan, 2002–2023); Cameron K. Andres, Ronald C. Smith & W. Ronald Woods, *Principles and Practices of Commercial Construction*, 10th ed. (Pearson, July 14, 2021); Cyril Chern, *The Law of Construction Disputes*, 3rd ed. (Informa Law from Routledge, Sept. 30, 2021); Richard Wilmot-Smith QC & Paul Darling, *Wilmot-Smith on Construction Contracts*, 4th ed. (July 20, 2021); Allen L. Overcash, Carol J. Patterson, Ross J. Altman & Stephen A. Hess, eds., *Construction Law*, 2d ed. (American Bar Association, May 7, 2020); Paul Levin (ed.), *Construction Contract Claims, Changes, and Dispute Resolution* (Asce Press, 3rd revised ed., American Society of Civil Engineers, Oct. 18, 2016); Will Hughes, *Construction Contracts*, 5th ed. (Apr. 20, 2015); Richard Wilmot-Smith QC, *Construction Contracts: Law and Practice, 2nd ed.* (OUP Oxford); Steven G.M. Stein, *Construction Law*, 8 vols. (Matthew Bender Elite Products); Justin Sweet, *Sweet on Construction Industry Contracts*, Wiley, 1987; McNeille Stokes, *International Construction Contracts*, 2nd ed. McGraw-Hill, 1981: Robert F. Cushman, Alan B. Stover, William R. Sneed, William J. Palmer, *Construction Management Form Book*, McGraw-Hill, 1983; William E. Combs, William J. Palmer, *The Handbook of Construction Accounting & Financial Management*, McGraw-Hill, 1984; William J. Palmer, Robert F. Cushman, *Businessman's Guide to Construction*, Dow Jones Books, 1980; Robert F. Cushman, James J. Myers, Stephen D. Butler and Lawrence N. Fisher, *Construction Dispute Resolution*, Formbook, Wiley, 1997; Stephen Hall, *Preparing Construction Claims*, Wiley-Blackwell 2020; John Stannard, *Delay in the Performance of Construction*

In some jurisdictions,[5] a distinction is made between construction contracts based on the nature of the contractor's involvement. Contracts where the contractor operates a business are classified as construction contracts, while those in which the contractor fulfills obligations personally (or with family members) are not considered construction contracts and are governed by different provisions.

Obligations, 2nd ed. 2018; Michael T. Callahan & John L. Hajob, *Construction Insurance: Coverage and Disputes*, 2nd ed. LexisNexis, 2023; Reg Thomas & Mark Wright, *Construction Contracts Claims*, Red Globe Press 2016. In Canada: R.E. Laidlaw, Christian Young & Angie Dick, *Engineering Law*, 5th ed. 1958; Immanuel Goldsmith, *Canadian Building Contracts*, 2nd ed., 1976; Donald L. Marston, *Law for Professional Engineers*, McGraw-Hill, 2nd ed., 1985; Douglas N. Macklem and David I. Bristow, *Mechanic's Liens in Canada*, 4th ed. 1978. In Switzerland: Peter Gauch, *Der Werkvertrag*, 3rd ed. Schulthess Verlag 1985; Carl Dürr, *Werkvertrag und Auftrag*, 3rd ed., Vira, 1983; Peter Gauch, *Der Unternehmer in Werkvertrag* 2nd ed. Zurich 1977; Peter Gauch, Justin Sweet, *Selected Problems of Construction Law*, International Approach, Freiburg/London, 1983; Werner Korinthenberg, *Erfüllung und Gewährleistung beim Werkvertrage*, Köln 1935; Mario M. Pedrazzini, *Der Werkvertrage*, Schweizerisches Privatrecht, Basel, 1977; Jean-François Perrin, *Le contract d'architecte*, Genève, 1970; Rudolf P. Schaub, *Der Engineeringvertrag*, Zurich, 1979. In Belgium: Maurice-André Flamme, *Le drois des constructeurs*, Bruxelles, 1984; André Brabant, *Le contrat international des construction*, Bruxelles, 1981. In Austria: Erich Feil/Walter Langer, *Der Werkvertrag in der Lehre und Rechtsprechung*, Prugg, 1974. In Germany: Walter Daub/Rudolf Piel/Carl Soergel/Wolfgang Steffani, *Kommentar zur VOB*, Wiesbaden 1976; Fritz Nicklisch, *Bau-und Anlagenverträge: Risiken, Haftung, Streitbeilegung (Technologie und Recht)* C.F. Muller 1984; Wolfgang Heiermann/Richard Riedl/Fritz Schwab, *Handkommentar zur VOB*, Weisbaden, 1975; Gisbert Kaiser, *Das* Mängelhaftungsrecht *der VOB/B*, Heidelberg, 1986; Horst Locher, *Das private baurecht*, 2nd ed., Munchen 1978; Werner Lorenz, *Contract for Work on Goods and Building Contracts*, in International Encyclopedia of Comparative Law, Vol. XVII Tübingen, 1980; Fritz Nicklisch/Günter Weick, *Verdingungsordnung für Bauleistungen: VOB*, Munchen 1981; Klaus Vygen/Edgar Joussen, *Bauvertragsrecht nach VOB und BGB: Handbuch des privaten Baurechts* Berlin, 1984; Ulrich Werner/Walter Pastor, *Des bauprozess*, Dusseldfor 1983. In France: Philippe Malinvaud, *Droit de la construction*, Dalloz 2000; Roger Saint-Alary, *Droit de la Construction*, Presses universitaires, 1977; Jacques Montmerle/Albert Caston, *Passation et exécution des marchés privés de travaux*, Moniteur, 1979; Georges Liet-Vaux, *Le droit de la construction*, 8th ed. Paris, 1986; Michel Huet, *Le droit de l'architecture*, Economica, 1985. In Italy: Domenico Rubino, *L'Appalto*, 4th ed, UTET, 1890; Domenico Rubino, (ed. Iudica) *Appalto*, Zanichelli 2007; Mauro Rubino-Sammartano, *Appalti di opere e contratti di servizi*, 2nd ed. Cedam, 2006; Carlo Giannattasio, *L'appalto*, 2nd ed. Giuffrè 1977; Angelo Luminoso, *Codice dell'appalto privato*, Giuffrè 2010; Maurizio De Tilla, *L'appalto privato e pubblico*, Giuffrè 1996; Vincenzo Cuffaro, *I contratti di appalto privato*, Utet 2011; Maria Costanza, ed. *L'appalto privato*, UTET, 2000. In Spain Fernando Díaz Barco, *Manual de Derecho de la Construccion* 5ª Ed. Enero 2024; Susana San Cristóbal Reales, Antonio Jesús, Alonso Timón, *Manual de derecho de la construcción*, La Ley 2023; Fernando Díaz Barco, *Manual de derecho de la construcción: adaptado al Código Técnico de la Edificación y a la Ley Reguladora de la Subcontratación en el sector de la construcción*, 3ª ed. Thomson-Aranzadi, 2008.

5 Lodovico Barassi, *Il contratto di lavoro nel diritto positivo italiano*, 2nd ed., Milan 1915–1917.

History of Construction Contracts

A construction contract is mentioned in the Code of Hammurabi, King of Babylon (2285–2242 B.C.). The Code addresses an architect who constructs a building based on consideration and specifies his liability if the building collapses. Architects, who at that time often served as both the designers and the builders, played a very important role in Egypt, Greece and in Rome and are remembered for their great achievements. Roman law has dealt with construction contracts qualifying them as *locatio operis*.[6] During the Middle Ages, the Edict of Rotary, a Longobard king, referred to construction contracts to be executed by a contractor known as magister comacinus. Construction contracts have been deeply analyzed in nineteenth century in France,[7] and addressed in the various legal systems through contract law and/or specific legislation.

The Parties

The Contractor and the Owner are the subject of other Chapters in this study, which explore not only their structure but also their respective duties.[8] This includes the various forms of cooperation among parties, such as joint ventures to bid for a project. That chapter will also address other entities closely connected to the project such as the Engineer,[9] the Designer[10] or Architect,[11] Construction managers or administrators,[12] the Resident Engineer, the Site Agent, their assistants, inspectors, and assessors. Forensic engineers and expert witnesses will also be included.[13] Based on these roles, their interrelationships must also be established.[14] The construction team deserves to be

6 Performance of works.

7 Robert Joseph Pothier, *Oeuvres, Traité du Contract de louage*, Paris, 1808.

8 Steven G.M. Stein, *Construction Law*, 8 vol. (Matthew Bender Elite Products).

9 Donald Charret, *Contracts for Construction and Engineering Projects*, cited.

10 Steven G.M. Stein, *Construction Law*, 8 vol. (Matthew Bender Elite Products); Will Hughes, *Construction Contracts*, 5th ed., April 20, 2015; Donald Charret, *Contracts for Construction and Engineering Projects*, cited.

11 Steven G.M. Stein, *Construction Law*, 8 vol. (Matthew Bender Elite Products).

12 Business Laws Inc. Editorial Staff, *Construction Contracts Deskbook*, 2024 ed. (Business Laws Inc. 2024); Will Hughes, *Construction Contracts*, 5th ed., April 20, 2015; Annette Davis Perrochet, *Construction Industry Formbook*, 2023–2024 ed. (Clark Boardman Callaghan, 2023).

13 Donald Charret, *Contracts for Construction and Engineering Projects*, cited.

14 Will Hughes, *Construction Contracts*, 5th ed., April 20, 2015; Cameron K. Andres, Ronald C. Smith & W. Ronald Woods, *Principles and Practices of Commercial Construction*, 10th ed. (Pearson, July 14, 2021).

thoroughly studied and may need to be revisited, if necessary, throughout the project's process.[15]

Distinction from Other Contracts

The construction contract is to be distinguished from various other types of contracts. This distinction is important even though, while governed by different statutory provisions, they may share some common aspects in certain circumstances.

Construction Contracts and Sale Agreements

In a construction contract, the contractor has an obligation to build for the Owner and to perform ("*facere*"), whereas in a sale agreement, the seller has a duty to transfer goods to the purchaser that he has either purchased or manufactured for sale, which constitutes a duty to give ("*dare*").[16]

Construction Contract and Periodical Supply of Goods

Construction contracts also must be distinguished from agreements whereby the parties agree on delivery of some items on a recurring or on a continuous basis.[17] This contract, which has been classified in such jurisdiction as a regular supply (*somministrazione*), is generally referred to as a supply contract.

Construction Contract and Agency

The construction contract must also be distinguished from an agreement in which one party undertakes to act on behalf of another. This type of agreement if often referred to as a *mandato*,[18] while in common law, it tends to fall under the definition of agency.[19]

---

15 Cameron K. Andres, Ronald C. Smith & W. Ronald Woods, *Principles and Practices of Commercial Construction*, 10th ed. (Pearson, July 14, 2021).

16 Domenico Rubino, *La compravendita*, 2nd ed. Giuffré, 1962; Philippe Le Tourneau, *Le contrat de vente*, Dalloz 2005; Henry Deeb Gabriel, *Contracts for the Sale of Goods*, 2nd ed. Oxford, University Press, 2022; Richard Christou, *Sale and Supply of Goods and Services*, Sweet & Maxwell 2nd ed. 2010.

17 Gastone Cottino, *Della somministrazione*, 2nd ed., in Comm. Cod. civ. Scialoja Branca, Vol. IV, 1966.

18 Benoit Kohl, *Le mandat dans la pratique*, Larcier 2014; Carlo Santagata, *Mandato. Disposizioni generali*. Zanichelli, 1998.

19 Roderick J.C. Munday, *Agency: Law and Principles*, Oxford University Press, 4th ed., 2022.

Contracts Related to Personal Estate

Another basic distinction is between works on real estate and works on personal estate. In the latter case, the object of the contract is a movable asset, and the overall framework of the contractor's work may differ significantly.

Contract for Services

Contract for services, which in many common law jurisdictions are a contract totally different from constructions contracts, belong in some other legal systems to the same type of contract, which may provide for the performance of works or of services.[20]

Various Construction Contracts

A broad term like "construction" naturally encompasses all types of construction projects. However, some of these construction contracts, despite sharing a common foundation with many others, may have a different objective. The nature of the object may indeed be different even within specific types of work.

Build, Repair, Correct, Maintain

The contract may involve either the construction of a new structure, the repair or modification of an existing one, or simply its maintenance. These contracts share the commonality of impacting the previous condition of the asset in question. While the legal framework for these various subtypes of construction contracts is generally basically the same, their differing objectives involve distinct works to be performed, which may be reflected in specific statutory, administrative, and/or or contractual provisions.

The Object of the Works

Works may pertain to construction for residential or commercial or public purposes. Major projects include skyscrapers, highways, dams, airports, factories, warehouses, and plants.

---

20 Mauro Rubino-Sammartano, *Appalti di opere e contratti di servizi*, 2nd ed. Cedam, 2006; Richard Christou, *Sale and Supply of Goods and Services*, Sweet & Maxwell 2nd ed. 2010.

The FIDIC General Conditions

Various standard contract forms are available.[21] The construction community is grateful to the International Fedcration of Consulting Engineer (FIDIC) for providing various standard conditions for contracts, which are widely used in international construction. Their series includes:

- Conditions of Contract for Construction, Building, and Engineering Works, designed by the Employer;[22]
- Conditions of Contract for Plant and Design Build for Electrical and Mechanical Plant and for Building and Engineering Works designed by the Contractor;[23]
- Conditions of Contract for EPC/Turkey Project;[24]
- A Short form of Contract: the Short Form;
- Conditions of contract for Underground Work;[25] and
- A Fidic Contract Guide.[26]

Differences Also as to the Consideration

Construction contracts must be distinguished also in respect of the consideration. Based on it, there are:

- contract for a lump sum;
- contract for measurement; and
- costs plus fees.

They are examined in another chapter of this study.

Contracts Where the Contractor Acts Exclusively Upon Owner's Request

In these contracts – which are quite different from construction contracts – the contractor has agreed to act merely as an agent of the owner with no margin of autonomy.[27] This may also be limited to a specific part of the work (or to certain hours that are to be compensated at the agreed price). The situation differs from

---

21 Donald Charret, *Contracts for Construction and Engineering Projects*, cited.

22 (The Construction Contract), 2nd ed. 2022 (the first edition of which is known as the Red Book).

23 The Plant and Design Building Contract, 2nd ed. 2022.

24 The EPC (Turkey Contract) 2nd ed. 2022.

25 1st ed. 2023.

26 2nd ed. 2022.

27 Court of Cassation (Italy) Chamber 3 Criminal division, March 28, 2018 no. 14359.

construction contracts where, although the contractor is entitled to organize and manage the construction autonomously and at his own risk, the Owner or Engineer may interfere with the contractor's autonomy rather than simply supervising and issuing directions. Contrary to construction contracts where some interferences may occur, in this specific situation, the Contractor has agreed in the contract that all decisions be made by the Owner or their Engineer. Nevertheless, the Contractor generally has a duty – even in this extreme case – to advise the Owner if, according to its knowledge and experience, the instructions it is asked to implement do not comply with regulations or construction rules.

Sale of Land Against Delivery of Part of What Has Been Built

The construction of real estate may also be just one of the objects of a broader agreement, called in France "*cession de terrain contre remise des locaux*,"[28] under which the owner of land sells it to a purchaser who, in consideration, undertakes to build – or have built – and to deliver to the seller a part of what is to be built on such land.

The Granting of the Right to Use What Will Be Built and Its Return to the Owner

A variation of this earlier scheme consists of the landowner authorizing a contractor to build on the land, use that building for a specified time period, either directly or by renting it out, and return it to the landowner at the end of that period. This less common contractual scheme is known in French as "*bail à construction*."[29]

Real Estate Development

Construction may also be the subject of an agreement through which a party decides to organize the construction of a large project, often on land belonging to a third party, with the purpose to sell it in parts to interested purchasers.[30] This project is defined in French as "*promotion immobilière*."[31]

Connected Contracts

The construction contract is just the peak of a contractual iceberg, under which various related contracts are submerged. Among them are not only the Owner's

---

28 Sale of land against delivery of apartments.

29 Lease of land to be built.

30 Mike E. Miles, *Real Estate Development*, 5th ed., Urban Land Institute, 2015.

31 Philippe Malinvaud, *Droit de la construction*, Dalloz 2000, at 107.

agreements with a designer and/or with an Engineer but also a wide range of other contracts, such as employment contacts with staff and labor, purchases of materials, the obtainment of rights, concessions from public authorities, transportation of materials, equipment and people, the hiring of equipment, consultancy agreements for the feasibility and for other studies, agreements for inspection, assessment, and other services.

Subcontracts play a vital role in many projects in particular when the contractor is a general contractor[32] when there are flow down provisions[33] from the main contract to the subcontract.

The Formation of the Contract

The meeting of the minds between the Owner and of the Contractor is often the result of a long series of steps.[34]

The Invitation to Tender

In international construction, the contract is typically initiated by an invitation to tender, which includes various documents such as the General Conditions, the Particular Conditions, Specifications, Drawings, Bill of Quantities, Schedules, information as to the underground conditions, as to access to the site, as to the location of quarries, and maps and documents from public authorities, sometimes on the basis of sealed bids. Depending on the project and local practice, the invitation is sent either to a large number of domestic and foreign contractors, or to those which have overcome a previous selection process, which limits their number. In some legal systems, the tender must be open and must be published.

Requests for Information

Many potential contractors will request clarifications or further information to prepare their bid.

32 Will Hughes, *Construction Contracts*, 5th ed., April 20, 2015.

33 Business Laws Inc. Editorial Staff, *Construction Contracts Deskbook*, 2024 ed. (Business Laws Inc. 2024); Reg Thomas & Mark Wright, *Construction Contracts Claims*, Red Globe Press 2016.

34 Reg Thomas & Mark Wright, *Construction Contracts Claims*, Red Globe Press 2016; Will Hughes, *Construction Contracts*, 5th ed., April 20, 2015.

Negotiations

While negotiation is a natural part of contracts aimed at finalizing the agreement,[35] in international construction contracts, there is generally very limited room to amend the invitation to bid. The contractor is typically aware that the more it attempts to modify a term of the invitation, the lower its his chances of success.

Criteria for Acceptance

Each Owner, particularly in the case of contract awarded by a public entity, has its own selection criteria. Generally, there must be no discrimination in favor of domestic contractors. It is common for the price to be the decisive element, a criterion that should be considered carefully (see the following sections).

Acceptance

Based on the criteria selected to assign the contract and on the bids it has received, the Owner issues its Acceptance of a Bid. In some situations, instead of a mere acceptance, the Owner first presents additional specifications or requirements.

The Meeting of the Minds

The existence of a construction contract depends on the alignment between the tender and its acceptance, which is not always clear and may require the interpretation of documents.

Negotiorum Gestio

Even if extremely rare, a construction may be built by someone who has taken the initiative to act in the interest of another party, without having been instructed by it what is referred to in Latin as "*negotiorum gestio*."[36] Such conduct is more likely to be accepted by the entity or person who has given no instructions if it produces a positive result.[37]

---

35 Philip Lane Bruner & Patrick J. O'Connor, *Bruner & O'Connor on Construction Law* (Clark Boardman Callaghan, 2002–2023).

36 Running a business of one's own initiative.

37 Aldo Cenderelli, *La negotiorum gestio*, Giappichelli, 1990.

The Form of the Contract

The requirements for the existence of a construction contract may be different in various legal systems. While generally the contract must be in writing, in some legal systems, it may be entered only verbally. A more formal document may be required, for example, in public procurements. The invitation to tender and the applicable law are the basis to establish the form of the contract. It is not infrequent that, while the basic elements of the contract are entered into in writing, other elements of it are only agreed on orally.

Acceptance by Conduct

When this is not prevented by the applicable law or by the terms of the invitation to bid, the acceptance of a bid may result from the conduct of the parties. A similar acceptance may open the door to disputes.

Documents Attached to the Contract

As a rule, the numerous documents covering the various aspects of a construction project, such as those listed above,[38] should be signed upon entering into the contract and attached to it. However, it is not uncommon for them to be finalized at different times, signed by the parties even after their acceptance, and not kept in an orderly manner.

Evidence of the Contract

Whenever the existence of a meeting of the minds is challenged, the applicable law will govern also the related burden of proof. The issue regarding the existence and contents of the construction contract – if it is a source of a disagreement – will be decided by the body appointed by the parties for such issues, or, in the absence of such a body, by the state court or arbitral tribunal with jurisdiction over the disputes between the parties.

The Applicable Law

The International Construction Contract

As discussed earlier, international construction contracts are characterized by a "foreign element" such as the nationality of a party, the performance of the

38 See *supra* under VII.

contract in a country different from the contractor's, or a foreign element related to other aspects of the contract.

The Applicable Law

Rather than one law, several laws apply to an international construction contract, such as:

- the law which governs the capacity of the parties;
- the law which governs the form of the contract;
- the contract law;
- local laws and regulations; and
- the law of tort.[39]

The existence of a "*lex constructionis*" has been suggested.[40]

The Law Which Governs the Capacity of the Parties

The law governing the capacity of each party, including the existence and status of that party, may often differ from the law which governs the contract. Typically, it is the national law of the respective party.[41] When the parties have different nationalities, the capacity of each of them will be governed by his/her national law.

The Law Which Governs the Form of the Contract

Unless otherwise regulated, the law governing the form of the contract tends to be the law of the place where the contract is entered into. This determination becomes less straightforward when that contract is formed at distance.

The Contract Law

The proper law of a construction contract, which governs the various aspects of the relationships between the parties,[42] includes:

- the construction of the contract;
- the rights and obligations of the parties;
- the performance of the contract;

---

39 Business Laws Inc. Editorial Staff, *Construction Contracts Deskbook*, 2024 ed. (Business Laws Inc. 2024).

40 Donald Charret, *Contracts for Construction and Engineering Projects*, cited.

41 P.M. North and J.J. Fawcett, *Cheshire and North's Private International Law*, Butterworths, 1987.

42 See *Cheshire and North's Private International Law*, cited.

- breaches of contract;
- the validity, voidability, and nullity of the contract;
- the various situations which bring that contract to an end; and
- the assessment of damages.

The Choice of the Parties

The fundamental rule as to the law of contract is the freedom of its choice by the parties. A choice that may also subsequently be modified by consent. The parties may either choose one law to govern the entire contract or decide that some parts of the contract be governed by different laws.[43] The parties may choose either the law of one of the parties, or the law of a third party, or the law of a different legal system, provided this choice does not breach any applicable mandatory provision. The parties may also provide that the contract be governed only by the UNIDROIT Principles,[44] or more vaguely, just by the general principles of law.

Criteria to Identify the Contract Law in the Absence of a Choice by the Parties

In the absence of a choice by the parties, the conflict-of-law rules of various legal systems offer different criteria for selecting the proper law governing that contract.

One must then look – for as held by von Savigny[45] – for the "seat of a legal relationship."

The Various Options

Among the criteria to select the proper law of a contract, the various legal systems include the law of:

- the place of performance of the contract;
- the place where the contract has been entered into;
- the domicile or residence of the party which carries out the characteristic performance of the contract; and
- the country which has the closest connection to the contract.

The Rome I Regulation

The Rome I Regulation[46] is in force in the member States of the European Union, except for Denmark. It has established specific criteria[47] to identify the applicable

43 Dismemberment referred to in French as "*dépeçage*."

44 Unidroit Principles of International Commercial Contract 2016.

45 Friedrich Carl von Savigny, *Vom Beruf unserer Zeit für Gesetzgebung und Rechtswissenschaft*, Heidelberg 1814 and *System Des Heutigen Römischen Rechts*, Heidelberg, 1841.

46 EC Regulation 593/2008 of June 17, 2008.

47 Art. 4.1.

law as to various contracts such as for the sale of goods, contracts for the provision of services, contracts related to a right in rem in immovable property, franchise contracts, distribution contracts, contracts for the sale of goods by auction, and contracts that bring together multiple third parties buying and selling interest in financial instruments.

Under this provision,[48] as to contracts not covered by the list above or those covered by more than one of the mentioned contracts, the contract shall be governed by, "the law of the country where the party required to effect the characteristic performance of the contract has his habitual residence."

However, when it is evident from the circumstances that the contract is more closely connected with a country different from those mentioned above, the Regulation stipulates[49] that the law of that other country shall apply. A final provision allows to reach a conclusion when the earlier provisions do not apply: "When the law applicable cannot be determined pursuant to paragraphs 1 or 2, the contract shall be governed by the law of the country with which it is most closely connected."[50] This in line with English law under which the preference is given to the country with which the transaction has the most real connection.

The Hague Convention on the Law Applicable to International Sales of Goods

The Hague Convention,[51] (the contracting parties of which are Italy, Belgium, France, Denmark, Finland, Norway and Sweden), provides, "In default of a law declared applicable by the parties under the conditions provided in the preceding article, a sale should be governed by the domestic law of the country where the vendor has his habitual residence at the time it receives the order."[52] The Convention further provides, "For the purposes of this Convention, contracts to deliver goods to be manufactured or produced shall be placed on the same footing as sales, provided the party which assumes delivery is to furnish the necessary raw materials for their manufacture or production."[53] This provision may be interpreted as also applying to construction contracts in which the contractor must provide the raw materials.

48 Art. 4.2.

49 Art. 4.3.

50 Art. 4.4.

51 The Hague Convention 15 June 1955.

52 Art. 3d.

53 Art. 1.

The United Nations Convention on Contract for the International Sales of Goods (CISG)[54]
A CISG provision that is similar to The Hague Convention's above-quoted provision states:

i. Contracts for the supply of goods to be manufactured or produced are to be considered sales unless the party who orders the goods undertakes to supply a substantial part of the materials necessary for such manufacture or production.
ii. The Convention does not apply to contracts in which the preponderant part of the obligations of the party who furnishes the goods consists in the supply of labor or other services.

As to International Constructions Contracts
It is suggested that the prevailing tendency is to hold that the closest connection of an international construction contract, wherein a foreign Contractor builds in the Owner's country, is with the country where the site is located. This conclusion might differ significantly if, for example, a plant is manufactured in the Contractor's country and is only assembled at the Employer's site.

The Mandatory Provisions
A natural reservation must be made for the situations involving a specific right, obligation, or other term in which the applicable law conflicts with mandatory provisions, such as those at the site. Given that this depends heavily on the specific circumstances, the assessment of the existence of such mandatory provisions must be conducted with care and expertise in each case.

Local Laws and Regulations

The performance of construction works is frequently the result of activities carried out in various locations, each of which may be subject to local laws or regulations, which "are regarded as crucial by a country for safeguarding its public interest, such as its political, social or economic organization."[55]

---

54 Vienna April 11, 1980, effective on 1 January 1988.

55 Art. 9 EC Regulation 593/2008.

Overriding Mandatory Provisions

Crucial provisions "are applicable to any situation falling within this scope irrespective of the law otherwise applicable to the contract under this regulation."[56] Such mandatory provisions may, therefore, override the proper law governing the contract.

The Law of Tort

Although contract law dominates the landscape of construction contracts, it is not the only body of law applicable to construction.[57] A party to such a contract may also be liable in tort to the other party for matters not covered by the contractual relationship, such as in cases of negligence. In some jurisdictions, tortious liability may be asserted together with contractual liability. This delicate aspect is to be checked within the context of each specific legal system. A party to a construction contract may also be liable in tort to third parties.

The Terms of the Construction Contract

The basic terms of a construction contract are:

- the Contractor's commitment to duly and timely execute the works;
- the Contractor's organizing the works and carrying the risks – with the due exceptions – related to them; and
- the Owner's obligations to pay the contract price.

The construction contract is bilateral, insofar as each party to the contract undertakes to perform its obligations under the agreement. In many legal systems,[58] the existence of a consideration is a requirement. In some of them,[59] construction contracts lacking a consideration are treated as a different type of contract. In other jurisdictions, it is statutorily required that payment be made in money rather than through alternatives such as the delivery of parts of the building. Not in all legal systems can the Contract Price – if not agreed upon – be determined according to custom and trade practices. The contract is characterized, as discussed earlier, by

---

56 Art. 9 EC Regulation 593/2008.

57 Richard Wilmot-Smit., C., *Construction Contracts: Law and Practice*, 2nd ed. (OUP Oxford, 2010).

58 See Hudson's, *Building and Engineering Contracts*, cited.

59 Maria Costanza, *L'appalto privato*, UTET, 2000 at 209.

the difference between the performance of the work, which is a "*facere*,"[60] and the obligation of the seller, in a sale of goods, which amounts to a "give" ("*dare*").[61]

In general, in the relationship between the obligations of the parties, each party undertakes to perform its duties on the premise that the other party has performed its part of the contract, which precedes the former's performance. This "reciprocity link" is referred to in civil law countries as a *synallagma*.[62]

The construction contact is also characterized by its performance having a long duration.

As discussed earlier, while the contractor is typically free to organize their operations at their discretion, in other situations – such as under the Italian legal system[63] – the contactor must establish a formal business structure. If the contractor operates solely by himself or with a few family members, the arrangement is no longer considered a construction contract and instead falls under different statutory provisions that govern personal labor in exchange for compensation.

The autonomy of the Contractor in carrying out the work, subject to supervision by the Owner or by his Engineer, is a basic requirement of construction contracts. Whenever the Contractor merely follows instructions from the Owner or the Engineer, in many jurisdictions, this relationship is not classified as a construction contract.

The contractor does not satisfy his commitment by using his best efforts. His obligation is to ensure that the results of his work – regarding quality, quantity, standards, and timelines – align with his contractual commitment. One consequence of this is that the contractor does not discharge his obligation by delivering only a portion of the work; his commitment to deliver must encompass the entirety of the project.

Implied terms add to the express written terms of the contract. An example of an implied term is the duty to act in cooperation with the other party.[64] Warranties by the contractor[65] may even be implied. Specifically, fitness for particular purpose of the work as well as warranty of design adequacy, if not expressly dealt with in the contract, is another implied term.[66]

---

60 Obligation to do.

61 Obligation to give.

62 The interdependence between the performances of the parties.

63 Sec. 2222 Italian Civil Code.

64 Samuel Williston and Richard A. Lord, *Williston on Contracts*, 4th ed. 2007.

65 Business Laws Inc. Editorial Staff, *Construction Contracts Deskbook*, 2024 ed. (Business Laws Inc. 2024). Philip Lane Bruner & Patrick J. O'Connor, *Bruner & O'Connor on Construction Law* (Clark Boardman Callaghan, 2002–2023).

66 Business Laws Inc. Editorial Staff, *Construction Contracts Deskbook*, 2024 ed. (Business Laws Inc. 2024); Philip Lane Bruner & Patrick J. O'Connor, *Bruner & O'Connor on Construction Law* (Clark Boardman Callaghan, 2002–2023.

In addition to express and implied terms, a construction contract is generally completed by custom and usage, unless such customs and usages have not been directly or indirectly excluded by the parties. An example of this includes compulsory regulations related to labor safety standards and material specifications.

The Contract Price

As it is well-known, there are various options available regarding the pricing of construction contracts.[67]

Lump Sum Contracts

Frequently the parties agree on a lump sum, *i.e.*, on a global amount to pay the entire works, without taking into account any measurement. This formula is not contradicted by the inclusion of a bill of quantities or similar documents among the contractual documents, provided that these documents are not binding on the parties. . In this case, the purpose is to provide the Owner with information to compare the various bids and to make sure the price proposed by the Contractor is reasonable. Such documents may contain data that assist in determining the price in event of variations.

The premise and requirement for a binding lump sum agreement is that the works to be performed have been exactly described, enabling the Contractor to accurately estimate the associated costs. Lump sum contracts are largely used when the contract is assigned on the basis of an invitation to bid. The tendency of many owners is – as mentioned earlier – to select the lowest price. It is suggested that the selection process should also consider a comparison between the standards and experience of the various bidders, relying on objective data related to elements such as their project organization and progress plans, as well as the quality of materials, equipment, and labor they have used in previous construction contracts. A lower bid is not necessarily the best choice when higher bids have been received from more experienced and successful bidders.

Contracts by Measurement

Contracts by measurement provide compensation consisting of a unit price for each type of work. That unit price will then be applied to the actual quantity of such type of work, which will then be measured for its purpose. In such contracts,

67 Philip Lane Bruner & Patrick J. O'Connor, *Bruner & O'Connor on Construction Law* (Clark Boardman Callaghan, 2002–2023.

the Contractor takes a risk not regarding the quantity of work rather concerning the adequacy of the unit price they have agreed upon. It inevitably follows from this that the Contractor will not be entitled to a higher price, if any materials or manpower have a higher cost than what the initial estimates are when agreeing to the unit price. Likewise, the Owner will not be entitled to a reduction of that unit price if the cost of any materials or labor is lower than the estimate they provided when agreeing to that unit price.

If the parties to a contract have not considered other work that becomes necessary during the execution of the project, the unit price for that additional work will typically be determined – either by the parties or through external guidance – often by referencing to the unit price of similar types of work.

Despite the clear distinction between choosing a lump sum and admeasurement, practical scenarios often arise where portions of the work are contracted on an admeasurement basis while others are based on a lump sum agreement. Even if the distinction between lump sum price and price by admeasurement is very straight forward, some construction contracts, particularly when parts of previous contracts are cut and pasted in it, are not sufficiently clear and may give rise to disputes regarding some items of the contract price.[68] It has been established that in cases of ambiguity, the lump sum formula would prevail only if it has been clearly expressed.[69]

Costs Plus Fees

Since both lump sums and contracts by admeasurement have positive effects for one party, but can also potentially cause adverse effects, an alternative approach has been sought: "costs plus fees" contracts. The advantage of this less common formula is that if the Engineer oversees the costs, the Owner is shielded from the risk of incurring excessive expenses. At the same time, the Contractor can anticipate covering its reasonable costs, with its remuneration predetermined in the calculation of the fees.

Quantum Meruit

Even if the rule is that the price is to be determined when entering into the contract, it is not rare that exceptions to this rule apply to construction contracts. This may be a source of serious conflicts. When the parties have not granted authority to a neutral body or person to make determinations, the price, in the absence of such

---

68 Cameron K. Andres, Ronald C. Smith & W. Ronald Woods, *Principles and Practices of Commercial Construction*, 10th ed. (Pearson, July 14, 2021).

69 See *Hudson's Building and engineering contracts*, cited and Donald Keating, *Building Contracts* cited.

appointment, will be determined by a state court or arbitrator – if so provided – on a quantum meruit[70] basis, frequently referring to tariffs and custom of that trade.

Pay When Paid

In various subcontracts, the subcontractor is required to accept payment only when contractor is paid. This provision has given rise to opposing views.[71]

Time in Constructions Contracts

Time is one of the top elements of the construction contract.[72] Apart from possible different local practices, there is typically not a basic difference between domestic and international construction contracts in regard to the timing.

Time Is of the Essence

Time conveys the notion that it is essential. This, however, is not accurate since even if Completion Time is not determined in the contract, the absence of it does not affect the validity of the contract and instead, will have to be determined as provided by the parties or, in the absence of any such arrangements, by the state court or by an arbitral tribunal, which has jurisdiction on such disputes. In the absence of a specific provision in the contract, it is generally the duty of the Contractor to complete the work with a reasonable time.[73] Aside from these remarks regarding the time factor as a contractual requirement, in practice, the Owner often stipulates in the invitation to bid that time be of the essence.

Fundamental Breach

If the time requirement is not complied with due to a breach by the Contractor, the consequence depends on the terms of the contract and on the seriousness of the breach. A very serious breach entitles, in common law as in civil law and other legal systems, to automatically discharge the entire contract when it deprives the

70 Business Laws Inc. Editorial Staff, *Construction Contracts Deskbook*, 2024 ed. (Business Laws Inc. 2024).

71 Philip Lane Bruner & Patrick J. O'Connor, *Bruner & O'Connor on Construction Law* (Clark Boardman Callaghan, 2002–2023.

72 Steven G.M. Stein, *Construction Law*, 8 vols. (Matthew Bender Elite Products).

73 Business Laws Inc. Editorial Staff, *Construction Contracts Deskbook*, 2024 ed. (Business Laws Inc. 2024).

other party of the main benefit of the contract.[74] Such a basic breach is frequently referred to as a "fundamental breach," as stated by Lord Denning, Master of the Rolls. The House of Lords has subsequently referred to this definition replacing it in *Suisse Atlantique*,[75] a landmark decision, with "repudiatory breach."

Damages

In the absence of a fundamental (or repudiatory) breach, the innocent party may avail itself of a claim for damages[76] caused by that breach.[77] Such damages are based on the benefit reasonably expected by the innocent party, or on the loss, the party incurs due to the breach. Damages may also be determined by the costs for the repair.[78]

Damages are distinguished in various legal systems[79] between general and special. General damages are those that the law presumes to be the consequence of that breach, while special damages are to be pleaded and proven.

Damages may be either liquidated or unliquidated. Liquidated damages are different from a penalty, since the latter is generally a high amount, determined to deter the parties from breaching the contract, while liquidated damages are an amount calculated to reasonably compensate the innocent party from the loss caused by the breach. Exemplary damages are not granted in the majority of jurisdictions. The breach involves liability for its usual consequences. This in general excludes remote damages.[80]

Specific Performance

A remedy that is not popular in common law jurisdictions is specific performance.[81] On the contrary, it is a well-accepted remedy in other jurisdictions,[82]

---

74 Joseph Chitty, *Chitty on Contracts*, Vol. I, Sweet & Maxwell, 28th, 1999.

75 *Suisse Atlantique Societe d'Armament Maritime* SA *v. NV* Rotterdamsche Kolen Centrale, [1962] 1AC361.

76 Will Hughes, *Construction Contracts*, 5th ed., April 20, 2015.

77 *Chitty on Contracts*, cited.

78 John Cartwrithg, *Remoteness of Damage in Contract and Tort: A Reconsideration*, Sweet and Maxwell 1996, CLJ 488 London.

79 Sir Edward Fry, *Treatise on the Specific Performance of Contracts* (6th ed.) Stevens and Sons, 1921.

80 Sec. 2932 Civil Code (Italy).

81 *Chitty on Contracts*, cited.

82 Sir Frederick Pollock, *Pollock's Principles of Contracts* (13th ed.) Stevens & Sons, 1950, *Chitty on Contracts*, cited 1274.

including civil law jurisdictions. In some jurisdictions, the innocent party is entitled to obtain a judgment, which orders that any unperformed work be performed at the costs of the party in breach.[83] In French law, *l'éxécution forcée en nature* allows the innocent party to either perform the uncompleted work themselves or to have it carried out by a third party.[84]

Necessary Extension of a Deadline

Some deadlines may become impossible to comply with for reasons other than issues with the contractor. The subsequent extension of time is then a matter of necessity.

Extension of a Term

Granting an extension to the defaulting party for the performance deadline is a common and practical solution, often provided in exchange for some benefit to the other party. An extension may be granted as to various deadlines, including completion of the work.

Program for the Execution of the Works

The Engineer monitors compliance with the time requirement by regularly checking the contractor's progress against the schedule outlined in his execution program. If necessary, notices are given to the Contractor to ensure compliance with progress.

Earlier Completion of the Works

Even if not common, the Contractor may complete the work before the time for completion. A premium is occasionally provided for this event.

Completion

Completion occurs when all the Works, including minor tasks, have been completed. At an earlier stage, when "substantial completion" is reached (meaning the project is functional even if minor works have to be completed), the Owner may take over part of or the entire the Works or parts of them.[85] Once the minor

83 Chitty on contracts, it. 1276, cited.

84 Art. 7.6. Fidic, Conditions of Contract for Construction (2nd ed. 2017).

85 Art. 10 Fidic Conditions of Contract for Construction (2nd ed. 2017).

and any outstanding works have been rectified, the completion of the work will occur within the timeframe specified in the taking over certificate. Any defects reported to the Contractor prior to the expiration of the defects notification period must be remedied before that period concludes. The FIDIC Conditions of Contract for Construction[86] stipulate that defects after taking over shall lead to the issuance of the Performance Certificate "within 28 days after the latest of the expiry dates of the Defects Notification Period." A period for the detection of defects after the contractor has completed the works is generally provided for. In general, construction contracts provide various warranties, which are discussed in another Chapter of this book, to guarantee the Contractor's performance.

Termination

Among the various situations that lead to termination, termination for convenience[87] and frustration[88] are to be mentioned.

Allocation of Risks – Unbalances

Higher Risks in International Construction

It is argued that there are the number of risks,[89] including those related to technology, that to be identified, allocated, and possibly mitigated is greater in international construction for both the contractor and the owner.

For the Contractor, this is primarily due *inter alia* to many factors, including nondomestic laws and regulations, environmental issues[90] tariffs, labor laws, custom and trade usages, to cross-border supplies and transnational transportation, and local regulations as to the importation of materials and of labor.

For the Owner, who often exerts influence over domestic Contractors beyond the actual contractual provisions, it frequently becomes apparent when dealing with a foreign contractor that such a contractor is much less amenable to that

86 Art. 11.9.

87 Business Laws Inc editorial staff, *Construction Contracts Deskbook*, 2024 ed. (Business Laws Inc., 2024).

88 Richard Wilmot-Smith QC & Paul Darling, *Wilmot-Smith on Construction Contracts*, 4th ed. (July 20, 2021).

89 Donald Charrett, *Contracts for Construction and Engineering Projects*, 2d ed. (Informa Law from Routledge, Sept. 25, 2023); Will Hughes, *Construction Contracts*, 5th ed. (Apr. 20, 2015); Philip Lane Bruner & Patrick J. O'Connor, *Bruner & O'Connor on Construction Law* (Clark Boardman Callaghan, 2002–2023).

90 Business Laws Inc editorial staff, *Construction Contracts Deskbook*, 2024 ed. (Business Laws Inc., 2024).

influence. Furthermore, the Owner may face local pressure regarding the use of foreign labor, the importation of foreign equipment and materials, and may be subject to political pressure by local contractors and opposition in Parliament, for instance, for awarding the contract to a nondomestic contractor.

Contractor's Risks

The contractor's risks are those related to the proper and timely performance of the works, the quality of materials, the efficiency of labor, the efficiency of the equipment, the fluctuation of costs, and to charges in local laws and regulations.

The Owner's Risks

The Owner's risks are inter alia related to the selection of the Engineer, his proper and timely supervision of the Contractor's Work and issue of directives and instructions, lack of cooperation by the Engineer with the contractor, interferences by itself and/or the Engineer with the Contractor's autonomy, ordering large variations or change orders[91] without carefully assessing their financial impact, as well as their impact on Time for Completion.

Unbalances

The philosophy of some Owners in their approach to construction contracts may, in addition to the above risks, give rise to unbalances. The tendency to award the contract to the lowest bidder, while seemingly obvious or even automatic, seriously risks overlooking the need to compare the tendered price with the standards and track record of the bidders. It has been suggested[92] that when we need a dentist, it would not be wise just to select the cheapest one.

Another concerning tendency is that when decisions are made by the Engineer, one should not later expect the Contractor to bear the consequences of those decision, even if they are incorrect. Any such approach is not in line with the duty of cooperation between the parties and frequently is not to the advantage of the construction project. Likewise, the frequent contractual duty of the contractor to continue the work, even if there has been a breach by the Owner or the Engineer, impacts the Contractor's right to refrain from performing if the other party has not fulfilled its prior commitments. More balanced solutions should be worked out. Furthermore, the authority granted to the Engineer to remove the Contractor's site agent may have a very negative impact on the site agent's ability

---

91 Business Laws Inc. Editorial Staff, *Construction Contracts Deskbook*, 2024 ed. (Business Laws Inc. 2024).

92 Mauro Rubino-Sammartano, *Appalti d'opere e contratti di servizi*, 2nd ed. Cedam 2006.

to discharge his duties when the Engineer disagrees with him. If the site agent insists, he risks being removed and this risk materially weakens his action.

Ultimately, on-demand performance bonds enable the Owner to exert pressure on the Contractor, who is compelled to reimburse the bank that issued the bond at his request. The Contractor is typically not allowed to demonstrate – except in cases of clear fraud – that he is not in breach of contract.[93] While it is indeed not suggested that the owner frequently tends to take advantage of this leverage, this may give rise to an additional imbalance.

All such imbalances often detract from achieving the best results in a construction project.

The Performance of the Works

Since the performance of the work serves as the foundation for various chapters of this book, a brief overview of some of its major steps will be provided:

- the delivery of the site, access to the site and control of it namely of its surface, subsurface and soil;[94]
- a very attentively drawn up program for the progress of the works;
- floats vis-a-vis the critical path;
- advance warnings;
- delays and their cause;
- tests by the Contractor;
- inspections and tests by the Engineer;[95]
- acceptance of parts of the works;
- retesting;
- rate of progress;
- extensions of time;
- notices by the Engineer to the Contractor;
- rejection of defective works;
- remedial works;
- variations;
- suspensions;
- substantial completion;
- taking over by the Owner (even of just a part of the works);

93 Mauro Rubino-Sammartano, *Performance Bonds in International Contracts*, Foro pad. 1979, I, 245.

94 Cameron K. Andres, Ronald C. Smith & W. Ronald Woods, *Principles and Practices of Commercial Construction*, 10th ed. (Pearson, July 14, 2021).

95 Philip Lane Bruner & Patrick J. O'Connor, *Bruner & O'Connor on Construction Law* (Clark Boardman Callaghan, 2002–2023).

- minor works to be remedied or completed(punch lists);
- completion of outstanding works;
- remedying defective works;
- final tests; and
- performance certificate.

The diverse range of claims[96] is addressed in another chapter of this book, beginning with claims for additional costs, extensions of time,[97] and defective works.

Ownership and Transfer of Title

While the rule is that the site remains in the ownership of the Owner during the performance of the work, the contractor is granted only possession of the site to carry out the work. The plant's items and materials, which are brought to the site by the Contractor, are usually owned by the Contractor, unless it has been agreed that they become – upon their arrival – the property of the Owner.[98] Equipment needed to execute the work, which is not intended to remain on the site upon the completion, is and remains the property of the contractor. If rented, it will be subject to the terms of the rental agreement.

Liability

Liability is another recurring theme in the various chapters of the book, and a brief mention will be provided here. Except when otherwise regulated by the applicable law, liability is governed, in domestic and in international construction contracts, by the law of contract or by the specific statutory provisions, which – under that law – govern construction contracts.

Each party to the contract may be liable to the other in contract or in tort, subject to the provisions of the applicable law.[99] Liability may also arise from misrepresentation and fraud.

Depending on the applicable law and contractual provisions, the Owner might also be liable to the contractor for the conduct of the Engineer, when the doctrine of vicarious liability applies. Each party to the contract may also be liable

96 Philip Lane Bruner & Patrick J. O'Connor, *Bruner & O'Connor on Construction Law* (Clark Boardman Callaghan, 2002–2023).

97 Reg Thomas & Mark Wright, *Construction Contracts Claims*, Red Globe Press 2016.

98 see art. 7.7 Fidic Conditions of Contract for Construction (2nd ed. 2017).

99 Will Hughes, *Construction Contracts*, 5th ed. (Apr. 20, 2015).

to third parties for actions or omissions on the site or related in general to the construction. The protection of persons and property is in part a requirement.[100]

In various jurisdictions, the contractor may be liable for many years – though the specific time periods differ across legal systems – not only to the Owner but also to their successors in title,[101] in the event of a building's collapse or if the building's condition significantly diminishes its enjoyment.[102] Since this liability is in the public interest, it is disputed whether it should be through contract or tort.[103] Limitation of liability is a frequent exercise in drafting such contracts.[104]

Nullity and Voidability of the Contract

The invalidity of a contract is a wide term, which includes nullity, the mere appearance of a contract, and voidability.

Nullity

Nullity must be distinguished from nonexistence of a contract. Nullity may be total or partial.

Subject to review in each single case, under the applicable law, a contract is null and void if there has been no meeting of the minds, when its object is in breach of a mandatory provision or may not be established or is impossible, it conflicts with public policy, or the parties have entered into in it exclusively for a common unlawful purpose.

Apparent Contract

The parties may choose to enter a contract while having no genuine intention of fulfilling it. In such cases, there is no contract. On other occasions, the parties may enter into a contract, with the intention to enter into a different one. If so, if the intention is to enter that, other contract may be proven under the applicable law, and the real contract will likely take the place of the apparent one. Apparent contracts may be entered into even in domestic and international construction contracts. A construction contract may, for example, conceal an employment relationship.

---

100 Philip Lane Bruner & Patrick J. O'Connor, *Bruner & O'Connor on Construction Law* (Clark Boardman Callaghan, 2002–2023).

101 Will Hughes, *Construction Contracts*, 5th ed. (Apr. 20, 2015).

102 *F.G. v. MU.MA. et al.* Court of Cassation (Italy) October 4, 2018 no. 24230.

103 Court of Cassation (Italy), November 17, 2017 no. 2017.

104 Philip Lane Bruner & Patrick J. O'Connor, *Bruner & O'Connor on Construction Law* (Clark Boardman Callaghan, 2002–2023).

Voidability

Voidability of a domestic or international construction contract may be due – under the applicable law of contract – to a mistake admitted by such law[105] (e.g., as to the contract price or as to the works to be performed or as to its design, or to its duration), misrepresentations,[106] negligence or fraud,[107] the lack of authority of the signatory for one of the parties, or pressure aiming to obtain an unlawful or unfair benefit.

Timing and Requirements to Assert the Invalidity

Legal systems vary significantly regarding the requirements, timing, and evidence needed to establish a contract's invalidity. Therefore, it is essential to refer to the applicable law in each individual case.

---

105 Richard Wilmot-Smith QC & Paul Darling, *Wilmot-Smith on Construction Contracts*, 4th ed. (July 20, 2021).

106 Business Laws Inc. Editorial Staff, *Construction Contracts Deskbook*, 2024 ed. (Business Laws Inc. 2024.

107 Business Laws Inc. Editorial Staff, *Construction Contracts Deskbook*, 2024 ed. (Business Laws Inc. 2024.

2

Legal Aspects of Pre-construction Activities*

Paolo Daino and Giovanni Minuto

Introduction

The preconstruction phase of a construction project is a crucial stage involving various activities, such as design development, procurement planning, risk management, and permitting, each characterized by legal aspects. These activities can significantly affect the success, cost, and quality of the project, as well as the allocation of responsibilities and liabilities among the parties involved. However, the preconstruction phase is often overlooked or under-regulated in the contractual arrangements between owners, contractors, subcontractors, and consultants. This chapter aims to provide an overview of the main issues and challenges parties face in regulating the preconstruction phase in construction projects and explore the different contractual and noncontractual options.

Typical Preconstruction Activities

The construction phase of a project is the final step following an extensive and costly preparatory process, which includes (i) the engineering and design of the works to be built and (ii) obtaining all necessary authorizations to begin the construction phase and carry out the planned works. Legal hazards and risks encountered by construction projects often originate in the early stages and are typically related to design errors and/or permit issues.

Parties considering entering into a construction contract should undertake the following preconstruction activities, which involve significant legal considerations.

* The authors would like to thank Giulia Vannoni, Martina Daniele, and Luca De Rosa of BonelliErede for their invaluable contribution to this work.

These activities should be performed regardless of any formal regulations pertaining to the preconstruction phase, which will be discussed in Section "Regulating Legal Aspects of Pre-construction Activities."

Design Phase

The concept of a construction project is transformed into a tangible technical design of a building, plant or infrastructure during the design phase. The design process moves from key initial steps like feasibility studies, site and ground assessments, and land acquisition to the subsequent steps resulting in the preliminary design and detailed design.

The designer is the core figure involved at this stage. The designer may be an individual or, more frequently, a firm[1] that provides the owner with drawings, measurements, calculations, and technical specifications that constitute the design of the works. It is not uncommon for the "engineering companies" – composed of engineers, architects, surveyors, directors of works, contract and project managers, and safety coordinators – to also provide supervision services on site and other services ancillary to the design activity, such as the preparation and filing of all necessary documentation to obtain authorizations, permits, licenses, and clearances to start the works.

The design of buildings and infrastructures is a complex and challenging activity, particularly regarding structural, mechanical, electrical, and plumbing design. Design errors are one of the main risks to consider during a construction project's development.

To elucidate the potential risks and liabilities of the parties involved in a construction project, it is imperative to examine two distinct scenarios: firstly, where the owner provides the project's design, and secondly, where the general contractor is entrusted with both the design and construction of the works. These two scenarios correspond to the procurement methods commonly employed in international construction projects, namely the Design-Bid-Build and the Design-Build methods.

Design-Bid-Build Method

The traditional procurement method necessitates that the owner assume responsibility for the preparation of an exhaustive tender design, encompassing drawings, specifications, bills of quantities, and calculations. This comprehensive task may be undertaken directly by the owner or with the assistance of a professional

---

1 Today the "engineering company" acting as designer prevails as design works are becoming very demanding and too large to be dealt with by an individual on their own. In this respect, see L. KLEE, *International Construction Contract Law*, Wiley Blackwell, 2018, p. 3.

design team. Following the tendering process, the owner appoints a contractor to execute the works based on the preprepared design provided during the tender phase.[2]

In accordance with the Design-Bid-Build (DBB) methodology, a substantial portion of the design must be completed prior to the selection of the contractor. During the construction phase, the project will be finalized by the contractor through detailed and constructive design efforts. The owner is required to enter into two distinct contracts: one with the designer to prepare the design documentation for the tender, and another subsequent contract with the selected contractor upon the completion of the tender stage. As the contractor is absolved from the responsibility of preparing the design and consequently does not assume the associated risks, the owner is likely to receive a more competitive quotation for the construction phase.

From the **owner's perspective**, it is self-evident that the DBB method entails a higher level of risk as the owner is responsible for the design prepared and provided to the contractor and for any errors contained therein that may result in project delays and/or cost increases.

To mitigate the risk of errors, liabilities, costs, and time inefficiencies during the design process and before the bid phase, the owner may wish to consider the following measures:

- if the design process is entrusted to several designers, appoint one of them with the role of coordinator, to avoid inconsistencies among the different design segments of the whole project;
- obtain consultancies by potential specialist contractors (such as mechanical, plumbing, electrical, etc.) to anticipate specific elements and components that will be needed for the constructability of the building or infrastructure concerned by the project and which shall be envisaged in the design; and
- request an overall review of the project by one or more contractors (not necessarily the one that will be awarded the construction contract) to ascertain the constructability and the likelihood of the costs and schedule forecasted.

From the **contractor's perspective**, the DBB method implies the obligation to comply with the design documentation furnished by the owner at the bid stage. Depending on the contractual terms of the relationship between the owner and the contractor, as well as on the type of breach, the failure by the contractor to comply with the owner's design may entitle the owner to ask the contractor to make good the defects and noncompliances at its own cost, to bear the consequences of the

2 A common example of this type of contracts is the FIDIC Conditions of Contract for Construction for Building and Engineering works designed by the Employer (the so called Red Book).

delay caused (e.g., by suffering the application of delay liquidated damages), to terminate the contract for default or to reject the works.

In the DBB method, the contractor is not typically responsible for the design. However, the contractor must notify the client and designer of any errors, defects, or inconsistencies in the technical documentation that a diligent and experienced contractor would be able to identify. If the design has defects that could not reasonably have been discovered by a competent contractor, the contractor generally will not be held liable for those defects and their negative impact on the works.

A diligent contractor would be generally expected to carefully review the design provided by the owner before signing the contract. Contractors should also notify the owner of any discrepancies in the tender documents to facilitate the implementation of the project.

Design-Build Method

This procurement method is characterized by the execution of an engineering, procurement, and construction contract (EPC) between the owner and contractor: the contractor typically assumes full responsibility for the entire project, encompassing both the design and construction phases.[3]

During the initial bid phase, the owner provides a set of requirements that outline the general purpose of the project. These requirements consist of technical criteria and performance standards, often based on a "fitness-for-purpose" approach. The contractor then submits a proposal in response to the owner's requirements. Upon being awarded the tender and signing the EPC contract with the owner, the contractor develops the project's design as part of its performance obligations according to the EPC contract.[4]

From the **contractor's perspective**, the Design-Build (DB) method involves a higher level of responsibility as the contractor is in charge of the work's design, procurement of materials, equipment, and labor, as well as the construction

3 Typical examples are the FIDIC model contracts for Plant and Design-Build by the Contractor (the so-called Yellow Book) and for EPC/Turnkey Projects (the so-called Silver Book).

4 In the design-build model, the owner's requirements are crucial documents and must be clear and unambiguous. This document must cross-refer to the conditions of the EPC contract while drafting. Attention must be paid in order to be consistent in the use of terminology and accurate in contractual references to paragraphs and sections of the owner's requirements. The owner's requirements must include all the definitions and purposes of the work, a definition of the site, of the quality and performance criteria (including testing) and of special obligations such as training, spare parts, and warranties. Furthermore, the specifications must be precise but not too much detailed to avoid that design responsibility being transferred back from the contractor to the owner. Indeed, the more the specifications are detailed, the higher the probability of a reduction of the contractor design responsibility. In this respect, see L. Klee, *International Construction Contract Law*, Wiley Blackwell, 2018, pp. 103–106.

phase, all while adhering to the timeline and budget agreed upon with the owner. Specifically, the contractor must produce a final product that meets the contract terms and applicable laws and regulations, and conforms to the purpose outlined in the tender requirements set by the owner.

The contractor is generally responsible for any problems that may arise due to design or construction defects, and bears the risk of the resulting increase in time and cost required to complete the work.

From the **owner's perspective**, the DB method involves concluding a single contract (typically a lump-sum contract) with the designer-contractor. The DB method facilitates an earlier start of the construction phase as some overlap between the design and construction phases is feasible. Additionally, the DB method provides the owner with more certainty regarding the anticipated costs and timeframe for the completion of the project. However, since the contractor develops the entire design after the tender is awarded, bearing the associated risks and costs, the price quoted during the bid phase is likely higher than that of the Design-Bid-Build (DBB) model.

Authorizations, Permits, and Other Regulatory Legal Issues

From a legal perspective, securing all necessary permits, authorizations, licenses, and regulatory clearances during the preconstruction phase is paramount for the commencement and execution of the project.

It is noteworthy that the permitting process often extends into the construction phase, as building and construction permits must remain valid until the project's completion and, frequently, beyond. Therefore, it is incumbent upon the involved parties to comply with any conditions imposed by regulatory authorities. Additionally, they may need to update permits if significant modifications to the project occur during construction.

The permit process is highly formalized and begins with an analysis of the legislative and regulatory framework to identify relevant procedures, government and local bodies to be addressed, and required documentation. Once these elements are identified, meetings with the competent authorities are scheduled before submitting the permit application to discuss the project, set up the process, and understand specific requirements. The responsible party then develops the necessary plans and documentation, often with the support of other parties, and proceeds with their submission.[5]

---

5 The documentation typically includes detailed project plans, environmental impact assessments, and other required information that varies depending on the specifics of the project and the jurisdiction in which it is located.

Following a thorough examination and review of the submitted dossier, including any necessary follow-up interactions, documentary integrations, and revisions, the relevant authorities issue the permits and clearances needed for the project to proceed.[6] These permits often include conditions and requirements that must be adhered to throughout the construction process under penalties such as sanctions, revocation of permits, or halting of construction. The permitting process may be assigned to the owner, the designer, or directly to the contractor, depending on the contractual scheme adopted for the project.[7]

The management of the permitting process may be challenging as the legislative and regulatory framework is often complex and the intervention of different authorities at a national, regional, and local level may be required. The degree of difficulty increases when the parties implement a project in unfamiliar countries with complex regulatory frameworks. For this reason, foreign contractors usually team up with local partners, either as co-contractors or subcontractors.

In this context, another common preconstruction legal activity for projects executed abroad or through **partnerships/joint ventures** between multiple contractors is the establishment of a **local branch** or a dedicated company in the country where the works will be performed. This is often a crucial legal step to facilitate the hiring of personnel, receipt of payments, leasing of spaces and vehicles, and for tax purposes. The legal activities involved primarily entail advice under the company, labor, and tax laws of the country where the project will be conducted. These activities will necessitate the assistance of legal professionals from the specified country.

Finally, it is worth mentioning that in some cases the design process, the bidding phase, and the contract negotiations between the owner and the contractor may conclude before the completion of the permit process. This may happen particularly in jurisdictions where the procedures are lengthy, complex, and subject to a slow bureaucracy. In this scenario, on the one hand, the parties are interested in signing the construction contract to commence at least with all preparatory activities: the owner to speed up the construction phase and the contractor to receive the first tranches of payments. On the other hand, they should be careful in providing all appropriate contractual protections in case the building permits are not obtained. To this end, the parties will typically extensively negotiate a

---

6 It may happen that the community inputs are also required during the review stage as long as a project affects the neighborhood.

7 In any case, all parties involved in the project are individually responsible for ensuring that the requisite authorizations in accordance with the applicable laws and regulations are in place to provide the services, the employment and work permits, and the health and safety of employees at workplace (especially with reference to construction sites).

detailed contractual regulation of the economic consequences for the untimely or total failure to obtain the required permits.

Regulating Legal Aspects of Preconstruction Activities

Parties in construction projects are free to determine how the preconstruction phase of their project should be governed. Very often there is no contract or binding understanding to discipline the preconstruction phase. However, complex and larger projects may require more sophisticated options. What follows is an overview of some of the solutions that parties in construction projects have adopted to define the legal aspects in the preconstruction phase.

Nonbinding Arrangements

The various subjects involved in a construction project (owner, potential contractor(s), designer, engineer, etc.) may opt to leave the preconstruction phase entirely unregulated, relying instead on discussions during meetings or through informal communications documented in minutes and correspondence. This approach may be particularly appealing in fast-paced preconstruction phases where the timetable of the forthcoming construction contract necessitates the early commencement of preliminary activities such as ordering materials and completing administrative formalities, despite the absence of a fully executed construction contract. These preliminary activities are not precluded by the lack of a formal contract and, in some instances, may be included among the conditions precedent for the contract's eventual execution.

Proceeding without formal commitments entails various risks, including potential disagreements and misunderstandings. These issues can result in wasted time and resources or even the abandonment of projects. Given the substantial investment required for construction projects, relying on unwritten and nonbinding agreements should be approached with caution, particularly during preconstruction activities.

Letter of Intent

When parties prefer to avoid mandatory preconstruction commitments, they typically opt for a Letter of Intent (LOI). This instrument is frequently utilized when parties are still engaged in negotiating a comprehensive contract or desire to initiate the construction program and other preconstruction activities. LOIs are not governed by standard contractual forms; their content and structure vary according to the specific requirements of the drafting parties. They may simply take the form of letters, but, more commonly, they present as a list of preconstruction activities to be undertaken pending the final award of a construction contract.

An LOI may authorize the prospective contractor to undertake certain activities while negotiations for the future contractual relationship are ongoing. The letter's content may specify the compensation for these activities and limit the owner's liability and the contractor's obligations should the final contract not be concluded.

In some instances, the LOI may explicitly reference the draft contract and stipulate that, should a contract be finalized, its terms – including the compensation for the identified preliminary activities – will apply. Consequently, the activities conducted under the LOI will receive the same contractual protection as other obligations covered by the contract.

Ad Hoc Agreements and Formularies

Complex projects may require more sophisticated solutions. In these cases, parties may enter ad hoc agreements to regulate the preconstruction phase or specific elements. While, as described above, LOIs have traditionally been employed as non-contractual instruments, in recent practice the scope of LOIs has widely expanded, transforming them, despite their title, into fully flagged preconstruction contracts.

The activities regulated under these ad hoc preconstruction contracts (sometimes still improperly named as LOIs) range from administrative tasks all the way to early construction activities. These may include definition of design criteria and submission of drawings, preliminary technical studies, as well as determination of technical specification for the purchase of necessary parts and materials. Not only the contractor but also the owner might be required to perform a set of activities, such as soil and geotechnical evaluations to be used in the design process. In addition to a simple list of activities, as stand-alone agreements, they include provisions on terms of payments, warranties, intellectual property rights, and indemnities as well as dispute resolution clauses.

Standard forms of construction contract also include some forms of agreement to govern the preconstruction phase, in particular elements such as design development, two-stage pricing, risk management and communication. An example are the Project Partnering Contracts (PPC), particularly the model **PPC2000**: it is a multiparty contract, that can replace the multiple bilateral contracts that the client, consultant, main contractor, and subcontractors would normally sign. It includes a preconstruction phase program called the "Partnering Timetable," regulating various aspects and activities of the preconstruction phase, including joint development,[8] joint selection of the main contractor's

8 PPC2000 Partnering Terms, clauses 8.3 to 8.12.

supply chain,[9] joint risk management,[10] and agreement on the construction phase program.[11]

Other binding agreements regulating preconstruction activities are more generally referred to as preconstruction service agreements (PCSA), often part of a two-stage procurement process. These procedures and contracts will be analyzed in more detail below in Section "Two-stage Tendering and Pre-construction Services Agreement (PCSA)".

Two-stage Tendering and Preconstruction Services Agreement (PCSA)

In case of larger construction projects requiring the participation of more sophisticated parties, owners may adopt a two-stage approach to govern both the preconstruction and construction phases of a project. This approach is often referred to as "two-stage procurement," or "two-stage tendering," involving owners and contractors in two different contracting phases. It is characterized by a first phase where a contractor is chosen for preconstruction activities, regulated under a pre-construction service agreement (PCSA) and a second phase characterized by the conclusion of the building construction contract, typically entered into at the dawn of the PCSA, often but not necessarily with the same contractor of the preconstruction phase.

Stage One

At the outset, the owner tenders the project with an incomplete design, program and price prepared by its team. At this initial stage, the tender documents may include pricing arrangements (for both stages), target costs, and a draft of the main contract conditions for the second stage (the construction phase). At the end of this preliminary tendering, the owner would have chosen a contractor to embark on the project's first stage. Their relationship will be governed by the PCSA, setting out the preconstruction activities.

A PCSA is akin to a consultancy contract, primarily encompassing nonconstruction activities. These activities may involve design development, such as architectural and engineering design, along with design review and revision. Additionally, it covers the selection of subcontractors, risk analysis and mitigation, project scheduling, and obtaining necessary permits to ensure compliance with all relevant regulations and codes. Occasionally, a PCSA may also include preliminary on-site work related to mobilization activities, such as site preparation,

9 PPC2000 Partnering Terms, clauses 10.3 to 10.9.

10 PPC2000 Partnering Terms, clauses 12.9 and 18.

11 PPC2000 Partnering Terms, clause 6.2.

mobilization of personnel, equipment, and materials, and the establishment of on-site facilities.

The contractor in Stage One is usually paid on a cost basis, although some pricing elements may be defined, like price caps for certain activities. PCSA may also include monetary incentives to ensure the contractor's commitment to the project, like a fee contingent upon awarding of the main contract in Stage Two tendering.

During Stage One activities, the owner and the contractor will usually negotiate the terms of the main construction contract, generally with a view to appointing the contractor selected during the preconstruction phase as the main contractor. Tendering for the main contract does not necessarily await completion of the pre-construction activities and will open depending on the project's need.

Stage Two

The preconstruction contractor in a two-stage tendering is usually also appointed as the main contractor for Stage Two. On the one hand, a long period of familiarity between the owner and the contractor, and between the contractor and the project, creates a better relationship, a faster learning curve and a performance program that is less likely to suffer from disruptive fluctuations. On the other hand, the owner inevitably loses leverage in Stage-Two tendering. With the preconstruction contractor already involved in the project team, other potential contractors may lose interest and consider Stage Two tendering more complicated to obtain. The ensuing negotiation power imbalance may drive up prices for Stage Two tendering. The owner may nonetheless appoint a different contractor, as evidenced in *Almacantar (Centre Point) Ltd v Sir Robert McAlpine Ltd.*[12]

While PSCAs have usually been customized agreements, various building contract suites – particularly those applied in the United Kingdom – include two-stage tendering and a PCSA option. Examples follow:

- **GC Works**: published in 1999 as part of the larger GC/Works suite of contract, designed for governmental projects, the *GC/Works/1/Two Stage Design and Build* was explicitly created for two-stage tendering. It envisaged the conditional appointment of a main contractor for participation in design, converting into an unconditional appointment when the design and other preconstruction activities had been sufficiently developed. The form is no longer in use and has been replaced by the *NEC suite of contracts.* It is nonetheless a testimony to the first attempt at two-stage tendering, useful to understand its basic components and mechanics;
- **Joint Contracts Tribunal – JCT**: JCT published two preconstruction service agreements, last updated in 2024: (i) the JCT Pre-Construction Services

12 The case is available at: https://www.bailii.org/ew/cases/EWHC/TCC/2018/232.html.

Agreement (General Contractor) (PCSA 2024), for use where the owner engages a general contractor; and (ii) the JCT Pre-Construction Services Agreement (Specialist) (PCSA/SP 2024), for use where the owner engages a specialist contractor. The JCT PCSAs are modeled on professional consultancy contracts. They cover all preconstruction activities leading up to the contractor's submission for Stage Two tendering, as well as the entry into operation of the main contract. The JCT PCSAs do not contain any obligation on the owner's side to conclude the main contract with the preconstruction contractor;
- **New Engineering Contract – NEC3**: in November 2015, NEC published a set of conditions for early contractor involvement (ECI). These clauses implement certain changes to the NEC3 Engineering and Construction Contract (ECC) to allow for two-stage tendering. With ECI, the parties can choose between two alternatives for performing preconstruction activities. The contractor can either (i) assist the client's consultant with design and other activities or (ii) carry out the design and other ECI activities by itself, with the assistance of the client's consultants (if any). The ECI contractor is appointed for a specific period, and, as it is usual for two-stage contracting, the client has either the option to confirm the contractor for the main contract or appoint a different contractor;
- **New Engineering Contract – NEC4**: in June 2017, NEC introduced the NEC4 suite of contracts, which, as NEC 3, included an ECI option, named Option X22. The renewed ECI conditions follow the experience built with NEC3 but provide more flexibility in the development of Stage One and greater clarity over the procedure in case the works do not advance to Stage Two. Moreover, NEC4 limits the Stage One contractor's liability to "reasonable skill and care" instead of a default rule of liability.

As previously noted, the transactional costs associated with two-stage tendering are justified in the context of larger and more complex projects. Conversely, simpler projects may not gain significant advantages from early construction involvement, thereby rendering the elevated costs of two-stage tendering unjustifiable.

Despite this, two-stage tendering can benefit both small and large projects. From the **owner's perspective**, engaging the contractor from an early stage can:

- improve the contractor's understanding of the project and reduce the risk of future variations and delays;
- result in better integrating the owner's project team and the contractor;
- ensure that the owner is involved in the selection of subcontractors, an activity that is usually covered under PCSA;
- let the project to commence before the scope of work is fully defined and all design-related studies are fully completed;

From the **contractor's perspective**, two-stage tendering can be an incentive to optimize the performance of preconstruction activities, develop a deep knowledge of the project, and put the contractor in a better position to secure the main contract. The expertise gained during the preconstruction phase may also be advantageous in negotiating the terms of the main contract and deterring other contractors from becoming interested in the project.

However, both the owner and the contractor remain at risk. From the owner's perspective, two-stage tendering results in higher costs and, possibly, a loss of competitive tension due to de significant advantage that the preconstruction contractor acquires over other interested contractors during the performance of Stage One. Owners may mitigate these risks by, for example, selecting more than one contractor for the preconstruction phase and then, based on performance, choosing the preferred contractor for the main contract. Alternatively, the owner may develop in detailed terms the PCSA to avoid re-negotiating specific provisions for the main contract.

The main risk from the contractor's point of view is performing all the activities under the PCSA and not obtaining the appointment for the main contract. There are no contractual avenues to avoid this risk, which should, however, be reduced to a minimum whenever the contractor can optimize the performance of preconstruction activities and gain valuable knowledge about the project.

Specific Preconstruction Legal Activities That a PCSA May Regulate

As outlined in Section "Regulating Legal Aspects of Pre-construction Activities", PCSAs could be a useful tool for both the owner and the contractor to deal with specific legal activities that should be carried out before the execution of the relevant construction contract and the performance of the works by the contractor. This section focuses on some of the critical preconstruction activities that could be covered by a PCSA, namely the activities related to (a) design, (b) procurement and subcontracting, (c) time and cost analysis, and (d) early works. Moreover, this section will highlight certain relevant issues that a PCSA should govern as well to mitigate the risk of future discussions between the parties, those being the price mechanism adopted to remunerate the activities carried out by a contractor under a PCSA and the provisions applicable if the relevant construction contract is not entered into.

Design Process

In a PCSA, the owner may take advantage of the contractor's design capabilities and share with the contractor the preliminary design documents developed at that stage by professionals directly retained by the owner. Engaging the contractor in the design process at a preconstruction stage could have multiple purposes. Namely, the contractor may:

- **review** the content of the **preliminary design documents** to identify and notify to the owner any errors, omissions, or other constructability risks and criticalities, which could impact the contractor's ability to perform the works expeditiously and economically. The review prepared by the contractor should then be compared with the analysis developed by the owner and its consultants, and any discrepancy should be discussed between the parties. This activity could be envisaged in every type of project, regardless of the complexity and value, to reduce the risks connected with the design of the works in the interest of all the parties involved. The foregoing is without prejudice to the mandatory provisions in force in certain jurisdictions stipulating the owner's obligation to obtain a positive constructability review by a panel of independent engineers as precondition for the performance of the works;
- based on the preliminary design documents it has received, develop **engineering proposals** to reduce the project's overall costs and/or anticipate the estimated completion date. If this activity, known as "value engineering," is carried out at the pretender stage, it may also be valued by the owner as a factor to be considered when selecting the contractor to carry out the work. It is important to bear in mind that if the design is still at its preliminary stages, the accuracy of the value engineering analysis will be limited, as the contractor will be unable to evaluate in detail the cost and time drivers of the project;
- be **involved in the design process** to expedite the completion of the relevant design and facilitate the commencement of the procurement phase, especially in circumstances where the owner intends to select the preconstruction contractor also for the performance of the works and/or time constraints related to the supply chain exist. It is self-evident that the aforementioned cooperation could be declined in multiple ways. Primarily, the contractor may start developing the executive design. Alternatively, the contractor may provide design assistance services to the design consultants already hired by the owner by sharing with them the contractor's specialized knowledge and skills, which could contribute to the overall evaluation of the buildability and affordability of the works.

From a legal standpoint, the PCSA must clearly state the objectives pursued through the owner's sharing of preliminary design documents with the preconstruction contractor. Depending on the selected objective (or combination of objectives), the PCSA must also define the respective obligations of the parties involved in the design development process. For instance, if the contractor is required to review the preliminary design, the PCSA could include time limits for the contractor to provide the owner with a written report of the findings of the review of the design documentation and/or for the owner to respond to the contractor's requests for clarification.

Furthermore, it is paramount to address the allocation of liability between the owner and the contractor in the event of errors or omissions in the preliminary design documents that are not identified during the preconstruction phase.

The **owner** will typically wish the contractor to assume full responsibility for any design documents it prepares, as well as for the preliminary design documents it reviews. Accordingly, the owner will require the contractor to declare that the preliminary design is complete and adequate to support both the detailed design and the execution of the works, except for any deficiencies that the contractor identifies and communicates to the owner prior to the signing of the contract governing the construction stage. As a result, the contract would exclude any claims by the contractor arising from errors or omissions in the preliminary design. In such a scenario, it is advisable to address whether the contractor's design liability will survive if the contractor is not engaged for the construction stage.

Conversely, the **contractor** will have an interest in avoiding any form of endorsement of the reviewed preliminary design, limiting its potential liability to the design activities directly carried out at the preconstruction stage (if any), such as the value engineering proposals or the executive design documents directly developed by the contractor ahead of the execution of the construction contract. In this scenario, the owner must ensure full recourse against the consultants employed to develop the preliminary design. The approach adopted under the PCSA regarding the allocation of liability for errors or omissions of the preliminary design will likely be reflected under the construction contract. Hence, the parties involved should give close attention to the negotiation and drafting of the relevant contractual provisions, given that their terms will significantly impact the design liabilities of the whole project.

In standard practice, if the contractor is involved in the design development at a preconstruction stage, it may propose to extend the participation of key subcontractors and suppliers to such a process. In this way, the contractor and the owner may rely on the specialized knowledge of the professionals who will directly undertake certain portions of the works. It may also be possible that the owner opts to deal only with subcontractors and suppliers, who will be treated as "nominated" subcontractors for the main contractor in the execution stage. For these reasons, the selection and appointment of key subcontractors and suppliers during the preconstruction phase plays an important role. The following subsection will address this legal activity.

Subcontracting

Under a PCSA, the owner and the contractor might deal with the early selection and appointment of subcontractors. Regarding the method of selection, the subcontractors could be (in whole or in part): (a) preselected by the owner and appointed by the contractor (the so-called nominated subcontractors); (b) identified and, subject to the positive outcome of a prequalification process managed by the owner, appointed by the contractor; or (c) jointly identified by the owner and the contractor, and finally appointed by the contractor.[13]

In the context of a structured preconstruction phase governed by a PCSA, opting for the joint selection of key subcontractors could be a viable solution to allow the subcontractor's selection based on criteria that are consistent with those applied to the selection of the contractor and avoid the need of nominated subcontractors. Furthermore, a joint selection can reasonably expedite the qualification. Careful consideration should be given to the subcontract agreement(s) to be entered into: specific reference is made to the scope of activities to be performed and the pricing method. Ideally, the regulation of these aspects should align – to the extent possible – with the terms and conditions of the PCSA, the characteristics of the project and the preferences of both the owner and the contractor.

The subcontractor appointed during the preconstruction phase may contribute to the design process and advise on the sequencing of works and identification of work packages. Furthermore, if the design documents are sufficiently developed, the subcontractor may begin placing orders for materials with long lead times and/or those whose prices are subject to fluctuations. The early commencement of the procurement phase may be advantageous in avoiding potential delays in the start of the works and ensuring cost certainty.

Cost and Time Analysis

In the event of involvement in the design process, under the PCSA, the contractor (directly or through any early subcontractors) may also be required to:

- develop a cost analysis of the works to be shared with the owner. A comparison of this analysis with that already developed by the owner's consultants could give the owner helpful information regarding the affordability of the project.

13 Either approach might have a different impact on the allocation of liability among the owner and the contractor with respect to subcontractors. For instance, opting for nominated subcontractors could give room to the contractor to claim the owner's liability for default and insolvency of the relevant subcontractor. For further considerations in this respect, see D. MOSEY, *Early Contractor Involvement in Building Procurement*, Wiley-Blackwell, 2009, p. 89.

The content of the cost analysis will necessarily vary depending on the pricing method adopted. If the parties have opted for a cost-plus fee basis, the contractor may be prepared to disclose to the client the cost of the works on an open-book basis.[14] Otherwise, the output will reasonably be a fixed and comprehensive sum estimated by the contractor as adequate to cover all the estimated costs, contingencies and profits of the project;

- analyze the schedule of the works and give inputs on the estimated completion date and the identification of the milestones for the completion of certain works packages. Once more, this activity is usually performed by the contractor under a construction contract before the commencement of the works within strict deadlines. Moving this activity to the preconstruction phase allows the contractor to allocate more time to the analysis of the relevant documentation, thus reducing delays in the commencement of the works.

Early Works

It happens very often that the owner, during the preconstruction phase, requires the contractor to carry out certain early works that are preparatory to the construction phase (e.g., excavation works, ground clearance works, strip out works). One of the reasons for this choice is that the execution of early works under a PSCA allows the works to start expeditiously upon execution of the main contract, as the site has already been prepared, thus reducing the risk of delays.

When early works are comprised within the scope of works of the contractor during the preconstruction phase, the PSCA shall envisage the main terms and conditions applicable to their performance, including the consideration (see Section Other Relevant Legal Aspects) and the final term of completion. It is also important to provide that, if the main contract is executed, the early works will be considered part of the works under the main contract and subject to the same terms and conditions.

Other Relevant Legal Aspects

As with any construction contract, one of the most complex issues when negotiating a PCSA is the **pricing mechanism** to be used to remunerate the work carried out by the contractor.

---

14 This means that the contractor, at the inception of the project, will share with the owner a detailed breakdown of all the estimated costs of the works, supported by the quotations received by the relevant suppliers and subcontractors, together with the applied margin and, during the performance of the works, it will report in detail the costs incurred.

From the contractor's perspective, the most convenient criteria would be a pricing mechanism based on a *time cost basis* for services and on a *cost plus fee* basis for early works. By adopting this pricing mechanism, indeed, the contractor would not face the risk of underestimating the effort required during the preconstruction phase as it will be paid for all the hours worked to perform the services and for all the documented costs incurred for carrying out the early works (plus a predetermined fee to cover general overheads and profit).

Nonetheless, the above-described pricing mechanism could present disadvantages for the owner, who will not have certainty of the costs associated with the preconstruction phase until its completion. Hence, the owner would be more willing to consider a lump sum method, by virtue of which the owner will pay to the contractor a preagreed and fixed sum against the performance of the works and activities, regardless of the actual costs incurred by the contractor. A balanced solution could be a time cost/cost plus fee criteria combined with a maximum guaranteed price.

Finally, it could happen that the construction contract is not entered into, because the **project is aborted**, or the owner decides to award the construction phase to another contractor or the parties do not reach an agreement on the terms of the main contract within a reasonable term. To avoid any liability on the part of the owner in this scenario, the PSCA should clearly provide that: (i) the awarding of the construction phase is only possible, but not necessary; and (ii) if the owner decides, for any reason, not to award to the preconstruction contractor the construction phase, the owner shall not have any obligation, liability or commitment, even of a precontractual nature, toward the contractor, other than the obligation to pay the consideration due under the PSCA and both parties shall be released from any obligation concerning the negotiation of the main contract. In addition to the foregoing, it is advisable to specify the contractor's obligation to cease the execution of the activities undertaken; deliver to the client all documents, projects, artifacts, rights or other utilities realized, executed, acquired, or in any case possessed in execution of or in connection with the PSCA; and clear the site (except as necessary for safety reasons or to preserve the usefulness of the work carried out).

Conclusion

The preconstruction phase of a construction project is a complex and dynamic process that requires careful planning, coordination, and communication among the parties involved. The regulation of the preconstruction phase can have significant implications for the project performance, risk, and dispute resolution. Therefore, parties should carefully consider the advantages and disadvantages of the different contractual and noncontractual options and tailor them to their specific needs and objectives. The choice of the most suitable method to regulate the preconstruction phase depends on various factors, such as the size, complexity, and duration of the project, the level of certainty and flexibility required, the degree of trust and cooperation among the parties, and the applicable legal and regulatory framework. Parties should also be aware of the legal activities that arise irrespective of executing a preconstruction contract, such as design, permitting, and other regulatory issues, and ensure that they are correctly addressed and allocated.

A well-regulated preconstruction phase can contribute to the project's successful delivery by enhancing the design's quality and efficiency, reducing the uncertainty and variability of the costs and time, and fostering a collaborative and constructive relationship among the parties.

3

Project Delivery Methods

Virginie Colaiuta and Costanza Mosca

Introduction

In the construction industry, project delivery methods refer to the approaches and contractual frameworks employed to plan, manage, and execute a construction project from inception to completion. These different approaches delineate the relationships among the parties, that is, the owners, the designers, the contractors, and the subcontractors, and define the allocation of their responsibilities.

This chapter will examine the different project delivery methods available to stakeholders for the successful completion of construction projects. Each method offers different levels of control, degree of collaboration, project risk allocation, speed of delivery, and distribution of responsibilities, allowing a project owner to choose the approach that best aligns with their objectives, costs and timeline.

The project delivery method is typically defined at the time of the procurement phase. The term "procurement" in a construction project refers to "*the process of the purchase of goods and services from conception to completion. It is a vital part of the overall construction process.*"[1] The contractual arrangements for the realization of a construction project will derive from the procurement terms.

The selection of the project delivery method will depend on various factors, including, among other things, the legal framework, the negotiating leverage of the party proposing a specific method, the level of risk that the proposing party is

1 Thomson Reuters Practical Law, 2025. *Construction: Regulatory Overview.* [online] Available at: <https://uk.practicallaw.thomsonreuters.com/Document/I114acbd9a93b11eabea3f0dc9fb69570/View/FullText.html?transitionType=SearchItem&contextData=(sc.Default)> [Accessed Feb. 7, 2025].

Major Issues in International Construction Law, First Edition. Mauro Rubino-Sammartano and Troy Harris.
Published 2026 by John Wiley & Sons, Inc.

willing to take, how the project is expected to be financed, and the nature of the project.

Project delivery methods are necessarily affected by the legal framework governing them.

In England and Wales, the Housing Grants, Construction and Regeneration Act 1996,[2] also known as the Construction Act 1996, and the Building Safety Act 2022[3] apply to construction projects located in the UK. For the purpose of this chapter, major focus will be placed on the Construction Act 1996.[4]

The Construction Act 1996[5] requires certain payment provisions to be included in construction contracts. If the contractual terms agreed by the parties do not comply with the requirements imposed by the Construction Act 1996, the terms in the Scheme for Construction Contracts will necessarily apply. The Construction Act 1996 also introduced a statutory right for contracting parties to refer disputes to adjudication, as well as the right to suspend the works in case of non-payment.

All construction and engineering projects, regardless of their size and technical particularities, involve a complex network of contractual relationships between multiple parties. The owner, also known as the employer or client, is the party for whose benefit the work is carried out. The owner is typically responsible for hiring the professional team that will overview the realization of the project and for appointing the multiple parties who will contribute to the realization of the project. The role of the owner varies depending on the nature of the project to be realized and the contract entered into.

The owner is often assisted by an architect whose typical role is that of the designer of the works. The role of the architect has evolved over time by extending beyond the mere development of the design. An example of activities the architect may be expected to carry out are:

- Reviewing and assessing the site;
- Advising the owner on the use of the land, such as planning and development issues, restrictive covenants, and access issues that may impact the performance of the works;
- Advising the owner on matters relating to the execution of the project, such as cost planning;
- Preparing the designs and the specifications for the works;
- Identifying project related risks;

2 Housing Grants, Construction and Regeneration Act 1996, c.53.

3 Building Safety Act 2022, c.30.

4 Housing Grants, Construction and Regeneration Act 1996, c.53.

5 Ibid.

- Preparing documents related to the tender process and providing recommendations on the submitted tenders;
- Monitoring the execution of the works.[6]

Another key role in a construction project is the engineer, who is typically responsible for supervising the technical aspects of the work. His role may overlap with that of the architect, especially when the project involves mechanical, electrical, and engineering works.

The owner may require the support of a quantity surveyor. The role of the quantity surveyor is to estimate the quantities required for the completion of the works. The quantity surveyor can be hired to support the owner in preparing the necessary tender documents or to review tender submissions where a bill of quantities has been prepared.

The contractor is responsible for carrying out the works. Depending on the type of contract, the contractor may perform the work on its own or rely on subcontractors to perform all or part of its scope of work. The nature and scope of the work may affect the role and responsibilities of the contractor. The subcontractors will undertake sections of the scope of the work delegated by the main contractor.

The suppliers will provide the materials that are specifically required for the project.[7]

Another key aspect in a construction project is the time for its realization. Time is of the essence in nearly all construction projects. In projects involving multiple contractors, the parties must agree on a timeline that specifies when the project should begin and be completed and the succession of its phases. Construction contracts often include express provisions specifying a completion date.

Most construction contracts require contractors to proceed with their work "regularly and diligently." *West Faulkner Associates v London Borough of Newham*[8] is

---

6 Stiegler, S. "Parties to a Construction Contract." Global Arbitration Review (GAR) – The Guide to Construction Arbitration, [online] Available at: https://www lexology.com/library/document?tk=eyJ0eXAiOiJKV1QiLCJhbGciOiJIUzUxMiJ9.eyJleHAiOjE3NTEzMDU1NTcsImRhdGEiOnsiRG9jdW1lbnRHdWlkIjoiMGQ4NDc3OTktMzBlOS00OTQwLThkYmUtOTQzMTBkYmM3NGVjIiwiQ29udGFjdEd1aWQiOiIwMDAwMDAwMC0wMDAwLTAwMDAtMDAwMC0wMDAwMDAwMDAwMDAiLCJCeXBhc3NMb2dpbiI6dHJ1ZX19.WUy6BaNYc5keOj8e9dvnyVpL_j4UgjnFU4HyhcNPKKhXBnJDOcokEWBny-LxgJVJqQwZ59WAlrsmGKj4dbH2Gg [Accessed Jun, 30 2025].

7 *Practical Law, Parties to a Construction Contract* | Thomson Reuters [online]. Available at: <https://uk.practicallaw.thomsonreuters.com/5-386-5860 originationContext=document&transitionType=DocumentItem&contextData=(sc.Default)&ppcid=a89ceb92e74d4450ab22e0a0487b16ed&comp=pluk> [Accessed Feb. 4, 2025].

8 [1994] 71 BLR 1.

considered the leading authority on this issue. The Court of Appeal held that "*regularly and diligently*" meant:

> "to proceed continuously, industriously, and efficiently with appropriate physical resources so as to progress the works steadily towards completion substantially in accordance with the contractual requirements as to time, sequence, and quality of work."[9]

The Court of Appeal held that providing useful guidance beyond this would not have been possible and that "*like the elephant,*" the failure to proceed regularly and diligently is "*far easier to recognize than describe.*"

The guidance in *West Faulkner*[10] was adopted and elaborated in the case *SABIC UK Petrochemicals Ltd v Punj Lloyd Ltd.*[11] In *SABIC UK Petrochemicals Ltd (formerly Huntsman Petrochemicals (UK) Ltd) v Punj Lloyd Ltd,*[12] the contract required the contractor to proceed "*with due diligence... and complete the Engineering Works in accordance with the Contract.*" The Court's interpretation was that the contractor had failed to execute the works in accordance with the plan and had failed to proceed with due diligence despite the warnings from the owner. The Court confirmed that the obligation of due diligence pertains to fulfilling the contractual duties related to completing the works. While delay may indicate a lack of due diligence, it does not, in itself, constitute conclusive proof. The obligation of due diligence is not diminished simply because achieving a particular contractual objective is, or becomes, impossible. In such circumstances, due diligence must be exercised to mitigate the continuing effects of the breach.[13]

The choice of the project delivery method can significantly influence the project's total cost. On the one hand, a specific delivery method that offers the client greater flexibility, such as allowing for design changes during construction, can lead to increased costs. Last-minute changes often result in additional expenses due to rework, procurement delays, and disrupted workflows. This approach, while adaptable, tends to reduce cost certainty and makes budgeting more difficult. On the other hand, a delivery method that limits the contractor's ability to deviate from the original design could allow the contractor to make only minor

9 Ibid paragraphs 26–29.

10 [1994] 71 BLR 1.

11 [2013] EWHC 2916 (QB).

12 *SABIC UK Petrochemicals Ltd (formerly Huntsman Petrochemicals (UK) Ltd) v Punj Lloyd Ltd* [2013] EWHC 2916 QB).

13 Lovells, H. 2017. *How does a contractor comply with a duty to proceed regularly and diligently with construction works?* Lexology. [online] Available at: https://www.lexology.com/library/detail.aspx?g=7caedb8a-3df9-49d8-ac21-0a999235101b [Accessed Jun, 30 2025].

adjustments while helping to have more control over the total costs of the project. By minimizing variations and defining scope and pricing early in the process, the latter method provides greater cost predictability and reduces the likelihood of budget overruns.

The project delivery method may be selected in function of the nature of the project and the related quality requirements. Quality can also be assessed in terms of the building's expected lifespan. Specifying certain quality standards in a contract may increase both the overall project cost and the time needed for completion.

The elements described above, including the nature, the value, and the risks of the project, are determining factors for the choice of the project delivery method.

Project delivery methods can be categorized into three main types: traditional, alternative, and collaborative, which will be examined in the sections below.

Traditional Project Delivery

Traditional project delivery methods are the most conventional, with distinct responsibilities assigned to each stakeholder. They are characterized by clearly defined responsibilities. The design and the construction phases of the project are kept separate and are carried out by different parties. These methods typically involve less collaboration among the contracting parties and can result in delays and cost overruns if the design and construction teams are not aligned.[14]

Design-Bid-Build

The design-bid-build ("DBB") is one of the most widely used project delivery method in the construction sector. As the name implies, the project is organized into three distinct phases: design, bidding, and construction. Each phase is built on the previous one, creating a straightforward process of successive steps.

This step-by-step approach ensures in principle efficient progression from the beginning of the project until its completion. This project delivery method is widely used because of its transparency and predictability. It clearly distinguishes the roles of the owner, designer, and contractor. In public projects where fairness and accountability are key to the performance of the contract, this structure is often preferred.

14 Thomson Reuters Practical Law, 2025. *Procurement: Common Construction Procurement Methods.* [online] Available at: <https://uk.practicallaw.thomsonreuters.com/Document/Ib9aa3f241c9a11e38578f7ccc38dcbee/View/FullText.html?comp=pluk&transitionType=SearchItem&contextData=(sc.Search)&OWSessionId=049e55ac422342e5904de5ab1870ef9d&skipAnonymous=true&firstPage=true> [Accessed Feb. 4, 2025].

The owner is the key figure in the process, as it is responsible for initiating the project, hiring the architect and the design team, and overseeing the entire realization of the project. The main responsibility of the owner is to ensure that every part of the project is carried out in compliance with defined requirements and in view of specific objectives.

The owner also serves as the bridge between all the parties involved, who may not interact directly with one another.

Designers are responsible for implementing the idea of the project that the owner intends to realize. Creativity, accuracy, and competency are fundamental for the realization of the project. Designers may also play a key role in supporting the owner in the realization of the project from a financial standpoint, in that their proposals need to fall within the owner's budget.

Contractors are key actors during the building phase as their job is to execute the project. Contractors are responsible for the choice and use of materials, implementation of the plans, and assessment of the risks.

The design phase lays down the foundation of the project. The procedure begins with the preliminary design by the designer or the architect establishing the structure, layout, and size of the construction project. Once the initial design is approved, a more detailed technical design is created, focusing on the specifications, plans, any technical requirements, and list of materials.

Once the design is approved, the contractors are invited to submit their bids, which outline the contractors' qualifications, a timeline for the completion of the work, and most importantly, the cost of the project. The owner is responsible for evaluating the bids and selecting the contractor on the basis of a combination of criteria, which may rely on the price, the credentials, and a history of successful performance and compliance with the requirements of similar projects.

The construction phase is the final phase of the project delivery method. The winning contractor manages the on-site operations and deals with the construction of the project. The role of the owner during this phase is to oversee the works, ensuring that they are carried out as previously planned and agreed.

In the DBB method, collaboration is mainly limited due to the separation of the various phases and the distinct roles of each stakeholder in such phases. Each party works independently by ensuring that its role is carried out as expected. Communication is as important as any miscommunication between the parties may lead to disruptions, delays, and cost increase.[15]

Advantages

The DBB project delivery method offers several advantages to the parties, making it a common choice in the construction industry.

---

15 Ibid.

One of the principal advantages is the clear division of the roles and responsibilities of all the stakeholders involved, supported by the three-phase structure of the project. This approach establishes well-defined boundaries and facilitates the allocation of tasks and responsibilities between the owner, architect/designer and contractor. Moreover, this method provides clarity in respect of the obligations each party has toward the completion of the project.[16]

The separation of the phases reduces to some extent the risk of any possible overlapping and ambiguity in the tasks of each party by establishing a clear framework, ensuring that everyone knows what to expect at each stage.

Multiple contractors can bid on the project and provide the owner with their pricing. This ensures transparency and enables the owner to select the best value for its investment by assessing and comparing multiple offers to determine the most cost-effective option.

Since the design must be agreed upon and finalized before selecting the contractor, the owner retains significant control over design decisions. This may in principle reduce the likelihood of alterations at later stages.

As mentioned, the design-bid-build method is a traditional approach, which has been tested in the construction industry for a long period of time. This traditional method is often selected, as owners and contractors are familiar with it.[17]

Disadvantages

However, despite its extensive use, the DBB method has faced significant criticism.

In particular, the process by which a contractor is selected during the bidding process mainly relies on pricing rather than on contractors' qualifications, their project-related experience, and past record of success in similar projects. As a result, the BDD approach may lead to the selection of contractors who may not be the most suitable for the nature of the project and for carrying out the requested tasks.

Furthermore, the separation of the design phase from the bidding and the building phases can potentially hinder effective communication and collaboration among the owner, the designer, and the contractors, which may lead to disagreements and misunderstandings. The limited role of the contractor and the reduced interaction with the engineer and the designer during the design phase

16 Almuhannadi, M.A. and Ghareeb, A.S., 2024. *Enhancing Design–Bid–Build Project Delivery: A Comprehensive Review and Framework for Contractor Selection and Project Optimisation in the Construction Industry.* Organization, Technology and Management in Construction, [online] 16(1), pp. 63–80. Available at: <https://www.researchgate.net/publication/381367174_Enhancing_design-bid-build_project_delivery_A_comprehensive_review_and_framework_for_contractor_selection_and_project_optimisation_in_the_construction_industry> [Accessed Feb. 5, 2025].

17 Studio Forma. (n.d.). *Design-Bid-Build vs Design-Build.* [online] Available at: <https://www.studioforma.ca/design-bid-build-vs-design-build/> [Accessed Feb. 5, 2025].

may create delays during the building phase, as the contractors may require clarifications on the technical plans and drawings. The reduced interaction between the stakeholders poses a limit to the use of their respective capabilities and to the increased effectiveness and cost-efficiency of the project. Increased communication between the parties would create more opportunities for improvement and innovations.[18]

With the DBB delivery method, the construction phase cannot commence until the design and bidding phases are completed. As previously mentioned, while the separate structure guarantees no overlap between the phases, it may delay the overall project timeline. The extension of the timeline of the project may lead to inefficiencies because of the sequential progression of the phases.[19]

Once the design is finalized, the project moves to the next phase, making adjustments difficult and challenging, thereby limiting the parties' flexibility to amend the project.

When to Use the DBB Method?

Although the DBB is a traditional method in the construction industry, it is not always appropriate for every type of project. The DBB method is widely used for public infrastructure projects as many governments and public bodies prioritize the structure, its transparent approach, and the use of competitive tendering which ensures value for money.

Alternative Project Delivery

Alternative project delivery methods are designed to streamline the construction process and minimize inefficiencies by fostering better coordination among project participants. They often seek to shorten project timelines, enhance cost-effectiveness, and promote greater collaboration between stakeholders throughout all phases of the project. We examine in the following paragraphs the aspects of some selected alternative project delivery methods.

Design Build

Design-Build ("DB") is a construction project delivery method that combines architectural and engineering design services with construction performance under one contract.[20]

18 Ibid.

19 Ibid.

20 National Institute of Governmental Purchasing (NIGP), 2019. *Selecting the Appropriate Construction Project Delivery Method: Best Practice.* [pdf] Available at: <https://www.nigp.org/resource/global-best-practices/Selecting%20the%20Appropriate%20Construction%20Project%20Delivery%20Method%20Best%20Practice.pdf?dl=true> [Accessed Feb. 6, 2025].

In a DB project, the owner may contract with a joint venture company, composed of a design engineering company and an in-house construction company, or with two separate companies collaborating to provide design and construction services together.[21]

After having decided on the procurement method, the owner will prepare a statement to provide the contractor with key information on the realization of the project. This statement is usually referred to as the "Owner's Requirements." The document can be both a general statement of the owner's needs and a more detailed statement incorporating the criteria for the project, drawings, and specifications. The Owner's Requirements are then sent to the selected contractor.[22] In the DB construction project delivery method, the contractor is selected through a Request for Proposals process based on the combination of qualifications of the contractor and the price charged. However, other owners may select the contractor by using the Request for Qualification process, whereby the selection occurs by evaluating the contractor's expertise and qualifications.

It is common for the client to instruct professional consultants to develop the initial designs. Once the designs are completed, the client, the consultants, and the contractor enter a novation agreement, appointing the contractor and allowing it to complete the construction of the project. All the parties to the original contract (the outgoing and continuing parties) and the incoming party must consent to the novation for it to be valid.[23] On construction projects, novation agreements usually substitute one party for another in respect of all the rights and obligations under a contract. In *Blyth & Blyth Ltd v Carillion Construction Ltd [2001]*, the owner engaged the company Carillion as a D&B contractor on a development project in April 1998 and appointed Blyth & Blyth as its consulting engineer in May 1998. The terms of the appointment entitled the owner to instruct and require Blyth & Blyth to enter a novation agreement. In June 1998, a tri-partite novation agreement was entered between the owner, Carillion, and Blyth & Blyth, with the effect that Carillion became Blyth & Blyth's owner and Blyth & Blyth became Carillion's subconsultant.

---

21 Mehany, M.H.M., 2018. *Claims and Project Performance between Traditional and Alternative Project Delivery Methods.* [pdf] Available at: <https://www.researchgate.net/profile/Mohammed-Hashem-M-Mehany/publication/326745396_Claims_and_Project_Performance_between_Traditional_and_Alternative_Project_Delivery_Methods/links/5efe056b299bf18816fa86d8/Claims-and-Project-Performance-between-Traditional-and-Alternative-Project-Delivery-Methods.pdf> [Accessed Feb. 6, 2025].

22 Ndekugri, I. and Turner, A., 1994. *Building Procurement by Design and Build Approach.* [pdf] Available at: <https://shaghool.ir/Files/1994-020.pdf> [Accessed Feb. 6, 2025].

23 Practical Law Construction. (n.d.). *Novation of Construction Documents* [online]. Available at <https://uk.practicallaw.thomsonreuters.com/1-315-1953?comp=pluk&contextData=(sc.Default)&transitionType=Default&firstPage=true&OWSessionId=39bd150ce55e47be8ac4a83ceb3dfe82&skipAnonymous=true> [Accessed Feb. 24, 2025].

Under Clauses 4 and 5 of the novation agreement, Blyth & Blyth agreed that:

- Its liability was to Carillion, whether the liability accrued before or after the date of the novation, as if Carillion had always been named as a party in place of the owner.
- It was liable to Carillion in respect of any breach of the appointment occurring prior to the novation as if Carillion had always been named as a party in place of the owner.

The dispute arose after Blyth & Blyth brought a claim against Carillion for unpaid fees. Carillion filed a counterclaim, alleging that Blyth & Blyth had performed the pre-novation services negligently causing it to suffer a loss. When considering that all pre-novation duties were to be regarded as having been performed for Carillion, the Court concluded that Carillion did not have the benefit of Blyth & Blyth's pre-novation duties in its own right. The Court agreed that Carillion was a "creditor" of the losses that the owner may have suffered as a result of the negligent performance of those duties.[24]

The *Blyth & Blyth Ltd v Carillion Construction Ltd* case created an element of uncertainty when drafting novation agreements. As a result, parties must ensure:

- The novation agreement does not contain any on-going post-novation rights owed to the original client.
- The contract to be novated makes practical legal sense for implementation.
- Drafting clauses that enable the contractor is entitled to recover its own losses in respect of pre-novation breaches by the consultant, in case of breaches by appointed consultants.

The client will have control over any design of the project that is included in the Owner's Requirements but will not have direct control over the contractor's detailed design once the responsibility of the design passes to the contractor.[25]

The contractor can appoint other professionals like subcontractors to carry out the detailed design, but the contractor remains liable toward the client for the completion of the project. The contract will typically confirm that even in the event of the appointment of subcontractors, the main contractor will continue to be liable for the performance of the works performed by the subcontractors. This is referred to as a "single point of responsibility."

24 *Blyth & Blyth Ltd v Carillion Construction Ltd* [2001] 79 Con. L.R. 142.

25 Thomson Reuters Practical Law, 2025. *Procurement: Common Construction Procurement methods*. [online] Available at: <https://uk.practicallaw.thomsonreuters.com/Document/Ib9aa3f241c9a11e38578f7ccc38dcbee/View/FullText.html?comp=pluk&transitionType=SearchItem&contextData=(sc.Search)&OWSessionId=049e55ac422342e5904de5ab1870ef9d&skipAnonymous=true&firstPage=true> [Accessed Feb. 6, 2025].

This streamlined structure encourages collaboration between the stakeholders involved from the beginning. Contrary to the DBB method, the parties can collaborate and communicate by addressing potential issues at an early stage thus reducing inefficiencies, technical problems, and the risk of delays.

The process involves planning and team selection and includes:

- A preliminary design and budget
- A collaborative design and development, and
- A streamlined construct and project execution.

A key strength of the DB method is that it simplifies the roles and responsibilities of the parties involved. Instead of separating the roles as in the DBB, all stakeholders collaborate under one single contract.

The owner's role is to define the grounds for the project, its goals, requirements, and budget. Rather than having multiple contracts and teams, in the DB, the owner builds a contractual relationship with one DB team.

The DB team includes architects, designers, contractors, and engineers all working together from the beginning up to the completion of the project.

In a DB project, the project manager handles the coordination between designers and builders, allowing the owner to focus on high-level decisions.

Advantages

The DB project delivery method provides multiple benefits to the parties and to the success of the construction project. The integration of design and construction activities under a single contract offers significant benefits to the overall management of the project. The DB team consists of a constructor that, pursuant to a prime contract with the project owner, is the "single point of responsibility" for both design and construction. The extent of any single point of responsibility depends on the terms of the DB contract. For this reason, a party should identify the risk of contractual and professional liability before signing the contract. For example, the architect or engineer must be cautious about the conflict between their duties to the contractor and to the ultimate owner of the project. On the one hand, the architect/engineer is instructed by the contractor and therefore acts as a client of the constructor not the owner. At the same time, the architect/engineer owe a duty also to third parties, for example, the owner, and must perform the work in accordance with reasonable skill and experience expected from individuals operating in the industry.[26] It is important to highlight that the engineer instructed by the contractor covers a completely different role with distinct

26 Smith Currie, n.d. *Who Owes What to Whom: Liability and Alternative Project Delivery.* [online] Available at: <https://www.smithcurrie.com/publications/common-sense-contract-law/who-owes-what-to-whom-liability-and-alternative-project-delivery/pdf/> [Accessed Feb. 6, 2025].

responsibilities with respect to the engineer appointed by the owner to monitor and control the work of the contractor.

Under the DB project delivery method, liability for non-performance of the work by subcontractors falls on the contractor.

The concurrent execution of design and construction phases can lead to reduced project timelines as construction can start before the design process is completed. As both phases are managed together from the outset, the parties can collaborate and have more control over the management of the budget and can limit the likelihood of extra costs.[27]

The parties' cooperation from the early phases of the project ensures that all the professionals involved can share their expertise, views and skills. This should ensure a seamless transition between design and construction, leading to better project outcomes and a holistic approach to decisions. Since the DB team is collectively responsible for both design and construction, there is a shared sense of accountability for the project success. Communication and collaboration are encouraged to ensure a coordinated effort and quicker resolution of any challenges. Integrated teams can more effectively manage quality throughout the project, leading to improved outcomes.

Disadvantages

However, with the same team managing both the design and construction phase, there's a risk that decisions may prioritize cost or convenience over the project's broader goals, giving rise to disputes between the parties.

Merging design and construction phases can result in fewer opportunities for the project owner to review the design proposals and make amendments later in the project.[28] Any proposed changes to the design would affect greatly the construction phase already underway and lead to potential cost increases.

In DBB projects, the owner, during the bidding phase, can compare a range of bids from multiple contractors deciding the best value for their investment after the design phase. In the DB project delivery method, the owner contracts with a single entity responsible for both design and construction, rather than engaging separate contracts. While this streamlines the execution of the project, it limits competitive bidding opportunities, making the selection process more critical for the owner. The owner therefore must evaluate the entire package, including design quality, team expertise and cost efficiency from the very start. In the DB

27 Royal Institution of Chartered Surveyors (RICS) 2018. *Developing a Construction Procurement Strategy.*[pdf] Available at: <https://www.rics.org/content/dam/ricsglobal/documents/standards/Developing-a-construction-procurement-strategy.pdf>[Accessed Feb. 6, 2025].

28 Ibid.

project delivery method, reliance on a single contract limits the opportunities for competitive bidding, leading the owner to evaluate the proposal appropriate for the project at the earliest phase of the process.

With emphasis being placed on cost efficiency and reduced project timelines, the DB project delivery method may disregard the importance of design, affecting the project's aesthetic or functional quality.[29]

Compromised design quality may also lead to serious consequences and to disputes regarding design responsibilities. This has been considered in *Costain v Haswell & Partners Ltd*[30] where Costain was contracted on a DB basis by United Utilities Water Ltd to design and construct a wastewater treatment plant. Consulting engineers were hired from Charles Haswell for the design. The consultants stated that the services were to be performed with reasonable skill and care, while the contract provided that the design had to meet the requirements set out in the design specifications. The conflicting criteria were construed as imposing a strict liability for design on the consultants, while other services had to comply with an obligation to exercise reasonable skill and care. The Court held that "*Clause 7.4 of the Consultancy Agreement does impose an obligation of strict liability upon Haswell in respect of its design if any part of the works based on such design is constructed by Costain to form part of the permanent works.*" The Court further concluded that "*it was not the intention of the parties that, potentially, Haswell would be under a strict liability obligation in respect of several successive designs but only in respect of the final design, which was incorporated into the permanent works by Costain*".

When to Use the DB method?

The DB method is very common in projects that have strict timelines, that is where the design and construction are complex and unique, and when staying within the budget is a key priority for the parties involved.

An example of the implementation of the DB method in an infrastructure project is the Luton DART, a fully automated double-shuttle cable-driven people mover at London Luton Airport (LLAL).[31] Completed in 2023 following an increased passenger capacity, that required LLAL to invest in developing and improving passenger experience. With a precise project timeline, established budget and the

29 Cambridge City Council, 2024.: Calford Seaden Procurement Strategy Report. [pdf] Available at: <https://www.cambridge.gov.uk/media/4ynhcu5h/appendix-4-calford-seaden-procurement-strategy-report.pdf> [Accessed Feb. 6, 2025].

30 *Costain v Haswell & Partners Ltd [2009] EWHC 3140 (TCC).* ([2009] EWHC B25 (TCC) *paragraph 63 of the judgment* - https://www.bailii.org/ew/cases/EWHC/TCC/2009/B25.html).

31 Kier Group (n.d) *Luton DART.* Available at: <https://www.kier.co.uk/discover-our-projects/luton-dart/>[Accessed Feb. 7, 2025].

need to have a team of experts working together, the DB method has resulted in the most appropriate project delivery method to adopt.

Construction Management

Under a Construction Management (CM) approach, the owner employs a design team including a construction manager, who would be responsible for managing, programing, and coordinating the design and construction activities. This occurs without the client allocating the risk and responsibilities to a main contractor.

The owner directly employs individual contractors, referred to as Trade Contractors, under individual contracts. Each trade contractor will be appointed throughout the construction of the project depending on the expertise and contribution required. The construction manager has no interest in the financial outcome of the project and carries no risk as his obligation is to act in the best interest of the owner. The construction manager supervises the construction process and coordinates the design team, but has no form of contractual relationship with the trade contractors.[32] The primary role of the construction manager is to offer professional expertise in managing the construction process.

The Construction Management method may lead to a form of overlap between the design and construction phases, and it is usually adopted by the owner when the primary objective is the completion of the project within a short timeline. This project delivery method is resource-intensive for the client as it requires the latter to instruct a dedicated team with experience in managing trade contract packages.[33]

On appointment, the Construction Manager will deal with any preliminary matters relating to the overall program of the contract, in particular estimate of costs, set realistic timelines, and flag potential risks. Collaboration and communication at this early stage will enable the parties to prevent any potential challenges.

The role of the contractors becomes pivotal at this stage. They are responsible for the review of the designs to ensure that the plans are implementable and practical. Their feedback may lead to some adjustments that will be potentially crucial in cost saving and technical issues. This step is fundamental to ensure that what initially planned aligns with the budget available and the Owner's goals.

---

32 Cambridge City Council, 2019. Appendix 4: Calford Seaden Procurement Strategy Report. [pdf] Available at: <https://www.cambridge.gov.uk/media/4ynheu5h/appendix-4-calford-seaden-procurement-strategy-report.pdf> [Accessed Feb. 7, 2025].

33 Royal Institution of Chartered Surveyors (RICS), 2024. Developing a construction procurement strategy. [online] Available at: <https://www.rics.org/content/dam/ricsglobal/documents/standards/Developing-a-construction-procurement-strategy.pdf> [Accessed Feb. 7, 2025].

Once the design is finalized, the project progresses into the construction phase. The role of the Construction Manager continues but shifts to a more practical level by coordinating the various teams of subcontractors involved.

The Construction Management delivery method is effective because of the clear division of roles and responsibilities amongst the parties involved. Each stakeholder has a specific role to play, in order to support the smooth delivery of the project.

The owner is the one that sets the vision of the project by choosing the Construction Manager at an early stage of the process. The owner is closely involved in each stage of the design and construction phase. It is key that the owner maintains his presence throughout the project and communicates with the experienced staff both off and on-site. This route is generally not suitable for inexperienced owners as they may encounter difficulties in understanding the technical aspects of construction projects.

The Construction Manager acts as a consultant at the initial stages of the process providing valuable input and once the project enters into the construction phase, he becomes involved with the day-to-day management of the project.

The designer/architect is responsible for developing the initial concept of the design based on the client's requirements. The designer/architect works closely with the Construction Manager to align design deliverables with the construction schedule. The responsibilities of each party involve also ensuring compliance with building regulations and planning permissions. The designer/architect is also responsible for conducting site visits and quality control checks during the construction phase.

The trade contractors are involved during the construction phase to handle specific tasks based on their expertise and work package. The packages are tendered individually for a lump sum price. They work under the supervision of the Construction Manager who ensures their work meets the requirements established with the owner.[34]

Advantages

Construction Management allows for the overlapping of design and construction phases, by reducing the overall project duration. In traditional procurement methods where each phase must be completed before moving to the next one, the Construction Manager enables early trade package procurement and site activities that accelerate the project timeline of completion.

---

34 Royal Institution of Chartered Surveyors (RICS), 2024, p. 15. Developing a construction procurement strategy. [online] Available at: <https://www.rics.org/content/dam/ricsglobal/documents/standards/Developing-a-construction-procurement-strategy.pdf> [Accessed Feb. 7, 2025].

Trade contractors and the Construction Manager actively contribute to the design process and project planning. Their involvement from the outset ensures practical and cost-effective solutions that help identify potential construction challenges from the beginning. This collaboration ensures that informed decisions are made and that the project is completed as requested by the owner.

The client directly engages and pays trade contractors, avoiding the need for a main contractor. This improves cash flow certainty for subcontractors, which can lead to lower prices and better value for money. Moreover, it ensures that the owner receives a project completed in line with their expectations and requirements.

Unlike other procurement routes that do not allow amendments throughout the construction phase, a Construction Manager provides greater flexibility by allowing design changes. This flexibility is especially beneficial for complex and evolving projects that may not always follow the initial plan and require adjustments during the construction phase.

Disadvantages

Unlike fixed price contracts used in the traditional procurement methods, Construction Management does not provide cost certainty from the outset. As trade contracts are awarded throughout the process, the cost of the final project remains uncertain until all the work packages are tendered and until the design is completed. This can result in problems for owners who have strict budgets and limited financial flexibility.

As previously mentioned, this project delivery method is not appropriate for clients who do not want to be involved or do not have the experience required. This is because throughout the design and construction phase a level of expertise and commitment. For inexperienced clients, having to manage multiple trade contractors can be overwhelming and may also lead to errors, disputes and increased costs.

The client does not have a single point of responsibility following a Construction Management approach. As multiple trade contractors are involved, each of them is liable for his contractual obligations under one's contract and for the performance of one's works. Determining liability may prove difficult when a dispute occurs.

Since the client holds direct contracts with all the trade contractors involved, any dispute must be addressed and resolved individually rather than through a single main contract. Delays in one package can affect the entire project, and if a trade contractor underperforms or goes into liquidation, the owner must find a replacement, which may disrupt the project timeline and cause further complications.

When Should Construction Management Be Used?

The Construction Management method stands out because it prioritizes collaboration and efficiency. It is often selected for large or complex projects where speed, flexibility, and client control are desired. It benefits projects that have a high degree of design innovation. This route is generally not recommended for clients with limited construction experience or those who require cost certainty from the outset.[35]

Construction Manager Multiple Prime

A Construction Manager Multi-Prime (CMMP) or Multi-Prime contract is an arrangement whereby the owner or client bypasses the traditional relationship with a single contractor and instead directly contracts with individual trade contractors, hence the term "multi-prime." The owner may choose to hire a construction manager to provide support throughout the project. Multi-Prime projects place owners directly in control, enhancing their oversight of subcontractors, project finances, and the overall project outcome.

In a construction project with a Multi-Prime contract, the owner hires a design firm to create construction drawings, a construction manager for project coordination, and directly manages individual contractors for various aspects of the project. The design team, comprising architects and engineers, collaborates directly with the owner to establish the project's design. The construction manager is involved early in the process to advise on design, coordinate efforts, and maintain schedule and budget, although they assume minimal control or risk for the project's outcome. Subcontractors in multi-prime contracts coordinate their work efforts and send updates and invoices directly to the owner or construction manager, ensuring streamlined communication and management.

Advantages

Multi-Prime contracts have several advantages for both owners and contractors. Owners maintain oversight of the project, allowing for direct management of individual contractors and potentially leading to improved quality and compliance with project requirements. By contracting directly with trade contractors, owners can avoid the burden of having to incur the costs of having a general contractor involved, which would in principle result in reduced overall costs. The involvement of multiple contractors allows for concurrent works on different packages, potentially expediting project delivery.[36]

---

35 Ibid.

36 Department of Finance (2010). *Construction Procurement Policy Framework*. Available at: <https://www.finance-ni.gov.uk/sites/default/files/publications/dfp/PGN%2006%2010%20%20Construction%20Procurement%20Policy%20Framework%20%5Bpdf%20version%5D.PDF> [Accessed Feb. 11, 2025].

Disadvantages

Managing multiple contracts requires significant coordination and administrative effort from the owner, necessitating robust project management capabilities. The involvement of multiple parties heightens the risk of disputes, often stemming from coordination difficulties or ambiguities in the division of responsibilities. Coordinating the schedules of various contractors may be challenging and time-consuming, potentially leading also to delays if not managed effectively by the owner.

Engineering, Procurement and Construction (EPC) and Turnkey Contracts

An Engineering Procurement Construction ("EPC") contract is *a form of building contract under which the builder will deliver a completed project, often on a turnkey basis.*

EPC contracts have been the most common forms of contracts used to undertake construction and engineering works by private investors in complex infrastructure projects across several industry sectors, including oil and gas, power, and transport.

In an EPC contract, a single contractor has responsibility for delivering a completed project according to the agreed specifications.

EPC contracts are often conceived as turnkey contracts. The expression 'turnkey' refers to the idea of a contractor being able to deliver a project to the owner upon completion in a ready-to-use state so that the project can be operated by simply "turning" the key.

In the construction industry, turnkey delivery simplifies the process by removing the need for clients to oversee multiple vendors or coordinate various phases of the project. This method is preferred by developers seeking to transfer most risks to the contractor in relation to the costs, the management, and the completion date. By delegating the entire project to a single entity, clients ensure accountability and minimize the risk of delays or unforeseen expenses.[37] There is a single point of contact.

In an EPC project, the contractor is responsible for the design as well as the construction.

Under an EPC contract, the contractor takes on more risks than under a Design-Build contract. An example of this is the different approach taken by ground conditions by the FIDIC Yellow Book (1999 and 2017 edition), referred to as FIDIC's D&B contract and FIDIC's Silver Book, which is the turnkey contract. The contractor takes the risk of the ground conditions under the Silver Book,

---

37 Mastt. (n.d.). *Turnkey Project.* [online] Available at: <https://www.mastt.com/blogs/turnkey-project> [Accessed Feb. 11, 2025].

whereas under the Yellow Book, the contractor is not liable for ground conditions that are "unforeseeable."

Moreover, the owner's involvement in the delivery of the project is very limited in a typical EPC contract. Additionally, under the EPC contract, there is little to no owner-design. The contractor is often provided with design specifications. The contractor is required to take full responsibility for the design due to the "single point of responsibility."

The contractor may also have the responsibility for testing and proving the performance and reliability of the facility.

In practice most EPC contracts are turnkey in nature. However, not all turnkey contracts are EPC contracts, as some turnkey arrangements may be simpler, and not all EPC contracts guarantee the full "ready to use" operational status that turnkey contracts imply. EPC contracts focus on how the project is delivered; turnkey contracts focus on what is delivered, that is, a fully operational facility ready for immediate use.

There are various standard forms of EPC contracts.

The *International Chamber of Commerce (ICC)* has published some forms of contract. Parties can use the ICC Model Turnkey Supply of an Industrial Plant Contract and the ICC Model turnkey Contract for Major Projects. Other examples of EPC contracts includes the Infrastructure Conditions of Contract, published by the *Association for Consultancy and Engineering (ACE).* Other standard forms of EPC contracts are those published by the International Federation of Consulting Engineers ("FIDIC"), known as the FIDIC Silver Book (https://fidic.org/books/epcturnkey-contract-2nd-ed-2017-silver-book). The EPC contractor holds full responsibility for all aspects of the project, including design, procurement, construction and commissioning. Typically, EPC contracts are based on a lump-sum price, providing cost certainty for the project owner. The contract specifies a completion date, with penalties often applied for delays attributable to the contractor. The contractor guarantees that the facility will meet defined specifications upon completion.

Procedure

The procedure under an EPC contract begins with project planning and feasibility, where the project owner conducts initial studies to define requirements and objectives. In the tendering and contractor selection phase, the owner issues a tender, inviting qualified contractors to submit proposals, which are evaluated based on technical capabilities, experience, and financial stability. Once a contractor is selected, contract negotiation takes place, defining the scope of work, pricing, schedule, and performance guarantees. The design and engineering phase follows, where the EPC contractor develops detailed designs in compliance with project specifications and regulatory requirements. During the procurement phase, the contractor acquires necessary materials and equipment while ensuring quality and adherence to the project timeline. The construction phase then commences with the contractor managing the construction process and conducting tests to verify operational performance. Finally, upon successful commissioning, the handover stage occurs, where the contractor transfers the fully operational facility to the owner.

If the contract is conceived simply as a turnkey contract, the procedure may be greatly simplified so that it starts with a contract agreement between the contractor and the owner, where the scope, budget, and timeline are established.

Allocation of Risks

Allocating design risk is a key element of every construction contract. In a typical EPC contract, the contractor takes on a greater part, if not all, of the design risk.[38]

One of the features of EPC contracts is that the contractor is expected to guarantee that the work will achieve a particular result. The extent to which the contractor's responsibility will be strictly objective with respect to the promised result or whether the contractor's responsibility will be limited to show that it exercised reasonable skill and care depends on the provisions of the contract and the law applicable to the contract. The owner may often require the contractor to be under an absolute obligation to deliver the promised result with the agreed specifications.

When should Turnkey contracts be used?

As explained above, the main difference between EPC and turnkey contracts is in the scope of the contractor's responsibility and in the allocation of risks between the contractor and the owner's project. In typical EPC contracts, the contractor is responsible for the design, procurement of materials, and construction of the project. Turnkey contracts comprise these same responsibilities plus the commissioning and handover of a fully operational project.

With turnkey contracts, more risks and responsibility are attributed to the contractors, who are to deliver a project ready for immediate use. Instead, in EPC contracts, there may be more involvement of the owner and more sharing of the project's risks.

Turnkey delivery is recommended for large-scale commercial developments like factories, retail centers, or hospitals. If an owner is looking for predictability and fixed budgets, turnkey is appropriate, as it provides for fixed-price contracts and clear timelines, providing the owner with a clear understanding of the expected outcome at project completion.[39] In cases where the owner/client lacks

38 Practical Law. (n.d.). *Engineer, Procure and Construct (EPC) Contracts: Key Issues.* [online] [Accessed Feb. 11, 2025].

39 Ibid.

construction expertise or seeks limited involvement in the project, the turnkey model enables delegation of full project responsibility to the contractor, who is accountable for delivering a fully operational facility.

Advantages and disadvantages of EPC and turnkey contracts:

Advantages

The advantages for the owner of having an EPC contract are that there is a single point of responsibility and management of the project and reduced risks because of the fixed price and fixed timeline. With an EPC contract, owners ensure accountability and minimize the risk of delays or unforeseen expenses. With a single contractor controlling all phases, the project timeline is often shorter than with traditional construction methods. Fewer delays due to miscommunication or coordination issues mean projects can progress smoothly and adhere to the schedule. As a result of such characteristics, there might be some potential cost savings.

Disadvantages

As the responsibility of the design and realization of the project is falling on the EPC contractor, the owner has limited control on those same aspects. There is also a reduced level of flexibility for possible changes or improvements. As the EPC contractor will have a limited budget for the realization of the project, there will be less emphasis on the quality of the features of the project that may be more costly. The success of the project is highly dependent on the performance of the EPC contractor. Reliance on a single contractor and reduced control by the owner may lead to a project not compliant with the owner's expectation and a dispute with higher costs.

Public–Private Partnerships

A Public-Private Partnership ("PPP") can be defined as the scenario where "*a government transfers the exclusive rights for developing and/or operating an infrastructure facility to a private operator or investor under certain conditions for a fixed period of time.*"[40]

The private investor agrees to provide public infrastructure in exchange for a specific financial benefit. Even though the private investor is responsible for the delivery of the service, the State plays a key role through the purchase of services provided, provision of public assets, and the creation of the social and economic environment for the PPP to be successful.

The private investor agrees to finance, build, and operate the project and provides the technical expertise and managerial skills required for the project.

40 HM Government. (2017). *Topic Guide: Public-Private Partnerships*. [pdf] Available at: <https://assets.publishing.service.gov.uk/media/5977576ee5274a289a000031/Topic_Guide_Public-Private_Partnerships.pdf> [Accessed Feb. 11, 2025].

The private investor is compensated according to its performance. PPPs can involve either the construction of a totally new environment called "greenfield projects" or the rehabilitation of an already existing asset, known as "brownfield projects." These partnerships rely on contracts that define specifically the roles and responsibilities of each party.

PPPs are usually relied upon because of a lack of public funds and in order to ensure better quality of services to the society. To the extent that PPPs rely on private investments, they in principle allow to complete projects on time and within budgets, increase market efficiency of the services provided, enhance market competition, exploit the flexibility of non-governmental entities, transfer in part the risk of a public project to private investors, encourage the use of technology and innovative solutions, may reduce maintenance costs, and optimize the use of resources.

A PPP project is achieved through the creation of a company created for the purpose of the project, known as a Special Purpose Vehicle ("SPV"). The role of the SPV is to collect toll money or receive payment stream over the life of the concession from the public sector as annual or unitary payments. This will be subject to the delivery of the services set out in the concession agreement. The private sector participants in an SPV may include contractors, service providers, insurance companies, financial institutions, bond and equity investors, facility managers and professional advisors.

The public sector participants will usually involve State departments, entities and local authorities depending on the nature of the project. A State entity will execute the contract with the SPV. The roles, rights, and duties of the State entity may vary depending on the arrangement in place and the jurisdiction.

The role of the Public Partner

The public partner may hold an equity stake in the SPV and may contribute capital to the project. Governments might provide existing assets such as land or infrastructure to the SPV, thus facilitating the project development. Most importantly, the public partner is responsible for ensuring that the project complies with applicable laws, regulations, and standards safeguarding the public interest.

The rights of the Public Partner

With equity participation, the public partner gains representation on the SPV's board, allowing it to contribute to the decision-making process. As an equity investor, the public partner is entitled to a share of the profits generated by the project, in line with the amount invested. The public partner has the right to be informed of the financial status and operational performance of the project.

The duties of the Public Partner

If holding equity, the public partner is responsible for providing its share of capital investment as agreed. The public partner will be key in assisting the SPV in obtaining necessary permits, licenses and regulatory approvals to ensure the project proceeds without delay. As previously mentioned, the public partner must oversee that the project adheres to contractual obligations, quality standards, and serves the public interest effectively.

The public partner is not responsible for the operational responsibilities in a PPP project, as these are often delegated to the private sector partner who is engaged in day-to-day operations. There are, however, arrangements where public interest is significantly impacted for the nature of the project. In this scenario, the public partner might take on a more active role in the operations whether directly or through other stakeholders.[41]

Most common types of Public-Private Partnerships project delivery methods

- Build-Operate-Transfer ("BOT"): The private sector company is responsible for the design, construction and operation of the project. After successful completion of the project, ownership transfers back to the government in charge. This method is commonly used in infrastructure projects like toll roads and power plants.
- Design-Build-Finance-Operate ("DBFO"): This type of PPP is where the private sector is responsible for the project in its entirety both from a construction point of view and operations side. This type is used in large-scale projects like schools and hospitals where long-term efficiency is crucial.
- Lease-Develop-Operate ("LDO"): A private company leases a public asset with the aim of improving it and managing the operations. This model is common when the public sector needs to bring improvements and upgrade existing infrastructures like airports.
- Operations and Maintenance Contracts ("O&M"): The private sector's role in management and operation contracts is relatively small. This type of contract can be beneficial for the maintenance of utilities that have been struggling financially. Here the private sector oversees the asset without owning or financing it. It is used for public facilities like water treatment plants.[42]

---

41 Asian Development Bank (ADB), *Public-Private Partnership*, 2012 [online]. Available at: <https://www.adb.org/sites/default/files/institutional-document/31484/public-private-partnership.pdf> [Accessed Feb. 25, 2025].

42 Mastt. (n.d.). *Public-Private Partnership.* [online] Available at: <https://www.mastt.com/blogs/public-private-partnership> [Accessed Feb. 11, 2025].

Procedure

The process starts with public authorities identifying infrastructure needs and conducting feasibility studies to assess a project's viability, considering factors such as economic benefits, risks, and alignment with public objectives. Once a project is identified, the scope, objectives, financial implications, and the rationale for adopting a PPP model are established. The public sector initiates a competitive procurement process to select the private sector party that will be involved in the project. This usually involves assessing potential bidders' capabilities and experience, issuing detailed project requirements and inviting bids, and finally reviewing the proposals based on a pre-established criterion. Once the private sector company is identified, the parties start negotiating the contractual terms of their relationship and the characteristics of the project. The public and private partners will negotiate, among other things, the performance standards, payment methods, and risk allocation. The PPP must establish how the project will be financed.[43] The private sector usually invests upfront by securing funds through equity, loans, or government bonds. The public sector may support the project through tax incentives, grants or guarantees. The private partner undertakes the design and construction of the project according to the projects' specifications and timelines negotiated by the parties. The public sector monitors the progress to ensure compliance with laws and regulations. After the construction, the private investor typically continues to be responsible for maintaining the facility, by overseeing the process and ensuring long-term functionality. The last stage of a PPP lies at the end of the contract term, where the completed asset is transferred back to the public sector.

Advantages

Many private companies invest in cutting-edge technologies and hire trained professionals to support their activities. By selecting innovative companies, the State is able to access the most innovative technologies available and use specialized skills and knowledge that may not be available within the public sector. These characteristics enhance the efficiency and effectiveness of public projects and services. The collaborative nature between the public and the private sectors in a PPP facilitates knowledge transfer between these two sectors.[44]

---

43 HM Treasury (2012) *A New Approach to Public Private Partnerships.* Available at: <https://assets.publishing.service.gov.uk/government/uploads/system/uploads/attachment_data/file/205112/pf2_infrastructure_new_approach_to_public_private_parnerships_051212.pdf> [Accessed Feb. 11, 2025].

44 HM Government. (2017). *Topic Guide: Public-Private Partnerships.* [pdf] Available at: <https://assets.publishing.service.gov.uk/media/5977576ee5274a289a000031/Topic_Guide_Public-Private_Partnerships.pdf> [Accessed Feb. 11, 2025].

Public projects require substantial investments, and by engaging private investors, financial responsibility is shared. Private investors bring in capital, reducing the financial burden on public budgets and taxpayers. This sharing arrangement allows governments to pursue critical projects that would be otherwise unfeasible without the support from the private sector. Also, risks are shared through the adoption of mitigation strategies and risks allocation to different parties. Both parties have an interest in the successful completion of the project, thus the risk sharing approach acts as an incentive to encourage better management of the challenges the project may face.

Private sector partners are to bring efficiency driven approaches that focus on customer satisfaction. Through their expertise, streamlined processes, and performance-driven mindset, private entities can optimize service provision and enhance the overall quality of the services.

PPPs lead to faster project completion timelines compared to solely public led projects. The private sector's ability to allocate resources efficiently, use cutting-edge technology, and employ innovative construction methods can contribute to quicker project completion.

The PPP model adapts more efficiently to changes than traditional public sector approaches. This adaptability ensures that the projects are responsive to the community needs and are able to adapt quickly in evolving circumstances by meeting the public needs.

Disadvantages[45]

The private sector's involvement in a PPP project is driven by the expectation of substantial profit and a guaranteed return on investment. The involvement of private investors can lead to higher costs, especially if the private investors contributing to the project have the expectation of high profits and use costly innovative technologies. As a consequence, the public sector may incur increased expenditures.

Due to the collaborative nature of the PPP, public entities and private investors share decision-making. This can result in reduced government oversight and can limit the level of control the public sector has over the project. Private investors involved in the project may end up acquiring significant influence on the management of the project and its operations, which may correspond to a reduced oversight by the public partner. Such a situation can raise concerns about the alignment of the project with public interests and priorities.

---

45 Snatika. (n.d.). *The Advantages and Disadvantages of Public-Private Partnerships.* [online] Available at: <https://snatika.com/single-blog/the-advantages-and-disadvantages-of-public-private-partnerships> [Accessed Feb. 11, 2025].

The private investors and the public partners may have distinct objectives. Private entities may be driven mainly by the financial gain and the commercial viability of the project, meanwhile the public sector may be more interested in addressing public needs. This difference in approaches may leave certain segments of the community underserved or marginalized and can result in unequal distribution of benefits or use of services provided on the basis of the PPP.

Government policies and regulations impact the business of private entities. When there are political changes in the government of a state, the legal and regulatory framework may change and impact the terms of existing agreements, introducing new requirements or restrictions on projects under construction or on the operation of projects already completed. The consequences may be delays, renegotiations, or even project cancellations. Change in the political regime represents a major risk that may affect a PPP.

An example of a PPP is the construction of tunnel linking southern England with northern France, called "The Channel Tunnel", delivered under a Build-Operate-Transfer model.

Collaborative Project Delivery

The collaborative contracts emphasize collaboration, transparency, and shared responsibilities between all key project stakeholders to achieve project goals. The terms "partnering" and "alliancing" are often used interchangeably to refer to collaborative contracts. They describe though procurement approaches which are quite different, particularly in the manner in which they address the distribution of both risk and reward.

Partnering

Partnering may be described as a strategic co-operative relationship between two or more companies, to promote and facilitate the operation of their respective businesses. It can take the form of either "*strategic partnering*" that relies on an incentive of developing an ongoing long-term relationship to the benefit of both parties or "*project partnering*," which is limited to short-term one-off schemes.

Partnering became more popular within the UK construction industry because of two distinct 1990 reports, the so called *Latham Report, published by Sir Michael Latham* in July 1994[46] and the *Egan Report,*[47] *published by Sir John Egan* in November 1998. The foundation of the partnering model lies in the

46 https://constructingexcellence.org.uk/wp-content/uploads/2014/10/Constructing-the-team-The-Latham-Report.pdf

47 https://constructingexcellence.org.uk/wp-content/uploads/2014/10/rethinking_construction_report.pdf

interdependence of the parties, whether for the successful execution of individual projects or the cultivation of a long-term collaboration over several projects. The partnering approach was viewed as an agreement that fostered collaboration, engendered mutual trust, and teamworking while creating a platform for sharing risks and rewards.

Partnering therefore involves a set of collaborative processes based on common goals and trust. Partnering can also be defined as a commitment by the parties involved in a project to work closely and cooperatively, rather than competitively or adversarially.

The role of the contract between the client and the contractor in the partnering process has a twofold aspect. On one hand, cooperation, dispute avoidance, and self-improvement are key to a successful project partnering agreement that can be implemented independently from the contract. On the other hand, the use of the contract reinforces the position of the parties in the partnering agreement.

Partnering, alliancing, and integrated teams are all different definitions and descriptive titles for collaborative contracts. Project partnering comprises free-standing binding or nonbinding "partnering charters" for single projects. Long-term partnering or alliancing operates based on an arrangement between a client and a consultant/contractor or a panel of consultants. These arrangements are usually in place for a period of between three to five years and enable the client to call the consultant(s) to provide services according to specific rates.[48]

No obligation is imposed on the client to award any specific work under the arrangement to the consultants, although the consultants have the obligation to provide any services requested by the client that fall within the arrangement according to the terms defined in the arrangement. Typically, the framework arrangements ensure that the consultants involved have a guaranteed income and workflow and in return the client receives lower prices, good performance, and enhanced project delivery.

At the commencement of the project, the risks associated with the arrangement are clearly identified and allocated to the party best equipped to manage and insure against them. At the center of the partnering approach are the duties of good faith, mutual cooperation, and trust between all the parties involved in the construction project.[49]

---

48 FIDIC, 2019. *Partnering, How It Works, Its Risks and Its Benefits.* [online] Available at: <https://fidic.org/sites/all/themes/fidic/pdf/policy/Partnering.pdf> [Accessible 10 February 2025].

49 Royal Institution of Chartered Surveyors (RICS), 2024, p. 20. Developing a construction procurement strategy. [online] Available at: <https://www.rics.org/content/dam/ricsglobal/documents/standards/Developing-a-construction-procurement-strategy.pdf> [Accessed Feb. 7, 2025].

Beyond the two approaches listed above, there are other contractual mechanisms to partnering:

- Framework agreements: This is operated on the basis of a bilateral agreement between the client and each member of the framework arrangement. The client selects the contractors and professional consultants to work for a defined period on individual projects. The pieces of work are separately awarded to the panel members. Each member retains individual risk and responsibility for delivering the schemes allocated to it. Work is not guaranteed, but in order to attract the best tenders for the framework, a framework agreement with no guaranteed work is very rare. Framework agreements are common in the public sector. In the UK, these types of agreements are governed by the Procurement Act 2023.[50]
- Multiparty Binding Contracts: Each party to the multiparty contract is in a direct contractual relationship with every other party.
- Construction consortium: Parties may come together to form a consortium to tender for a specific project, in particular major infrastructure projects. A construction consortium is common when different parties with different skill sets come together to work in a group, and every individual specialist skill is required for the successful tender. The group that works together improves its position in the market or wins a series of projects.

In a partnering agreement within the construction sector, each stakeholder plays a vital role in ensuring the successful outcome of the project through collaboration and shared objectives. The project's owner is responsible for defining the project's vision and objectives, ensuring alignment with the main goals. Moreover, the owner is responsible for providing the funding and resource allocation required for the execution of the project.

The contractor's main responsibility is to oversee the day-to-day operations, manage the coordination of activities and carry out any risk assessment to identify potential future risks. Another responsibility of the contractor is to manage the performance of subcontractors. The contractor ensures that the construction process is aligned with the project's goals set by the owner. Additional stakeholders, such as the engineer, the subcontractors, and the suppliers contribute with their expertise to the project, ensuring their tasks meet the required standards and they actively participate in the decision-making process.

There are a number of potential advantages to partnering.[51] Due to the collaborative nature of the delivery method, partnering projects may encourage faster

---

50 2023 c. 54.

51 Constructing Excellence. (2015). *Partnering.* [online] Available at: <https://constructingexcellence.org.uk/wp-content/uploads/2015/03/partnering.pdf> [Accessed Feb. 10, 2025].

project completion. The early involvement and interaction of all stakeholders and continuous communication create an environment where identifying and resolving potential problems is less challenging. Partnering projects are more likely to remain within the budget agreed at the beginning, therefore the parties can better predict the costs of the project. The shared risk and reward structure incentivizes all the parties to come up with strategic decisions that minimize costs and maximize efficiency. The project delivery method fosters a collaborative environment where all the stakeholders involved work together under a single contract. This approach should ensure clear communication and coordination throughout the various phases of the project. The parties can identify and assess the risks at an early stage of the project. This ability minimizes the issues that can arise at later stages of the project that could delay the timeline of the construction. The reduction in disputes in partnering agreements stems from a shift away from traditional adversarial contracts to a more integrated, open, and solution-oriented approach where stakeholders collaborate to achieve the same objectives.

Partnering also presents a number of potential disadvantages. Partnering tends to need a high level of management and involvement of the parties to be a successful method. Under a project partnering arrangement, the owner does not promise any further work, but a series of partnered projects. If project partnering does not lead to any repeated activities and business, the "investment" from all parties in collaborative teams and ways of working may be lost.[52] The need to adapt to the necessities of the project may lead to the elimination of "layers" and consequently the end of some relationships between parties. For example, if a client can obtain materials directly from one supplier, he or she may not need to source it relying on a specialist supplier or on the main contractor. This, on the one hand, can reduce duplication and bureaucracy, leading to reduce costs, but on the other hand, may put businesses under financial strain.

Examples of standard forms of partnering contracts include:

- NEC contracts: If the parties adopt a NEC contract, they agree to "act in a spirit of mutual trust and co-operation."[53]
- JCT contracts: Although the JCT's main contracts are not partnering contracts, there are two documents that can be used by parties that relate to

52 Thomson Reuters. (n.d.). Partnering or Alliancing on a Construction or Engineering Project. [online] Available at: <https://uk.practicallaw.thomsonreuters.com/Document/I68c680c5e5a911e398db8b09b4f043e0/View/FullText.html?comp=pluk&transitionType=SearchItem&contextData=(sc.Search)&OWSessionId=ab03a57705df45ca93378d351d7a0c12&skipAnonymous=true&firstPage=true> [Accessed Feb. 10, 2025].

53 (*NEC3 clause 10.1 and NEC4 clause 10.2.*)

partnering, one being a nonbinding partnering charter[54] and the other, the JCT Constructing Excellence contract.[55]
- PPC2000 and TPC2005 contracts: The Association of Consultant Architects (ACA) provides for Standard Form of Contract for Project Partnering (PPC2000), which requires the parties to sign-up to a single multiparty contract to procure the project, as well as the Standard Form of Contract for Term Partnering 2005 (TPC2005) amended in 2008, which provides a term partnering contract that may be used on different projects for a particular set of period.
- Framework alliance (FAC-1) and term alliance contracts (TAC-1): Published in 2016 by ACA, FAC-1 is a new type of contract, mapping the strategy that connects and integrates any number of related two-part contracts for works and/or services and/or supplies. It is endorsed by the Construction Industry Council and by Constructing Excellence. TAC-1 instead supports and integrates the provision of any type or scale of works and/or services and/or supplies.[56]

Alliancing

In alliancing arrangements, the parties enter into an agreement to work cooperatively and to share risk and reward measures against the performance indicators.[57] The owner and other professionals work as a single team to deliver a specific project under a contractual framework where their commercial interests are aligned with actual project objectives. A formal contract is entered into between the parties that undertake to act in the best interest of the project. In a partnering agreement instead the parties do not undertake to act in such a manner as it is voluntary. In an alliancing contract, the main idea lies in transforming the interests of each party involved encouraging one approach, where the commercial interests of all parties are aligned toward a common goal.

One of the main components of this approach is "risk sharing." Each party will share the success or failure of the project and collectively will take responsibility for the decision-making process and its impact on the project's outcome.

---

54 JCT Partnering Charter Non-binding, 2016 Edition (PC/N 2016).

55 JCT Constructing Excellence Contract, 2024 Edition (CE 2024).

56 Thomson Reuters. (n.d.). Partnering or Alliancing on a Construction or Engineering Project. [online] Available at: <https://uk.practicallaw.thomsonreuters.com/Document/I68c680c5e5a911e398db8b09b4f043e0/View/FullText.html?comp=pluk&transitionType=SearchItem&contextData=(sc.Search)&OWSessionId=ab03a57705df45ca93378d351d7a0c12&skipAnonymous=true&firstPage=true> [Accessed Feb. 10, 2025].

57 MacDonald, C. (2005). *What are the Important Differences Between Partnering and Alliance Procurement Models and Why are the Terms so Seldom Confused?* Alliance Contracting Electronic Law Journal. [online] Available at: <https://citeseerx.ist.psu.edu/document?repid=rep1&type=pdf&doi=8ed6162a549b1b12b9a772713b7e987883f71ccb> [Accessed Feb. 10, 2025].

Alliancing contracts are structured to enable the parties to begin work immediately, offer flexibility in the activities performed, and provide clients with the capacity to deliver substantial workloads within tight timelines, even in resource-constrained markets.[58]

In order to set up an alliance, the parties can adopt two different legal corporate structures. The first one is by incorporating a Special Purpose Vehicle ("SPV") where all the parties become shareholders or by forming a quasi-alliance, where instead the parties to the contract behave as they were part of an alliance but without formally forming an SPV.

The way an alliance contract is structured is very similar to the structure of a private limited company:

- The Owner is the entity that will procure the project, and which will hold the assets in the long-term.
- The Nonowner Participants ("NOP") are all the stakeholders involved in the project delivery, like contractors, designers, engineers, consultants and subcontractors.
- The owner and the NOPs are the main parties of the alliance contract. They represent the shareholders.
- The Alliance Leadership Team is what in a private limited company we would call the board of directors. They are responsible for the strategic management and direction of the project. Usually, the alliance leadership team is made up by one representative for each party.
- The Alliance Manager is responsible for overseeing the alliance in accordance with the contract.
- Different representatives from the owner and NOPs can form the Alliance Management Team responsible for the coordination of the delivery team.
- The Delivery Team is the workforce of the project and is responsible for the day to day delivery of the construction project under the alliance.

Alliancing is often viewed as a "risk-embracing" project delivery method. The commercial framework creates a mechanism referred to as "gain share/pain share." The contract members share the rewards of exceptional performance but also bear the consequences of poor performance.[59] Examples of risks that the members share are those in relation to design errors and latent conditions.

58 CMS Law. (2005). *Contract alliancing in construction. What is alliancing?* [pdf] Available at: <https://cms.law/en/int/publication/contract-alliancing-in-construction/what-is-alliancing?format=pdf&v=3> [Accessed Feb. 11, 2025].

59 MacDonald, C. (2005). *What are the Important Differences Between Partnering and Alliance Procurement Models and Why are the Terms so Seldom Confused?* Alliance Contracting Electronic Law Journal. [online] Available at: <https://citeseerx.ist.psu.edu/document?repid=rep1&type=pdf&doi=8ed6162a549b1b12b9a772713b7e987883f71ccb> [Accessed Feb. 11, 2025].

From a practical point of view, the contract is executed with a Target Outrun Cost ("TOC"), a cost benchmark used to govern the gain-pain mechanism. If the alliance members successfully deliver the project below the cost benchmark, the members will typically share a pre-agreed ratio where 50% is allocated to the owner and 50% is allocated to the NOPs. If the project delivery is unsuccessful and therefore above the TOC, the parties will share the pain, a loss amount usually set under the contract for which the NOPs are liable.

Potential advantages of alliancing include the following:[60]

- Enhanced Collaboration: alliance fosters collaboration and encourages communication and teamworking between the parties.
- Improved risk management: due to the risk sharing mechanism, the members are more likely to manage from the outset the potential risks under the project, particularly the financial exposure should the project be unsuccessful. Risk management reduces the financial exposure and liability of the parties to the alliance.
- Cost savings: the alliance structure helps the parties optimize the costs, minimize delays, and ensure efficient resource allocation.
- Flexibility: allows for adaptability in the project design and execution to accommodate changing circumstances.
- Transparency: with a co-operative framework made of relationships between the members, the alliance enhances transparency throughout the project operations, facilitating mutual performance.
- Innovative problem-solving: the collaborative environment promotes creative solutions to the challenges faced by the members throughout the execution of the project. This is because individuals from different backgrounds, with strong expertise, and qualifications support each other and strive to find efficient and cost-effective solutions to issues.

On the other hand, there are also potential disadvantages to alliancing:

- Complex governance structure: alliance contracts require complex legal structures that must be defined prior to the commencement of the project. At the beginning, this may result in challenging situations and difficulties in the decision-making process.
- Resource intensive: setting up and maintaining the alliance contract requires significant resources in terms of time, effort and selection of personnel.
- Dependency on trust: the key to successful alliance contracts is reciprocal trust amongst the members of the alliance. Distrust in a member's capabilities

60 Construction Front. (n.d.). *Alliance Contract.* [online] Available at: <https://constructionfront.com/alliance-contract/> [Accessed Feb. 11, 2025].

or conflict between members can lead to adversarial behavior that undermines the performance of the project.

- Limited use in small projects: due to the complicated nature of setting up from a contractual and legal prospective an alliance, this method may not be practical for small-scale projects.

Alliancing is recommended when the parties are dealing with a complex construction project that requires a long-term relationship between the members, and the members are willing and capable of working in a co-operative environment. From an owner perspective, this method is appropriate also when it is difficult to accurately define the finished "product." Moreover, the owner must be capable of selecting the best deal for the delivery of the works and services and therefore be able to efficiently pool together a group of individuals who have the appropriate knowledge, skills and resources.[61]

An example of a project where the alliance model has been adopted is for the Ministry of Justice construction project of up to 1200 new prison spaces across the United Kingdom in collaboration with the three contractors Kier, Laing O'Rourke and Wates. The aim of the project is prioritizing quality and sustainability by reducing carbon emissions and enhancing social impact in the prison community.[62]

Conclusion

The choice of a project delivery method significantly impacts the efficiency, cost, management, and allocation of risks of construction projects. Traditional methods like DBB offer clear roles and responsibilities but may lack flexibility and speed. Design & Build models provide cost and time certainty but can limit the client's control over design. Construction Management offers flexibility and early involvement of contractors but demands significant client coordination and responsibility. Turnkey contracts further simplify project execution by providing a single point of responsibility, though they limit owner control over design changes. Multi-prime contracting offers cost-savings and enhanced control but requires careful coordination and a robust project management approach to avoid disputes and delays. PPPs allow a participation of private investors in large-scale public

61 PwC Australia. (2016). *Investing in Infrastructure: Alliancing.* [pdf] Available at: <https://www.pwc.com.au/legal/assets/investing-in-infrastructure/iif-21-alliancing-feb16-3.pdf> [Accessed Feb. 11, 2025].

62 Wates. (2023). *Wates appointed to deliver £225 m Ministry of Justice contract.* [online] Available at: <https://www.wates.co.uk/news/construction/justice/225 m-ministry-of-justice-contract/> [Accessed Feb. 11, 2025].

infrastructure projects but require complex financing and risk-sharing arrangements. Finally, alliancing promotes collaboration between the client and multiple stakeholders, ensuring shared risks and rewards, but relies heavily on trust and effective relationship management. Each project delivery method offers distinct strengths and weaknesses, and selecting the most appropriate approach depends on factors such as the complexity of the project, the client's preferred level of involvement, and the acceptable allocation of risks. Ultimately, a thorough understanding of these elements is essential to ensure successful project outcomes and effective collaboration among all parties.

4

Payment Issues

Marco Padovan, Michele Bonomini, and Fabiola Viccei

Pricing Models

One of the first, and likely foremost significant, issues an owner and a contractor face when considering starting a project is the decision on which pricing model shall be used.

This aspect is very important as it has impacts on the entire performance of the project. It impacts the preliminary study of the project, the allocation of risks between the owner and the contractor and it shapes the accounting systems of the work, the payment methods and procedures, etc. While the parties to a construction contract retain certain freedom to establish pricing models and payment systems, in practice there are four models that are most commonly used and recognized by legal systems and contractual practice.

Lump Sum

Lump-Sum (or Fixed-Price) pricing model is one of the most commonly used pricing method in the construction industry, especially for "turn-key" or "EPC" projects. This method consists in the ex-ante establishment of a predetermined price for the entire project. The lump-sum price is agreed upon at once and is thereafter, in principle, fixed and invariable (save for price review provisions and variations, if any).

In legal terms, the Lump-Sum contract is often described as a contract in which the contractual consideration refers to the overall works and services as performed and as described in the contract. By way of example of contractual clauses on lump-sum method,[1] although with some differences among them, we may refer

1 Baker, E., Mellors, B., Chalmers, S., Lavers, A, *FIDIC Contracts: Law and Practice.* (Abingdon: Informa Law – Routledge, 2009), 163.

Major Issues in International Construction Law, First Edition. Mauro Rubino-Sammartano and Troy Harris.
Published 2026 by John Wiley & Sons, Inc.

to the FIDIC Yellow Book[2] and the FIDIC Silver Book.[3] Clause 14.1 [The Contract Price] of the FIDIC Silver Book Second Edition 2017 provides: "*Unless otherwise stated in the Particular Conditions: (a) payment for the Works shall be made on the basis of the lump sum Contract Price stated in the Contract Agreement.*" Clause 14.1 [The Contract Price] of the FIDIC Yellow Book Second Edition 2017 provides: "*Unless otherwise stated in the Particular Conditions: (a) the Contract Price shall be the lump sum Accepted Contract Amount.*"

In principle, the main advantage of this pricing method is that it provides cost certainty for the entire project. This is true especially when the drawings of the project are in a quite advanced stage, and the quantities are, to a great extent, certain.[4] When this is not the case, the risk of cost of completion is generally on the contractor. This is why this method is often the preferred method proposed by owners, especially when it comes to complex projects. Moreover, setting an overall fixed price at least allows owners to prepare with some comfort a suitable business plan and to bank the project if needed, provided that they are ready to pay a premium to achieve greater certainty on the final price.

This pricing method shifts the risk of costs increases on the contractor. The contractor in fact takes the risk of the quantities, and if the design is to be prepared/developed by it, of the qualities, necessary to complete the works. That's why, when a fixed-price contract is used for a project, contractors need to carefully assess the project scope, materials, labor, and other factors (such as taxes, their general expenses and overheads and expected profit) in order to be able to determine, with the highest possible degree of certainty, the overall cost of the project.

The main risk a contractor faces in this case is to underestimate the actual costs of the project. And, when a project is complex and likely extended over multiple years, it is clear that the number of risks that may be faced by the contractor increases. The most common risks that may put at risk the contractors' estimate of the price are usually related to unforeseen site conditions, changes in the design or in the legal framework applicable to the project (e.g., a new law is introduced, which requires a reduction of environmental impact and therefore the need to use different, more expensive, materials or working methods, etc.), or the increase in prices of materials, transport, labor, and related taxation. In all these circumstances, unless the contract provides for some "review" or adjustment" mechanism, the contractor may have to incur additional expenses, reducing the profit or even turning the project into the red.

---

2 See FIDIC dedicated website page: https://fidic.org/books/plant-and-design-build-contract-2nd-ed-2017-yellow-book.

3 See FIDIC dedicated website page: https://fidic.org/books/epcturnkey-contract-2nd-ed-2017-silver-book.

4 Baker, Mellors, Chalmers, Lavers, 2009, 164.

But also the contrary may be true. In fact, if the price decreases, or the contractor overestimated the project costs (which were nevertheless accepted by the owner), unless there is a project benefit sharing clause, the contractor will make a higher profit.

Mitigations to the Fixed Nature of Lump-sum Prices

While in principle the lump-sum amount is meant to be fixed for the entire duration of the project, therefore shifting the economic risk on the contractor, the law or the contract may provide certain mitigations to such principle.

The parties to a construction contract often negotiate clauses to mitigate the possible consequences of the fixed and invariable nature of the lump sum in case of occurrence of subsequent, unforeseen events that affects prices of materials, workforce and utilities or unforeseeable natural events or geological conditions, to supplement or derogate relevant legal provisions, if any. Certain legal systems also directly provide for derogation to the fixed nature of the lump sum in certain cases.[5] The most common mitigations, subject to the specificalities of each jurisdiction, are indicate below.

Principle of Fairness and Good Faith

The first mitigation to the fixed and invariable nature of the lump sum derives from the principles of fairness and good faith in the relationships between the parties. These principles may come into play both at precontractual and contractual stage and are based on the need to correct the asymmetrical availability of information between the owner and the contractor. According to these principles, the owner is required to disclose to the contractor all information regarding the project that may affect the performance and the completion of the project itself.

Accordingly, the owner shall provide the prospective contractor with geological studies of the site, access routes, availability of utilities near or at site, whether conditions, etc. to allow the contractor to establish an as much as possible accurate evaluation and estimate of the project costs, and therefore of its proposed price.

The contractor nevertheless remains responsible for properly analyzing and elaborating such information, so that it cannot transfer on the owner any mistake it may incur in analyzing such documents when preparing its offer. Moreover, if the contractor deems that there are errors in the documents made available to it, it shall promptly notify the owner of such errors, and in lack thereof, it is at risk of not being able to recover possible increases in costs. This "duty to warn" the client is generally known in many civil law jurisdictions and applies to any construction professional (a designer, a contractor, etc.). The basis

5 Baker, Mellors, Chalmers, Lavers, 2009, pp. 164, 166.

of this duty is the principle of good faith, but it is also a consequence of the duty of reasonable skill and care and the duty to achieve a result, which often applies to contractors.[6] The extent of such duty to warn varies in different jurisdictions. It is debated whether such duty may be extended so as to impose to contractors the duty to make disclosures that may have an impact on price determination in favor of the owner.

While the principle may be underestimated at the beginning of a project, it is very often relied upon by contractors during the works execution phase, especially in case of unforeseeable conditions claims.

Variations Required by the Owner

The law or the contract usually provides that the owner is entitled to request changes in the work, and that the contractor is obliged to perform such changes without the right to refuse or withdraw from the contract, at least for changes up to a certain threshold (e.g., up to 20% of the initial contract price).

The rationale behind this exception is that if the owner, within its freedom to request the work that he likes, wishes to make a change to the work, he may do that; however, since the contractor was not aware of such different works when it made its quotation, it is entitled to an adjustment in the price.[7] Moreover, unless otherwise provided for, the owner may perform such variations through other contractors after completion of the works.

Certain legislation further introduces quantitative and/or qualitative limitations to the right of the owner to order variations in the works, allowing the contractor to refuse to make such variations in case their amount exceeds a percentage of the original price or in case they involve activities of a different nature than the original ones or require different construction methods,[8] unless a specific agreement is reached between the contractor and the owner on the variation and the ensuing adjustment to the lump sum.

The reason behind these exceptions is that when making the price, the contractor has evaluated certain costs and expected profits on the basis of certain work requirements, which are afterword unilaterally changed by the owner, and therefore the contractor is entitled to a consequent adjustment of its price.[9] Nevertheless, legal provisions on variations are often nonmandatory, and the parties may agree different terms in the contract. This is an issue to be carefully considered during the negotiation of the price and the contractual terms and conditions.

---

6 Aleksandrov, G., 2024, *Duty to Warn in Construction Contracts*. Available at: https://www.ibanet.org/clint-april-2024-feature-2.

7 Rubino, D., Iudica, G, *Appalto*, (Bologna: Zanichelli, 2007), 284.

8 See for instance art. 1661 Italian Civil Code.

9 Rubino, Iudica, 2007, 284, 287.

Necessary Variations

Necessary variations are variations that are not a consequence of a mistake of any of the party (e.g., the architect appointed by the owner/developer) but that become necessary, during the performance of the contract, to complete the project on a workmanlike manner in accordance with the prevailing legal and technical requirements. A typical example is a variation that is required by the entry into force of new law prescribing different materials or design parameters.

In certain jurisdictions, if necessary variations are beyond certain quantitative or qualitative thresholds, the contractor or the owner may terminate the contract. The reason behind a similar legal provision[10] is that an owner may not be obliged to complete a work that is not anymore of its liking or to incur excessive costs, due to changes required by external factors.[11] In these cases, normally he may withdraw from the contract and pay to the contractor a relief (e.g., based on the contractor's expected profit). On the other hand, the contractor may not possess the necessary technical or financial capacities to perform the new work that would result from the necessary variation, and in this case, it may be entitled to withdraw from the contract and receive an equitable relief.[12]

However, the parties are free to reach a specific agreement on the performance of the necessary variation, with the ensuing changes to the lump sum.

Excessive Supervening Onerousness in the Completion of the Project

Especially in projects where the performance of the works is to last long, it may well be that during performance of the works certain circumstances arise which may render the performance considerably more onerous than what was foreseeable. In these cases, the law or the contract may provide for a mechanism of review or adjustment of the price to restore the economic balance of the contract, even if the price is a lump sum.[13]

Usually, such revision provisions activate upon the occurrence of particular conditions of an objective nature, and which were unforeseeable by the parties at the time of execution of the contract.[14] The evaluation of the foreseeability is based on the professional diligence to be expected by the parties involved, which, in turns, depends on the size and capacities of the parties and the size and nature of the project.

---

10 See for instance Italian law, Art. 1661 Civil Code.

11 Rubino, Iudica, 2007, 283.

12 Rubino, Iudica, 2007, 280.

13 Rubino, Iudica, 2007, 284, 285.

14 Rubino, Iudica, 2007, 326, 327.

Moreover, all the prevailing circumstances (legal, economic and political, technical) at the time of execution of the contract shall be taken into account when assessing the foreseeability of a change. Relevant circumstances that may be relevant to trigger the review of the lump sum are those that determine a variation in the cost of the work, supply or service, up or down, exceeding a certain percentage (e.g., 5 or 10 percent) of the total contract amount. The review may operate to the full extent or to a more limited extent (e.g., 80 percent) of the variation itself, in relation to the services yet to be performed.

Construction contracts usually include detailed clauses to cover the foregoing situations. Often such clauses provide for a derogation to the relevant legal provisions, when and to the extent permitted by the applicable law.[15]

Parties' Negligence

Changes to the lump sum may occur also as a consequence of the acts or omissions of the parties. For instance, when inaccurate drawings and contract documents prepared by the owner do not describe the work properly, and the contractor was not in a position to discover such defects upon execution of the contract, the lump sum may be reviewed accordingly.

Lump-sum vs Turn-key Contracts

A particular case of Lump-Sum contract is the Turn-Key contract. This may be defined as an "extreme version" of the lump sum, in so much as it shifts even more risks and responsibilities on the contractor. Turn-Key is sometimes referred to also as "EPC Contract." Traditionally, under a "Turn-Key" contract, the contractor's obligations include also activities (and risks) that under standard lump-sum build or design and build contracts are on the owner. The contractor is in fact entrusted with the obligation to design, procure, supply, install, execute, commission, test and hand over the works in accordance with the contract and the owner's requirements, on time and within the agreed contract price.

With a Turn-Key contract, the contractor assumes all risks related to construction. It depends on local laws whether such different allocation of risks is such that it changes also the nature of the contract, from a commutative agreement to an aleatory agreement, or not. Moreover, unless otherwise expressly agreed upon, in principle only reasonably foreseeable obligations not expressly provided for in the contract are covered, and contractor is entitled to additional compensation (and therefore review of the lump-sum price) for unforeseeable activities and hardship.

Turn-key contracts are often used in complex international construction projects that need to be financially independent and viable. In these cases, the contractor

---

15 Rubino, Iudica, 2007, 324, 325.

will put a higher contingency for risks in its price. On the other hand, it may be more acceptable to pay a premium on a project to ensure price (and return) certainty rather than finding out later on that the project is not viable because the costs have increased significantly.[16]

Specific model contracts have been prepared by international professional and business organizations for these projects. An example of a Turn-Key model contract is constituted by the "Conditions of Contract for EPC Turnkey Projects" published by FIDIC, otherwise known as the Silver Book.[17] Another example is the Turn-Key model contract proposed by the International Chamber of Commerce (ICC), which, according to ICC itself, diverges form other Turn-Key contracts as it is meant to be more balanced taking into account recent trends in international construction industry toward "*more collaborative relationship between contractor and Employer in terms of rights and obligations.*"[18] The ICC model puts an emphasis on the "Assumptions," thus putting pressure on the owner to disclose as much information as possible on the prevailing conditions of the project in order to allow the contractor to better study its offer and reducing the risk premium for uncertain events.

Cost-plus Contract

The cost-plus contract is on the other end of the spectrum compared with the Lump-Sum contract. Cost-plus contracts in fact involve reimbursement of costs incurred by the contractor along with a predetermined fee or percentage for contractor's overhead and profit. Under a cost-plus contract, the contractor is required to provide detailed records of all expenses, including labor, materials, equipment, and subcontractor costs. The owner then reimburses the contractor for these expenses, typically with an agreed-upon fee or percentage added to cover overhead and profit.[19]

The main advantage of this model is that it offers transparency to the parties, as the owner pays for the actual costs incurred during construction (plus the contractor's predetermined fee). This method also allows for flexibility, as it accommodates changes and unforeseen circumstances that may arise during construction. On the other hand, the cost-plus contract lacks the cost certainty provided by the fixed-price method. Owners may be concerned about potential cost overruns,

16 Baker, Mellors, Chalmers, Lavers, 2009, 164.

17 See FIDIC website dedicated page: https://fidic.org/books/epcturnkey-contract 2nd-ed-2017-silver-book.

18 See ICC dedicated page on its website: https://iccwbo.org/business-solutions/model-contracts-clauses/icc-model-turnkey-contract-for-major-projects/.

19 Van der Puil, J., van Weele, A, *International Contracting. Contract Management in Complex Construction Projects*. (London: Imperial College Press, 2014), 139.

especially if the project encounters unexpected challenges or if the contractor's expenses are not closely monitored.

To mitigate this risk, owners often negotiate a fixed date within which the works must be completed, sometimes combined with incentive for early completion, and a cap on the contractor's fee, or establish a maximum budget for the project or a guaranteed maximum price (GMP). Cost-plus contracts are usually based on open-book procedure, to allow the owner to verify costs incurred by the contractor and to be paid to it. These contracts require a high level of confidence on the side of the owner on the procurement method of the contractor, to be satisfied that it is fair, on an arm's length basis and efficient.

Unit Pricing

Unit pricing is another common method in construction. It involves pricing activities based on individual units of measurement, such as per square meter or per linear foot.[20] It is mainly used for standardized, repetitive and common activities which however are difficult to estimate ex-ante in terms of quantity and time. In the construction industry, unit pricing is used for activities such as excavation, concrete work, or electrical installations.[21] Under this pricing method, the contractor is usually paid on a monthly basis, on the basis of the relevant measurement and evaluation of the works performed in such period duly approved and certified by the engineer.[22]

This method enables owners to compare bids more accurately, as they can assess the cost per unit and determine the total cost based on the project's requirements. At the same time, the owner takes the risk of the quantities and of the qualities of the activities/materials necessary to complete the works.

Sometimes unit price contracts also provide for an indicative contract amount. This amount is not the price that will be paid to the contractor and is meant to be purely indicative, as it may change during performance of the contract, and the parties are aware of it. This "indicative contract amount" is useful, for example, to establish the amount of the performance bond (whose amount is usually quantified in a percentage, *e.g.*, 10%, of such contract value), or to set thresholds for variations that the contractor is bound to perform.[23] The effective price to be paid to the contractor shall only be the price resulting from the application of the unit

20 Rubino, Iudica, 2007, 228, 229.

21 Van Der Puil, Van Weele, 2014, pp. 140, 141.

22 Baker, Mellors, Chalmers, Lavers, 2009, 161.

23 Baker, Mellors, Chalmers, Lavers, 2009, 161, 162.

prices to the quantities of the items actually used in the works. The unit prices may be fixed and invariable, or subject to adjustments.[24]

One of the key benefits of unit pricing in construction is its ability to accommodate changes in project scope. As the scope changes, the pricing can be adjusted based on the additional or reduced units, usually without the necessity for the contractor to make a specific claim.[25] Moreover, unit pricing method also provides owners with a clear breakdown of the costs involved in each unit. This transparency allows owners to understand the pricing structure and make informed decisions. It also helps in identifying any potential areas of cost savings or optimization. Finally, unit prices schemes may lead to reduced "risk costs" incorporated in the contractor's offer, resulting in more accurate unit rates.[26] At the same time, contractors need to carefully consider all cost factors and ensure that the unit prices align with market rates and industry standards to submit a competitive offer.

However, unit pricing may not capture the full complexity of a project. On the owner's perspective, certain factors, such as site conditions, project specifications, and design complexity, can significantly impact the overall cost. Therefore, it is essential for owners to provide detailed information and specifications to ensure accurate unit pricing.

Moreover, there is a risk that the contractor can take advantage of its (better) ability to estimate quantities to manipulate the rates in the bill of quantities to its advantage. The contractor in fact could identify certain quantities that are likely to increase in the performance of the works and submit a schedule of rates so that at the end the price turns out to be more favorable to the contractor.[27]

Retainage

Retainage or retention constitutes a form of security in international construction contracts, aiming at mitigating the risks of lack of completion of the project or of poor-quality completion. Through retainage, an amount of money is deducted by the owner or the contractor from payments respectively due to the contractor or the subcontractor, in respect to either progress payments (usually set out on a monthly basis) or milestone payments. The sum so withheld ensures that the contractor or subcontractor is incentivized to finish uncompleted work and correct any defects in adherence with the defects liability period. Should the contractor/subcontractor fail

24 Rubino, Iudica, 2007, 319.

25 See for example for FIDIC contracts Baker, Mellors, Chalmers, Lavers, 2009, 161.

26 Baker, Mellors, Chalmers, Lavers, 2009, 162.

27 Baker, Mellors, Chalmers, Lavers, 2009, 162 and case law quoted therein.

to complete the work to correct the defects, the retention may be used to have the works corrected by third-parties contractors or subcontractors, pursuant to the relevant contract provisions and the governing law of the contract. Retainage shall be indeed released, according to the relevant contract provisions, to the extent that the defects liability period expires and no defects in the works is uncovered.

The retainage is usually set out as a percentage of the total value of the contract, including variations and changes (thus, in case of re-measurement contracts, the contract sum initially agreed is not considered as the parameter to calculate the retention). Retainage is commonly set forth between 5% and 10% and contracts may put a cap to limit the maximum percentage of deductible retention.

FIDIC and NEC model forms of contract provide specific regulation on retainage. FIDIC Red Book 1999 Edition includes retainage in the definitions: "*Retention Money*" means the accumulated retention moneys which the Employer retains under Subclause 14.3 [*Application for Interim Payment Certificates*] and pays under Subclause 14.9 [*Payment of Retention Money*]. Accordingly, the contractor shall submit its Statement (i.e., an application for payment) at the end of each month, including: "*any amount to be deducted for retention, calculated by applying the percentage of retention stated in the Appendix to Tender to the total of the above amounts, until the amount so retained by the Employer reaches the limit of Retention Money (if any) stated in the Appendix to Tender.*" Retention release occurs in two moments: the first half of retainage is released upon substantial completion by the contractor and acceptance of the works by the Owner certified by the Taking-Over Certificate, while the contract administrator, the Engineer, shall certify payment of the balance of retainage after the latest expiry of the defects notification periods. The release of the retention can be possibly withheld if there are outstanding activities, having regard to the relevant defects liability provisions; the sum so withheld shall be equal to the estimated cost of the work to be executed and final release shall be made only once the outstanding work is completed. FIDIC Yellow Book and Silver Book contain analogous conditions in relation to retention money, which have been substantially maintained under the FIDIC 2017 Editions.

The NEC Engineering and Construction Contract provide that retention shall be formally introduced in the contract by selecting the option clause X16, which should clearly be set out in the contract data. In contract data part 1, the owner should establish either the retention free amount (being the amount payable to the contractor without any applicable retention) and the applicable retainage percentage. The NEC guidance notes set out just indicative figures, while the applicable values shall be chosen and inserted by the owner. The retention is a debt of the owner vis-a-vis the contractor, so it is subject to the normal rules applicable to it – for example, it can be seized by the contractor's creditors or, conversely, its repayment is subject to any possible bankruptcy proceedings involving the owner.

Practical Issues

In some jurisdictions, limits are expressly provided in respect either to the amounts possibly subject to retainage and to the retention mode and/or the kind of projects where retainage may be applied.

Some construction contracts provide that the retainage can be applied only after certain payments have been made to the contractor/subcontractor, who thus shall receive certain cash flow before retention money is withheld. If the contract provides that the works shall be executed by sections, the retainage is usually released upon completion and acceptance of the relevant section of the works. The release of the balance may occur either upon expiry of the defects liability period for each section or for the whole project works. In case of contracts providing for partial taking over of the works, the reimbursement of the retainage is to be made upon occurrence of the partial possession by the owner.

Termination of the construction contract can originate critical issues as to retainage, as the contract provisions on release of retention may cease to be applicable in consideration of the ground of the termination and of the then relevant contract clauses. Therefore, the contractual path for the release of the retainage should be carefully analyzed, to avoid that retainage is lately repaid or its release is unexpectedly challenged. Financing charges may also accrue in the event of late repayment of retainage, thus also clauses on interest accrual deserve specific attention.

Further, worth considering certain possible drawbacks in applying retainage. Firstly, as in the industry concerned profit margins and cash flow are essential for contractors and subcontractors, the application of a retainage may contribute in reducing actual financial availability for such subjects, affecting the whole process as a result thereof. Also bookkeeping and administrative requirements may create additional requirements while applying the retainage. Lastly, it appears that in certain cases retainage may give rise to delay in payments and even abuses; in the 2018 case *United Riggers & Erectors v. Coast Iron & Steel*, where a general contractor retained payments without any grounds, the California Supreme Court stated that retainage could be withheld exclusively to the extent that the dispute was in good faith.

Alternatives and Additional Mechanisms

Further tools to strengthen or to meet analogous purposes of the contractual retainage are:[28] (i) project bank accounts; (ii) retention bonds; (iii) performance bonds; (iv) parent company guarantees, and (v) retentions trust funds. The first two schemes are specifically intended to ensure that retention – and not the

28 https://www.fenwickelliott.com/research-insight/annual-review/2023/retentions-get-rid-or-retain.

performance of the obligations in general – is risk-preserved and timely released according to the contract provisions.

The project bank accounts imply asset segregation and thus security for the money in case of insolvency; funds are in fact held on trust for identified beneficiaries. The NEC4 provides the specific option Y(UK)1 for project bank accounts, where a trust deed is required and funds are held on trust according to the provisions regulating the trust and the project bank account.

Retention money can also be held in a trust scheme with separate accounts. New Zealand[29] has recently approved an Amendment Act with the specific aim of granting protection of the retention funds from the contractor's insolvency.

Retention bonds are an alternative to retention deductions and are expressly provided for in the NEC4 as well as the JCT Forms of Contract. By this kind of security, no deduction is made from the payments due, while a bond is issued to cover the obligation that would otherwise have been granted through percentage amounts deducted. Retention bonds are thus meant to cover correction of defects emerging after completion and are frequently set up on an on-demand basis. The bond issuance implies upfront costs, which are, however, mitigated by the increased availability of cashflow. The text of the guarantee should be clearly connected with the aim of retainage – that is, correction of defects – as the retention bond secures not the contractor's general performance of obligations but specifically the retainage amount.

Statutory Holdbacks

In the construction industry, statutory holdback (or statutory retention) is a legal mechanism designed by the law to protect the interests of subcontractors, suppliers, and laborers by ensuring they get paid for their work and materials supplied in a specific project. The underlying purpose is to provide a fund to protect subcontractors and suppliers in the event of insolvency or payment default by others involved in the project. In practice, the owner will withhold a predetermined percentage of the payment due to the contractor until certain project milestones are achieved, primarily to cover any potential claims or liens that may arise. The owner must abide by the holdback requirements to avoid the risk of making double payment, the first time to the contractor and the second time to the subcontractors and suppliers that should have been paid by the contractor, but they were not.

---

29 The Construction Contracts (Retention Money) Amendment Act 2023 has been passed by the Government of New Zealand in April 2023, to give certainty to the retention money regime by making it easier for subcontractors to access retention money without the requirement of a court order, in the event of a head contractor's insolvency.

The specific elements of statutory holdbacks can vary significantly from one jurisdiction to another, reflecting different legal traditions and construction industry practices. Without going into the details of each legal system, it is possible to briefly outline the main characteristics of certain jurisdictions.

United States

In the United States, statutory holdback practices are regulated at the state level, with each state having its own rules. Statutory holdbacks are usually accompanied by construction lien laws, which provide additional security for payment to the subjects involved in the project. Typically, the holdback amount ranges from 5% to 10% of the contract value. Moreover, there is a specified period after project completion during which subcontractors and suppliers can file liens. The holdback is released only after this period expires and any liens are resolved. Noncompliance with statutory holdback requirements can result in the owner being liable for unpaid subcontractor claims, even if the owner has already paid the general contractor.

An interesting feature of the US construction industry is that contractors are required to provide owners with a 100% surety bonds to guarantee that the contractor properly performs its contractually agreed obligations and, in case it fails to do so, to compensate the owner. The surety bond may be mandated by legislation (e.g., when public funding is involved in a project, in order to ensure the public authority is protected for the work that it has contracted) or by contractual agreement. At the federal level, the Miller Act is a federal statute that requires prime contractors on most federal construction projects to furnish payment and performance bonds, each with a value equivalent to the contract price. Such bonds, issued by sureties, protect the government (the bond's obligee) in case the contractor (the bond's principal) defaults on the work, while payment bonds protect both the government and downstream subcontractors and suppliers in the event the contractor defaults on its payment obligations. Most states have their own variation of the Miller Act, commonly referred to as "Little Miller Acts", which may apply in public contracts involving state and local government entities. Although similar in concept and general outline to the Miller Act, the substance and procedure of Little Miller Act can vary greatly.[30] Such ample surety coverage – especially if compared with European countries, where surety bonds are usually in the region of 5% to 20% of the contract amount – provides the owner with greater security, at least in financial terms, even though the owner continues to bear the corporate risk of the of the contractor and of the surety company.

---

30 Cohn, J., 2022, *Miller Act Payment Bond Claims: The Basics*. Available at: https://www.americanbar.org/groups/construction_industry/publications/under_construction/2022/winter2022/miller-act-payment-bond-claims/.

Canada

In Canada, construction lien acts govern statutory holdback practices. Typically, holdbacks amounts are equal to 10% of the value of the work done or materials supplied. Owners must withhold the statutory amount until the end of the lien period, which is generally 45 days after project completion. The holdback can be released upon substantial performance of the contract, but a portion may be retained until the project is fully completed and any deficiencies are corrected.

United Kingdom

In the UK, statutory holdback practices are influenced by the Housing Grants, Construction and Regeneration Act 1996 (HGCRA), though they are less common than in North America. The use of retention sums (akin to holdback) is prevalent, often governed by contract terms rather than statute. The HGCRA provides for a quick adjudication process to resolve disputes over retention payments. The contracts further include specific percentages and conditions for retention, reflecting industry standards.

Australia

In Australia, statutory holdback practices are regulated by security of payment legislation, which varies by state. Typically, the percentage of the retention ranges from 5% to 10% of the contract value. Legislation further mandates prompt payment and outlines strict timelines for the release of holdback funds. Security of payment laws provide mechanisms for rapid dispute resolution, ensuring that holdback funds are released promptly once claims are resolved.

European Union

In the European Union (EU), statutory holdback practices vary widely among member states, often influenced by local construction laws and industry practices. In many civil law jurisdictions, statutory holdback is governed by specific provisions contained in civil codes or construction laws. Parties often have significant freedom to negotiate holdback terms, subject to overarching legal principles of fairness and good faith. For public construction contracts, statutory holdback is governed by a combination of EU directives and national legislation. The primary objective of these EU rules is, among others, to ensure fairness, transparency, and the protection of subcontractors and suppliers in public construction projects. The EU's public procurement framework is primarily based on two key directives:

1. ***Directive 2014/24/EU of the European Parliament and of the Council of 26 February 2014 on public procurement and repealing Directive 2004/18/EC Text with EEA relevance***: This directive sets out rules for the procurement processes used by public sector bodies. It emphasizes transparency, competition, and nondiscrimination.
2. ***Directive 2014/25/EU of the European Parliament and of the Council of 26 February 2014 on procurement by entities operating in the water, energy, transport and postal services sectors and repealing Directive 2004/17/EC Text with EEA relevance***: This directive governs procurement procedures in the utilities sectors, with similar principles of fairness and transparency.

While these EU Directives do not explicitly mandate statutory holdback percentages, they require member states to ensure that subcontractors and suppliers are paid promptly and fairly. The specifics of statutory holdback are therefore implemented through national legislation, which must comply with the overarching principles of the EU directives.

EU Member States National Legislation

Each EU member state incorporates the requirements of the EU Directives into its national laws, which must comply with the overarching principles of the EU Directives, leading to variations in how statutory holdback is applied. The percentage of statutory holdback typically ranges from 5% to 10% of the contract value. This amount is retained to cover potential defects, ensure project completion, and protect against subcontractor claims. The holdback period usually extends until the project reaches substantial completion and for a specified period thereafter, often to allow for the filing and resolution of any claims or liens. Holdback funds are released once the project is substantially completed, any defects have been rectified, and no valid claims or liens are outstanding.

Germany

In Germany, public procurement is governed by the *Vergabeverordnung* (VgV) and the *Vergabe- und Vertragsordnung für Bauleistungen* (VOB/B). The Construction Contract Procedures Part B (VOB/B) dealing with security are mostly supplemented and detailed in the construction contract. Generally, the contractor must provide a performance guarantee amounting to 10% of the gross contract value. The amount of any payment/prepayment guarantee is also generally 10% of the gross contract value. Where agreed upon in the contract, the owner may withhold an amount of up to 10% of each installment and the final payment as security. Generally, the contract provides that at the time the final invoice is issued, the retention is reduced to 5%. This is then paid into a blocked account unless the contractor provides a contractually agreed warranty guarantee in order to release the full amount. Pursuant to German case law, an on-demand warranty guarantee cannot be agreed on in general terms and conditions. The holdback is usually released after the completion of the project and the expiration of the defects liability period, provided there are no outstanding claims.

France

In France, public procurement is regulated by the *Code de la commande publique*. The specific holdback percentage is provided in the contract; generally, it amounts to 5% of the contract value. As an alternative to the holdback, contractors can provide a bank guarantee or insurance bond. The holdback is released after project completion and a defined period for latent defects.

Italy

Italy's public procurement is governed by the *Codice dei contratti pubblici*, currently contained in Legislative Decree no. 36 of 31 March 2023, as supplemented and integrated from time to time. Typically, Italian law does not provide for a statutory holdback but rather for a performance bond, in the amount of 10% of the contract value. However, before execution of the contract, the contractor can ask the contracting authority to substitute the performance bond with a retainage, in the same amount of 10%, to be applied on each interim payment certificate (also called SAL).

The holdback period extends until the issuance of the provisional testing certificate or the certificate or regular completion of the works, and in any case no longer than 12 months from completion of the works resulting from the certificate. Holdback funds are released in stages with the progress of the works, with a portion (usually 20%) released upon issuance of the provisional acceptance certificate. Payment of the final installment is conditional upon the contractor providing the contracting authority with a bank or insurance bond in the same amount and with validity until the testing certificate becomes final.

Statutory Holdback for Noncompliance in Payment of Social Security Contributions

A specific feature of the Italian Public Contracts Code is that it allows contracting authorities to provide in the tender documents a specific statutory holdback with the specific aim of protecting laborers. In the event of noncompliance in payment of social security contributions relating to personnel employed by the contractor or the subcontractors in the execution of the contract, the contracting authority withholds from the payment certificate the amount corresponding to the default for the subsequent direct payment to social security and insurance institutions, including, in the works, the construction fund. In any case, a withholding of 0.50% is applied to the progressive net amount of the services. Such withholdings can only be released at the time of final payment, after approval by the contracting authority of the testing or conformity verification certificate, following issuance of the single document of contributory regularity (DURC).

Direct Actions of Contractor's Employees vis-à-vis Owner

As an alternative to statutory holdbacks and liens (see below), in order to protect those employed in a project, some legislations grant to the employees of the contractor a direct action vis-à-vis the owner in case of unpaid amounts (including social security contributions, safety insurance premiums and/or tax retentions) by the contractor to its employees.

Such direct actions may be limited to certain levels of the contracting and subcontracting chain, or in amounts and/or in time. For instance, the action can be available only to the employees of the contractor vis-à-vis the owner, but not to the employees of the subcontractors vis-à-vis the owner; or the direct action can be limited to amounts due and not paid yet by the owner to the contractor at the time of the filing of the direct request by the contractor's employees to the owner. Local laws or case law may further specify if the request of the employee must be judicial or simply an out-of-court notice to the owner.

The purpose of these provisions is generally to freeze, after the request of the employee, the credit of the contractor vis-à-vis the owner, to protect the employees and ensure they are paid by the owner. The effect of such notices is that in case the owner pays the contractor after it has received the notice from the employee, the payment is not effective vis-à-vis the employee, and the owner will be required in any case to pay the employee. Usually, these provisions are mandatory and cannot be derogated or excluded contractually by the parties.

This direct action vis-à-vis the owner is usually different from the action the employee has against its owner, *i.e.*, the contractor. An owner in such jurisdictions may wish to protect itself against the risks: (i) that it pays the contractor after it has received a notice from the employee (in this case an indemnification, or better, a reimbursement, is usually agreed upon), and/or (ii) the legal cost and fees incurred in case the employee opts for the court request.

Liens

In the US system and other Western jurisdictions, contractors who have not been paid by the owner may become lienholder on the property where the construction project is being executed; however, similar rights are not recognized under many other legislations – such as in South Asian countries.

Liens are specifically ruled by the laws applicable to the international construction contract. The regime of liens may vary significantly in each jurisdiction, so understanding the specifics of the applicable law system is paramount to figure out their scope and possible effects. 1999 and 2017 FIDIC model contracts mention liens at Subclause 7.7 [Ownership of Plants and Materials]. The provision establishes that Plant and Materials shall become the property of the Owner, "free from liens and other encumbrances" at the earlier of: (i) their delivery on Site or (ii) when the Contractor is entitled to payment of their value pursuant to the clauses regulating payments. The mentioned Subclause specifies, however, that the provision applies "to the extent consistent with the Laws of the Country," meaning and implying that the laws of the country where the works are being executed may provide differently.

The Scope of Construction Liens

Construction liens (or contractor's liens or mechanic's liens) are claims that the contractor-creditor may file against the property of the owner-debtor to secure a debt. In the construction context, liens grant the contractor the right to sell the owner's property, in case the owner fails to pay any amount due for materials or labor provided by the contractor to the project.

Enforcing Construction Liens

The enforcement of a construction lien is a process starting with a formal notification to the owner-property owner, then implying the submission of a claim to the competent authorities and, possibly, the filing of a lawsuit in courts.[31] Actually, after being filed, a lien turns into a public record matter, which is attached to the property's title. As a result thereof, the property may not be anyhow disposed, unless and until the lien has been properly addressed: therefore, recourse to liens motivates the owner-property owner to pay the owed amounts, considering that the property's disposal is somehow frozen as a consequence of any unresolved liens – this is mostly the case of real estate transactions. To the extent that the owner- property owner challenges the debt or is not able to pay, the contractor-creditor may file a lawsuit to enforce its right, which is expected to conclude with a mandatory order of sale of the property, in order to get the relevant proceeds to settle the debt. The lienholder is generally required to comply

31 https://www.mellorlawfirm.com/construction-law-education/construction-liens-vs-mechanics-liens and https://pbl.legal/surety-bonds-and-liens-for-construction-projects-understanding-your-rights-and-obligations

with specific terms for the procedure, commonly referred to a number of months after the work completion or the providing of labor or materials. Failure to respect these terms can lead to withdrawal of the lienholder's right.

Mechanic's Liens

Mechanic's liens are a kind of construction liens aiming at securing all the professionals participating in a project, particularly if it implies a "vertical" improvement, including either construction but also upgrading of lands (e.g., landscaping) or underground utilities. Contractors and suppliers are protected in case of lack of payment in respect to services and materials provided within these wider property upgrade. As a result of a mechanic's lien, the property's title and the related rights are substantially affected, as the possibilities of sale or refinancing the property are in fact nullified as long as the lien is in place and effective. These hindrances clearly have an impact on the property owner, but also on all the perspective buyers as well as on the value of the property itself.

The US Scenario

The US regulations may help to better understand the picture. Actually, there are certain differences between state laws but a written contract identifying the scope of the work, the envisaged materials and the contract price is a common prerequisite to file a lien. In some jurisdictions, the regulation changes between liens against residential property and against commercial property.

Under the laws of the State of New York, construction liens can be filed by operators performing labor or providing material for the property improvement, at any time during the work or up to eight months after project completion. Then, the lien is effective up to one year. Indeed, according to the laws of New Jersey: (i) for commercial projects, a lien must be filed with a local clerk within 90 days after services or materials were furnished; (ii) for residential projects, a *Notice of Unpaid Balance and Right to File Lien* must be filed within 90 days after the last day of service. The property owner must get a copy of that notice and the further step for a lien in respect to a residential property is an arbitration hearing.

A Final Balance

In the light of the foregoing, four main pros of construction liens may be envisaged:

- Payment security: Due to the abovementioned legal and financial consequences, construction liens give a critical remedy to the contractor-creditor and constitute an inducement for property owners to duly pay for materials and performances provided within the project.

- Legal acknowledgment: Holding an interest in the property until the debts are paid grants a legal acknowledgment of the work/contribution given by contractors and suppliers, clearly stating their credit rights.
- Financial enhancement: Placing a lien on the property means having at disposal a remarkable negotiation leverage, helping settling of the debts in favor of the lienholder.
- Risk mitigant: The use of liens by participants in the construction projects mitigates the risk of lack of payment and facilitates their engagement in construction commitments.

Payment Process

When a Contractor Becomes Entitled to Payment of the Price?

Before entering in the details of payment mechanisms, one of the main legal issues relating to payment is to establish when the contractor becomes entitled to payment of the contract price.

Under certain civil law jurisdictions, this entitlement rises only at the time when the work is accepted by the owner. This principle shall not be overestimated. No contractor would be able to finance a project, nor it will be willing to do so on behalf of the owner. To face such issues, advance payments and interim payments come into play. They provide the contractor, at an early stage of the project and during its performance, with the necessary financing to set up the site, perform preliminary activities, pay for performance bonds and insurances, and pay the subcontractors and suppliers as they are involved in the project.

It is often specified, in the law or in the contract, that such advance payments are only accounts of the overall price, and that the contractor becomes entitled to payment (and therefore to finally retain the amounts received so far) only when the works have been finally accepted by the owner.

The situation is different when projects are to be performed (and paid) by sections. In this case, the contractor becomes entitled to payment of each section when the relevant section is completed and accepted by the owner. If the section is divided in interim milestones payment, the same principle on payments in account applies to each section.

Terms of Payment: Legal and Contractual Aspects

The contracts usually provide for the relevant terms of payment. While certain jurisdictions may have specific terms for payment under construction contracts, in other jurisdictions, there are general principles that apply to contracts in general, and therefore also to construction contracts.

An example is the European Union Directive 2011/7/EU of the European Parliament and of the Council of 16 February 2011 on combating late payment in commercial transactions, which has been transposed into local legislation by EU Countries. According to this legislation, standard terms of payment are 30 days from:

i. receipt of invoices; or
ii. where the invoice's date is uncertain or subsequent, supply of goods/services;
iii. inspection of goods/services, which should not require more than 30 days from receipt/performance.

The parties may agree on different terms of payment. However, terms that are longer than 60 days, provided that they are not grossly unfair in view of the prevailing circumstances, require specific agreement in writing. Terms that are grossly unfair can be declared null and void upon request of the party or *ex officio*, and are substituted by lawful terms. In public works, public procurement legislation often provides specific terms of payment, also in respect of the final payment. Within the limits established by applicable legislation, the parties are usually free to establish specific terms and methods of payment applicable to their project.

The most common contractual terms of payment are:

- Interim progress payments, e.g., on a monthly basis; they can be with or without a downpayment at commencement of the works; the downpayment is normally set at around 10% of the contract price, but in certain cases (including in public works), it can be raised up to 30% of the contract price; the downpayment is normally associated with the issuance by the contractor of an advance payment bond in the same amount of the advance payment, to be released at the same pace as repayment of the advance payment with subsequent interim payments;
- Payment for single parts of the works, which applies if the works are to be completed in various single parts qualified as such; and
- In one installment at the agreed time (prior, upon or after completion of the works).

Payments' Administration

Typically, and subject to the specific provisions of the relevant contract or model contract, in construction contracts, there are four main moments relating to payment of the contract price: (i) payment of the advance payment (this of course apply only if the contract provides for and advance payment to be paid); (ii) payment of the contract price (less the advance payment) through interim progress payments; (iii) release of the moneys retained on each progress payment; and (iv) payment of the final payment certificate.[32]

Construction contracts usually provide a flow of administrative activities preliminary to payments that the parties must perform. Assuming the owner has appointed an engineer to supervise the performance of the project, such activities can be summarized as follows:

- (If relevant:) collection, by the main contractor, of payment applications/ documents from subcontractors;
- Submission of contractor's interim application for payment to the engineer; the interim payment application shall indicate retainages and withholdings, if any;
- assessment of the interim payment application by the engineer against the relevant contract documents;
- If, and to the extent that, it is satisfied with the interim payment application, the engineer informs the owner, asking it to issue the relevant payment certificate for the same amount or the different amount that the engineer will certify as due to the contractor;
- Issuance of the relevant payment certificate by owner, on the basis of and for the amounts certified by the engineer;
- Notification of the payment certificate by the engineer to the contractor, with authorization to the latter to issue the invoice for the amount indicated in the payment certificate;
- Issuance by contractor of the relevant invoice for the certified amount; and
- Payment of the contractor's invoice, within the contractual/legal terms, by the owner.

It is not uncommon that diverging opinions may arise between the contractor and the engineer (or the owner) as to the amounts that are actually due and payable to the contractor. Given the relevance of the payment matters on the prosecution of the works, usually construction contracts for large projects provide for expedite dispute resolution procedures specifically designed to quickly resolve disputes regarding payments.

Payment Delays and Failures

In case of delays in issuance of interim payment certificates and/or payment of the contract price, the contractor is entitled to interests on late payments. Interests are due at the legal or contractually agreed upon rate. If the owner fails to make

32 For FIDIC contracts see, for instance, Baker, Mellors, Chalmers, Lavers, 2009, 182.

payment when due, and unless otherwise agreed upon in the contract, the contractor may: (i) suspend performance of the works; (ii) upon completion, refuse to deliver the works; (iii) apply to court to get payment of the price; in this case, it will be up to the contractor to provide satisfactory evidence of the works performed and for which it claims payment; (iv) if the unpaid amount is material, apply for termination of the contract and damages.

In case the final accounts of the works show a credit of the owner, the latter may in principle and unless otherwise agreed upon: (i) retain from the balance of the price due to the contractor, if any, the amounts due to it; (ii) enforce the performance bond, if available; (iii) pursue the ordinary remedies on the assets of the contractor to satisfy its credit.

Payment Security

Payment dynamics in international construction contracts constitute the most important issues, as they imply the value for the contractor's work and can be undermined by significant payment risks. Disputes in construction projects are in fact very often originated by financial and/or payment problems, which cannot be avoided as such but can be clearly identified, taking into account the hierarchy of the participants in the project chain – owner, contractor/s, subcontractors, suppliers, designers, consultants and, further, financing institutions/banks. The structure of payments within an international construction project follows this chain, therefore the failure to pay by the owner has a waterfall effect on all the other participants and result in delays in the project completion, loss of productivity and/or poor quality of the work. The owner's insolvency indeed constitutes the higher payment risks for the contractor, which can prejudice the swift progress and completion of the project as originally envisaged in the construction contract.

Standard conditions of contract like FIDIC, NEC, and JCT provide certain suitable rights and remedies for the contractor in case of failure to pay by the owner. Nevertheless, securities of payment can be affected by the governing law of the contract, thus some securities of payment may eventually be not available in certain jurisdictions. Different tools in international construction contracts are set out to control and secure payments.

Evidence of Financial Arrangements

In developing its model contracts, FIDIC has significantly enhanced the level of control of the funds destinated to payments to the contractor. In fact, under Subclause 2.4 *[Employer's Financial Arrangements]* of the 1999 Red Book, the Owner is required to provide, within 28 days of the Contractor's request, "*reasonable evidence that financial arrangements have been set up and are being maintained which will enable the Employer to pay the Contract Price"* and shall further notify the Contractor with relevant details, in case it intends to make "*any material change to his financial arrangements.*" The same Subclause 2.4 in the 2010 Pink Book (to be used for projects financed by MDBs-multilateral development banks) sets forth that the Owner is required to submit reasonable evidence of the financial availability "*before the Commencement Date*" as well as within 28 days of the Contractor's request and also give evidence that the payments may be made "*punctually.*" In addition, Subclause 8.1 of the 2010 Pink Book includes, under its letter (b), the "*delivery to the Contractor of reasonable evidence of the Employer's Financial arrangements (under Subclause 2.4 [Employer's Financial Arrangements])*" as one of the condition precedents to be met to trigger the agreed Commencement Date. The 2017 FIDIC editions require that the Owner's arrangement for financing shall be inserted in the Contract Data – thus, also before signature of the contract. The Contractor is entitled in any case request clear evidence of the financial arrangements in the event of: (i) receipt of an instruction for variations exceeding certain thresholds; (ii) lack of payment according to the contract; (iii) becoming aware that a material change occurred in respect to financial arrangements.

Advance Payment or Down Payment

Subclause 14.2 of the FIDIC Red, Yellow, Silver Book (either 1999 and 2017 editions) allow the Contractor to receive, before commencement of the works, an amount which is expressly aiming at granting cash flow at the initial stage of the contract execution, being "*an interest-free loan for the mobilization.*" Also NEC provides advance payment under Option X14 and various JCT Contracts as well (Intermediate, Standard, D&B). The advance payment is generally associated to the contractor's obligation to submit a counterpart advance payment bond, which will be progressively reduced as long as payments due under the contract are made by the owner. Repayment of the advance payment is commonly set up by deducting percentage amounts from the payments envisaged in the contract. Through the advance payment, the contractor may dispose, in the course of the construction, of cashflow already paid even if not actually earned yet, which means hedging against a possible lack of payment. The advance payment will thus enable the contractor to mitigate the risk of the owner's insolvency.

Security from a Third Party

Payment in construction contracts is also commonly secured through a guarantee released by a third-party in respect to the contract/project, such as banks or insurance companies. The range of options here may include bank guarantees (often on-demand) and different types of bonds and/or credit letters. FIDIC model contracts provides a specific Annex, envisaging a "Form of Payment Guarantee by Employer." The cons to this option are the upfront and actual costs to be sustained by the owner.

Suspension (and Possible Termination)

Since the 1999 editions, FIDIC model contracts (Red Book, Yellow Book, Silver Book) allow the Contractor to suspend the execution of the works or to reduce the rate of work, under Subclause 16.1 [*Suspension and Termination by Contractor*], in the event of a payment delay or a failure to pay or in case of lack of evidence of the Owner's financial arrangements. In case one of these circumstances persists beyond certain terms, the Contractor is entitled to terminate the construction contract. The underlying principle to this provision is known by most jurisdictions, based on the Romanistic principle of "*inadimplenti non est adimplendum.*"

Direct Payment by Lenders

In project finance transactions, the agreements denominated "Direct agreements" are instrumental to provide lenders with a direct relation with the main counterparties of the project. Indeed, direct agreements are entered into to strengthen the lenders' protection of their interests, enabling them to enforce their rights under the contracts if a financial hardship or a default by the special purpose vehicle jeopardizes the deal. In this regard, there are step-in rights in favor of the lenders, who can take the place of the project company and assume its contractual rights and obligations vis-a-vis the other key project counterparts. As a result, the lenders may also directly make certain payments whenever required for the smooth proceeding of the project.

Litigation

Disputes relating or connected to payment issues in international construction contracts are highly frequent; the choice of the best dispute resolution system is therefore crucial in this regard. The option of local courts is often disregarded by international contractors, due to the uncertainty variables and risks that may affect litigation in a specific country. Payment and security payment disputes in international projects are thus commonly submitted to arbitration; all FIDIC general conditions provide this option as applicable – unless otherwise agreed by the parties in the contract – and address the choice to arbitration under the rules of the International Chamber of Commerce (ICC).

Liens

Contractors that do not receive duly payment are entitled to place a lien on the various properties relating to the project, thus materials, machines, equipment and construction products may be subject to liens. By the lien right, the contractor is granted with a security interest in the project, allowing to mitigate the owner's insolvency risk, but particular attention should be paid in respect to local laws applicable to the contract, as not all the jurisdictions recognize the lien rights in favor of the contractor. In respect to liens, see also the preceding paragraph.

Bibliography

Masons Thelen Reid LLP, *Managing Payment Risks on International Construction Projects*, 2006.

Musolino, G, *L'appalto internazionale*. (Milano: Giuffré, 2003).

National Construction Projects "*Securing high Performance through Cultural awareness and Dispute Avoidance*" SHANGHAI, CHINA, November 21–23, 2008.

Patroni Griffi, U., Gli appalti internazionali di costruzioni e gli ordinamenti di *civil law*. *Diritto del Commercio Internazionale*, 825, 1998.

Rubino Sammartano, M. *Gli appalti internazionali*, in Appalto Pubblico e Privato a cura di Iudica. Padova, Cedam, 1996.

Wenfa Hu, Jianguo Chen, Zhaomin Ren, George Wang. *Identification and control of payment risks in international construction projects* from International Conference on Multi-national Construction Projects "Securing high Performance through Cultural awareness and Dispute Avoidance", Shanghai, China, November 21–23, 2008, available at: https://www.irbnet.de/daten/iconda/CIB12202.pdf.

5

The Parties

Shona Frame and Keith Robertson

The identity and number of parties involved in any construction project will vary depending on the nature, size, and complexity of the project and which form of contract is being used. This chapter sets out details of the types of party typically involved in construction projects and their roles as part of the delivery team.

What this illustrates is that while there are common names for parties, the role they in fact play can differ from contract to contract and it is important to understand the nature of their role in the particular contract being used and not make assumptions as to what this is based only on their title.

Employer/Owner/Client

In every contract, there will be a party for whose benefit the work is being carried out. Terminology varies from country to country and within different contract forms but they tend to be known as the Employer, Owner, or Client.

Their obligations would typically involve providing access to the site to the Contractor to allow work to be carried out and making payments to the Contractor. For example, in the NEC4 Engineering and Construction Contract ("NEC"), the Employer allows the Contractor access to the Site which is necessary to perform the work included in the contract.[1]

In relation to payment, the Employer's obligation is often to pay a series of interim payments as the work progresses either based on the value of the work undertaken to that point in time or milestone payments linked to either dates or progress such as completion of certain activities.

1 Clause 33 NEC.

Major Issues in International Construction Law, First Edition. Mauro Rubino-Sammartano and Troy Harris.
Published 2026 by John Wiley & Sons, Inc.

The International Federation of Consulting Engineers (FIDIC) Conditions of Contract for EPC Turnkey Projects (2nd Edition, 2017; Reprinted 2022) (the Silver Book) contains an additional requirement for the Employer to provide the Contractor with reasonable evidence of financial arrangements to pay the contract price. This evidence must be provided upon the Contractor's request and if there is a significant change in the Employer's financial situation.[2] The Silver Book also allows the Contractor to terminate the contract if there is a change in the Employer's financial situation and the Employer fails to provide reasonable evidence of financial arrangements to pay the Contract Price.[3]

The Employer will have to provide details of the project to be constructed. Depending on the form of contract, the Employer may provide the whole of the design or, if a design and build contract, that will be restricted to the Employer's Requirements from which the Contractor can develop the more detailed design.

In FIDIC, this is reflected by an obligation on the Employer to clearly specify the intended purpose of the facility in the Employer's Requirements.[4]

In NEC, the Employer provides the Scope which is information which specifies and describes the works to be undertaken or specifies any constraints on how that is to be done.[5] This includes any design elements that the Client is to provide.

The Employer may also be involved in assisting with obtaining any necessary licenses and permissions although often this obligation sits with the Contractor.

Employer's Professional Team

Architect

In building projects, the Architect will have a role as Designer but may also have a role as a contract administrator in some forms of contract. They will generally be appointed by the Employer under a form of Appointment. They will be responsible for inspecting the work as it progresses and monitoring it for quality and compliance with the contract documents.

If there is a design and build form of contract whereby the Contractor is taking on responsibility for both design and construction, then it is often the case that the Architect will initially be employed by the Employer but their Appointment will then be novated to the Contractor so that they become a subcontractor of the Contractor.

2 Clause 2.4 Silver Book.

3 Clause 16.2 Silver Book.

4 Clause 4.1 Silver Book.

5 Clause 11.2(16) NEC.

It will be important to consider the terms of the novation, particularly in relation to responsibility to the Contractor for work already carried out for the Employer.

Neither the FIDIC or NEC forms of contract, being primarily for engineering projects, specify the Architect as a party to the contract.

In the UK, both The Joint Contracts Tribunal (JCT) and Scottish Building Contracts Committee (SBCC) suites of contracts refer to the role of Architect/ Contract Administrator and in that role they have duties related to the administration of the contract.

Engineer

The Engineer is another party with a potentially dual role. The Engineer can have a design role in either building (where it would usually be structural or mechanical & electrical engineers who are involved) or engineering projects (which are likely to have civil engineers).

However, they may also have a role as a contract administrator. For example, in the FIDIC Construction Contract 2nd Ed (2017 Red Book, Reprinted 2022 with amendments) ("Red Book"), the Engineer has a range of duties and has wide authority within the contract conditions. These include that the Engineer shall act as a skilled professional and will act for the Employer.[6] The Engineer's role includes the ability to instruct variations,[7] facilitating Agreement discussions and providing Determinations on matters in dispute[8] and issuing Interim Payment Certificates.[9]

In NEC, the Engineer is not referenced as a named party and the contract administrator type role sometimes carried out by the Engineer will be fulfilled by the Project Manager and Supervisor.

Contract Administrator/Employer's Agent/Employer's Representative

This is a role that again is referred to by various different terminologies. It may be undertaken by the Architect (on a building project), Engineer (on a civil project), or by another party.

The Contract Administrator will effectively have conduct of the project in terms of being the party, which certifies payments, issues instructions, receives contractual notices and makes decisions on, for example, extension of time awards to the Contractor and whether practical completion has taken place. They will

6 Clause 3.2 Red Book.

7 Clause 13.1 Red Book.

8 Clause 3.7 Red Book.

9 Clause 14.6 Red Book.

be involved in site meetings, monitoring progress of the works and flow of information.

In that role, they are intended to act impartially although they are employed (and paid for) by the Employer.

In the Silver Book, the Employer's Representative holds full authority from the Employer, except in relation to termination.[10] The Employer's Representative is responsible for performing assigned duties, exercising delegated authority, and acting on behalf of the Employer. The Employer's Representative must act as a skilled professional and be fluent in the ruling language of the contract's location.[11]

In NEC, the Project Manager fulfills the role of the Contract Administrator. NEC gives the Project Manager wide-ranging duties related to the management of the project. They are obligated to act according to the contract and in the spirit of mutual trust and cooperation between the parties.[12] The Project Manager is responsible for using the contractual mechanisms to accept or reject the project program and updates to it, changes to the Price, changes to the Completion Date, and changing the Scope and other matters as set out in the contract.

Quantity Surveyor

The Quantity Surveyor's role relates to the costs of the project. They may be involved in preparing bills of quantities, where those are being used as a way to ask Contractors to price the project at tender stage, in valuing work carried out by the Contractor and in ascertaining any claims submitted by the Contractor.

Neither the FIDIC or the NEC forms of contract provide for the role of the Quantity Surveyor but in JCT/SBCC, this is a defined role.

Specialist Consultants

Depending on the nature of the project, there can be a number of specialist consultants involved. They will not necessarily have defined roles within the contract between Employer and Contractor but will have important roles nevertheless.

10 Clause 3.1 Silver Book.

11 Ibid.

12 Clauses 10.1 and 10.2 NEC.

Civil Engineer

The civil engineer's role would tend to relate to infrastructure projects. The Institution of Civil Engineers (ICE) in the UK describes it as centering around "the design, construction and maintenance of infrastructure, from bridges to flood defences, wind turbines to tunnels, and more."[13] Infrastructure itself covers a wide range of types of project such as transport, water/waste water, sanitation, power generation, telecommunications.

Structural Engineer

The Institution of Structural Engineers (IStructE) in the UK describes the role of structural engineers as designing the strength and stability of buildings and structures to ensure they are strong enough to support the loads on them and stable against elements such as wind or, in some parts of the world, hurricanes, earthquakes, and other natural disasters.[14]

Mechanical and Electrical Engineer

The Mechanical and Electrical (M&E) Engineer would get involved in the design, installation, and maintenance of building systems like heating, ventilation, air conditioning, plumbing and fire protection (mechanical elements) and electrical systems and equipment such as control systems, fire alarms, and emergency lighting.

Other Specialists

A number of other specialist consultants can be involved depending on the nature of the project. These could include specialists in Planning, Environment & Sustainability, Fire, Landscape Architect, Lifts, Interior Designers, Acoustics/ Lighting or many other specialist disciplines.

These specialists would tend to be engaged through Appointments by either the Employer or the Contractor depending on which party has the responsibility for the design of the particular element of work. The duties owed by the specialist will vary based on their role, and these terms will be specified in their Appointment.

---

13 https://www.ice.org.uk/what-is-civil-engineering/civil-engineering-explained.

14 https://www.istructe.org/become-a-structural-engineer/what-is-an-engineer/.

Health and Safety

Health and safety is an important consideration on construction projects. There tend to be obligations incumbent on parties as a matter of law and regardless of any contractual provisions. Obligations would tend to sit with the Contractor and the Subcontractors (and other parties) for their own employees. These might relate to provisions for the supply and use of personal protective equipment as well as to their methods of working. In the UK, there is a general duty on employers (in the sense of a party who employs workers as opposed to in the construction sense described above) to ensure the health, safety, and welfare at work of their workers.[15]

The designer will also have obligations related to avoiding health and safety risks in terms of how the building or structure is to be built as well as in relation to the operation and maintenance of the building or structure thereafter.

In addition there are often specific contractual provisions related to health and safety obligations and sometimes accompanied by termination rights in the event of breach.

In FIDIC, the Employer is responsible for ensuring its personnel or other contractors adhere to health and safety obligations.[16] The Contractor must comply with all health and safety regulations and any specific health and safety requirements stated in the contract.[17] Additionally, the Contractor must submit a project-specific health and safety manual to the Employer.[18] The Contractor must take all necessary precautions to maintain the health and safety of its Personnel, including appointing a health and safety officer at the Site.[19] In NEC, the Contractor must adhere to the health and safety requirements specified in the Scope.[20] The Contractor includes provisions for health and safety within the programme.[21]

Project Time Manager

This is a role referenced in the CIOB (The Chartered Institute of Building) Contract for use with Complex Projects. The Project Time Manager (PTM) role is to monitor and manage the program. The PTM reports to the Employer or Contractor on the progress of all procurement, engineering and construction activities.

15 Section 2(1) Health and Safety at Work etc. Act 1974.

16 Clause 2.3 Silver Book.

17 Clause 4.8 Silver Book.

18 Ibid.

19 Clause 6.7 Silver Book.

20 Clause 27.4 NEC.

21 Clause 31.2 NEC.

They monitor contract milestones, assess risks related to meeting the program, and propose mitigation measures to ensure key milestone dates are met.

Project Manager

In some forms of contract such as NEC, the Project Manager appointed by the Employer has a role, in effect as a contract administrator in the same way as described above. Project Managers can also be appointed by Employers to take on tasks such as handling of the tender process, preparation of contract documents and monitoring of progress.

Clerk of Works/Supervisor/Technical Adviser

A Clerk of Works/Supervisor/Technical Adviser may be appointed as the "eyes and ears" of the Employer on site to monitor the works as they are carried out and check for compliance with contractual standards. They would tend to have a more extensive inspection obligation than the Architect as they would have a more frequent site presence and have access to works as they progress rather than only a snapshot of progress obtained through periodic inspections.

In NEC, the role is referred to as Supervisor and they have obligations to carry out tests and inspections, instruct the Contractor to search for defects, notify defects and issue the defects certificate.[22]

Contractor/Main Contractor

The Contractor is the party responsible for undertaking the work which is the subject of the contract. They can have design as well as construction obligations.

Their obligations would typically involve carrying out and completing the work (which could be build only or design and build), complying with instructions issued for or on behalf of the Employer, providing documentation such as as-built drawings, any certifications and information for the Operation & Maintenance Manuals.

Some obligations apply to all contracting parties. For example, in NEC, the duty to act in a spirit of mutual trust and cooperation[23] applies equally to the Employer, the Contractor, the Project Manager, and the Supervisor. In Alliancing forms

22 Clause 14 NEC.

23 Clause 10.2 NEC.

of contract, similar obligations would apply among all parties in a potentially multiparty arrangement.

The Contractor will usually have duties imposed on them related to quality of the work including compliance with regulatory and statutory requirements as well as any contract specific requirements.

Within Contractors there will be a raft of different roles – Commercial, Project Manager, Quantity Surveyor, Site Manager, Supervisors, Buyers, Estimators, BIM Manager, Health and Safety, and so on.

The interplay between the Contractor's team on site (Project Manager, Site Manager, Site Agent) and the Employer's team (Engineer/Architect) can be important in ensuring the successful execution of projects. A constructive relationship can assist, whereas when that relationship deteriorates, this can start to contribute to conflicts developing.

Subcontractor

The majority of Contractors do not have the ability to undertake work on every trade required in a project. They therefore employ Subcontractors to undertake work on their behalf. On a complex project, there can be very large numbers of subcontractors involved.

Domestic Subcontractor

As between the Employer and Contractor, the Contractor remains responsible for all work undertaken by Subcontractors but in the event of there being delay or any defects then the Contractor would have recourse against the relevant Subcontractor(s).

Often contracts require the Employer to grant permission to the Contractor to subcontract elements of the work. For example, NEC requires the Contractor to submit the Subcontractors' names for acceptance.[24]

In FIDIC, the Contractor is responsible for the Subcontractor's work and must give the Employer 28 days' notice before appointing a Subcontractor.[25]

Named Subcontractor

Where there are particularly specialized elements of work or if the Employer has a preference as to which party should undertake certain elements of work, the Employer may name a Subcontractor as its preferred choice to be appointed by the

24 Clause 26 NEC.

25 Clause 4.4 Silver Book.

Contractor. The Contractor would remain responsible for the work undertaken by that Subcontractor in the same way as for a Domestic Subcontractor.

Nominated Subcontractor

It is possible in some cases for the Employer to require certain Subcontractors to be used, known as Nominated Subcontractors.

Where that is done, because the Contractor has had no choice, in some cases, the Employer takes responsibility for the work of the Nominated Subcontractor.

FIDIC allows the Employer to nominate a Subcontractor in the Employer's Requirements or to instruct the Contractor to employ a specific Subcontractor.[26] The Contractor is not obligated to employ a nominated Subcontractor if it raises reasonable objections, such as that the Subcontractor is not competent to carry out the work.[27] If, however, the Contractor accepts the nominated Subcontractor, they are regarded as a Subcontractor under the Contract.[28] Consequently, the Contractor is responsible for the work of the nominated Subcontractor.

Suppliers

Suppliers of materials, plant, and equipment used to undertake the build are important parties in the construction process. They may supply to Contractors, Subcontractors, or direct to the Employer in some circumstances if the Employer is providing "free issue" materials to the Contractor.

Insurers

Construction projects require a number of insurance products including for loss or damage to the works, plant, materials and equipment, loss of or damage to property other than the works and liability for death or personal injury. In relation to design elements, those with responsibility for design would take out professional indemnity insurance cover.

There can be other more specialist insurance products such as Latent Defect Insurance or insurance related to business interruption or pollution.

---

26 Clause 4.5 Silver Book.

27 Clause 4.5.1 Silver Book.

28 Clause 4.5 Silver Book.

In Alliancing forms of contract, there is the potential for Integrated Project Insurance (IPI). This is a single insurance policy for the entire project and all participants. IPI insures risks and outcomes rather than focusing on liabilities and causes. In the NEC4 Alliance Contract, each member is required to insure against death or bodily injury to its employees but the Employer provides the Insurance stated in the Insurancc table, except where otherwise identified as the responsibility of the other parties.

In FIDIC and NEC, the Employer is responsible for insuring against property risks and any risks classified as force majeure. The Contractor is obligated to obtain insurance policies for the execution of the works. Typically, the Contractor must provide proof of insurance to the Employer or Project Manager. In FIDIC, the Contractor must provide the Employer with evidence of the necessary insurance policies and copies of receipt of payment, whenever required.[29] If the Contractor fails to comply with the insurance obligations, the Employer may effect and keep in force any such insurances and pay a premium as necessary which can be recovered from the Contractor.[30]

In NEC, the Employer must supply the insurances specified in the Contract, while the Contractor is tasked with securing the insurances detailed in the Insurance Table (such as for loss of or damage to the works, plant and materials, loss of or damage to equipment, loss of or damage to property and liability for bodily injury or death arising from the works and death of or bodily injury to employees of the Contractor.[31] Should the Contractor fail to provide the necessary insurance certificate, the Employer has the right, at the Contractor's cost, to obtain insurance for the event that the Contractor was obligated to cover.[32]

Sureties

If the Contractor is asked to provide performance security against the possibility of its breach of contract, in the form of a Bond or Guarantee, that would tend to be provided by a surety (a bank or other financial institution).

In the UK, conditional performance bonds are the most common type of Bond provided by a bank or insurance company, on instruction of the Contractor, in favor of the Employer. A conditional bond would require proof of the breach of Contract by the Contractor and generally requires losses and damages sustained

29 Clause 19.1 Silver Book.

30 Ibid.

31 Clause 83 NEC.

32 Clause 85.1 NEC.

by the Employer to be established and ascertained in accordance with the Contract as a prerequisite to payment by the surety.

The amount of the Bond will be based on a percentage of the total contract value. In the UK, bonds are typically issued for 10% of the contract value, whereas in Canada, bonds are typically issued for 50% of the contract value, though bonds can be issued for up to 100% in some instances.

In some cases, the Contractor may ask for a Bond or other form of security for payment in the event the Employer fails to pay.

Bonds can also be provided in lieu of retention meaning that instead of the Employer deducting a retention percentage from payments to the Contractor, they will instead pay the Contractor in full in return for the Contractor providing a bond.

The same may happen if there is an advance payment made to the Contractor where a Bond may be provided as security until the Contractor has carried out sufficient work to earn the sum paid.

More commonly used on international projects and in non-UK jurisdictions, on demand bonds may be provided by the Contractor to the Employer. On demand bonds create a primary obligation of the bondsman to pay immediately upon a compliant written request from the Employer, usually on supply of the required documents outlined in the bond.

In FIDIC, there are specific obligations for contractors' sureties, particularly in relation to performance security[33] and advance payment guarantees.[34]

In NEC, the secondary option clauses contain the optional obligation for a Contractor to provide a performance bond,[35] an advance payment bond,[36] a retention bond,[37] and ultimate holding company guarantees.[38]

Funders

Where the Employer needs to secure additional finance to allow it to proceed with the works, Funders can be involved. They will ordinarily want a security package in return for providing funding, which often includes a requirement for collateral warranties from the Contractor, Designer(s), and key Subcontractors.

---

33 Clause 4.2 Silver Book.

34 Clause 14.2 Silver Book.

35 Option X13 NEC.

36 Option X14 NEC.

37 Option X16.3 NEC.

38 Option X4 NEC.

The purpose of collateral warranties is to create a direct contractual relationship between the Funder and Contractor, certain Subcontractors and/or Designer to give the Funder a direct route to pursue them should it not be possible for it to pursue the Employer.

Prospective Purchasers/Tenants

On completion of certain projects, there will be a future purchaser of the building or other asset and/or a future tenant of a property. In the same way as for Funders, they will ordinarily require collateral warranties from the Contractor, Designer(s), and some key Subcontractors. These collateral warranties can be useful to these parties as it gives them a recovery route against these parties, primarily in the event of latent defects coming to light.

Dispute Resolution Procedure Parties

Most construction contracts include a dispute resolution procedure. Often this is a multistage process meaning that a number of parties may be involved.

Senior Representatives

The dispute resolution process will often start with a reference to Senior Representatives of the parties with escalation to Chief Executives or other senior officers. These individuals are often named in the contract, although subject to being updated as required through passage of time.

In NEC, disputes may initially be referred to the Senior Representatives. If they are unable to resolve the issue, it is then referred to the Adjudicator for a decision (although in the UK, there is an ability to adjudicate at any time so the Senior Representatives stage may be bypassed).[39]

Mediator

There is an increasing trend of Mediation being included within some contracts as part of the multistage process for dealing with disputes. Mediators are independent third parties who would be appointed to work with the parties to explore whether a resolution can be reached. Often the contract provision is simply to encourage parties to consider Mediation as an option as opposed to being a mandatory requirement.

39 Option W1.1/W2.1 NEC.

Dispute Board Members

Some dispute resolution processes involve reference to a Dispute Board. Often Dispute Board members will be appointed at the outset of the project and called upon as necessary throughout the life of the project. Failing that, an appointment mechanism will be set out in the contract.

In FIDIC, the parties jointly appoint members to the Dispute Avoidance/Adjudication Board (DAAB) to help resolve and prevent disputes during the project.[40] The parties may appoint either one or three suitably qualified members. The parties may jointly request the DAAB to provide assistance or informally discuss and attempt to resolve any issue or disagreement.[41] If a dispute arises, either party has the right to refer a dispute to the DAAB for a decision.[42] If a dispute cannot be resolved amicably or if parties are dissatisfied with the DAAB's decision, it is possible to refer the matter to Arbitration for determination.[43]

In NEC, there is an option to select a Dispute Avoidance Board as the method of dispute resolution. It can consist of one or three members who are to act impartially and to assist the parties in resolving potential disputes.[44]

Adjudicators

Some contracts include a contractual Adjudication option – generally a fast-track form of dispute resolution resulting in a binding decision. NEC includes this option.[45]

In some countries, including the UK, there is a statutory right to Adjudicate as a fast-track form of dispute resolution. In NEC this is provided for by a UK-specific Adjudication provision, which complies with the requirements of the statute allowing for Adjudication.[46]

Contracts will either include reference to named Adjudicators or will specify the Adjudication nominating body which will be asked to appoint an Adjudicator.

---

40 Clause 21.1 Silver Book.

41 Clause 21.3 Silver Book.

42 Clause 21.4 Silver Book.

43 Clause 21.6 Silver Book.

44 NEC Option W3.

45 The parties can include Option W1 to include Adjudication where the contract is not considered a "construction contract" under the United Kingdom Housing Grants, Construction and Regeneration Act 1996.

46 The parties would include Option W2 to include Adjudication where the contract is considered a "construction contract" under the United Kingdom Housing Grants, Construction and Regeneration Act 1996.

Arbitrators

Where Arbitration is the specified form of dispute resolution, the contract may contain named Arbitrators who may be instructed or specify the appointing body or arbitral institution to be used to make an appointment. The contract may also provide for specific arbitral rules to apply to the Arbitration.

Expert Determiner

Some contracts provide for an Expert Determiner to be appointed to resolve disputes or issues between the contracting parties. This role is typically established in the contract or can be agreed upon later if a dispute arises and parties consider that to be appropriate. The Expert Determiner's decision is usually binding on the parties, and the procedure followed tends to be less formal than Arbitration or litigation.

The Institution of Chemical Engineers (IChemE) forms of contract contain provision for expert determination of certain disputes. In the Engineering, Procurement and Construction Management (EPCM) Contract (The Blue Book) 2023, there is an optional clause C53 that allows the parties to include the option to appoint an Expert Determiner and refer disputes under certain clauses for determination by the Expert Determiner.[47] The parties will be bound by the decisions of the Expert Determiner[48] and would not then be able to refer the dispute to Arbitration.[49]

New Recommended Roles in the UK

Following the Grenfell Tower fire on 14 June 2017, a Public Inquiry took place. Phase 2 of the Inquiry's Report[50] was published in September 2024. It contains wide-ranging recommendations, which include some new parties to be involved in construction. These include:

- Construction Regulator – a central, independent authority headed by the Construction Regulator and responsible for overseeing and ensuring the safety and compliance of the construction industry. This regulator would consolidate various functions currently dispersed among multiple bodies including

47 Clause C53 The Blue Book.

48 Clause C53.7 The Blue Book.

49 Clause C53.9 The Blue Book.

50 https://www.grenfelltowerinquiry.org.uk/phase-2-report.

the regulation of construction products, development and certification of fire testing methods, oversight of building control, and licensing of contractors wishing to construct/refurbish higher-risk buildings.

- Chief Construction Adviser appointed by the Secretary of State, to provide advice on all matters related to the construction industry.
- A body of professional fire engineers accredited by a professional regulator and accredited fire risk assessors.

6

The Designer

Jennifer Varley

Introduction

This chapter explores design liability in construction and engineering projects. By design liability, we mean the party that has legal responsibility in the event that the design of the works does not meet the required standard. Typically, many different parties are involved in the design process for construction projects, including architects, engineers, and contractors from both general and specialist disciplines. The liability of such parties for defective design will depend on the standard of care that they are required to exercise in producing the design. Such a standard will be determined by what has been agreed contractually. If the contract is silent, then the standard will be determined by statute and common law. The standard of care that designers are required to exercise will depend on the nature of the contract and whether it is governed by statute,[1] what has been contractually agreed or, in the absence of a specified standard of care, the common law. It is possible to exclude or vary common law standards[2] and in certain circumstances, statutory provisions.[3] This chapter will focus on the standards commonly included in construction standard form contracts.

There are a range of design standards to which a designer may be subject, ranging from reasonable skill (requiring the least input) and fitness for purpose (constituting an absolute obligation, the product either meets the agreed criteria

1 For example, the Defective Premises Act 1972, Sales of Goods Act 1979, Supply of Goods and Services Act 1982, the Consumer Rights Act 2015, and the Unfair Contract Terms Act 1977. In-depth discussion of such legislation is outside the scope of this chapter.

2 The Mercini Lady [2010] EWCA 1145.

3 For example, §11 Supply of Goods and Services Act 1982, although the Defective Premises Act 1972 cannot be excluded.

Major Issues in International Construction Law, First Edition. Mauro Rubino-Sammartano and Troy Harris.
Published 2026 by John Wiley & Sons, Inc.

or it doesn't). This chapter explores the range of design obligations found in construction standard form contracts.

This chapter is written from an English law perspective. It is not intended to be comprehensive, nor does it constitute legal advice. Specific legal advice should always be sought before taking or refraining from taking any action.

The Impact of Procurement Routes

The contractual obligations and standard of care that the designer is held to will be one of the matters for commercial and contractual negotiation between the various parties involved in any construction project. However, the procurement route that is selected for the project will also influence the allocation of design responsibility between the parties at the outset of commercial and contractual discussions, as demonstrated by the summaries of the common English law procurement routes as follows:

Traditional Procurement Route

The owner appoints a professional designer to prepare the design and engages a contractor for the construction of the design. The contractor is responsible to the owner for the construction only and the designer is responsible for the design followed by the contractor.

Design and Build Procurement Route

A design and build contractor is responsible for the design and construction of the works and is a single point of responsibility for the owner when defects or issues arise in the works. Under the design and build procurement route, the owner usually appoints a design professional to set out the owner's design requirements (e.g., in the owner's/employer's requirements document) and develop an initial design before the project is tendered to contractors.

The contractor generally provides details of a proposed design for the project at the tender stage (this document is often referred to as the contractor's proposals). While the exact details will differ from project to project, upon signature of the construction contract, the contractor may assume responsibility for the owner's initial design (prepared by the design professional), by express words in the contract and assume responsibility for developing the design.

It is typical for the design professional's appointment with the owner to be novated to the contractor. Novation involves the discharge of the initial contract between the owner and design professional, and the creation of a new contract in its place between the designer and the contractor. Following a novation, the design professional is liable to the contractor for its duties and responsibilities under the terms of its appointment with the owner. The design professional, under the terms of its appointment, will normally be required to provide a collateral warranty to the owner, where its appointment is novated to the contractor.

Engineering Procurement and Construction Route

Engineering Procurement and Construction (EPC) contractors deliver a complete facility to the project company. It is often called a turnkey project on the basis that the owner can simply turn the key to start the facility. The EPC route is frequently used for large-scale and complex infrastructure projects. Depending on the standard form of contract adopted, it is more common in an EPC contract procurement structure for the contractor to warrant the accuracy and/or sufficiency of the design set out in the owner's requirements document.

Appointing a Professional Designer

Appointing a professional designer is usually an essential step for an owner in a traditional procurement method, where the professional designer prepares the design and the contractor constructs the design. The owner will often also appoint a professional designer as part of a design and build procurement route to produce an initial design and a set of owner's requirements, specifying the parameters of what the contractor is required to design and build and which are provided to the contractor as part of the tender process. In an EPC procurement route, the owner will likely appoint an engineer to provide the front-end engineering design (FEED), which the contractor is later asked to verify. In all procurement routes, the owner may wish to have a consultant designer appointed to review the design and progress of the works.

Design Duty and Standard of Care – Professional Designer

A professional designer has a common law and (where applicable) a statutory duty[4] to exercise a reasonable degree of skill and care in carrying out the design services in connection with the construction or engineering project, unless the contract clearly excludes such standard or specifies a different standard. Therefore, it must be shown that the designer has been negligent (i.e., fallen below the required

4 This duty is also an implied term under the Supply of Goods and Service Act 1982, Section 13.

standard of reasonable skill and care), before damages can be recovered from the designer either by virtue of a contract claim for breach of contract or in tort of negligence.

The common law test of whether a professional designer has exercised reasonable skill and care is judged against what is reasonable for a reasonably competent member of that profession. The case of *Bolam v Friern Hospital Management Committee*[5] established that where special skill and competence are involved, the test for negligence is not to be measured against that of an ordinary man, but of someone who possesses the same special skill set: "*it is sufficient if he exercises the ordinary skill of an ordinary competent man exercising that particular art.*"[6] The professional designer does not need to demonstrate greater than average skills to discharge its common law duty of care. The court in *Eckersley v Binnie and Partners*[7] considered that, "*A professional man...must bring to any professional task he undertakes no less expertise, skill and care than other ordinarily competent members of his profession would bring but need bring no more. The standard is that of the reasonable average.*" Therefore, in carrying out the design services, the professional designer must live up to the average standard of the profession. In a claim for professional negligence, it is usual for a court to require expert evidence as to the standards ordinarily observed within a profession before it will find that a professional's conduct amounts to negligence.[8]

A professional designer can agree to a higher standard of care through express terms in the contract. In *Costain Ltd v Charles Haswell & Partners Limited*,[9] the consultancy agreement contained both reasonable skill and care and absolute obligations: clause 7.2 of the consultancy contract stated "*in the provision of the services the consultant shall exercise all reasonable professional skill, care and diligence.*" Clause 7.4 provided "*any part of the works designed pursuant to this agreement if constructed in accordance with such design, shall meet the requirements described in the Specification or reasonably to be inferred from the Tender Documents or the Contract or the written requirements of Costain and designed in accordance with good up to date engineering practice and with all applicable laws, by laws, codes or mandatory regulations and in all respects with the requirements of the contract.*" The reference to the "Contract" was to Costain's design and build

5 [1957] 1 W.L.R 582.

6 *Bolam v Friern Hospital Management Committee* [1957] 1 W.L.R 582 at 586.

7 [1955-95] P.N.L.R 348 at 383.

8 *Avondale Exhibitions Limited v Arthur J. Gallagher Insurance Brokers Limited* [2018] EWHC 1311 (QB), at 119. See also, *Sansom v Metcalfe Hambleton and Co* [1998] P.N.L.R 542 and *Pantelli Associates Ltd v Corporate City Developments Number Two Ltd* [2010] EWHC 3189 (TCC).

9 [2009] EWHC 3140 (TCC).

contract with the owner to provide a new water treatment works and the reference to the "Specification" was to the specification within that "Contract." The judge found that while it is true that usually a professional man in the field of construction only undertakes an obligation to exercise reasonable professional skill and care, it is perfectly normal, in any given case, for a professional to give express warranties that impose strict liability or a performance obligation such as the building will be reasonably fit for a specified purpose. The judge concluded that there is nothing wrong in principle or unusual in finding such provisions (i.e., absolute obligations) in professional engagements.

The interaction between any absolute obligations in the designer's appointment and the reasonable skill and care obligations will depend upon the express terms of the appointment. In *MW High Tech Projects UK v Haase Environmental Consulting*,[10] Haase was appointed by the contractor, "MW", to provide the process engineering design for a waste-to-energy plant. In its appointment, Haase had obligations to design in accordance with the EPC Output Specification and EPC Delivery Plan. These obligations were expressed to be "independent obligations" but "*subject to the Consultant's overriding obligations to exercise reasonable skill and care.*" The court found that in the hierarchy of principal obligations, the obligation to exercise reasonable skill and care is paramount.[11] For example, if complying with the EPC Output Specification would render Haase negligent, then Haase was not obliged to comply with that part of the EPC Output Specification. However, if they could comply with the EPC Output Specification and produce a design that was not negligent, then they were obliged to take reasonable skill and care to do so. In this case, the court found that Haase could be liable to MW for the cost consequences of modifications to the design which did not comply with the EPC Outline Specification.

A designer may impliedly agree to exercise a higher standard of care if it warrants that the design, once constructed, will be fit for a purpose made known to the designer. In *Greaves & Co v Baynham*,[12] the consultant structural engineers had designed a factory complex. The consultant engineers were made aware that the first floor was to be used for moving heavy loads. The floor cracked and the design and build contractor brought a claim against the consultant engineer. It was held that the structural engineer breached an implied warranty that the design of the floor would be fit for the purpose for which they knew it was required. The consultant engineer had knowledge of the requirements for the first floor and, as

---

10 [2015] EWHC 152 (TCC).

11 *MW High Tech Projects UK v Haase Environmental Consulting* [2015] EWHC 152 (TCC), at para 46.

12 [1975] 1 W.L.R 1095.

such, these "*special circumstances*" meant that there was a higher duty imposed on the consultant engineer than the law in general imposes on a medical or other professional man.[13]

Professional Designer – Standard of Care – Standard Form Contracts

The RIBA Standard Professional Services Contract 2020 requires the designer to exercise "the reasonable skill care and diligence to be expected of an architect/consultant experienced in projects of a similar size, nature and complexity."[14] The RIBA Standard Agreement 2010 had a slightly different standard of care, relating the standard of care to the Architect's profession rather than relating it to similar projects. For large, particularly complex or high value construction and engineering projects, a standard of care that requires the designer to act as a designer experienced in projects of a similar size, nature, and complexity arguably requires a higher standard of care from the designer than a standard of care that requires that designer to act in accordance with the normal standards of its profession. The RIBA 2020 contract, in respect of large, complex, or high-value projects, requires a higher standard of care from professionals appointed under its terms than its 2010 predecessor.

The FIDIC White Book Client/Consultant Appointment Model Services Agreement (fifth edition, 2017) requires the consultant to exercise not just reasonable skill, care, and diligence but "reasonable skill, care and diligence to be expected from a consultant experienced in the provision of such services for projects of a similar size, nature and complexity." The 2006 FIDIC White Book had a lower threshold standard of care, which was expressed in the exclusive and was applicable to all contractual obligations not just the performance of the professional's obligations under the agreement: "the Consultant shall have no other responsibility than to exercise reasonable skill, care and diligence in the performance of its obligations under the Agreement." In the FIDIC White Book 2017, the standard of care warranty is now appliable to the performance of the services, rather than the consultant's contractual obligations in their entirety and is expressed as a specialist standard of care (i.e., the standard expected of a consultant providing services for projects of a similar size, nature, and complexity).

The NEC3 Professional Services Contract (PSC) requires the consultant to provide the services in accordance with the scope (clause 21.1) and provides that the consultant's obligation is to use the skill and care normally used by professionals

---

13 *Greaves & Co. (Contractors) Ltd. V Baynham Meikle & Partners* [1975] 1 W.L.R 1095.

14 Clause 3.1.

providing services similar to the services (clause 21.2). The details of the services provided by the consultant, and any constraints on how the services are provided, are set out in the scope. Clause 21.1 potentially imports into the contract an absolute obligation for the consultant to comply with, subject to what is included in the scope. It is possible for the consultant to be in breach of the contract by failing to adhere to the scope, even if everything that is done is done with the requisite skill and care.[15] The requirement to exercise skill and care does not diminish the duty to comply with the specific details of the scope, but it is an additional requirement.[16]

NEC4 PSC contains an additional subclause, clause 21.3, providing that the consultant is not liable for a defect unless it failed to carry out the service using the skill and care normally used by professionals providing similar services. The commentary[17] around the amendment is that, at least in the context of a defect, the consultant will not be liable unless it has failed to exercise reasonable skill and care. As such, it removes the ambiguity and potential under NEC3 PSC, for a consultant to have absolute obligations imposed on it.

In the Institute of Chemical Engineers (IChemE) standard form professional services contract (known as the silver book), there is an obligation to exercise reasonable skill and care and an obligation that the consultant shall regularly and diligently carry out and complete the services in accordance with the contract. This could present a more onerous standard of care than a standalone reasonable skill and care obligation if, for example, the contract specified a design life that ought to be achieved and may, in such circumstances, impose an absolute obligation on the design professional.

Varieties of Design Obligations

Design Obligations – Traditional Contractor – Standard of Care

In a traditional procurement route, the owner engages a professional design team to prepare the design. The owner provides the completed design to the contractor for construction. The design professionals are responsible to the owner for any design defects, and the contractor has no responsibility for the design of the works and is responsible for the construction of the works only (unless the contractor undertakes, in the contract, to design part of the works). There are two points of responsibility for the works: the design professionals for the design and the

15 Keating on NEC at page 697, paragraph 14-006.

16 Keating on NEC at page 697, paragraph 14-006.

17 See, for example, Keating on NEC para 14-009 at page 704.

contractor for the construction. The traditional procurement approach can cause difficulties when defects arise, and it is not clear whether the defect stems from a defect in the design or construction because there is no single point of responsibility for defects. The traditional contractor has a duty to ensure that the goods it is delivering (being the building or plant it is constructing) are reasonably fit for their intended purpose. This duty comes from the Supply of Goods and Services Act 1982.[18] Fitness for purpose is an absolute obligation to achieve a specified result or outcome and it imposes a higher standard than reasonable skill and care.

Design Obligations – Design and Build Contractor – Standard of Care

In design and build and EPC procurement routes, the contractor fulfills the dual role of designing and constructing. A contractor may also be required to provide a design for a portion of the works under a traditional procurement route, with the owner retaining primary responsibility for the design. This usually occurs when a particular contractor or specialist subcontractor has the specialist skills needed to execute an element of the design. In the JCT suite of contracts, for example, the owner can designate certain works for the contractor to design. Where the contractor assumes some or all of the design responsibility, the standard of care that the contractor owes in respect of the design, in the absence of any express contract provision otherwise, is that the design must be reasonably fit for its intended purpose.[19] In the case of *Independent Broadcasting Authority v BICC Construction Ltd*,[20] the court considered: "*In the absence of any term (express or to be implied) negativing the obligation, one who contracts to design an article for a purpose made known to him undertakes that the design is reasonably fit for the purpose*," the court also noted that this position was consistent with the statutory sale of goods law.

A fitness for purpose is an absolute obligation to achieve a specified result or outcome and is considered to be a higher standard to satisfy than a reasonable skill and care obligation. The default position is therefore that a design and build contractor owes a higher standard of care in respect of the design (who must achieve fitness for purpose) than a professional designer (who must exercise reasonable skill and care), unless the contract provides otherwise. Most construction contracts encompassing a design element will expressly set out the contractor's design obligations and will determine whether the contractor owes a fitness for

18 Section 4(5).

19 See, for example, *Independent Broadcasting Authority v BICC Construction Ltd* (1980) 14 B.L.R. 1.

20 (1980) [1955] P.N.L.R. 179.

purpose obligation in connection with its design or whether its design obligations are limited to a standard of reasonable skill and care. Owners, developers, and project funders will usually require the contractor to meet certain absolute obligations while contractors would prefer to exclude or limit absolute obligations. It is likely to be the subject of negotiations and parties may address those tensions by negotiating on other areas of the contract, for example, a limit on the contractor's liability, rather than a blanket exclusion on the contractor achieving absolute obligations in connection with the design.

Within a construction contract, a fitness for purpose obligation may be expressed as a generic obligation that the works must be fit for purpose when complete, for example "*the works, when complete, will be fit for their intended purpose*" or "*the completed works will be fit for the purpose stated in the Specification/Owner's Requirement's/Contract.*" The owner's requirements or the specification may contain particular results that the completed works should meet or achieve. For example, a power plant might be required to achieve performance parameters relating to availability, output, emission, and wattage. A process plant might require a certain level of throughput. Other construction projects might specify a certain design life that ought to be achieved. Absolute obligations may be imposed on the contractor without the contract expressly using the words "fitness for purpose" by requiring the contractor to "warrant" the completed works will comply with the specification or the owner's requirements.

Disputes have arisen where construction contracts contain obligations for the designer to exercise reasonable skill and care and absolute obligations (e.g., the completed works will be fit for the purpose stated in the owner's requirements). The courts have established that reasonable skill and care obligations and absolute obligations to achieve a particular result or output are not mutually incompatible, and these obligations can exist side by side. *In MT Højgaard A/S v E.ON Climate & Renewables UK Robin Rigg East Limited and another,*[21] the court found, on the facts of that case, that the two standards in the contract that were in dispute could be met by the party fulfilling the more onerous standard of the two.

In the Højgaard case, MTH was contracted to design, fabricate, and install foundations for proposed wind turbines for an offshore wind farm. The foundation structures failed shortly after completion of the project. The parties were in dispute as to who was responsible for the cost of the remedial works. MTH assumed, in a complicated and extensive suite of contractual documents, various obligations in connection with the design.

21 The Supreme Court considered these issues in *MT Højgaard A/S v E.ON Climate & Renewables UK Robin Rigg East Limited and another* [2017] UKSC 59. In its decision, the Supreme Court restored the order made at first instance by the Technology and Construction Court, at paragraph 54.

The main contract had a section titled "general obligations" which stated that: "*the Contractor shall, in accordance with the agreement, design, manufacture, test, deliver and install and complete the Works*" and at subclause (i) it continued "*with due care and diligence expected of appropriately qualified and experienced designers, engineers and constructors (as the case may be).*" It is a similar clause to the standard skill and care obligation that can be found in most professional designer appointments. At subclause (x), it stated "*so that each item of Plant and the Works as a whole shall be fit for its purpose as determined in accordance with the Specification using Good Industry Practice.*" Fit For Purpose was defined in the contract as "*fitness for the purpose in accordance with, and as may be properly inferred from, the Employer's Requirements.*"

The "Employer's Requirements" included various technical requirements relating to the design basis. Notably, the detailed design for the foundations was to be designed to an international standard, J101, for the design of offshore wind turbines. It is also relevant that the contract referred to J101 as defining the "minimum requirements." MTH warranted that the wind turbine foundations would have a service life of 20 years in every aspect without planned replacement. MTH designed to the international standard set out in J101, but the standard contained an error which significantly reduced the service life of the foundations, and the foundations failed. The dispute revolved around who was responsible for the remedial works.

The Supreme Court overturned the Court of Appeal decision finding that the obligation to comply with standard J101 and the design life requirement were not inconsistent because J101 was expressed as a minimum requirement to be taken into account in the design. The correct analysis was that the more rigorous or demanding of the two standards should prevail, as the less rigorous can be treated as a minimum requirement and that MTH would be liable for the failure to comply with the required criteria, as it was MTH's duty to identify the need to improve on the design. The court noted that:

> *While each case must turn on its own facts, the message from decisions and observations of judges in the United Kingdom and Canada is that the courts are generally inclined to give full effect to the requirement that the item as produced complies with the prescribed criteria, on the basis that, even if the customer or owner has specified or approved the design, it is the contractor who can be expected to take the risk if he agreed to work to a design which would render the item incapable of meeting the criteria to which he has agreed.*[22]

---

22 *MT Højgaard A/S v E.ON Climate & Renewables UK Robin Rigg East Limited and another* [2017] UKSC 59, at para 44.

In the Højgaard case, the purpose for which the foundations had to be fit for was determined by the specification (in addition to good industry practice). This is not an uncommon contractual approach and the contract's specification is often used to set out the performance requirements or other purpose of the project: the fitness for purpose obligation needs detailed information or requirements to define and make clear the purpose. The emphasis, particularly for the owner (or contractor in contracts with its supply chain) is therefore on furnishing the specification with the required performance requirements and other specific details as to the intended purpose of the project. In the absence of such information, an owner may be reliant on expert evidence as to whether a particular construction project or process plant can fulfill its purpose or not, with the existence of what is alleged to be a defect. For example, in the absence of an express performance requirement, is a bus depot with water pooling on its forecourt fit for purpose? Alternatively, are foundations that are supportive of a structure fit for purpose if they clash with existing services?[23]

The Supreme Court in Højgaard also found that there was nothing improbable or unbusinesslike in the design life warranty (that the design of the foundations shall ensure a lifetime of 20 years in every aspect without planned replacement) and that there was nothing wrong or surprising with the warranty being put in the technical requirements – it was not tucked away in an unexpected location. The Supreme Court also considered[24] whether MTH had a liability to warrant that the foundations would last 20 years or would be designed so as to achieve a lifetime of 20 years. While the Supreme Court determined that it was not necessary to decide this point, it considered that the warranty required MTH to provide a design that could be expected to have a lifetime of 20 years without guaranteeing that the product will last that long.

In *LDC (Portfolio One) Limited v George Downing Construction Limited*,[25] the court considered whether an obligation in a design and build subcontract to not put the contractor in breach of the main contract, where an obligation in the main contract is to comply with all statutory requirements, was superseded by a skill and care obligation in the subcontract. The court found that the subcontractor had an absolute obligation to comply with the statutory requirements in carrying out the works. Referring to Lord Neuberger in Højgaard, the court noted that: "*if there are two clauses imposing different standards or requirements, treating the*

23 For a discussion on this point, see Keating Chambers Podcast TCC150 Series: Episode 2 – *MT Højgaard A/S v E.ON Climate & Renewables UK Robin Rigg East Limited and another* [2017].

24 *MT Højgaard A/S v E.ON Climate & Renewables UK Robin Rigg East Limited and another* [2017] UKSC 59, at 32.

25 [2022] EWHC 3356 (TCC).

clause imposing the lesser standard as a minimum requirement makes more sense. Treating it as qualifying a more onerous obligation effectively renders the more onerous obligation redundant."[26] The court also considered that if the reasonable skill and care obligation was to take precedence over the strict obligation to comply with the statutory requirements, the parties' attempts to make the requirements in the subcontract "back-to-back" with those in the main contract would have failed, even though this was clearly the intention.

Design Obligations – Design and Build Contractor – Allocating Responsibility for Design

The scope of the design contractor's design responsibility is influenced by whether its design obligations are absolute or one of reasonable skill and care. It is also influenced by what design responsibility, if any, is retained by the owner. If the contractor is fully responsible for design (as is common in an EPC contract procurement route), the owner's requirements document will usually contain only design parameters for the contractor to meet. The contractor will then subsequently propose the design solution in its contractor's proposals document, for which it takes full responsibility. It is common in design and build contracts for owners to provide a design that has been progressed to a certain level so that a contractor can complete and develop the design. The contractor may be asked, within the contract, to take full responsibility for the owner's design, with the owner passing the full design responsibility to the contractor. If there is no such warranty, or equivalent contractual provision, the responsibility for the initial design remains with the owner.

In *Workman Properties Ltd v Adi Building & Refurbishment Ltd*,[27] there was a dispute as to whether the contractor or the owner was responsible for completing the design of the works up to RIBA stage 4. The contract was an amended JCT 2016 design and build contract. In the standard form of the JCT 2016 design and build contract, the contractor is not responsible for the contents of the owner's requirements or verifying the adequacy of the design contained within them, and any correction made due to any design inadequacy in them is to be treated as a contract change in the contractor's favor.[28] However, the bespoke amendments to the contract in this case were materially different from the standard wording and

---

26 [2022] EWHC 3356 (TCC) at 46.

27 [2024] EWHC 2627 (TCC), for further discussion on this case, generally see: BCLP Insight, Design Responsibility: The need for clarity: https://www.bclplaw.com/en-US/events-insights-news/design-responsibility-the-need-for-clarity.html

28 [2024] EWHC 2627 (TCC), at 82.

imposed a far more wide-ranging design responsibility on the contractor for the whole of the owner's requirements.

The owner's requirements contained the following statement: "*significant design has been developed to date which has been taken to end of RIBA Stage 4 with some part of the contractor specialist design elements together with Services design to Stage 4... ...*" The contractor claimed that this statement amounted to a contractual warranty that the owner had completed the design to Stage 4, that the owner had breached this warranty by not having completed the design to Stage 4, and that the contractor was entitled to claim damages, additional time, and costs as a result of that breach. The owner's position was that the contractor had inherited responsibility for the existing design contained in the owner's requirements and had accepted responsibility for completing the design. The owner relied on wording in the contract stating that: "*the contractor has examined the Employer's Requirements and has agreed to accept full responsibility for any design contained in them and acknowledges that the Employer's Requirement's form part of the Contractor's Design Documents*," which was a bespoke amendment to the JCT 2016 design and build contract standard form. The owner also relied on an amended clause 2.17.1 under which the contractor was to be "*Fully responsible in all respects for the design of the Works including all design work proposed by or on behalf of the Employer on or before the date of this Contract forming part of the Employer's Requirements*," and (by subsequent sub-clauses) this included design work carried out by consultants and sub-consultants, whether before or after the date of the contract.

In Part 8 proceedings in the Technology and Construction Court, the judge rejected the contractor's claim observing that the alleged warranty was "*nowhere near sufficient to require the other unequivocal contract provisions to be read as so heavily qualified*," and found that this did not negate the contractor's contractual responsibility to satisfy themselves that what was in the existing design was sufficient in all respects.[29] The decision was consistent with the approach in *Co-operative Insurance Society Ltd v Henry Boot Scotland Ltd*[30] where it was held that the contractor, in fulfilling its duty to exercise reasonable skill and care, is required to examine and ensure that the conceptual design of the owner is appropriate so that they can develop the conceptual design into a completed design capable of being constructed.

In the EPC procurement route, the owner may provide a FEED to the contractor. This is typical of many energy and process plant projects. The contractor is typically expected to assume responsibility for the FEED and its accuracy.

29 [2024] EWHC 2627 (TCC), at 97.

30 [2002] EWHC 1270 (TCC), at 68.

This is dealt with in the contracts by providing the contractor with a period of time to check the accuracy of the FEED and to verify it either before the contract is signed or within a short period after the contract has been signed. Otherwise, if the FEED is provided as part of the owner's requirements document or relying upon information provided by the owner, the owner will retain responsibility for it.

Design Obligations – Design and Build Contractor – Standard Form Contracts

In the JCT 2024 Design and Build standard form contract (JCT 2024 DB), which made several changes to the JCT 2016 design and build standard form contract (JCT 2016 DB), clause 2.17 elevated the standard of care to be exercised by the contractor to that of "*a qualified and experienced architect (or other appropriate professional designer) undertaking the design of works similar in scope and character to such design of the Works.*" The standard of care in the JCT 2016 DB is that the contractor shall have the same liability "*in respect of any inadequacy in the design*" as would "*an architect or other appropriate designer who holds himself out as competent to take on work for such design,*" and effectively holds the designer to a standard of reasonable skill and care that would be expected of a professional designer.

The JCT 2024 DB expressly states that the contractor has "*no greater duty, obligation or liability in respect of design than to exercise reasonable skill and care as provided in clause 2.17.1.1 in respect of such design and under no circumstances shall the contractor be subject to any duty, obligation or liability which requires that any such design shall be fit for its purpose.*"[31] This clause specifically excludes fitness for purpose in connection with design. The newly added clause is a reaction to case law such as the Højgaard case, where the contractor was held liable for breaching fitness for purpose obligations set out in technical specification documents, which were not qualified by the reasonable skill and care clause set out in the JCT 2016 DB.

In the FIDIC suite of contracts, where the contractor is required to undertake design, the contractor has a fitness-for-purpose obligation. The FIDIC 2017 Yellow Book and Silver Book require the contractor to design and build the works. Under the Yellow and Silver Books, the works completed by the contractor (including the design) must be fit for the purpose intended as defined in the owner's requirements.[32]

31 JCT Design and Build Contract 2024, Clause 2.17.1.2.

32 See the Contractor's General Obligations, Clause 4.1.

Under the FIDIC 2017 Yellow Book, the contractor has a period to review the owner's requirements document and report any errors or inaccuracies in it.[33] If an error isn't detected during this process but is indeed present, and the contractor incurs delays or additional costs as a result, it is entitled to relief if an experienced contractor exercising due care and diligence would not have discovered the error.[34] In the FIDIC 2017 Silver Book, the contractor is also expressly responsible for all errors, inaccuracies, or omissions in the owner's requirements, except for: (i) any information or data which is stated in the contract as being the responsibility of the owner; (ii) definitions of intended purposes of the works; (iii) criteria for the performance of the completed works and any information or data that cannot be verified by the contractor.

In the NEC Engineering and Construction Contract (ECC) suite of contracts, there is no separate design and build standard form contract and it is intended to be suitable for use irrespective of the contractor's level of design responsibility, with the requirements as to design, including what design is to be carried out by the contractor, set out in the works information[35] or scope documents.[36] The NEC4 guide to preparing an ECC gives examples of the type of information that might be included by way of requirements. There is no express standard of care that the contractor has to meet in respect of the design it is engaged to carry out. This means that the contractor is under an implied fitness-for-purpose obligation in relation to its design. However, the NEC ECC provides an optional secondary clause X15, which introduces a reasonable skill and care provision for the contractor's design obligations.

In the IChemE Red Book, there is a standard skill and care obligation and an obligation that the completed plant shall comply with the contract and shall be in every respect fit for the purpose for which it is intended, as specified in the specification or in any other provision of the contract. This type of fitness for purpose obligation is not uncommon in the process industry, where contracts are often engaged for the technology that they have access to or experience of.

33 See the Contractor's General Obligations, Clause 5.1.

34 See the Contractor's General Obligations, Clause 1.9.

35 See NEC3.

36 See NEC4.

Conclusion

It is key for a designer to know and understand the design obligations that it is required to meet. Those provisions should be clearly outlined in the parties' contract. Ambiguities (and consequently disputes) can arise where the contract terms are ambiguous and the parties must revert to statute or common law in order to define the standard of care that applies to the design. The parties should also ensure that the contract reflects the standards that were agreed during contract negotiations. Different standard form contracts will provide different starting points, in terms of negotiations, for the standard of care that the designer is required to meet. It is not uncommon for the standard terms to be amended; however, it is important to note that, in the event of conflicting provisions, English law has established that the higher standard normally prevails. The designer will also want to ensure, to the extent possible, that the obligations it is required to meet are consistent throughout its supply chain.

7

The Engineer

Jeremy Glover and Giuseppe Giancarlo Franco

> *Engineers have to wear many hats. They have great technical expertise and can move mountains, bridge rivers, construct skyscrapers, design and build power stations, and they generally create the modern world with which we are familiar. However, the demands on engineers do not end there. They are often required to be project managers and to administer complex contracts .... In these various roles, engineers ... often have to deal with disputes.*
>
> – Robert Gaitskell, *Engineers' Dispute Resolution Handbook* (2006)

Introduction

In the construction industry, the role of the engineer is traditionally defined as the professional who designs and oversees the construction of projects. In a legal context, however, the term is more difficult to define, especially if one goes beyond borders and looks at the engineer's role under a comparative perspective.

One difficulty in establishing a universally accepted legal definition of "engineer" arises from the different experiences of the common and civil law traditions. In the common law tradition, which in modern times has assumed the dominant position worldwide, the engineer's domain has significantly expanded, combining their conventional functions with sophisticated consulting services. Nowadays, on major infrastructure projects, the engineer commonly provides expert assistance during the preconstruction phase, supplying the design and other services related to the technical and financial feasibility of the project. Then, during the execution of the works, it becomes responsible for the administration of the construction contract, assuming a duty to act fairly and impartially between the parties.

Finally, under certain standard forms of contract, the engineer is also entrusted with decision-making functions for the assessment of the parties' claims.

On the contrary, in civil law countries, the practice of appointing an engineer that encompasses the functions of multiple professionals – such as designer, contract administrator and adjudicator – is relatively uncommon.[1] Therefore, the legal characterization of the *modern* engineer remains a subject of ongoing debate among civil law practitioners.

The broadening scope of the engineer's responsibilities has also reshaped the industry. Individual engineers or small professional firms often lack the capacity to deliver the full range of services now expected of them. Consequently, owners are increasingly turning to large, multiservice engineering firms capable of meeting these diverse demands.

With this in mind, this chapter examines the role of the modern engineer in construction projects through the lens of the civil-common law dichotomy. First and foremost, while acknowledging that the exact scope of the engineer's role is defined by the terms of their engagement, it is equally important to recognize that these terms do not exist in isolation. Rather, they are affected and constrained by the governing laws of the jurisdiction in which they are meant to operate. Therefore, our analysis will inevitably incorporate a degree of *localisation* to assess the role of the modern engineer within the specific legal frameworks of different jurisdictions. For the civil law tradition, the authors will anchor the analysis on the French and Italian systems; for the common law tradition, the analysis will primarily focus on the laws of England and Wales.

As a note from the authors, the chapter will use the lowercase ("engineer") when generally referring to the construction consultant appointed by the owner to provide the design and/or supervise the execution of the works. The authors will use the uppercase ("Engineer") when specifically referring to the engineer in the context of the FIDIC standard forms of contract.[2]

---

1 Seppälä C, *The FIDIC Red Book Contract: An International Clause-by-Clause Commentary* (Kluwer, 2023), 79.

2 The FIDIC contracts include standard forms that are tailored to different construction procurement routes. The most notable ones are: the Conditions of Contract for Construction which follows the traditional route (hereinafter "Red Book"); the Conditions of Contract for Plant and Design-Build which follows the design and build route (hereinafter "Yellow Book"); the Conditions of Contract for EPC/Turnkey Projects which follows the turnkey route (hereinafter "Silver Book"). The FIDIC contracts also include the Client/Consultant Model Services Agreement (hereinafter "White Book"), which is the standard form for the appointment of the Consultant acting as engineer, owner's representative, project manager, or similar.

Civil Law

Overview

In civil law jurisdictions, the special part of the civil codes contains specific sets of rules, each tailored to the regulation of a particular type of contract that was regarded as significant at the time of drafting (or reforming) the code. Civil law practitioners refer to these model contracts as the "nominate contracts" – e.g., the sale, lease, construction, and agency contracts – and endeavor to subsume the specific parties' arrangement under one of these models. Once identified, the relevant model contract governs the parties' ad hoc agreement.

The special part of the civil code is distinct from the general part, which applies indistinctly across all the arrangements that parties devise to regulate their transactions. In particular, the general part covers matters such as contract formation, interpretation, execution, termination as well as the assessment of damages in the event of breach.

Legal Characterization of the Relationship Between the Engineer and the Owner

At first glance, the contract binding the engineer to the owner appears to match with three contract types commonly regulated by most civil codes: (i) the employment contract, (ii) the agency contract, and (iii) the contract for hire of work. The first option may be discarded outright because it implies a relationship of subordination, whereby one party (i.e., the employee) agrees to perform work under the authority of another (i.e., the employer) in return for payment of a wage. This does not apply to contracts in which the promisor undertakes to perform its obligations independently of the promisee.

This leaves agency and the contract for hire of work, and the legal relationship between the engineer and the owner appears to fit either classification. Under the French and Italian civil codes, agency is a contract whereby one party, the agent, undertakes to perform one or more legal acts on behalf of the principal.[3] On the other hand, under a contract for hire of work, one party undertakes to perform certain works or services for the benefit of the promisee. The key distinction between the two lies in the nature of the activity performed for the benefit of the promisee. Whereas the agency contract involves the performance of a *legal* activity, such as entering into a contract, the contract for hire of work involves the performance of material or intellectual work, i.e., a *nonlegal* activity.[4] Given that

3 Articles 1984 ff. French civil code; Articles 1703 ff. Italian civil code.

4 Mainguy D, *Contrats spéciaux* (Dalloz, 2022) 561 and 651; Barchiesi L, Bottoni F, "Il mandato" in *Trattato di Diritto Commerciale* ed. R Costi (Giappichelli, 2024), 7.

the engineer's obligations generally pertain to technical or professional services rather than legal acts, the contract between the engineer and the owner is most appropriately characterized as a contract for hire of work.

However, when the owner appoints the engineer, it also defines the scope of the engineer's authority to act on its behalf. Therefore, the engineer-owner relationship may also be characterized as an agency contract to the extent that the terms of the engagement confer upon the engineer the authority to act as the owner's agent. This is the case, for example, in the FIDIC contracts. The FIDIC White Book states: "*when acting as the engineer ..., the Consultant shall have the authority to act on behalf of the Client to the extent provided in the Works Contract.*"[5] This is consistent with the wording in the FIDIC construction contracts, where the Engineer is deemed to act for the Employer.[6]

Article 1710 of the French civil code very broadly defines the contract for hire of work[7] as the "*contract by which one party undertakes to do something for another, in return for a price agreed between them.*" Pursuant to Article 1779, this contract type is further divided into three subtypes: the contract for hire of service; the contract of carriage; and a third category, which specifically refers to the services performed by architects, contractors and technicians in relation to construction works ("*Celui des architectes, entrepreneurs d'ouvrages et techniciens par suite d'études, devis ou marches*"). The contracts falling under this latter category have two essential features that distinguish them from other contract types. First, unlike the employment contract, the promisor undertakes to perform the agreed services with full independence from the promisee. Second, unlike the agency contract, the promisor does not have the authority to act on behalf of the promisee. According to the French Court of Cassation, the contract entered into between the owner (*maître d'ouvrage*) and the engineer (*maître d'œuvre* or *architecte*) is generally not characterized as an agency contract. Instead, it qualifies as a *contrat de louage d'ouvrage* regulated by Articles 1787 to 1799-1 of the French Civil Code.[8] Notably, these same provisions also regulate the construction contract.

Italian legal scholarship refers to the contract for hire of a multiservice engineering company as the "engineering contract," which is described as an arrangement by which the engineer undertakes to perform services instrumental to the execution of construction works. Historically, two forms of engineering have been identified. The first is "consulting engineering," whereby the engineering company only provides the project design and all services related to the

5 Subclause 3.9.3.

6 Subclause 3.2(1), 2017 Red and Yellow Books; Subclause 3.1(5)(a), 1999 Red and Yellow Books.

7 *Contrat de louage d'ouvrage* or *contrat d'entreprise*.

8 Cass. civ. 1, 21 January 1963; Cass. civ. 3, 17 February 1999, No. 95-21.412; P Malinvaud, *Droit de la construction*, 7th ed. (Dalloz, 2018) para. 101.81.

preconstruction phase (such as buildability studies and cost evaluation). The second is "commercial engineering," whereby the engineering company not only provides the design but also undertakes to administer the execution of the works carried out by a third party. The first form of engineering is characterized as a contract for hire of *services*;[9] the latter is regarded as a contract for hire of *work*.[10] Although the terminology may differ, both contracts are governed by Articles 1655 to 1677 of the Italian civil code. Furthermore, as under French law, these provisions also apply to construction contracts between the contractor and the owner.

Duties of the Engineer

In civil law jurisdictions, parties are free to agree on the content of their contracts, subject to the limits imposed by the law.[11] Therefore, since neither the French nor the Italian civil code expressly defines the specific duties of construction consultants, parties enjoy broad discretion in shaping the terms of the engineer's engagement. At the same time, however, through the adjudication of construction matters and the application of the statutory duty to execute contracts in good faith, courts have developed a body of collateral duties, which consultants owe to other project participants.[12] This process of judicial development, which complements and elaborates on the provisions of the civil codes, is not too dissimilar to the evolution of the common law, the main difference being that judgments do not constitute binding precedents. The most influential role in shaping the development of the law is played by supreme courts, namely the *Cour de Cassation* in France and the *Corte di Cassazione* in Italy. In the context of French construction law, the landmark decisions are typically issued by the *troisième chambre civile* of the Court of Cassation. In Italy, the second chamber fulfils the same function within the Italian Supreme Court.

As anticipated above, in civil law systems, the modern engineer encompasses the functions of a wide range of consultants. In a typical French construction setting, the engineer is likely to combine the roles of *architecte* and *maître d'œuvre*. The *architecte* is the regulated professional entrusted to develop the design and obtain the building permit, whereas the *maître d'œuvre* is tasked with administering the execution of the works. However, French courts tend to use these terms interchangeably, reflecting the frequent practice of appointing the *architecte* as the contract administrator.

9 Cass., sez. II, 6 December 1986, No. 7263; P Montalenti, "Società di progettazione e società di intermediazione progettuale: nuovi profili interpretative" in Giur. Comm., 1983, I, 485, 502.

10 Rubino D, Iudica G, *Appalto*, 4th ed. (Zanichelli, 2007), 73.

11 Article 1102 French civil code; Article 1322 Italian civil code.

12 Lavers A, Shorter R, "A contractor's duty to warn of defects: common law and civil law approaches compared," *ICLR* 195 (2024), 204.

The *architecte* is subject to an extensive duty to advise and inform the owner (*obligation de conseil*) both during the preconstruction and construction phase.[13] As part of this overarching duty, the *architecte* must also enquire (*obligation de se renseigner*) about the owner's financial capabilities[14] and the purpose of the commissioned work.[15]

Secondly, in developing the design, the *architecte* must comply with all applicable planning and building regulations, exercising the standard of reasonable skill and care expected of a qualified professional experienced in works of similar size, scope, and nature.[16]

Thirdly, if retained as *maître d'œuvre*, the *architecte*'s scope of work extends to supervising the execution of the works and issuing instructions to ensure the contractor's compliance with the contract.[17] However, the *maître d'œuvre* does not typically have the authority to issue variation orders, and their certifications of the contractor's interim payment applications are usually not binding on the parties. Moreover, the *maître d'œuvre* may not subcontract work unless expressly authorized to do so.[18]

Finally, the *architecte* must assist the owner during the taking-over process ("*réception*"). In particular, the *architecte* must draw the owner's attention to any apparent defects and warn against the consequences of taking over without preparing a punch list to be attached to the taking over certificate.[19] In any case, the *architecte* does not retain the authority to accept works on the owner's behalf unless expressly authorized to do so.[20]

A similar set-up can be encountered on Italian projects, where the modern engineer is likely to encompass the roles of *progettista* and *direttore dei lavori*. Again, since the Italian civil code does not specify the duties of construction consultants, guidance is derived from case law and legal scholarship. The *progettista* is the professional responsible for developing the design, whereas the *direttore dei lavori* is primarily tasked with ensuring that the works comply with both the construction contract and the approved design. However,

---

13 Malinvaud, *Droit de la construction*, para. 100.22.

14 Court of Appeal Pau, 21 November 1991, No. 1936/90.

15 Cass. civ. 3, 2 June 2016, No. 15–16.981.

16 Muller F, Aquino V, "French Law (Napoleonic)" in *International Construction Law: An Overview* ed. W Breyer (Routledge, 2024) para 5.206.

17 Malinvaud P, Jestaz P, Jourdain P, Tournafond O, *Droit de la promotion immobilière*, 9th ed. (Dalloz, 2014) 137.

18 Malinvaud, *Droit de la construction*, paras. 511.12 and 511.151.

19 Muller/Aquino, "French Law (Napoleonic)," para. 5.218; Cass. civ. 3, 30 October 1991, No. 90–12.993.

20 Cass. civ. 3, 17 February 1982, No. 80–13.761.

unless expressly authorized by the owner to verify the buildability of the design, the *direttore dei lavori* is not liable for defects in the works resulting from flaws in the design.[21]

Liability Regime

The French and Italian systems slightly differ in respect to the liability regime. Under Article 1792 of the French civil code, the *architecte* or *maître d'œuvre* warrant to the owner, and together with the contractor, that the work's structural integrity will not be impaired and that the works will remain fit for purpose. This warranty extends for ten years from taking over and is governed by the fitness for purpose standard. The *architecte*'s liability regime therefore varies at taking over. Before taking over, the *architecte* is only subject to the standard of reasonable skill and care, and thus evidence of their fault, as well as of causation, are required to establish liability. After taking over, during the 10-year warranty period, liability is presumed and proof of fault is not required to hold the *architecte* liable. Indeed, pursuant to Article 1792, the *architecte* or *maître d'œuvre* can avoid liability only by proving force majeure or that the risk has been taken on by the contractor.[22]

In any case, the *architecte*'s liability cannot exceed the scope of its engagement. In other words, if the *architecte* is not retained as *maître d'œuvre*, it cannot be liable for damages that stem, for example, from a breach of the obligation to oversee the works.

Contrarily, the Italian *progettista* and *direttore dei lavori* are only held to the standard of reasonable skill and care. Yet the standard required of these professionals is higher than usual in consideration of their qualified and skilled capacity. This means that the *direttore dei lavori* is under an obligation to proactively ensure, through the issuance of instructions, that the contractor complies with both the contract and the design. In performing this task, the contract administrator must assess the works against the design and promptly notify the contractor and owner of any potential nonconformities.[23] In any event, unless expressly authorized to do so, the *direttore dei lavori* does not have the authority to issue variation orders or to accept works on behalf of the owner.[24]

---

21 Cass., sez. II, 19 September 2016, No. 18285.

22 Muller/Aquino, "French Law (Napoleonic)," para. 5.219; Saint-Alary-Houin C, Poumarède M, *Droit de la construction*, 12th ed. (Dalloz, 2020) 219, 222 and 225; Malinvaud, *Droit de la construction*, paras. 100.91 and 100.101; Malinvaud/Jestaz/Jourdain/Tournafond, 137.

23 Cass., sez. II, 3 September 2020, No. 18289.

24 Trapuzzano C, *L'appalto privato* (Giuffrè Editore, 2020), 66 and 67; Caredda V, "Autonomia dell'appaltatore e ingerenza del committente" in *Codice dell'appalto private*, ed. A Luminoso, 2nd ed. (Giuffrè Editore, 2016), 313.

It must be recalled, however, that under Italian law, the contract for hire of a multiservice engineering company is likely to be characterized as a contract for hire of work (or services), governed by Articles 1655 to 1677 of the Italian civil code. Under this legal framework, the engineer would be equated with the contractor and would therefore be subject to the same liability regime. In particular, under Article 1667, the contractor warrants to the owner that the works will be free from defects for a period of two years following taking over, and the applicable standard of liability is that of fitness for purpose.[25] Additionally, under Article 1669, the contractor is liable to compensate the owner for any damage the works may cause to the owner or third parties within ten years of taking over. Again, the contractor's fault is presumed, and to avoid liability, the contractor must demonstrate that it promptly warned of any design defects and that the owner (or its agent) instructed the contractor to proceed despite these warnings.[26]

As can be observed, both French and Italian law substantially limit the authority of the contract administrator, who, unlike their common law counterpart, is normally not authorized to order variations, certify work, subcontract, or accept the works. Instead, case law emphasizes the consultant's duty to advise and warn the owner. In this way, it is the owner that assumes responsibility for final decisions, which it takes on the basis of the information received from the consultant.

Common Law

Overview

The UK regulatory body for the engineering profession, the Engineering Council, says on its website:

> *The words engineer and engineering have both been in common use for centuries in the UK. Neither is legally defined and in everyday language the term engineer is very often taken to mean anyone who is in some way associated with engineering, including the design, manufacture, maintenance or operation of a technical product or system.*[27]

---

25 Trapuzzano, 598; Rubino/Iudica, 424; Rubino-Sammartano M, *Appalti di opere e contratti di servizi* (CEDAM, 2006), 440.

26 Trapuzzano, 706.

27 https://www.engc.org.uk/glossary-faqs/frequently-asked-questions/status-of-engineers/#:~:text=Commonplace%20use%20of%20the%20word,Engineering%20Council%20as%20totally%20impractical.

That said, in many respects, the position under the common law is potentially simpler than its civil law counterpart. This is largely because the engineer's responsibilities are primarily defined by the terms of their contract with the owner. Importantly, when the engineer is required to perform tasks under the construction contract between the owner and contractor, it is essential that the duties outlined in the construction contract align with those in the engineer's own contract.

However, the term "potentially simpler" is warranted because, in the common law, the contract does not always expressly set out all the obligations the engineer may be subject to; indeed, terms can also be implied.

Implication of Terms

In *Marks & Spencer v BNP Paribas*,[28] the UK Supreme Court addressed the test for the implication of terms, identifying two types of contractual terms that may be implied into a contract. The first type is implied based on the express terms, commercial common sense, and the facts known to both parties at the time the contract was made. The second type arises from statutory provisions that, unless expressly excluded, automatically impose certain terms into contracts. For instance, the *Supply of Goods and Services Act 1982* imposes design obligations where the contract is silent on the matter.

In relation to the first type of implied terms, the Supreme Court went back to the Privy Council case of *BP Refinery (Westernport) Pty Ltd v Shire of Hastings*, where Lord Simon affirmed that for a term to be implied, the following five conditions must be satisfied:

1. *it must be reasonable and equitable;*
2. *it must be necessary to give business efficacy to the contract, so that no term will be implied if the contract is effective without it;*
3. *it must be so obvious that "it goes without saying";*
4. *it must be capable of clear expression;*
5. *it must not contradict any express term of the contract.*[29]

However, the implication of terms is subject to strict limitations, ensuring that they are not inserted arbitrarily or without clear justification. As affirmed by Lord Neuberger in *Arnold v Britton*: *"there is no principle of interpretation which entitles a court to re-write a contractual provision simply because the factor which the parties catered for does not seem to be developing in the way in which the parties may well have expected."*[30]

---

28 [2015] UKSC 72.

29 [1977] UKPC 13.

30 [2015] UKSC 36, [2015] AC 619, [41].

Good Faith Obligation and the Duty to Warn

Good faith obligations are a well-established principle in civil law systems, where the provisions of civil codes require parties to negotiate, conclude, and perform contracts with fairness and honesty.[31] In contrast, the approach under the common law is more fragmented and varies across jurisdictions. In England, as well as in Singapore and Hong Kong, the general position is to refrain from implying a universal obligation to act in good faith, with courts holding that parties wishing to impose such a duty must do so expressly in their contract.[32] In Australia, a duty of good faith is more readily implied,[33] although the High Court of Australia has rejected the existence of a broad, overarching obligation to act in good faith.[34]

The approach to good faith in standard forms of contract is similarly variable. For instance, Subclause 1.16.1 of the White Book expressly requires: "*In all dealings under the Agreement the Client and Consultant* [i.e., the Engineer] *shall act in good faith and in a spirit of mutual trust.*" Likewise, Core Clauses 10.1 of NEC3 and 10.2 of NEC4 mandate: "*The Parties, the Project Manager and the Supervisor act in a spirit of mutual trust and co-operation.*" Such obligations are notably absent from the FIDIC Red and Yellow Books, which govern the relationship between the contractor and the owner. As a result, under English law, an Engineer operating under the White Book and a Project Manager[35] operating under NEC3 and NEC4 would be bound by duties of good faith. On the contrary, contractors operating under the Red and Yellow Books would not share this obligation, unless it was expressly included in the particular conditions of the contract.

Although a duty of good faith is rarely implied in common law jurisdictions, the engineer's duty to warn is treated differently. Under English law, this duty is typically tied to the broader obligation to exercise reasonable skill and care, and is readily implied in cases where it becomes clear that the works pose a risk of loss or damage to the owner.[36] As a professional engaged to provide services in connection to construction work, the engineer is also expected to fulfil this duty, which arises: "*when a careful professional ought to have known of such danger,*

31 Article 1104 French civil code; Articles 1337 and 1375 Italian civil code.

32 *Compass Group UK and Ireland Ltd (t/a Medirest) v Mid Essex Hospital Services NHS Trust* [2013] EWCA 200.

33 *Renard Constructions (ME) v Minister for Public Works* (1992) 26 NSWLR 234.

34 *Royal Botanic Gardens and Domain Trust v South Sydney City Council* [2002] HCA 5.

35 Under the NEC standard forms, the Project Manager serves as the contract administrator, a role similar to that of the Engineer under FIDIC contracts.

36 *Equitable Debenture Assets Corporation Ltd v William Moss Group Ltd* (1984) 2 Con LR 1.

having regard to all the facts and circumstances."[37] The duty to warn acts as a safeguard, ensuring that all parties remain informed and able to make decisions based on accurate and timely information.

In *Hart Investments Ltd v Fidler*, the court confirmed that the engineer's duty to warn extends to the inadequacy of temporary works. The court stated:

> *If an Engineer employed by an owner in respect of permanent works observes a state of temporary works which is dangerous and causing immediate peril to the temporary works in respect of which he is employed, he is obliged to take such steps as are open to him to obviate that danger.*[38]

The court further reasoned that this obligation arises not only from "*common sense*" but also from the engineer's overarching responsibility in relation to the suitability of the permanent works. It noted:

> *The engineer is, after all, instructed in relation to the permanent works as a whole. It would appear strange if he is under a duty to take such steps as he can to see that they survive for say, the next 25 years, or whatever the design life for the building is, but is not obliged to take any steps to warn of an immediate danger to those works caused by an imperilling act by the contractor.*[39]

Can the Engineer Be an Employee of the Owner?

The answer to this question strictly depends on the contract wording. In *Scheldebouw BV v St James Homes Ltd*,[40] Jackson J held that an owner cannot assume the role of construction manager unless there is an express contractual provision permitting them to do so. The issue for the court was that the construction manager was also acting as the decision maker on matters where the contractor and the owner might have conflicting interests.

The same question may arise in the context of the FIDIC contracts. Strictly, there is nothing in the Red and Yellow Books, which expressly prohibits the owner from acting in this way and assuming the Engineer's role. However, under Subclause 3.6, the contractor has the right to object to the proposed appointment and one might reasonably question how an owner-engineer could maintain the neutrality and fairness required when making determinations under the FIDIC contracts.[41]

37 *Goldswain v Beltec Ltd (t/a BCS Consulting)* [2015] EWHC 556 (TCC), [47].

38 [2007] EWHC 1058 (TCC), [22].

39 Ibid.

40 [2006] EWHC 89 (TCC).

41 Subclause 3.5, 1999 Red and Yellow Books; Subclause 3.7, 2017 Red and Yellow Books.

Design Duties

Similar to civil law jurisdictions, under the common law, the contract will govern the extent of the engineer's authority, as contract administrator, to order variations, certify work, subcontract, and accept the works. Although again, unless otherwise specified in the contract terms, it is the owner who assumes final responsibility with the contractor for the actions of the engineer.

The contract should also specify the nature of the engineer's first (and some might say primary) duty, namely design. There are two common design obligations. First, to design works with reasonable skill and care, in which case the engineer only needs to show that the works were designed to the standard of the average member of the consultant's profession. Second, to ensure that the works once completed are fit for their stated purpose. A less common alternative, which sits between these two, is the obligation to design in accordance with good industry practice. In this case, the standard is higher than that of the average professional, but its precise meaning remains somewhat ambiguous.

Under English law, if no standard is set under the contract, Section 13 of the *Supply of Goods and Services Act 1982* will supply the implied term, providing as follows:

> *Implied term about care and skill*
>
> *In a contract for the supply of a service where the supplier is acting in the course of a business, there is an implied term that the supplier will carry out the service with reasonable care and skill.*

In *Eckersley v Binnie*, Bingham LJ provided this explanation of the reasonable skill and care obligation:

> *A professional man should command the corpus of knowledge which forms part of the professional equipment of the ordinary member of his profession. He should not lag behind other ordinarily assiduous and intelligent members of his profession in knowledge of new advances, discoveries and developments in his field. He should have such awareness as an ordinary competent practitioner would have of the deficiencies in his knowledge and the limitations on his skills. He should be alert to the hazards and risks inherent in any professional task he undertakes to the extent that other ordinarily competent members of his profession would be alert. He must bring to any professional task he undertakes no less expertise, skill and care than any other ordinarily competent members of profession would bring, but need bring no more. The standard is that of the reasonable average. The law does not require of a*

professional man that he be a paragon, combining the qualities of polymath and prophet.[42]

Engineers should note the need to keep abreast of emerging technologies and advancements, a task of no little importance given the rapid pace of innovation of recent years, not least in the field of artificial intelligence, digital design, and sustainable development.

Under the FIDIC construction contracts, the requirement in respect of contractor-designed works is that, once completed, they shall: "*be fit for such purposes for which the part is intended as are specified in the Contract.*"[43] This imposes an absolute duty. In other words, the absence of negligence in the design will not be a defense for the contractor. In *MT Højgaard A/S v E.ON Climate*,[44] the UK Supreme Court held that the contractor had agreed that the design of the foundations would be fit for their purpose, and last for 20 years. It did not matter that the foundations had been designed according to the latest standards. That was no defense. The foundations failed so they were not fit for their purpose.

In contrast, Subclause 3.3.1 of the White Book provides that the consultant shall "*have no other responsibility than to exercise the reasonable skill, care and diligence to be expected from a consultant experienced in the provision of such services for projects of similar size, nature and complexity.*"

One reason for the stricter "fit for purpose" obligation being imposed on contractors is that, under FIDIC contracts, the contractor typically undertakes design work in a design-and-build capacity. Unlike contracts where design and construction responsibilities are separate, this model places full responsibility on the contractor to ensure that the works meet the owner's specified purpose, and not merely the industry standards. Consequently, the contractor is held to a higher level of accountability, requiring the final product to function as intended, even if the design complies with professional standards.

The Engineer as Part of the Dispute Resolution Process

Common law-based construction contracts commonly also entrust the contract administrator with the authority to determine the parties' claims. This is commonly referred to as the engineer's "dual role," which aligns with the principles of

42 [1988] 18 Con LR 1.

43 Subclause 4.1, 1999 and 2017 Red and Yellow Books.

44 [2017] UKSC 59.

the English legal system, where contract administrators are subject to a duty to act fairly and impartially between the parties. A series of English cases have outlined the nature of this role and the meaning of such a duty.

In *ICI Ltd v Merit Merrell Technology Ltd*, Fraser J noted that the majority of the commonly used standard forms of construction and engineering contracts had express provisions within them: "*for the involvement of a third party entity who is central to the operation of the contractual provisions.*"[45] The judge further explained that the duties of these entities, as we have explained above, can broadly be considered as falling into one of two functions:

> *One is the "agency function"; an example of this is instructing variations, acting on behalf of the employer. The other is the "decision-making function." This is quite different and is where the decision-maker has to reach decisions on matters where, at least potentially, the contractor and the employer have opposing interests.*[46]

Fraser J referred to the *Scheldebouw* case, where Jackson J described the role of the engineer as a decision-maker as follows:

1. *The precise role and duties of the decision-maker will be determined by the terms of the contract under which he is required to act.*
2. *Generally the decision-maker is not, and cannot be regarded as, independent of the employer.*
3. *When performing his decision-making function, the decision-maker is required to act in a manner which has variously been described as independent, impartial, fair and honest. These concepts are overlapping but not synonymous. They connote that the decision-maker must use his professional skills and his best endeavors to reach the right decision, as opposed to a decision, which favors the interests of the employer.*[47]

FIDIC incorporated this philosophy into their standard forms, where the Engineer's determination today acts as the threshold of the dispute resolution mechanism. The Engineer must first try and seek the agreement of the parties, failing which their determination will provisionally bind the parties until revision by the dispute board or, subsequently, by the arbitral tribunal.[48]

---

45 [2017] EWHC 1763 (TCC), [128]. In this case, the parties utilized the NEC3 Engineering and Construction Contract.

46 *Ibid.*

47 *Scheldebouw BV v St James Homes (Grosvenor Dock) Ltd* [2006] EWHC 89 (TCC), [35].

48 Subclauses 3.5 and 20.1, 1999 Red and Yellow Books; Subclauses 3.7 and 20.2, 2017 Red and Yellow Books.

The evolution of FIDIC contracts has seen a number of changes to the Engineer's role. In the 1999 edition, the express reference to the Engineer's impartiality[49] was removed, as it was recognized that in performing their duties or exercising their authority, the Engineer would be deemed to act on behalf of the Employer. By contrast, the 2017 edition introduced a requirement under Subclause 3.7 for the Engineer to act "*neutrally between the Parties*" when making determinations. The accompanying language of the provision – "*shall not be deemed to act for the Employer*" – confirms that, during the determination process, the Engineer is not acting as the Employer's agent.

The 2017 FIDIC edition broadened the role of the Engineer during the determination process. The Engineer is now required to proactively try and seek an agreement between the parties on any claim, which may promote adherence to dispute avoidance principles by both the Engineer and the parties involved. This is further underpinned by the requirement that the Engineer maintain a record of the consultation process.

By contrast in civil law jurisdictions, there is no duty to act fairly and impartially between the parties, and the concept of empowering the engineer to act neutrally between the parties is largely unknown.[50]

Consequently, the dual role of the engineer has faced longstanding criticism, not least because, in most construction projects, the contract administrator is appointed and paid by the owner.[51] As noted earlier, the engineer is often vested with the authority to act on the owner's behalf when performing administrative functions. This creates a potential conflict of interest for two reasons. First, it is regarded as unrealistic to expect that an individual appointed and paid by the owner can impartially switch roles to resolve the disputes between the parties. Second, the engineer is frequently required to adjudicate claims arising from their own prior determinations (e.g., certifying one of the contractor's interim

---

49 Contained in the 1987 edition.

50 Seppälä, 79; Howard C, Lavers A, Hanna P, "English Common Law" in *International Construction Law: An Overview*, ed. W Breyer (Routledge, 2024) para. 3.38; Jaeger A, Hök G, *FIDIC – A Guide for Practitioners* (Springer, 2009) 225.

51 Jenkins J, *International Construction Arbitration Law*, 3rd edn. (Kluwer 2021) 30; Baker E, Mellors B, Chalmers S, Lavers A, *FIDIC Contracts: Law and Practice* (London, 2009) para. 9.11; Feld G, "One step forward, two steps backward! A critical review of the FIDIC conditions of contract for plant and design-build (2017 edition)," *International Business Law Journal* 433, (2020): 441; Shorland R, "Role of the engineer and settlement of disputes under FIDIC conditions of contract for works of civil engineering construction," *Arbitration* 65(2), (1995): 92; Stipanowich T, "Reconstructing Construction Law: Reality and Reform in a Transactional System," *Wisconsin Law Review* 463, (1998): 499; Pike A, "Dispute Review Boards and Adjudicators," ICLR 157, (1993): 159; Rubino-Sammartano M, "The Role of the Engineer – Myth and Reality," *International Business Lawyer* 14, (1986): 81.

payment applications). In effect, this places the engineer in the problematic position of reviewing and re-evaluating their own decisions.

However, it is important to recognize that construction consultants, including engineers and architects, are bound by the ethical obligations set by their professional bodies. For instance, FIDIC has established a Code of Ethics for its members, outlining key principles considered fundamental to the conduct of consulting engineers.[52] These include that the consulting engineer shall be impartial in the provision of professional advice and judgment, shall act at all times in the legitimate interest of the client, shall provide all services with integrity and faithfulness, and shall maintain knowledge and skill at levels consistent with development in technology.

Additionally, despite the absence in civil law systems of a formal duty for engineers to act fairly and impartially, nothing prevents the contracting parties from exercising their contractual freedom to include such obligations in their agreement. There are obvious advantages to entrusting the contract administrator with the power to determine the parties' claims. Given their familiarity with the project, its complexities, and the parties' respective positions, the engineer is particularly well-placed to make timely and informed determinations. Moreover, engineers – like other professionals – too have their own position and market reputation to protect. Therefore, in theory, an engineer appointed to act impartially between the parties, as under the FIDIC regime, would be less likely to show bias, as doing so could harm their professional credibility.

In this context, there is an emerging trend in which the engineer is increasingly viewed as the first part of a dispute *avoidance* process rather than a *resolution* one. This shift requires, on one hand, that all involved parties, including the engineer, are educated about the engineer's role, and on the other, that the engineer actually acts impartially between the parties and issues determinations as promptly and fairly as possible, in compliance with the tight timeframes specified in the contract.

A word of caution is warranted here. In some cases, contracts stipulate that the engineer can grant extensions of time or approve claims above a certain value only with the prior consent of the owner. This can undermine the engineer's role and independence, both in perception and reality. It may also lead to disputes, as the lack of owner's approval prevents the engineer from addressing a claim and issuing a determination, leaving the contractor with no choice but to escalate the matter to a formal dispute.

Finally, when examining the engineer's decision-making function, an important consideration concerns the legal nature of their determinations. In performing

52 https://fidic.org/sites/default/files/fidic_codeofethics_201510.pdf.

this function, the engineer does not act as a judicial body, as this would necessitate a truly independent and impartial position between the parties.[53] Accordingly, the engineer's determinations are not afforded the same legal standing of courts' judgments and arbitral awards, and thus they cannot be enforced in the same way.[54] Instead, the engineer's determination constitutes the first, provisional step of a contractually agreed dispute resolution mechanism. If either party is dissatisfied with such determination, it has the option to refer the dispute to an independent adjudicator (*i.e.*, the dispute board), whose decision will eventually be enforced through arbitration.

The Role of the Engineer in the Future

Looking ahead at the role of the engineer in the future, from a contract management perspective, it is clear that the engineer will increasingly have the opportunity become a valuable partner in the dispute (or even conflict) avoidance process. Consequently, if engineers reject this opportunity to be a part of the collaborative process increasingly offered by standard forms of contract, their role in this aspect of project management will gradually diminish over time.

The engineer's toolbox is equipped with many powerful dispute avoidance mechanisms. One essential tool for managing risk is early or advance warning. The idea standing behind early warning is rather simple: the sooner an issue is identified, the easier it is to resolve. Both the NEC and FIDIC forms require not only the contractor and owner but also the Project Manager or Engineer to promptly notify all parties of any issues that may adversely impact the quality of the works, delay the project, or increase the contract price.

Building on this proactive role, the FIDIC contracts reinforce the Engineer's responsibility in dispute avoidance. In the first stage of the determination process, the Engineer is tasked with facilitating an agreement between the parties. It is only if this effort fails that the Engineer issues a formal determination. This early intervention aims to resolve disputes at an earlier stage, thereby avoiding the need to escalate the dispute to the dispute board or arbitration.

FIDIC contracts also vest the Engineer with the power to convene meetings. Direct communication, particularly through in-person discussions, is a highly effective method for resolving conflicts and keeping the project on track.

53 Gaillard E, Savage J (eds), *Fouchard Gaillard Goldman on International Commercial Arbitration* (Kluwer, 1999) para. 24; Crivellaro A, "Appalti internazionali: il ruolo dell'engineer," Rivista dell'Arbitrato 207, (1987): 222.

54 ICC Case No. 10619 (2002), Interim Award; ICC Case No. 3790 (1983), Partial Award; Rome Court of Appeal, 12 December 1994; Baker/Mellors/Chalmers/Lavers, para. 9.6.

Engineers can also take a more proactive role by guiding contractors to maintain records in a specific format or, even better, encouraging both parties to agree on how to document payment applications and claims. This approach not only reassures the owner by reducing the likelihood of exaggerated claims but also fosters transparency with the contractor, significantly minimizing the potential for disputes.

Predicting the engineer's role from a technical perspective is more challenging. The sheer pace of digital advancement will continue to reshape the industry, presenting new opportunities and challenges. These are exciting times for the next generation of designers and engineers, as technology redefines the boundaries of what is possible.

Table of Cases

France

Cour de Cassation, Chambre Civile 1, 21 January 1963

Cour de Cassation, Chambre Civile 3, 17 February 1999, No. 95-21.412

Cour d'Appel de Pau, 21 November 1991, No. 1936/90

Cour de Cassation, Chambre Civile 3, 2 June 2016, No. 15-981

Cour de Cassation, Chambre Civile 3, 30 October 1991, No. 90-12.993

Cour de Cassation, Chambre Civile 3, 17 February 1982, No. 80-13.761

Italy

Corte di Cassazione, sez. II, 6 December 1986, No. 7623

Corte di Cassazione, sez. II, 19 September 2016, No. 18285

Corte di Cassazione, sez. II, 3 September 2020, No. 18289

Corte d'Appello di Roma, 12 December 1994 in [1996] Diritto del Commercio Internazionale 949

England and Wales

Marks & Spencer Plc v BNP Paribas Securities Services Trust Co (Jersey) Ltd [2015] UKSC 72

BP Refinery (Westernport) Proprietary Limited v The President Councillors and Ratepayers of the Shire of Hastings [1977] UKPC 13

Arnold v Britton [2015] UKSC 36, [2015] AC 619

Compass Group UK and Ireland Ltd (t/a Medirest) v Mid Essex Hospital Services NHS Trust [2013] EWCA 200

England and Wales

Equitable Debenture Assets Corporation Ltd v William Moss Group Ltd (1984) 2 Con LR 1

Goldswain v Beltec Ltd (t/a BCS Consulting) [2015] EWHC 556 (TCC)

Hart Investments Ltd v Fidler [2007] EWHC 1058 (TCC)

Scheldebouw BV v St James Homes (Grosvenor Dock) Ltd [2006] EWHC 89 (TCC)

Eckersley v Binnie & Partners [1988] 18 Con LR 1

MT Højgaard A/S v E.ON Climate and Renewables UK Robin Rigg East Ltd [2017] UKSC 59

Imperial Chemical Industries Ltd v Merit Merrell Technology Ltd [2017] EWHC 1763 (TCC)

Australia

Renard Constructions (ME) v Minister for Public Works (1992) 26 NSWLR 234

Royal Botanic Gardens and Domain Trust v South Sydney City Council [2002] HCA 5

Arbitral Awards

ICC Case No. 10619 (2002), Interim Award in (2008) 19 ICC International Court of Bulletin, No. 2, 85–90

ICC Case No. 3790 (1983), Partial Award in ICCA Yearbook Commercial Arbitration 1986 – Volume XI, pp. 119–126

Table of Statutes

Code civile (France)

Codice civile (Italy), Regio decreto of 16 March 1962, No. 262

Supply of Goods and Services Act 1982

Bibliography

Dr Robert Gaitskell QC, *Engineers' Dispute Resolution Handbook* (Thomas Telford 2006).

Engineering Council. https://www.engc.org.uk/glossary-faqs/frequently-asked-questions/status-of-engineers/#:~:text=Commonplace%20use%20of%20the%20word,Engineering%20Council%20as%20totally%20impractical (last accessed 1 October 2024).

FIDIC Code of Ethics: https://fidic.org/sites/default/files/fidic_codeofethics_201510.pdf (last accessed 1 October 2024).

Standard Forms of Contract

FIDIC (1999), *Conditions of Contract for Construction*. Geneva, Switzerland: International Federation of Consulting Engineers.
FIDIC (1999), *Conditions of Contract for Plant and Design-Build*. Geneva, Switzerland: International Federation of Consulting Engineers.
FIDIC (2017), *Conditions of Contract for Construction*. Geneva, Switzerland: International Federation of Consulting Engineers.
FIDIC (2017), *Conditions of Contract for Plant and Design-Build*. Geneva, Switzerland: International Federation of Consulting Engineers.
FIDIC (2017), *Client/Consultant Model Services Agreement*. Geneva, Switzerland: International Federation of Consulting Engineers.
NEC3 Engineering and Construction Contract, June 2005.
NEC4 Engineering and Construction Contract, June 2017.

8

Public Procurement

Marco Padovan, Michele Bonomini, and Fabiola Viccei

Introduction

Central governments, governmental agencies, and local public authorities often need to purchase goods and services with public resources and for public purposes to fulfill their functions. Such purchases are generally referred to as government/public procurement.[1] It has been calculated that government procurement accounts for 10–15 per cent of the GDP of an economy on average. In the EU, in 2017, the public purchase of goods and services was estimated to be worth some €2 trillion or 13.3% of GDP.[2] Public procurement therefore constitutes a significant market and an important aspect of international trade. Of this, a large portion is constituted by works and related technical services.

The main objective of public procurement instruments is to achieve the best "value for money." The principles of open, transparent, and nondiscriminatory procurement are generally considered to be the best tool to achieve this goal, as they increase competition among suppliers. However, more recently, other principles emerged, aimed at achieving other, less monetary, objectives.

So, for instance, certain countries use the instruments of government procurement to achieve other domestic policy goals, such as the promotion of specific local industry sectors or products or the promotion of social groups. Examples are several governments' rules that require the contractors who are awarded public tenders or the contracting authorities to procure locally at least a threshold of

1 See https://www.wto.org/english/tratop_e/gproc_e/gproc_e.htm.

2 See European Commission, Public procurement indicators, 2017. Available at: https://policy.trade.ec.europa.eu/help-exporters-and-importers/accessing-markets/public-procurement_en.

Major Issues in International Construction Law, First Edition. Mauro Rubino-Sammartano and Troy Harris.
Published 2026 by John Wiley & Sons, Inc.

certain services or products (e.g., asphalt, steel etc.).[3] The aim is clearly to foster local industrial development and economic welfare.[4] Other instruments, such as the Indian Public Procurement Policy for Micro and Small Enterprises (MSEs) Order, 2018, require every Central Ministry/Department / PSUs to set an annual target for 25% procurement from MSE Sector.[5]

In other cases, public procurement is used as a tool to promote social and environmental goals. For example, the awarded party, or any participating party in a bidding procedure, may be required to apply collective bargain agreements approved by the most representative Trade Unions in the relevant industry sector, or to comply with certain social standards established in international labor conventions, or again with the most advanced environmental standards and label certifications. Similarly, the EU is pushing its green public procurement, through which public authorities are encouraged to seek to procure goods, services and works that have a reduced impact on the environment throughout the lifecycle of these goods, services, and works.[6]

On the other hand, we are witnessing a resurgence of protectionism, in that certain jurisdictions require bidders to declare any foreign subsidies they have received, which may lead to distortions in the competition.

On this aspect, probably the most striking example is the Regulation (EU) 2022/2560 of the European Parliament and of the Council of 14 December 2022 on foreign subsidies distorting the internal market. The purpose of this Regulation is to contribute to the proper functioning of the EU internal market by establishing a harmonized framework to address distortions caused, directly or indirectly, by foreign subsidies, with a view to ensuring a more uniform level playing field. This Regulation lays down rules and procedures for investigating foreign subsidies that distort the internal market and for redressing such distortions with respect to concentrations and public procurement procedures. This Regulation addresses foreign (i.e., non-EU) subsidies granted to an undertaking, including a public undertaking which is directly or indirectly controlled by the State, participating to

---

3 An example of this trend is the Public Procurement (Preference to Make in India) Order of 2017 and subsequent amendment and integrations. The purpose of this Order is to promote manufacturing and production of goods and services in India with a view to enhance income & employment, even though this is at the cost of more limited (international) competition.

4 For a comprehensive list of such measures, see: https://trade.ec.europa.eu/access-to-markets/en/barriers/details?isSps=false&barrier_id=10863.

5 See for more info: https://msme.gov.in/public-procurement-policy#:~:text=Every%20Central%20Ministry%20%2FDepartment%20%2F%20PSUs,mandatory%20from%201st%20April%202015.

6 See "EU-U.S. Trade and Technology Council Working Group 2 – Climate and Clean Tech. Joint EU-US Catalogue of Best Practices on Green Public Procurement," available at https://policy.trade.ec.europa.eu/help-exporters-and-importers/accessing-markets/public-procurement_en.

a public tender procedure in the EU. It imposes a notification obligation to bidders and grants the European Commission wide investigating powers in respect of the existence, or not, of any foreign economic subsidy.

Additionally, the growing number of sanctions related to international political crisis situations has recently led to restrictions in the participation to public procurement procedures or in the award of public contracts to certain entities based on their origin or controlling influence.

An example of this limitations to participation to public tender procedures is given, once again at the European level, is Regulation (EU) No 833/2014 of 31 July 2014 concerning restrictive measures in view of Russia's actions destabilizing the situation in Ukraine.

Pursuant to article 5*k* of the same Regulation, it shall be prohibited to European public authorities to award or continue the execution of any public or concession contract falling within the scope of the public procurement Directives to or with: (a) a Russian national, a natural person residing in Russia, or a legal person, entity or body established in Russia; (b) a legal person, entity or body whose proprietary rights are directly or indirectly owned for more than 50 % by an entity referred to in point (a) above; or (c) a natural or legal person, entity or body acting on behalf or at the direction of an entity referred to in point (a) or (b) above, including, where they account for more than 10 % of the contract value, subcontractors, suppliers, or entities whose capacities are being relied on.[7]

---

7 The same EU Regulation provides for certain exemptions, though, related to:

a. the operation, maintenance, decommissioning and radioactive waste management, fuel supply and retreatment and safety of civil nuclear capabilities, and the continuation of design, construction and commissioning required for the completion of civil nuclear facilities, as well as the supply of precursor material for the production of medical radioisotopes and similar medical applications, critical technology for environmental radiation monitoring, as well as civil nuclear cooperation, in particular in the field of research and development;
b. intergovernmental cooperation in space programmes;
c. the provision of strictly necessary goods or services which can only be provided, or which can only be provided in sufficient quantities, by the persons referred to in paragraph 1;
d. the functioning of diplomatic and consular representations of the Union and of the Member States in Russia, including delegations, embassies and missions, or international organizations in Russia enjoying immunities in accordance with international law; or
e. unless elsewhere prohibited, the purchase, import or transport of natural gas and oil, including refined petroleum products, as well as titanium, aluminum, copper, nickel, palladium and iron ore from or through Russia into the Union.

The Member States concerned are required to inform the other Member States and the Commission of any authorizations granted within two weeks of the authorization.

Competitive Bidding Laws

Before moving to examine the most relevant legislative texts, some basic concepts of public procurement law and practice need to be at least enucleated.

Public Procurement

The expression "Public procurement" covers substantially the assessment of need of the public authority, the rules to be followed when issuing invitation to pre-qualify or to enlist or to bid, procedures for the selection of the bidder, and the contracts executed by public authorities aimed at the purchase of goods and services with public resources and for public purposes, in general with the aim of obtaining the best value for money.

Public Authority

One of the most relevant concepts that are relevant for public procurement is the concept of "public authority." Usually, competitive bidding laws provide for a definition of the subjects which are required to apply the competitive bidding procedures. While each jurisdiction may have its own notion and list of what it considers public authorities, usually the concept of "public authority" is construed in a broad way so as to cover the State, regional or local authorities, bodies governed by public law or associations formed by one or more such authorities or one or more such bodies governed by public law. Legislation usually categorizes such authorities depending on their level (national/central or local), functions (public or economic) and level of autonomy.

Among the public authorities, "Central government authorities" usually include governmental ministries and departments. "Subcentral contracting authorities" usually refers to all those contracting authorities which are not central government authorities. Finally, "bodies governed by public law" usually refer to those bodies that have all of the following characteristics: (a) are established for the specific purpose of meeting needs in the general interest, not having an industrial or commercial character; (b) have legal personality; and (c) are financed, for the most part, by the State, regional or local authorities, or by other bodies governed by public law; or are subject to management supervision by those authorities or bodies; or have an administrative, managerial or supervisory board, more than half of whose members are appointed by the State, regional or local authorities, or by other bodies governed by public law.[8]

---

8 At the EU level, the notion of "contracting authorities" and in particular that of "bodies governed by public law" have been examined repeatedly in the case-law of the Court of Justice

Similar to the concept of "public authority" is the concept of "contracting authority" or "awarding authority." This notion is broader than the one of "public authority" and includes also private entities, such as commercial companies incorporated under private law and having a traditional business purpose, which however become entrusted of a public service or use public funds on the basis of certain individual rights granted to them by the public authorities or by the law. A typical example are the concessionaires of public services. While the concessionaire may be a private law company, it is entrusted by a public authority of certain powers, which are relevant for public purposes. Examples of concessionaires include the subject entrusted with the management and maintenance of the road system or the railway system of a country. When these entities need to procure works or services for the purpose of obtaining their scope, they act as public authorities and are therefore subject to public procurement laws.

Another example of private entities that are subject to public procurement procedures is the holder of a building permit for a private real estate development initiative which is required to include also certain works that are meant to be of public use, such as access roads and bridges, green areas, parking slots etc. (such works also referred to as "infrastructure works"). In many jurisdictions, in fact, construction is subject to the control of the public authorities and the execution of the works is conditional upon permission being granted by the municipality. Laws concerning the suitability of land for development may also provide that any activity involving the urban development of municipal land and building works on such land entails liability, for the applicant for the building permit, to contribute to the related urban development and construction costs (also referred to as "infrastructure contribution" to be paid to the municipality).

Those infrastructure works are usually meant to be executed, through public procurement, by the relevant municipality that grants the building permit. However, local legislation may provide that, by way of total or partial set-off against the amount due as infrastructure contribution, the holder of the permission may

of the European Union. In this regard, it should be mentioned that according to such EU case law, a body which operates in normal market conditions aims to make a profit, and bears the losses resulting from the exercise of its activity should not be considered as being a "body governed by public law" since the needs in the general interest, that it has been set up to meet, can be deemed to have an industrial or commercial character. Similarly, the condition relating to the origin of the funding of the body considered has also been examined in the case-law, which has clarified inter alia that being financed for "the most part" means for more than half, and that such financing may include payments from users, which are imposed, calculated, and collected in accordance with rules of public law.

undertake to execute the infrastructure works[9] directly, in accordance with the procedures and standards laid down by the municipality.

In such cases, the granting and validity of the building permit is conditional upon the execution, contemporary with the real estate development initiative works or within a short period thereafter, of (at least) primary infrastructure works, often formalized through an agreement with the municipality itself. In such cases, while the developer of the real estate initiative is not *in se* a public authority, it is bound by public procurement rules, as the infrastructure works that it agreed to perform on behalf of the municipality transform it into a contracting authority.[10] In such a case, in fact, the developer must be regarded, by virtue of the agreements concluded with the municipality exempting him from the infrastructure contribution in return for the execution of public infrastructure works, as the holder of an express mandate granted by the municipality for the construction of that work. As such, it is subject to the application (and to apply to its procurement) the rules concerning public procurement.

Public Procedures

Competitive bidding laws regulate in detail the public procedure. The management of such procedures is usually informed by the principles of freedom to participate in public tender procedures, equal treatment, nondiscrimination, mutual recognition, proportionality, and transparency.

This means that in principle, contracting authorities shall treat economic operators equally and without discrimination and shall act in a transparent and proportionate manner. The design of the procurement shall not be made with the intention of excluding the bidding procedure form the application of public procurement rules or of artificially narrowing competition, for instance by way of

---

9 Infrastructure works are usually divided in "primary" and "secondary." "Primary infrastructure works" comprise residential streets, leisure areas, parking spaces, sewers, networks for the distribution of water, electricity and gas, street lighting and formal parks and gardens. "Secondary infrastructure works" usually comprise preschool facilities; primary and secondary schools; buildings and campuses to accommodate higher and further education facilities; local markets; municipal branch offices; churches and other religious buildings; local sports facilities; community centers; cultural and health and fitness facilities; and local parks and gardens.

10 See for instance *Judgment of the Court (Sixth Chamber) of 12 July 2001. – Ordine degli Architetti delle province di Milano e Lodi, Piero De Amicis, Consiglio Nazionale degli Architetti and Leopoldo Freyrie v Comune di Milano, and Pirelli SpA, Milano Centrale Servizi SpA and Fondazione Teatro alla Scala. Case C-399/98*, available at https://eur-lex.europa.eu/legal-content/EN/ALL/?uri=CELEX:61998CJ0399.

designing the procurement in a manner that can lead to unduly favoring or disadvantaging certain economic operators.

Bidding laws provide detailed rules on: (i) communication of the contracting authority with tenderers and economic operators; (ii) criteria for the choice of the relevant awarding procedure; (iii) the conduct of the procedure (including the technical specifications, that shall be nondiscriminatory; variants; division of the tender into lots; form and manner of publication of notices); (iv) choice of participants and award of contracts; (v) contracts award criteria; and (vi) contracts performance and subsequent amendments. Of particular interest for the economic operator participating to public tender procedures are the bidding systems and the award criteria.

Bidding Systems

Bidding systems are designed to achieve an appropriate balance between the needs of the public authority to ensure it gets the appropriate quality from the appropriate source (provider) and the appropriate price in relation to specific needs specified in the relevant technical requirements. In certain critical and complex requirements, the technical and financial capability of the prospective supplier/contractor becomes an important element to achieve value for money.

Bidding systems include rules that require an initial assessment of the financial, professional, and experience capabilities of the bidder, which are also evaluated against the tender documents, and may lead immediately to the exclusion of bidders that do not comply with the prescribed tender requirements.

The following types of procedures are often provided for:

- Open procedures. In open procedures, any interested economic operator may submit a tender in response to a call for competition. A time limit for the receipt of tenders is set in the tender notice (usually around 30 days from the date on which the contract notice was sent). The tender is usually accompanied by the information for qualitative selection that is requested by the contracting authority.
- Restricted procedures. In restricted procedures, any economic operator may submit a request to participate in response to a call for competition by providing the information for qualitative selection that is requested by the contracting authority. The minimum time limit for receipt of requests to participate is usually around 30 days from the date on which the contract notice or, where a prior information notice is used as a mean of calling for competition, the invitation to confirm interest was sent. Only those economic operators invited to do so by the contracting authority following its assessment of the information provided may submit a tender. Contracting authorities

may limit the number of suitable candidates to be invited to participate in the procedure.

- Competitive procedure with negotiation. In competitive procedures with negotiation, any economic operator may submit a request to participate in response to a call for competition by providing the information for qualitative selection that is requested by the contracting authority. In the procurement documents, contracting authorities usually identify the subject-matter of the procurement by providing a description of their needs and the characteristics required of the supplies, works or services to be procured and specify the contract award criteria and also indicate which elements of the description define the minimum requirements to be met by all tenders. The information provided shall be sufficiently precise to enable economic operators to identify the nature and scope of the procurement and decide whether to request to participate in the procedure. The authority sets also the time limit for receipt of requests to participate (usually around 30 days from the date on which the contract notice or, where a prior information notice is used as a means of calling for competition, the invitation to confirm interest was sent).
- Negotiated procedure without prior publication. In certain specific cases and circumstances,[11] contracting authorities may award public contracts by a negotiated procedure without prior publication. Only those economic operators invited by the contracting authority following its assessment of the information provided may submit an initial tender which shall be the basis for the subsequent negotiations. Contracting authorities may limit the number of suitable candidates to be invited to participate in the procedure. Unless otherwise provided for, contracting authorities shall negotiate with tenderers the initial and all subsequent tenders submitted by them, except for the final tenders, to improve the content thereof. The minimum requirements and the award criteria shall not be subject to negotiations. Contracting authorities may award contracts on the basis of the initial tenders without negotiation where they have indicated, in the contract notice or in the invitation

11 For instance: (a) where no tenders or no suitable tenders or no requests to participate or no suitable requests to participate have been submitted in response to an open procedure or a restricted procedure, provided that the initial conditions of the contract are not substantially altered; (b) where the works, supplies, or services can be supplied only by a particular economic operator (e.g., because the aim of the procurement is the creation or acquisition of a unique work of art or artistic performance; or competition is absent for technical reasons; or because of the protection of exclusive rights, including intellectual property rights); (c) in so far as is strictly necessary where, for reasons of extreme urgency brought about by events unforeseeable by the contracting authority, the time limits for the open or restricted procedures or competitive procedures with negotiation cannot be complied with. The circumstances invoked to justify extreme urgency shall not be attributable to the contracting authority.

to confirm interest, that they reserve the possibility of doing so. During the negotiations, contracting authorities shall ensure the equal treatment of all tenderers. To that end, they shall not provide information in a discriminatory manner which may give some tenderers an advantage over others. They shall inform all tenderers, whose tenders have not been eliminated, in writing of any changes to the technical specifications or other procurement documents, other than those setting out the minimum requirements. Following these changes, contracting authorities shall provide sufficient time for tenderers to modify and re-submit amended tenders, as appropriate.

Contract Award Criteria

The most common criteria for award of public works contracts are (i) the most economically advantageous tender and (ii) the lowest price.

The most economically advantageous tender from the point of view of the contracting authority shall be identified on the basis of the price or cost, using a cost-effectiveness approach, such as life-cycle costing, and may include the best price-quality ratio, which shall be assessed on the basis of criteria, including qualitative, environmental and/or social aspects, linked to the subject-matter of the public contract in question. Such criteria may comprise, for instance: (a) quality, including technical merit, aesthetic and functional characteristics, accessibility, design for all users, social, environmental and innovative characteristics and trading and its conditions; (b) organization, qualification, and experience of staff assigned to performing the contract, where the quality of the staff assigned can have a significant impact on the level of performance of the contract; or (c) delivery conditions such as delivery date, delivery process and delivery period or period of completion.

The cost element may also take the form of a fixed price or cost on the basis of which economic operators will compete on quality criteria only.

The lowest price criteria are usually reserved for simpler, more standardized activities. In such a case, the contracting authority sets out a list of activities and quantities, and requires the bidder to indicate an individual price in addition to the bidder's own business safety costs.

Contracting authorities may require economic operators to explain the price or costs proposed in the tender where tenders appear to be abnormally low in relation to the works, supplies, or services.

The explanations may in particular relate to: (a) the economics of the manufacturing process, of the services provided or of the construction method; (b) the technical solutions chosen or any exceptionally favorable conditions available to the tenderer for the supply of the products or services or for the execution of the work; (c) the originality of the work, supplies, or services proposed by the tenderer; (d) compliance with environmental, social, and labor obligations; (e) prices offered by subcontractors; (f) the possibility of the tenderer obtaining State aid or other form of foreign subsides.[12]

If the tenderer fails to provide adequate evidence to justify the economic sustainability of the tender, it is excluded from the tender and the tender awarded to the first subsequent sustainable offer.

Public Contracts

"Public contracts" means contracts for pecuniary interest concluded in writing between one or more economic operators and one or more contracting authorities and having as their object the execution of works, the supply of products or the provision of services.

Among the public contracts, "public works contracts" usually define public contracts having as their object one of the following:

- the execution, or both the design and execution, of works related to activities such as: (i) construction of new buildings and works, restoring and common repairs; demolition and wrecking of buildings; earth moving; test drilling and boring; (ii) general construction of buildings and civil engineering works; erection of roof covering and frames; construction of highways, roads, airfields and sport facilities; construction of water projects; other construction work involving special trades; (iii) installation of electrical wiring and fittings; insulation work activities; plumbing; other building installation; (iv) plastering; joinery installation; floor and wall covering; painting and glazing; (v) other building completion such as building completion and finishing works;
- the execution, or both the design and execution, of a work; and
- the realization, by whatever means, of a work corresponding to the requirements specified by the contracting authority exercising a decisive influence on the type or design of the work.

On the other hand, "public supply contracts" means public contracts having as their object the purchase, lease, rental or hire-purchase, with or without an option to buy, of products. A public supply contract may include, as an incidental matter, siting and installation operations.

12 See on this aspect, at EU level, Regulation (EU) 2022/2560 of the European Parliament and of the Council of 14 December 2022 on foreign subsidies distorting the internal market.

Finally, "public service contracts" means public contracts having as their object the provision of services other than those referred to in reference to works. In certain legislations, however, "services" include construction works and services.

Other forms of contracts include lease agreement (especially for real estate complexes), framework agreements and other forms of agreement.

Of great importance in the construction industry are also the concessions agreements and other forms of Public-Private Initiatives ("**PPI**").

Generally speaking, concession contracts are used by public authorities to deliver services or construct infrastructures. Concessions involve a contractual arrangement between a public authority and an economic operator (the concession holder) by which the latter provides services or carries out works and is remunerated by being permitted to exploit the work or service for a certain period of time.[13] One significant aspect of concessions contracts is that the right to exploit implies the transfer to the concession holder of the economic operating risk involving, in particular, the possibility that the concession holder will not recover the investments made.

Concessions are a particularly attractive way of carrying out projects in the public interest when state or local authorities need to mobilize private capital and know-how to supplement scarce public resources. In the EU, they underpin a significant share of economic activity and are especially common in network industries and for the delivery of services of general economic interest. Concession holders may, for example, build and manage motorways, provide airport services, or operate water distribution networks.[14]

Public-Private Initiatives include arrangements between the central, statutory entity, or any other government-owned entity, on the one side, and a private sector entity, on the other side, for the provision of public assets or public services or a combination thereof, through investments being made or management being undertaken by the private sector entity, for a specified period of time, where there is predefined allocation of risk between the private sector and the public entity and the private entity receives performance-linked payments that conform (or are benchmarked) to specified and predetermined performance standards.

General Contractor

Particularly relevant in the context of complex international construction projects is the "General Contractor" scheme. The assignment of global services to the

13 See for instance, at the EU level: https://single-market-economy.ec.europa.eu/single-market/public-procurement/legal-rules-and-implementation/concession-contracts_en

14 *Ibidem.*

general contractor is achieved through the conclusion of a contract that obliges the economic operator (i.e., the general contractor) to carry out a work and to pursue a specific administrative result indicated in the tender notice and in the contract, with the organization of the necessary means and with management at its own risk in exchange for a fee determined on the basis of the result obtained and the services rendered.

The assignment to the general contractor is decided by the public authority taking into account the complexity and heterogeneity of the services requested and the need to pursue a high-quality and effective administrative result. It is used for huge complex projects such as the realization of highways, railways, and similar projects.

The relationships between the awarding entity and the general contractor are regulated by the tender notice, the contract, and the relevant provisions of the public procurement and concession contracts law.

The general contractor may be required, among other things:

- to draw up the detailed project of the works, in accordance with the technical-economic feasibility project drawn up by the contracting authority, and to carry out the activities that are instrumental to its approval;
- to ensure the prefinancing, in whole or in part, of the work;
- to constantly communicate to the contracting entity the development of the project, including any information regarding anti-corruption issues.

The contract may further provide that:

- the economic operator be vested with the status of expropriating authority, and that it will carry out expropriation of lands for public utility, with the power to expropriate and to supervise the related procedure;
- the economic operator identifies the methods of management of the work and of selection of the entities to whom such management can be entrusted.

The public authority usually draws up the technical-economic feasibility project and approves the executive project and its variants.

The tender notice and the contract usually also establish:

- the criteria for determining and reducing the fee due to the general contractor based on the result obtained and the services rendered;
- the methods and times for payment of the fee (usually after testing);
- the general contractor's own resources;
- the methods for attributing to the parties any subsequent charges, affecting the fee and deriving from regulatory provisions or from measures of other authorities; and
- the measures suitable for preventing attempts of corruption and criminal infiltration.

Economic Operators

One important aspect to consider when participating in a public tender procedure is the possibility to meet, alone or with third party, the required qualification criteria. Especially for huge construction projects, it may well be more appropriate to pursue the bid and the project execution (if awarded) allying with other contractors in order to satisfy the participation requirement if the single operator is lacking in any respect, and share the risks. In this regard, public tender rules usually set out the rules for individual and joint participation to the public tender. This second aspect is particularly relevant and deserves careful consideration.

As we have seen, one of the main purposes of competitive bidding rules is to enhance competition, also by way of increasing the possibility of participation in different legal forms. In this context, an economic operator can be either a single entity or a grouping of different entities. Public tender rules usually provide for a broad definition of economic operator,[15] which may cover any natural or legal person or public entity or group of such persons and/or entities, including any temporary association of undertakings, which offers the execution of works, the supply of products or the provision of services on the market.

Among the most frequent forms of association are contractual or corporate joint ventures ("**JV**"), consortia and other form of association such us "reliance agreements."[16]

---

15 Concepts similar to economic operator are the concepts of "tenderer" and "candidate." "Tenderer" usually refers to an economic operator that has submitted a tender; "candidate" is generally used to designate an economic operator that has sought an invitation or has been invited to take part in a restricted procedure, in a competitive procedure with negotiation, in a negotiated procedure without prior publication, in a competitive dialogue or in an innovation partnership.

16 In order to satisfy criteria relating to economic and financial standing and/or criteria relating to technical and professional ability, certain competitive bidding laws provide that an economic operator may, where appropriate and for a particular contract, rely on the capacities of other entities, regardless of the legal nature of the links which it has with them. Sometimes such laws also provide that the economic operators may only rely on the capacities of other entities where the latter will perform the works or services for which these capacities are required. The economic operator that wants to rely on the capacities of other entities is usually required to prove to the contracting authority that it will have at its disposal the resources necessary, for example, by producing a commitment by those entities to that effect. The tender rules may also require that the economic operator replaces an entity which does not meet a relevant selection criterion, or in respect of which there are grounds for exclusion. Where an economic operator relies on the capacities of other entities with regard to criteria relating to economic and financial standing, the tender rules may require that the economic operator and those entities be jointly liable for the execution of the contract. Especially in the case of works contracts, service contracts and siting or installation operations in the context of a supply contract, tender rules may require that certain critical tasks be performed directly by the tenderer itself or, where the tender is submitted by a group of economic operators, by a specific participant in that group.

In case of joint pursuit of a public tender, from a contractual point of view, it may be useful that the parties start their cooperation by way of execution of confidentiality and exclusivity agreement whereby the parties agree not to participate in the tender with others and to keep confidential the information exchanged for the study of the tender and preparation of the offer. The exclusivity ad confidentiality agreement is usually followed by the execution of a pre-bid joint venture agreement (“**PB-JVA**”) where the parties establish the basic criteria of their prospective collaboration in the preparation of the tender, the submission of the offer and the main principles regarding the execution of the works in case the tender is awarded to them. Such PB-JVA agreement may provide in this regard the prospective participating interests of the parties, the apportionment of the scope of works, the rules regarding the decision-making process of the JV or consortium, the rules on procurement from third parties, cases of exclusion of one or more parties form the JV or consortium and termination of the PB-JVA. In the event of award of the tender, the PB-JVA is usually superseded by a full joint venture agreement (“**JVA**”) that regulates in detail the collaboration of the parties in the project performance.

Finally, tender rules may additionally require that economic operators will assume, after award of the project, a specific legal form for the performance of the project, such as the incorporation of a special purpose vehicle in certain public-private initiatives.

Bonds and Guarantees

One particular aspect to consider when an economic operator decides to participate in a public tender procedure is also the type and amount of participation bonds that are required. Such bonds, usually referred to as “bid bonds”, shall usually be in the form of first-demand guarantees issued by banks, financial or insurance institutions duly authorized to carry out their activities in the jurisdiction where the project has to be carried out and shall remain valid for the entire duration of the tender procedure.

In case of award of the project, the successful bidder is usually required, as condition for the effectiveness of the contract, to provide the contracting authority with an advance payment bond (if an advance payment is provided for by the tender rules or the contract), in the same amount as the advance payment, a performance bond (usually in the amount of 10%[17]) and a warranty bond after testing of the works.

17 The performance bond may be replaced in certain cases by a retention money on interim payments.

Remedies

One of the foremost interests of economic operators taking part in public tender procedures is the availability of judicial remedies to ensure the tender rules are drafted and applied in a fair way and there are no infringements thereof. This, in turn, maximizes competition among economic operators.

In order to pursue such results several public procurement laws provide for a set of rules aimed at guaranteeing transparency and nondiscrimination in the course of tender procedures and ensure that rapid and effective means of redress are available if a bidder believes that the procedure has been run in an improper way in breach of public procurement laws.[18] Laws on remedies in public procurement usually provide for a standstill period (around 10 days) between the award and the signature of a public contract and require the contracting authority to inform all tenderers about the outcome of tender procedures. Since procedures for the award of public contracts are often of very short duration, competent review bodies are often authorized to take interim measures aimed at suspending a procedure or the implementation of any decisions which may be taken by the contracting authority. Other remedies available may be related to the granting of interim measures to correct alleged deviations or prevent further damages, the setting aside of awarding authority's decisions taken unlawfully, the award of damages, as well as time limits for precontractual remedies, sanctions for "ineffectiveness" and the possibility to introduce alternative penalties via fines or the limitation of contract duration.

In civil law jurisdictions, the competent court for such remedies are the administrative courts, which run the case on the basis of "fast-track" procedures.[19]

International Rules on Public Procurement

International Level

As public procurement plays a significant role in today's integrated world economy, an attempt to adopt internationally shared rules on public procurement has been put forward in several instances.

In fact, it has been considered that the liberalization of government procurement markets holds the potential to generate benefits both in terms of procurement

18 See for instance at European level Council Directive 89/665/EEC of December 21, 1989, on the coordination of the laws, regulations and administrative provisions relating to the application of review procedures to the award of public supply and public works contracts, as amended from time to time.

19 On the other hand, issues regarding performance of the contract etc. may be referred to ordinary (nonadministrative) courts or arbitral tribunals.

efficiency and commercial interests. Therefore, at the World Trade Organization (WTO) level, members have worked on this issue on three fronts, namely via: (i) the plurilateral Agreement on Government Procurement (GPA); (ii) the Doha Development Agenda (DDA) Working Group on Transparency in Government Procurement (which is currently inactive); and (iii) the GATS negotiations on government procurement.[20]

Among these three areas, work on the GPA is certainly the most active and has produced substantial trade liberalization. On 6 April 2014, the GPA 2012 came into force and marked a significant milestone of the WTO.

The GPA is a plurilateral agreement within the framework of the WTO. This means that not all WTO members are parties to the GPA. At present, the GPA has 22 parties comprising 49 WTO members.[21] Thirty-five WTO members/observers participate in the Committee on Government Procurement as observers. Several WTO members have initiated accession negotiations.

The fundamental aim of the GPA is to mutually open government procurement markets among its parties. As a result of several rounds of negotiations, the GPA parties have opened procurement activities estimated to be worth more than US$ 1.7 trillion annually to international competition (i.e., to suppliers from the GPA parties offering goods, services, or construction services).

The GPA is composed mainly of two parts: (i) the text of the Agreement and (ii) the Parties' market access schedules of commitments.

The text of the GPA Agreement establishes rules requiring its Parties that open, fair and transparent conditions of competition be ensured in government procurement. Particular importance and attention are given to technical specifications, in order to avoid discrimination and limitation of competition among bidders.[22]

Other legal instruments at international level include Regulation (EU) 2022/1031 for International Procurement (Regulation (EU) 2022/1031 of the European Parliament and of the Council of 23 June 2022, on the access of third-country (non-EU) economic operators, goods and services to the European Union's public procurement and concession markets and procedures supporting negotiations on access of EU economic operators, and goods and services to the public procurement and concession markets of third countries (also referred to as International Procurement Instrument – IPI). The IPI Regulation aims to promote reciprocity in access to international public procurement markets. In particular, it lays down procedures for the European Commission to investigate alleged measures or practices negatively affecting the access of EU businesses,

---

20 WTO website: https://www.wto.org/english/tratop_e/gproc_e/gproc_e.htm.

21 *Ibidem.*

22 See GPA, article X.

goods and services to non-EU procurement markets, and consult with the non-EU countries concerned, and to impose, as a last resort, measures to restrict access to EU public procurement procedures for businesses, goods, and services from the non-EU countries concerned.

Regional Level

At regional level, the EU has since long time adopted a set of rules regulating public procurement contracts and concessions agreements. Those rules are now contained in the following Directives:

- Directive 2014/23/EU on awarding concession contracts;[23]
- Directive 2014/24/EU on public procurement in ordinary sectors; and
- Directive 2014/25/EU on public procurement in the utilities sectors (water, energy, transport and postal services).

EU member states are required to implement EU Directives on public procurement and concessions through local (national) legislation. Such local (national) legislation must be in compliance with EU directives.

National Level

It may be useful to take a very brief look at how public procurement is regulated in the main countries of the European Union.

In **Germany**, most of these contracts are awarded by public authorities at federal, state or local level. However, contracting authorities are not only public bodies, but also some private companies in some areas of public infrastructure, such as drinking water supply, transport or energy supply.

---

23 This Directive sets out the EU rules for procurement by public sector contracting authorities and by contracting entities in the utilities sector by means of a concession (e.g., the right to operate infrastructure, such as a motorway, or a service, such as a bus route). Concessions are contracts for payment whereby one or more contracting authorities or contracting entities entrusts works or the provision and management of services to one or more companies. For the purposes of this Directive, contracting authorities (as defined in Article 6) are national, regional, or local authorities or bodies governed by public law; contracting entities (as defined in Article 7) are authorities or operators in the utilities sector exercising one of the relevant activities and awarding a concession for carrying them out.

The award of the works or services consists in the right to exploit them or that right together with payment. The right to exploit implies the transfer to the concession holder of the economic operating risk involving, in particular, the possibility that the concession holder will not recover the investments made.

The rules on the award of concessions aim to set out a clear legal approach. They also aim to guarantee effective access to the concessions market for all European businesses, including small- and medium-sized enterprises, and provide them with possibilities for investment in major public services in the future.

Public procurement law in Germany is very fragmented and is not collected in a homogeneous *body of* law due to the federal structure. The activity of the contracting authority is considered to be inherent to fiscal administration (*fiscaliche Verwaltung*) and is generally brought under private law. In fact, a segment of private law is identified as the private law of the administration, consisting of private law overlaid with legislation to safeguard the pursuit of interests with a public purpose. Ordinary (nonadministrative) courts have jurisdiction on public procurement matters involving the contracting authority,[24] although the rules of such judicial proceedings are similar to those of administrative judicial proceedings.

The transposition of the EU procurement directives was done centrally by the Act against Restrictions on Competition, *Kartellvergaberecht*, which places a dividing line between above- and below-threshold procurement. Above the threshold, the law against restrictions on competition applies, while below it the tax laws of the Federation and the Länder continue to apply.[25]

As for **France**, which has always shown its propensity for a centralized public administration, the system is simpler. French legislation on public contracts is currently summarized in the '*Code de la commande publique*,' which entered into force on 1 April 2019. This legal text regulates the entire 'life' of the contract, from its formation to its execution.[26]

The system in force in **Spain,**[27] in turn, presents some interesting peculiarities. Unlike other civil law jurisdictions in which a special procedural discipline has been set up for the relevant disputes that take place before the administrative courts, in the Iberian country a specific remedy, called *recurso especial en materia de contratation*, has also been provided for, with the corresponding creation of a system of "special administrative justice," alternative to the *Jurisdicciòn Contencioso-Administrativa*, but not replacing it. The structure under consideration entailed the creation of a complex of bodies, established by the central State, but decentralized in the various Autonomous Communities, called upon to settle, as

---

24 For higher-value contracts, there is also a preliminary procedure entrusted to a specialized administrative body.

25 M. Noccelli, *Diritto degli appalti pubblici in Germania*, in *Studio comparato della disciplina in tema di contenzioso sugli appalti pubblici nei principali paesi europei*, Ufficio Studi, massimario e formazione del Consiglio di Stato, 2020, 15 et seq.

26 G. TULUMELLO, *Diritto dei contratti pubblici in Francia*, in *Studio comparato della disciplina in tema di contenzioso sugli appalti pubblici nei principali paesi europei*, Ufficio Studi, massimario e formazione del Consiglio di Stato, 2020, 2 et seq.

27 For this part, see I. RAIOLA, *Diritto dei contratti pubblici in Spagna*, in *Studio comparato della disciplina in tema di contenzioso sugli appalti pubblici nei principali paesi europei*, Ufficio Studi, massimario e formazione del Consiglio di Stato, 2020, 27 et seq.

an alternative to the ordinary system of administrative litigation, disputes arising in the matter of the awarding of public contracts.

There are three elements of certain interest that characterize this system:

- the recognition, in Spanish law, of wide-ranging legal standing to bring an action, in the wake of the well-established Iberian administrative jurisprudence on *legitimatio ad causam*;
- free *recurso special en materia de contratación pública*;
- the automatic suspensive effect attached to the submission of the *recurso especial* in the event of a challenge to the award decision.[28]

In **Italy**, Legislative Decree No. 36 of 31 March 2023, approved the new Public Procurement Code (the "Code"), which entered into force on 1 April 2023, and will apply from 1 July 2023. The aim of this reform is to bring national legislation in line with European law and to rationalize, reorganize and simplify the discipline of public contracts for works, services and supplies (objectives which were in fact also proclaimed by the previous public procurement codes), as well as to adapt it to the principles expressed by case law over the years. The Code also has the specific function of completing one of the reforms required of Italy by the European Commission for the implementation of the National Recovery and Resilience Plan (PNRR). The Code is divided into five books: Principles, Digitalization, Planning and Design (I); Procurement (II); Procurement in Special Sectors (III); Public-Private Partnerships and Concessions (IV); Litigation, National Anti-Corruption Authority (ANAC), and Final and Transitional Provisions (V).

Anti-corruption Laws

The anti-corruption legislation applicable also to the public procurement sector is spread across international, supranational and state legal sources. The reason for this multilevel regulation lies in the transnational dimension of corrupt phenomena, affecting all countries and economies.[29] The framework constitutes an articulated system, which is mainly represented by the following sources:

---

28 The automatic suspensive effect is excluded in the case of procedures relating to contracts based on a framework agreement or to specific contracts under a dynamic purchasing system.

29 Council of the European Union, Proposal for a Directive of the European Parliament and of the Council on the fight against corruption, replacing Council Framework Decision 2003/568/JHA and the Convention on the fight against corruption involving officials of the European Communities or officials of Member States of the European Union, and amending Directive (EU) 2017/1371 of the European Parliament and of the Council-General Guideline, 24 May 2024, 10247/24, 2023/0135(COD), Recital (4), p. 5.

- *International sources:*
  - the United Nations Convention against Corruption (UNCAC) adopted by the UN General Assembly on 31 October 2003 by Resolution No. 58/4;
  - the OECD (Organization for Economic Cooperation and Development) Convention on Combating Bribery of Foreign Public Officials in International Business Transactions, signed in Paris on 17 December 1997;
  - the Council of Europe Criminal Law Convention on Corruption, opened for signature on 27 January 1999, in force internationally since 1 July 2002;
  - the Additional Protocol to the Council of Europe Criminal Law Convention on Corruption, concluded in Strasbourg on 15 May 2003;
  - the Council of Europe's Civil Law Convention on Corruption, opened for signature on 4 November 1999, in force internationally since 1 November 2003;
  - Council of Europe Resolution (97) 24 laying down the Twenty Guiding Principles against Corruption, adopted by the Committee of Ministers on 6 November 1997;
  - Council of Europe Recommendation (2000) 10 on codes of conduct for civil servants, adopted by the Committee of Ministers on 11 May 2000.
- *Supranational (EU) sources:*
  - the Council Decision of 25 September 2008 on the conclusion, on behalf of the European Community, of the United Nations Convention against Corruption (2008/801/EC);
  - Council Framework Decision 2003/568/JHA of 22 July 2003 on combating corruption in the private sector;
  - the Council Act of 26 May 1997 drawing up, on the basis of Article K.3(2)(c) of the Treaty on European Union, the Convention on the fight against corruption involving officials of the European Communities or officials of Member States of the European Union;
  - the Convention on the basis of Article K.3(2)(c) of the Treaty on European Union on the fight against corruption involving officials of the European Communities or officials of Member States of the European Union;
  - the Recommendation of the European Parliament of 17 February 2022 to the Council and the Vice-President of the Commission/High Representative of the Union for Foreign Affairs and Security Policy concerning corruption and human rights (2021/2066(INI)) (2022/C 342/23);
  - the Council Conclusions of 9 March 2023, which set out the EU's priorities in the fight against organized and serious crime for EMPACT 2022–2025;
  - the anti-corruption package adopted by the EU Commission on 3 May 2023, consisting of: a joint Communication from the Commission and the High Representative of the Union for Foreign Affairs and Security Policy; a proposal for a Directive on the fight against corruption; the strengthening of

Common Foreign and Security Policy (CFSP) sanctions with the proposal for a new framework for CFSP sanctions against corruption.

The multilevel system of anti-corruption acts and instruments is enacted by state regulations.

The United Nations Convention Against Corruption

The United Nations Convention Against Corruption (UNCAC) adopted by the UN General Assembly on 31 October 2003, by Resolution 58/4, known as the Merida Convention, is, to date, "*the only legally binding universal anti-corruption instrument*,"[30] and 191 states and supranational organizations (including the European Union) have so far adhered to it (the most recent ratification dates back to 7 August 2024). The UNCAC outlines a system of "minimum standards" and expressly provides that each State Party may adopt stricter or more stringent measures to prevent and combat corruption (Art. 65).

With regard to public procurement, Article 9 of the Convention on "Public procurement and management of public finances" sets forth the main principles for action:

1. *Each State Party shall, in accordance with the fundamental principles of its legal system, take the necessary steps to establish appropriate systems of procurement, based on transparency, competition and objective criteria in decision-making, that are effective, inter alia, in preventing corruption. ... Such systems, which may take into account appropriate threshold values in their application, shall address, inter alia:*
   a. *The public distribution of information relating to procurement procedures and contracts, including information on invitations to tender and relevant or pertinent information on the award of contracts, allowing potential tenderers sufficient time to prepare and submit their tenders;*
   b. *The establishment, in advance, of conditions for participation, including selection and award criteria and tendering rules, and their publication;*
   c. *The use of objective and predetermined criteria for public procurement decisions, in order to facilitate the subsequent verification of the correct application of the rules or procedures;*
   d. *An effective system of domestic review, including an effective system of appeal, to ensure legal recourse and remedies in the event that the rules or procedures established pursuant to this paragraph are not followed;*

30 *See* https://www.unodc.org/corruption/en/uncac/index.html.

e. *Where appropriate, measures to regulate matters regarding personnel responsible for procurement, such as declaration of interest in particular public procurements, screening procedures and training requirements.*

2. *Each State Party shall, in accordance with the fundamental principles of its legal system, take appropriate measures to promote transparency and accountability in the management of public finances. ...*
3. *Each State Party shall take such civil and administrative measures as may be necessary, in accordance with the fundamental principles of its domestic law, to preserve the integrity of accounting books, records, financial statements or other documents related to public expenditure and revenue and to prevent the falsification of such documents.*

EU Anti-corruption Legislation

The EU legal framework includes both the ratification of the UNCAC (by Council Decision of 25 September 2008) and the abovementioned dedicated regulations – supranational EU sources.

In public procurement, corruption is relevant both with regard to the public administration and the economic operator. With reference to the first aspect, it has been considered that the traceability and transparency of decision-making in procurement procedures is essential for ensuring sound processes, including efficiently fighting corruption and fraud.[31] For this purpose, contracting authorities should keep copies of concluded high-value contracts, in order to be able to provide access to those documents to interested parties, in accordance with applicable rules on access to documents. Furthermore, the essential elements and decisions of individual procurement procedures should be documented in a procurement report.

As to the second aspect, corruption is relevant at the admission and selection of the tenderer stage. In fact, contracting authorities shall exclude an economic operator from participation in a procurement procedure where they have established (or are otherwise aware) that that economic operator has been the subject of a conviction by final judgment for one of the following reasons:

- participation in a criminal organization, as defined in Article 2 of Council Framework Decision 2008/841/JHA (32);

31 Directive 2014/24/EU of the European Parliament and of the Council of 26 February 2014 on public procurement and repealing Directive 2004/18/EC Text with EEA relevance, *considerandum* 126.

- corruption, as defined in Article 3 of the Convention on the fight against corruption involving officials of the European Communities or officials of Member States of the European Union (33) and Article 2(1) of Council Framework Decision 2003/568/JHA (34) as well as corruption as defined in the national law of the contracting authority or the economic operator;
- fraud within the meaning of Article 1 of the Convention on the protection of the European Communities' financial interests.[32]

The public administration shall also exclude an economic operator where the person convicted by final judgment is a member of the administrative, management or supervisory body of that economic operator or has powers of representation, decision or control therein.[33]

Mandatory Contractual Provisions

As discussed briefly in the Introduction to this chapter, public procurement is currently being used as an instrument to promote certain social collective goals, such as the promotion of social rights and the protection of the environment. One way to achieve this goal is to put mandatory clauses in public procurement tender documents and contracts. In such context, the setup of public procurement and concessions has been particularly improved by operation of the EU Directives on public concessions 2014/23/EU, on public procurement 2014/24/EU and on utilities 2014/25/EU, which have introduced mandatory clauses with the aim of fostering the development of the underlying instances.

Social Clauses

Under the abovementioned EU Directives, public purchasers shall not be only interested in buying at the lowest price or best value for money, but also in ensuring that social benefits are achieved through procurement and that adverse social impacts are avoided or mitigated during the execution of the contract.

EU Directives commit Member States, *inter alia*, to strengthen the consideration of social and environmental aspects in public contracts. In particular, EU Directive 2014/24/EU sets forth that all Member States must take appropriate measures to ensure compliance with social criteria in the execution of public contracts (Art. 18), integrate these criteria into the principles for the award of

32 Directive 2014/24/EU of the European Parliament and of the Council of 26 February 2014 on public procurement and repealing Directive 2004/18/EC Text with EEA relevance, Art. 57.1.

33 *Ibidem*.

public contracts and concessions (Art. 30), and provide for the exclusion from participation of those economic operators convicted of serious violations of human rights and workers' health and safety standards (Art. 80).

In order to promote compliance by economic operators with social obligations, in many jurisdictions such compliance is retained as a criterion for the exclusion of economic operators. For instance, at the EU level, it is a general principle of public procurement that member States shall take appropriate measures to ensure that, in the performance of public contracts, economic operators abide by applicable obligations in the fields of environmental, social and labor law established by Union law, national law, collective agreements, or by international environmental, social and labor law provisions.[34]

Moreover, an economic operator shall be excluded from participation in a procurement procedure where the contracting authority is aware that the economic operator is in breach of its obligations on payment of taxes or social security contributions and where this has been established by a judicial or administrative decision having final and binding effect in accordance with the laws of the country in which it is established or with those of the Member State of the contracting authority.[35]

Environmental Clauses (Minimum Environmental Criteria – CAM)

Protection of the environment has been significantly pursued in the EU context through the implementation of the Green Public Procurement (GPP),[36] which regulation has been harmonized to overcome the individual approaches of the member States by means of the quoted directives 2014/23/EU, 2014/24/EU and 2014/25/EU, with the aim of fulfilling half percent of the GPP target. The framework provides the inclusion, in the design and tender documentation, of the technical specifications and contractual clauses in adherence with certain minimum environmental criteria (so-called CAM). As a result, the construction sector has been greatly affected by the EU requirements.

34 Directive 2014/24/EU of the European Parliament and of the Council of 26 February 2014 on public procurement and repealing Directive 2004/18/EC Text with EEA relevance, Art. 18.2.

35 Directive 2014/24/EU of the European Parliament and of the Council of 26 February 2014 on public procurement and repealing Directive 2004/18/EC Text with EEA relevance, Art. 57.2.

36 Fabio Cinti, *The Uptake of Green Public Procurement in the EU in Light of New Directives, Policies and Expectations: Towards a Modern Holistic Society, or Just an Integrated Economic Governance for the EU*? in Trento Student Law Review, Vol. 2:1 (2020); Stockholm Environmental Institute – Report February 2023, *Green Public Procurement: a key to decarbonizing construction and road transport in the EU.*

Certain minimum procurement values are set out as criteria for the application of the Directives and, since public procurement is considered strategic for environmental purposes, the evaluation of the bidders must be based on the specific principle of the most economically advantageous tender – "MEAT" principle. Due to this, the bidders' evaluation shall be oriented not only by the lowest cost, but also on environmental footprint, energy performance and life-cycle associated costs.

Environmental considerations come into the public procurement procedure at different stages:

- Selection of candidates: subjective qualification requirements to prove the candidate's technical capacity to perform the contract in a way that causes the least possible damage to the environment;
- Contract clauses: the base contract proposed by the administration shall include clauses that require the contractor to perform the contract in the best environmental manner;
- Technical specifications: through technical specifications the administration can *define the characteristics envisaged for the works, services or supplies.* These characteristics *may refer to the process or method of production or performance of the requested works, supplies or services, or to a specific process for another stage of their life cycle even if these factors are not part of their substantial content, provided that they are linked to the subject-matter of the contract and proportionate to its value and objectives*;
- Awarding criteria: tender documents can provide requirements aimed at selecting products/services with better environmental performance than that guaranteed by the technical specifications, to which a technical score is to be attributed for the purposes of awarding the contract according to the offer at best value for money. Award criteria are not compulsory, but the Public Contracts Code (Art. 57 c. 2) stipulates that they must be taken into account, even when defining the "*contract award criteria*."

Political Risk

An international construction project implies that foreign companies are committed to operate in another Country, where political risk invariably results much enlarged. Political risks are associated with "*government actions which deny or restrict the right of an investor/owner (i) to use or benefit from his/her assets; or (ii) which reduce the value of the firm. Political risks include war, revolutions, government seizure of property and actions to restrict the movement of profits or other revenues from within a country.*"[37]

Although the political risk is not always adequately considered in international construction projects – particularly in the bidding phase – unpredictability of a political system and its related developments may materially affect the successful completion of construction works.

Due to the scale of the projects concerned, political risk has a greater impact on Engineering Procurement Construction (EPC) turnkeys, Public-Private Partnerships (PPP), Build Operate and Transfer (BOT), and design, build and operate (DBO) contracts, where an efficient risk- weighting must include a careful analysis of the documentation, which covers a multitier set of provisions covering both legal and technical aspects.

From the most recent analyses on political risk in international construction projects[38] a risk categorization is set out on the basis of the likelihood of occurrence. The main items are identified with reference to: (i) expropriation; (ii) applications for local property; (iii) corruption and bribery; (iv) labor unrest; (v) repudiation; (iv) contract problems; (vii) permit delays; (ix) demonstrations; (x) local ownership issues; (xi) political violence; (xii) change of law.

Besides business-related considerations regarding the opportunity to pursue a project in a specific political context, from a legal standpoint it is necessary to pay attention to the legal framework and contractual clauses aimed at regulating the political risk.[39]

First of all, in case the contractor/developer is a foreign entity in respect of the Country where the project is to be located, and depending also on the nature of the project, it is useful to check if the State where the project is located is party to a bilateral (or multilateral) investment treaties (briefly referred to as "**BITs**") or to even more complex Free Trade Agreements ("**FTAs**") with the Country of the contractor. The main purpose of BITs is twofold: (i) they create obligations for the government of the economy in which an investment is made (the "host economy") to treat foreign investors according to particular standards which are set out in the treaty; and (ii) they almost always grant foreign investors the right

37 MIGA, *Glossary of Terms Used in the Political Risk Insurance Industry*, available on MIGA website.

38 *Risk identification techniques for international contracting projects by construction professionals using factor analysis* in Ain Shams Engineering Journal Volume 15, Issue 4, April 2024, 102655.

39 For detailed reference see for instance Rubins. N., Kinsella N. S., *International Investment, Political Risk and Dispute Resolution. A Practitioner's Guide*, Oceana Publications, Inc. 2005,

to submit claims for compensation against the host economy for violation of the standards of treatment contained in the treaty.[40]

At the contractual level, it is necessary to establish the proper set of project documents (such as concession agreements, take-or-pay agreements, construction agreements, etc.) in such a manner to establish adequate protection for the investors.[41] The most relevant clauses for this purpose are those relating to force majeure, change in law, stabilization clauses, duties of cooperation by local authorities/owner, damages, choice of law and dispute resolution.

Finally, a prospective contractor/investor may reduce (the consequences of) political risk by way of execution of suitable political risks insurance coverages on the private market or from state-sponsored national or international insurance agencies.[42]

40 See APEC Committee on Trade and Investment (CTI), *Handbook on Obligations In International Investment Treaties*, 11, 2020.

41 Rubins. N., Kinsella N. S., *International Investment, Political Risk and Dispute Resolution. A Practitioner's Guide*, cit., 36, 37.

42 Rubins. N., Kinsella N. S., *International Investment, Political Risk and Dispute Resolution. A Practitioner's Guide*, 69 and ff.

9

Project Financing

Lauren Davies and Nonkululeko Zondo

Introduction

This chapter will examine the role of a financing structure known as "project finance," in particular looking at the importance of a project's construction strategy and contractual structure in the successful closing of a project-financed transaction. At a high-level, project finance is a well-established method of raising long-term debt for major capital-intensive energy and infrastructure projects. It has been central to the development of sectors and assets, including upstream oil and gas, pipelines, refineries, petrochemical plants, power plants, solar, wind, airports, roads, bridges and tunnels, telecommunications, and a host of other assets and, looking ahead, project finance will continue to play a significant role in the transition away from fossil fuels to renewable and lower carbon energy sources. Without access to long-term financing from external debt providers or other forms of investment, it would not be possible for many project owners or shareholders (typically called the "sponsors") to develop energy and infrastructure projects because of the sheer magnitude of the capital costs required. Project finance is therefore a vital tool in the development of important projects across the world.

Due to the limited-recourse nature of project financed transactions (explained below), a significant amount of work is undertaken by a project's sponsors and its advisers during the predevelopment phase to ensure that all elements of a project are structured in a manner that will be accepted by the lender and investor market (an approach commonly referred to as "bankability"). Lenders are only willing to lend to a project where they are comfortable that the project will reach completion, become operational, meet the projected performance levels, generate revenue and repay their debt. In all cases, the borrower must satisfy lender requirements and meet conditions to draw down on the debt,

Major Issues in International Construction Law, First Edition. Mauro Rubino-Sammartano and Troy Harris.
Published 2026 by John Wiley & Sons, Inc.

as well as ensure that ongoing obligations are met throughout the loan tenor. Moreover, each individual lender may also have requirements that are specific to the use of the funds they are providing and the general implementation of the project. Because the costs associated with a project's construction approach represent a substantial portion of a project's completion costs, and due to the importance of the construction strategy in ensuring that a project achieves completion, lenders will (in particular) scrutinize the construction contracts and project completion strategy before agreeing to lend amounts to a project. It is therefore imperative that contractors understand the requirements of their project owners in circumstances in which the project is seeking financing and demonstrate willingness and flexibility to ensure that the project meets stringent lender requirements.

In this chapter, the term "borrower" should be read as the entity that acts as the project owner, project company and borrower of the loans. "Lenders" includes the financiers and finance parties (some of which entities may not actually lend to the project but may provide financial products such as insurance or guarantees or otherwise assume an agency role).

The Fundamentals of Project Finance

Key Concepts

In a project financing structure, a project's sponsors incorporate a project company as a special purpose vehicle (SPV). In most cases and subject to the corporate structure implemented, this SPV entity is also the borrower of the loans from the external lender group. These SPVs have no existing cashflows or operating history and they are established for the sole and exclusive purpose of acting as the borrower to construct, develop and operate the project. This entity also enters into all of the contracts and agreements required to develop and operate, as well as finance, a successful project, including the construction contract(s). Lenders take security over the assets of the SPV, as well as over the equity interests held by the sponsors in the SPV and security assignments over the SPV's rights under the key project agreements. By creating a separate legal entity, project assets are ring-fenced and there is (subject to any sponsor completion or other support available to the lenders) limited or no recourse to the assets of the sponsors and any other equity investors. Instead, the primary, and typically sole, source of income for the repayment of the debt provided by the lenders is the revenue generated by the project. The result is that the lenders will undertake substantial diligence and conduct a very detailed risk analysis to ensure that the borrower will achieve completion, become operational, and repay their debt in accordance with the agreed repayment schedule.

From the perspective of a project's sponsors and other equity investors, financial indebtedness incurred by a borrower will customarily be off-balance sheet for a project's equity holders, subject to the terms of any agreed sponsor guarantees or support, as well as the relevant accounting treatment, as the repayment of the loans will depend on the project's cashflow and on the collateral value of the project's assets. While lenders will scrutinize the creditworthiness of a project's sponsors and their ability to implement a project of the nature of the one under consideration, the customary project financing structure may be attractive to potential sponsors because they have greater flexibility to use their debt capacity for other investments. While a significant portion of a project's costs to complete may be provided by external debt providers, the sponsors will be required to fund a minimum amount of equity. The debt:equity ratio applicable to the project is negotiated between the sponsors and the lenders, and will depend on a number of factors including the debt capacity and economics of a particular project. The sponsor and any other equity investors also have their own requirements (from shareholders, their own debt providers, or otherwise) and desired investment returns that need to be factored into the structuring of a project and reflected in its contractual framework.

The following diagram reflects a customary legal and contractual structure for a project financed transaction. Significant work is undertaken by the project's sponsors and its advisers at the outset of a project financing to ensure that the project is structured in a bankable manner i.e., the structure must reflect a structure which project lenders are willing to lend against. Project parties develop the key terms and conditions for a project's agreements in a manner that reflects lender requirements, and agreements customarily considered at this early stage include (among others) the engineering, procurement, and construction (EPC) arrangements, licenses, any concession agreement, operation and maintenance agreements, project management contracts, supply arrangements, offtake agreements, and utilities arrangements. In developing such a commercial and contractual structure, the sponsors and their advisers consider potential project risks and seek to implement a structure that mitigates such risks, where possible, so that lenders are comfortable that the project will be completed on time, generate revenues and repay their debt. Section "Legal and Contractual Framework" provides a high-level overview of the key finance documents and their typical terms in project financed transactions, including how these documents address various project risks and lender requirements throughout different stages of the project lifecycle.

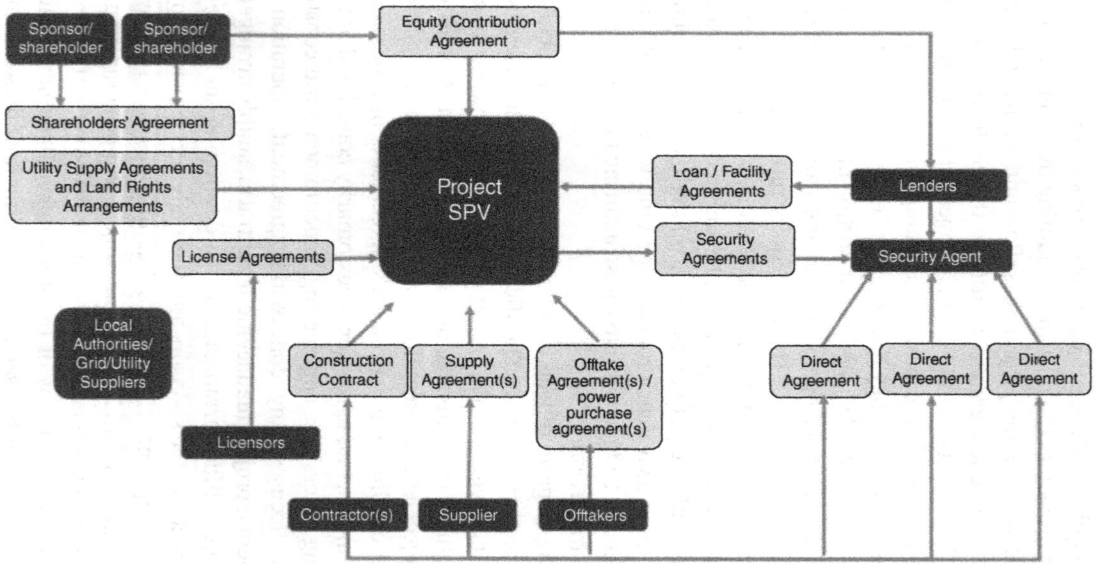
Sponsor/ shareholder
Sponsor/ shareholder
Equity Contribution Agreement
Shareholders' Agreement
Utility Supply Agreements and Land Rights Arrangements
License Agreements
Project SPV
Loan / Facility Agreements
Lenders
Security Agreements
Security Agent
Local Authorities/ Grid/Utility Suppliers
Licensors
Construction Contract
Supply Agreement(s)
Offtake Agreement(s) / power purchase agreement(s)
Direct Agreement
Direct Agreement
Direct Agreement
Contractor(s)
Supplier
Offtakers

Project Finance vs Corporate Finance

When determining the financing approach that a project takes, it is useful for a sponsor to analyze the distinctions between the project finance structure and a classic corporate financing model. One of the key differences between project financing and corporate financing lies in the recourse that lenders have to the assets of the borrower. As discussed above, in project financed transactions, an SPV is incorporated and the lenders lend against an identifiable pool of ring-fenced assets. By contrast, in corporate financing, lenders typically lend to an existing operating entity and have recourse to all the assets of that entity (to the extent such assets have not been charged to other lenders). As a result, in corporate financing, debt and borrowing costs are determined based on the assets, creditworthiness, and risk associated with the borrowing entity, whereas in project financing, the amount of debt that can be raised is customarily sized on projections of the project's ability to repay debt through the cashflows of that project alone. The project's sponsors negotiate and agree a financial model with the lenders (and their respective advisers) to determine a project's cashflow projections. The assumptions used in developing such financial model may include (among other things) projections as to the capacity, operational ability and production levels of the project, pricing of key inputs and outputs (including capital costs, other project costs, financing costs, costs of feedstock and raw materials, utilities, and product sale prices), market share, taxation rates, legal regime, and foreign exchange costs. Lenders only sign off on a financial model that demonstrates that a project will generate sufficient revenue, taking into account all of the assumptions, inputs and outputs, to repay their debt in accordance with an agreed repayment schedule.

Project finance is often more expensive than raising corporate funding due to its complexity and the extensive time required to develop and implement a bankable structure. The use of the project financing structure can result in additional management time and expertise to implement and administer throughout the project's life. Project financing is further complicated by the reliance on the project cashflows as the primary source of repayment, rather than the broader financial health of the sponsoring company. This requires detailed financial modeling and risk assessment to ensure the project's viability and the ability to meet debt obligations.[1] In connection with this, project finance necessitates a robust legal and contractual framework to manage and mitigate the various risks and ensure successful project execution. This includes negotiating complex agreements with multiple stakeholders, which can be time-consuming but nevertheless remains vital for ensuring that the project's contractual structure supports the project's ability to

---

1 Lauren Davies and Vanessa McLellan, "Don't look back in anger – operational considerations in a project financing," Project Finance International (June 30, 2021).

attract financing and external investment. Bankable construction contract(s) are key to achieving this robust contractual framework.

It is often necessary for a project to attract financing from a syndicate of lenders in order to raise the substantial volumes of debt required to implement a major project. The lender group may consist of any combination of development financial institutions, export credit agencies, financial institutions, bondholders and other forms of capital providers. Due to the broad range of potential lenders that participate in the project finance market, a project may be able to attract a higher level of debt, on longer tenors and offering more competitive (and therefore favorable) financial terms than would be the case with other forms of financing; however the size of the lender group may prolong negotiations and result in a longer financing and project implementation timetable. For certain projects, specific elements of the project may dictate the financing strategy and the identity of the debt providers. The financing strategy may also be closely aligned with the construction strategy, particularly where export credit agencies are involved to support their country's contractors and goods and services.[2] An understanding by a project's contractors of the nuances required by specific lenders, as well as flexibility in agreeing to their requirements, is key in ensuring the successful financing of a project.

Project Risks, Risk Allocation, and Mitigation

Summary

As described in Section "The Fundamentals of Project Finance", a project's lenders conduct detailed diligence and risk analysis before providing any funds towards the development of a project. It is prudent for a borrower and its sponsors (together with advisers) to ensure that such potential risks have been considered at the outset and appropriately mitigated, to the extent possible. Because the borrower is an SPV holding only the project assets, parties seek to ensure that key risks are allocated to other project parties where appropriate and therefore the project agreements form a critical element of a project finance structure. Risk allocation is primarily dealt with through due diligence (specifically, the appointment of independent consultants and delivery of due diligence reports), the contractual structure and terms and conditions (i.e., the project agreements and finance documents), insurance, political risk cover, and hedging. Ultimately (and ideally), risks should be borne by the party best able to foresee, manage and mitigate such risk and its associated consequences.

---

2 Lauren Davies and Vanessa McLellan, "The role of export credit agency financing," *Construction Law* (Aug/Sept 2021).

The borrower and its sponsors appoint their own advisers to assist in developing a project financing commercial and contractual structure that is closely aligned with the risk allocation and mitigation strategy. Advisers may include legal (both international and local counsel), financial, technical, insurance, market, and environmental and social consultants, among others. The project's lenders will also require their own independent consultants, and as a condition to advancing any funds to the project the lenders need to be satisfied as to the findings outlined in the due diligence reports prepared by those independent consultants as well as more generally being satisfied as to the risk profile of the project. The timing for the appointment of each consultant will depend on the needs and particulars of the project, but it is key for a borrower to ensure that some advisers (such as legal and financial) are appointed at an early stage in the process given the complexities involved in a project financing transaction. By way of example, it is customary in more complex projects for a lenders' legal adviser to review the major project agreements (as well as any term sheet of key provisions) before these are executed between the borrower and the counterparties to ensure that any critical concerns are identified and dealt with in an acceptable manner. The project finance structure tends to gravitate toward projects where risks can be identified, mitigated, and allocated in a manner that is acceptable to both lenders and the key project participants.

Potential project lenders will evaluate various types of risk, including (but not limited to) the following.

Completion Risk

For the reasons described above in this chapter, completion risk is a key focus of the lenders and construction-related risks are central to the success (or failure) of a project. If a project has not achieved completion by a certain point and met minimum performance levels, the lenders face significant risk that they will not be repaid. As such, lenders to project financed transactions do not customarily take completion risk. Given the scales and timelines typical of project-financed projects, project completion can be hampered by many factors including cost overruns, delay, inadequate performance and/or specification failure, and shortcomings or failures of relevant technology. For example, sponsors and lenders must consider how any cost overruns under the construction budget will be met. The size of the debt financing and the initial equity contributions from the sponsors are agreed at financial close based on detailed analysis and verification of the construction budget and other project costs. If there is a cost overrun during the construction phase (i.e., if amounts required to be paid to complete the project exceed the amount of available funding), the unbudgeted amounts need to be paid for (otherwise construction may need to be halted, risking completion). Lenders are unlikely to agree to fund these from additional debt commitments and sponsors may be reluctant or unable to offer (or will want to cap) any form of cost overrun or other additional equity contribution support.

Delays to the project schedule under the construction contract(s) impact a borrower's ability to achieve completion and therefore to generate revenue to service the debt. Lenders are particularly concerned about delays because these typically result in a shortfall in revenue towards the end of the project, when the lenders have lent significant volumes of debt and are at an increased risk of nonpayment.[3] Delays also result in additional interest during construction which again must be funded from somewhere. Lenders also assess the risk that a project will not meet required specifications and minimum performance levels.[4] As described in Section "The Fundamentals of Project Finance," the financial model agreed between the sponsors and the lenders will assume a certain level of revenue, based on agreed assumptions as to performance levels and product specifications (among other things). If the project does not produce the quantity or quality of the product projected to be sold in that financial model, the project is unlikely to generate the level of income required to service the debt.

The construction contract(s) form an important element of the completion risk analysis because it is usually the biggest component of aggregate project costs. The approach that a borrower takes in negotiating the terms of the construction contract(s) is closely aligned with the financing approach, including the level of completion support (if any) available from the sponsors and the borrower's completion strategy more generally. For example, if a sponsor is willing (and able) to guarantee to the lenders the payment of all financing and other amounts due from the borrower to the lenders in circumstances in which these remain outstanding and project completion does not occur by a longstop date, then the lenders may accept a weaker form of construction contract(s) with less risk transfer to the contractor(s). On the other hand, if the sponsors are not able to provide a debt guarantee and instead agree limited undertakings to provide additional equity to the borrower in the event of cost overrun, delay or any other agreed trigger (or no undertakings at all), then the lenders are likely to require a lump-sum, turnkey, single-point-of-responsibility EPC contract with a creditworthy and reputable contractor. Moreover, the sponsors may be called on to provide completion credit support where, in particular, there is unproven technology or a particularly complex or large-scale construction process. The concept of project "completion" is also heavily negotiated in the finance documents and includes, among other

---

3 Scott L. Hoffman, *The Law and Business of International Project Finance: A Resource for Governments, Sponsors, Lawyers, and Project Participants*, 4th ed. (Cambridge: Cambridge University Press, 2020), ISBN: 9781108427021, 6.

4 John D. Finnerty, "Chapter 8: Designing Security Arrangements" in *Project financing: Asset-based financial engineering* (New Jersey: John Wiley & Sons, 2013), 122.

things, conditions such as: mechanical completion and acceptance of the project under the construction contract, minimum performance, quality and reliability testing requirements, confirmation of receipt of all permits and authorizations, no funding shortfall, and compliance with environmental and social requirements.

Identifying and selecting an experienced, creditworthy, and reputable contractor, and having robust project budgets and schedules that factor in contingencies are key mitigants to completion risk. There are various other ways that project sponsors can mitigate the completion risks associated with a project, some of which will be explored in the context of how these risks are mitigated through the contractual framework. The risk mitigation strategies entail (but are not limited to): carrying out feasibility studies to assess the viability of a project, having a robust fixed price via a turnkey EPC package (EPC wrap) with a single point of responsibility for construction risk, using proven technology (at scale) and ensuring that there is independent verification of the construction budget and project design and specification as discussed above. Certain projects also necessitate the appointment of a project management consultant to coordinate between the borrower and the contractor(s) and manage any interface or other related risks.

Contractually, parties may manage the risk through the turnkey EPC contract, which is the most prevalent form of construction arrangement in project finance. In this model, the turnkey contractor assumes full responsibility for on-time completion and takes on all risks associated with the design, construction and installation of the project, provides a significant sense of security to lenders for a fixed price. It offers a single point of responsibility should the project fail to meet the guaranteed performance standards, thereby reducing the risk for all stakeholders.[5] The turnkey arrangements will typically include a regime of liquidated damages and indemnities for construction and any delays or defects in performance. The turnkey EPC contracts also include monitoring, reporting, and testing provisions throughout the project's construction phase to ensure that the project is hitting the required milestones within requisite timeframes.

Legal and Political Risks

Legal and political risks may hinder production (as well as the ability of a project to repay debt) in cross-border projects in the form of, for example, sanctions, changes in law and taxation, and the revocation or nonavailability of consents and permits. Political violence, the outbreak of war and outright hostile actions such as discriminatory actions towards foreign investors (in the form of, for example, the

5 Edward R. Yescombe, *Principles of Project Finance*, 2nd ed. (Amsterdam: Academic Press, 2014), 109. ISBN: 9780123910585, 189.

imposition of high taxes and royalties or the imposition of inconvertibility events) are also factors. Lenders require comfort that the project does not face expropriation and nationalization risks, particularly if the project involves natural resources which are vital for the economy of the host jurisdiction. All of these factors can hinder a project's economic viability and affect its ability to secure financing. In developing jurisdictions, lenders may seek host government support in the form of, among other structures, investment agreements, concession agreements, and favorable taxation and consent arrangements, as well as comfort that the local legal regime and political climate provide sufficient protection. In some projects, the provision of favorable tax, foreign exchange, and customs privileges (along with stabilization mechanisms) may be a key element of a project's economics and ability to attract financing. Lenders take some comfort from the availability of bilateral investment treaties, which may include, among other benefits, guarantees against discrimination and expropriation and protections in favor of fair and equitable treatment, the right to repatriate profits and the right for any disputes to be dealt with by way of international arbitration. In some project-financed transactions, the borrower will procure political risk insurance to provide comfort to the lenders of their debt repayment on the occurrence of a defined set of political risk events.

Supply and Offtake Risks

Project lenders require certainty that the feedstock, raw materials, fuel or other commodities required as input to the project will be available as soon as the project becomes operational, at the specifications, volumes, and prices required to meet the project's projected operational requirements.[6] Likewise, lenders require certainty as to the sale and offtake of relevant product from the project, again at prices, volumes and specifications that are consistent with the project's projected operational levels. Execution of long-term agreements for the supply and offtake provides comfort to the lenders that a project will have access to key supplies and the sales market at a preagreed price (or subject to a preagreed pricing formula) for the duration of the tenor of their debt, as well as meeting the quality and quantity required for the project at hand. Existence of a long-term supply agreement provides comfort that a borrower can better fulfil its obligations under any long-term offtake agreements, and the availability of long-term offtake arrangements mitigates risks such as demand risk, market or price risk, and/or currency risk.[7]

---

6 Stefano Gatti, *Project Finance in Theory and Practice: Designing, Structuring, and Financing Private and Public Projects*, 3rd ed. (Amsterdam: Academic Press, 2018), 414, 417. ISBN: 9780128114018.

7 Yescombe (2014), 222–223.

Moreover, the lenders require that the counterparties to these long-term arrangements are creditworthy and reputable, and capable of performing such roles. Lenders may require suppliers or offtakers to demonstrate their ability to meet payment commitments under the supply or offtake agreement (as applicable) by meeting certain minimum credit ratings and/or through the provision of letters of credit, guarantees or other assurances. In addition, it is customary for supply and offtake agreements to be concluded on a "deliver or pay" or "take or pay" (as applicable) basis, which obliges the supplier or the offtaker to pay the set contract price regardless of whether such seller has supplied, or offtaker has taken delivery of, the agreed quantity and quality of applicable product and materials.

Prior to financial close, the project sponsors develop and agree a financial model with the lender group and their advisers. This financial model determines projected cashflows and debt repayment terms based on (among other inputs) volumes and prices of the supply and offtake arrangements. It is therefore necessary for the terms of the supply and offtake agreements to align with these financial model inputs. While the exact requirements depend on the industry and project under consideration, customarily lenders require that long-term supply and offtake agreements are executed at financial close, reflecting terms acceptable to the lenders and entered into between the borrower and creditworthy, reputable and experienced counterparties.

Technology Risks

The response to technological risks in project financing turns on whether the project relies on proven or unproven technology. The choice of technology, particularly in respect of scalability and proven use cases, may result in lenders requiring more equity contributions from the sponsors, larger reserves for debt servicing or maintenance and specialized insurance products to protect against underperformance or technological failures. In addition to these risk mitigation strategies, where unproven technology is used, projects may face enhanced due diligence enquiries, higher financing costs and have a smaller pool of lenders with the risk appetite to fund projects that rely on unproven technology. Some sectors – such the green technology and alternative energy space – are more vulnerable to technology risk. Proven technology, therefore, is usually more desirable to a project financing. As a result, project financing in an established sector is generally more attractive to potential lenders and equity investors because well-tested and proven technology decreases project and operational (and therefore overall transactional) risks. Despite some technology having an established track record, risk mitigation strategies will still be required to respond to technological failures during the testing and operational phases of a project.

Operational Risks

Once a project is completed, its continued profitable operation and maintenance is critical for the borrower to meet its repayment obligations. To manage this risk, lenders would expect that (unless the sponsors possess such skills in-house) the borrower engage service providers responsible for managing and maintaining a project through an operation and maintenance (O&M) agreement. Similarly to the offtake and supply agreement expectations, lenders would expect that the borrower appoints a competent operator that will comply with the often extensive reporting and inspecting obligations. The O&M agreement's extensive reporting obligations provide critical early warning signals that empower the lenders to act pre-emptively to resolve any issues with the project.

O&M agreements significantly impact the project's financial feasibility by influencing operating costs, cashflow and exposure to performance risks and liabilities. Standards of care are critical in these agreements, setting benchmarks for the quality of services provided. These standards are often defined by industry or regulatory guidelines or mutual agreement between parties, ensuring the services meet the project's technical and commercial requirements. Pricing for services under these agreements can be structured in various ways, including fixed fees, cost reimbursements or performance-based incentives. Fixed fees typically cover core operations, while reimbursement models pass actual costs to the borrower. Incentive payments encourage providers to meet or exceed performance benchmarks, aligning their interests with those of the borrower.

Key Sponsor and Shareholder-level Agreements

Where a borrower is owned by two or more shareholders, the relationship between these shareholders is typically governed by a shareholders' agreement. This agreement outlines critical aspects such as the injection of share capital, detailing how much capital will be contributed, the timing and the form it will take. It also addresses the funding of the borrower, specifying the financial commitments required from each shareholder.[8] Additionally, the agreement sets forth voting requirements for certain decisions, shareholder dispute resolution procedures, and the management structure of the borrower. Shareholders agreements may also include provisions for the disposal of shares, pre-emption rights, drag-along and tag-along rights to protect existing shareholders' interests.

---

8 Carla Milani do Prado Rossi, "Step-In Rights Mechanisms in Project Finance Transactions and Lenders' Liabilities–The English and Brazilian Legal Approaches," *FGV Direito SP Research Paper Series* 160 (2018): 12–13.

In some project financings, as discussed in the risk mitigation section above, shareholders may also enter a shareholder or sponsor support agreement with the borrower and lenders. This agreement contains several commitments, including the obligation to provide funding, which may be through equity subscription or loans. These loans are typically unsecured and subordinated to the project's primary loans, ensuring that lenders have priority in repayment.[9] The agreement may also require shareholders to provide management and technical assistance to support the project's development. Restrictions on the ability of shareholders to dispose of their shares in the borrower are often included, safeguarding the project's stability and continuity. Additionally, the agreement may outline any completion guarantees or cost overrun guarantees, as well as security requirements that support the shareholders' commitments to provide equity at a later stage.

Construction Phase of a Project Financing

As described above, lenders to project financed transactions rely primarily on revenue generated by the project for the repayment of their debt. Project finance loans customarily comprise long availability periods, during which the borrower utilizes the loans to pay project costs (including construction-related costs), with the repayment period commencing once the project has been completed and becomes operational (subject to long-stop dates). Large-scale energy and infrastructure projects can take years to be developed, financed and completed, and the scale of these projects as well as the nuances of the project finance model results in competing and complicated workstreams throughout a project's lifetime. For example, the construction phase is typically divided into several distinct stages, each with its own set of activities and objectives. These stages work alongside the financing process, with the aim of ensuring that a project is financially viable, environmentally compliant and operationally successful. It is also necessary to align these processes so that financial close has been achieved and debt is available to make any necessary downpayments and milestone payments under the construction contracts as and when due. An example timeline demonstrating key construction and financing workstreams in a large-scale, multisource project financing is set out below. It should be noted that each project is unique and the construction and financing approaches must be aligned to the project under development.

9 Richard Tinsley, *"Equity" Advanced Project Financing Structuring Risk* (London: Euromoney Institutional Investor PLC, 2022).

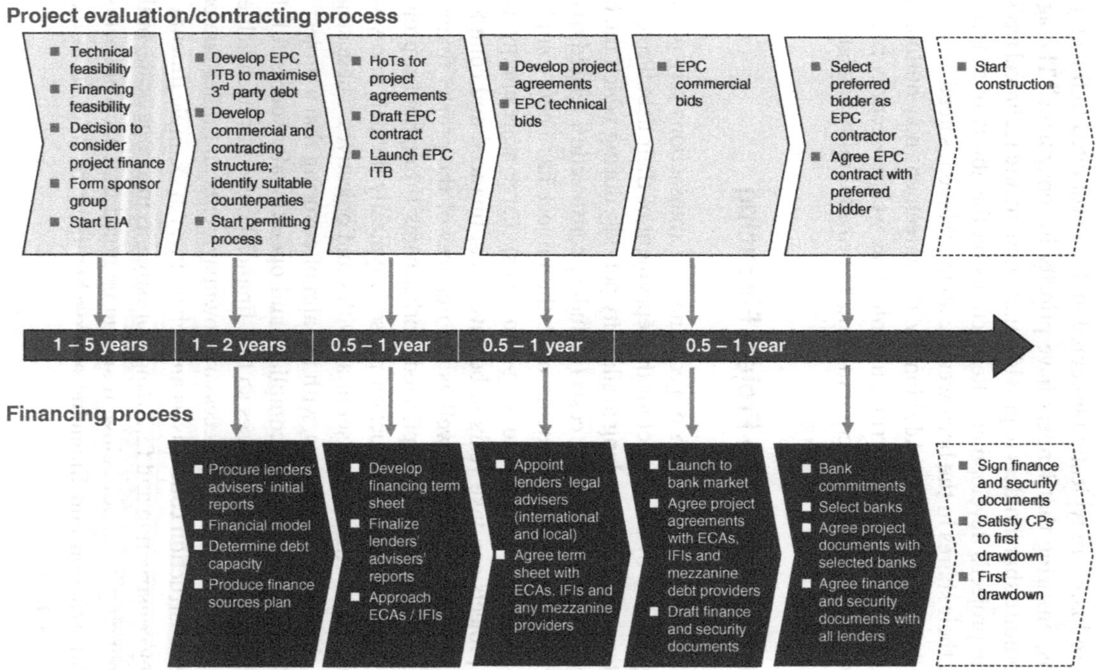
Project evaluation/contracting process
Technical feasibility
Financing feasibility
Decision to consider project finance
Form sponsor group
Start EIA
Develop EPC ITB to maximise 3rd party debt
Develop commercial and contracting structure; identify suitable counterparties
Start permitting process
HoTs for project agreements
Draft EPC contract
Launch EPC ITB
Develop project agreements
EPC technical bids
EPC commercial bids
Select preferred bidder as EPC contractor
Agree EPC contract with preferred bidder
Start construction
1 – 5 years
1 – 2 years
0.5 – 1 year
0.5 – 1 year
0.5 – 1 year
Financing process
Procure lenders' advisers' initial reports
Financial model
Determine debt capacity
Produce finance sources plan
Develop financing term sheet
Finalize lenders' advisers' reports
Approach ECAs / IFIs
Appoint lenders' legal advisers (international and local)
Agree term sheet with ECAs, IFIs and any mezzanine providers
Launch to bank market
Agree project agreements with ECAs, IFIs and mezzanine debt providers
Draft finance and security documents
Bank commitments
Select banks
Agree project documents with selected banks
Agree finance and security documents with all lenders
Sign finance and security documents
Satisfy CPs to first drawdown
First drawdown

Predevelopment Stage

The predevelopment stage is crucial for laying the groundwork for an energy or infrastructure project. It involves, as part of the risk mitigation strategy, conducting feasibility studies and environmental assessments. Feasibility studies are comprehensive analyses that evaluate the technical, economic and financial viability of the project. These studies help in identifying potential risks and determining whether the project can be successfully completed within the proposed budget and timeline. Environmental assessments, on the other hand, evaluate the potential environmental and social impacts of the project and ensure compliance with environmental and social regulations. This process of environmental assessments is essential for securing necessary permits and gaining public support, as well as attracting financing from the project finance lender market.[10]

Development Stage

Once the predevelopment stage is complete, the project moves into the development stage. This stage is characterized by the conclusion of key agreements (discussed below) and reaching financial close. Concluding these agreements involves negotiations and formalizing contracts with various stakeholders, including contractors, suppliers, sponsors and lenders. Financial close is the point at which all financial arrangements are finalized, and funds are secured to begin construction. This stage is critical as it marks the transition from planning to execution, ensuring that the project has the financial backing needed to proceed.

Construction Stage

During the construction stage, the focus shifts to monitoring and reporting. Effective monitoring ensures that the construction progresses according to the plan, stays within budget and meets quality standards. Regular reporting provides stakeholders, including the project lenders, with updates on the project's status, helping to identify and address any issues promptly. The construction stage's focus on monitoring and reporting is directly reflected in the key finance documents. Lenders typically include specific covenants and conditions in the loan agreement and other finance documents to ensure they receive regular updates on the project's progress. These contractual obligations, as discussed in Section "Legal and Contractual Framework," serve to formalize the monitoring process and provide lenders with the necessary oversight during the construction phase.

---

10 Alan Bond, Jenny Pope, Monica Fundingsland, Angus Morrison-Saunders, Francois Retief, and Morgan Hauptfleisch, "Explaining the political nature of environmental impact assessment (EIA): A neo-Gramscian perspective" *Journal of Cleaner Production* 244, (2020).

Operational Stage

The final stage is the operational stage, where the project becomes fully functional. Revenue generation begins as the project starts to deliver its intended services or products. This stage also involves regular maintenance to ensure the project's longevity and efficiency. Proper maintenance is crucial for sustaining revenue streams and maximizing the return on investment. These stages of a project financed transaction are interdependent, with each stage building on the previous one to ensure the project's overall success. By meticulously planning and executing each stage, developers can mitigate risks and achieve their financial and operational goals.

Legal and Contractual Framework

The sponsors develop a financing strategy during the initial phases of a project. Prior to commencing any formal financing approach, the sponsors must be comfortable that the results of any feasibility study and preliminary input from their financial and legal advisers support the project's bankability and ability to meet the sponsors' investment goals. The make-up of the proposed lender group, identity of individual lenders, and the nature of the debt provided will depend on the specifics of the project under consideration. For example, the geographical location, sector, anticipated timeline, financing costs, required debt tenors, key financial terms and creditworthiness and reputation of the sponsors and other key project counterparties, among other things, help form the financing strategy. This process takes place alongside the structuring and development of the project's contractual structure and the development (and negotiation) of the terms and conditions of the key project agreements. In light of the number of workstreams, tasks, and parties involved in a project-financed transaction, each process is managed in parallel to minimize the risk of delay to the timeline.

The project sponsors and lenders (and their advisers) draft and negotiate a large and complex suite of finance documents. In large-scale, multisourced finance transactions the parties may negotiate short- and/or long-form financing term sheets at the earlier stages of the financing process. These are customarily nonbinding but ensure that the sponsors and potential lenders are aligned as to the key financing concepts before project participants dedicate extensive time and resource to detailed due diligence, negotiations, and development of comprehensive long-form definitive documentation. Key provisions included in financing term sheets include, among other things, the proposed values of debt and equity to be contributed and timings and conditions for contribution, the projected project cost figure (including the amount anticipated for the construction budget and any contingency), the nature and level of sponsor support offered, key financing terms such as the repayment terms, final maturity date and interest payment provisions, the structure for the borrower's project accounts and payment waterfall, the dividend and distribution policy, ownership requirements, key borrower/borrower and sponsor undertakings, and the project's security package.

At the relevant point of time in the process, parties then develop and negotiate the full suite of definitive financing documents. These customarily include a common terms agreement which sets out the key terms and conditions of the financing common to each member of the lender group. The terms included in the common terms agreement reflect customary requirements in financing transactions and include a common set of definitions, conditions precedent to drawdown and ongoing utilizations of the debt, representations and warranties, debt repayment and prepayment mechanics and requirements, covenants, reporting and budget obligations, undertakings, and events of default. The financing documentation also includes any facility agreements, fee letters, the sponsor support agreement, accounts agreements, security documents, direct agreements, intercreditor agreement, and hedging agreements.

For the reasons described above in this chapter, project lenders require significant levels of control over their borrowers. Prior to execution of the finance documentation, lenders conduct detailed due diligence and risk analysis to ensure that they are comfortable with the risk profile of the project that they are financing. Once the financing documentation has been signed, project lenders require comfort that the borrower will not take any action, or inaction, that may alter such risk assessment or otherwise harm the ability of the project to service the debt. Similarly, lenders control the ability to make amendments to the project agreements (and the risk allocation structure reflected in those documents) that they have diligenced and approved prior to financial close. Lenders are alert to early warning signs that there are issues with the project, such as cost overruns and delays or technical issues, and require the ability to exercise their customary enforcement rights in the event of default pursuant to the terms of the financing documents. For that reason, project finance documentation contains, among other things, detailed undertakings and covenants on the borrower in respect of the development, construction, and operation of the project. In respect of the construction arrangements, provisions can include:

- Conditions to utilizations of the debt. There is a long list of conditions precedent prior to the initial utilization by the borrower of the loans. These conditions precedent include, among other things, execution of the key project agreements including the construction contract(s), delivery of the agreed construction budget and schedule, delivery of the agreed financial model, receipt of all necessary permits, consents and approvals, receipt of satisfactory due diligence reports, corporate authorizations, approvals and customary legal opinions for the borrower, sponsors and each key project participant (including the contractor), financial statements for such entities, evidence of placement of construction phase insurances, and evidence of perfection of security. The lenders will not lend the entire debt amounts in the initial disbursement, but lend in a number of separate disbursements throughout the availability period. Lenders also insist on the satisfaction of a number of conditions and documentary requirements before they lend further amounts. For example, a borrower may be required to submit copies of invoices received from the contractor for the provision of goods and services under the construction contracts, along with evidence of the performance of such services or delivery of the goods, copies of insurance policies applicable to such goods and services, confirmation from the borrower that there is (and after the utilization will be) no project completion funding shortfall, and certificates from the lenders' technical consultant confirming that the proceeds of the utilization will be used toward approved construction costs and in accordance with the agreed construction budget.

- Restrictions on amendments, waivers, and modifications to material project agreements, including the construction contracts. This includes restrictions on a borrower's ability to approve change orders without the lenders' prior written consent, subject to preagreed tolerances and carve-outs.
- Restrictions on the borrower from exercising certain rights under the construction contract(s) or agreeing certain requests, in each case without the prior written approval of the lenders, as well as obligations on the borrower to take certain actions or give notices on the instruction of the lenders. For example, the borrower is prohibited from agreeing to any assignment, transfer, or novation by the contractor of its rights and/or obligations under the construction contract(s) without lender preapproval. The borrower may also be restricted from agreeing to a contractor's request to subcontract certain services under the construction arrangements.
- Mandatory prepayment requirements that are triggered on the occurrence of certain events or payments. For example, proceeds received by the borrower from any performance liquidated damages to be used towards prepayment of the debt, subject to negotiated thresholds.
- Reporting undertakings throughout the construction phase. Lenders expect to receive copies of regular progress and other reports given by the contractor to the borrower, as well as regular reporting by the borrower setting out the costs incurred to date and indicators as to whether such costs are in line with the construction budget and whether the construction progress is in line with the construction schedule. The borrower also delivers copies of any

change orders to the lenders as well as information on the anticipated dates for commencement of any performance or other material tests under the construction contracts. Lenders also require delivery of financial statements for the borrower, the sponsors and the key project counterparties, including the contractor during the construction phase.

- Restrictions on the borrower approving any material tests under the construction contract(s) without the presence and approval of the lenders' technical consultant. This includes tests such as mechanical completion, provisional acceptance, and final acceptance (among other tests).
- Notification of any material claim or dispute (or potential claim or dispute) under the construction contract(s), and restrictions on settling or compromising any claims.
- Requirements on the borrower to exercise, maintain, protect and enforce its rights under the construction contract(s).
- Events of default linked to the solvency of the contractor, litigation or other proceedings against the contractor, and compliance by the contractor with the terms and conditions of the construction contract subject to cure rights. It is also an event of default if project completion (including satisfaction of all of the conditions thereto, such as provisional acceptance and satisfaction of performance and reliability tests) is not achieved by a preagreed long-stop date.

The lenders are entitled to exercise enforcement rights on the occurrence of an event of default, which rights include the ability to ensure their security package, call on any credit or other forms of support and generally take over the project. In addition to specific events of default, breach of any other provision of a finance document (such as an undertaking or a misrepresentation) may result in an event of default if not cured within a preagreed time period.

Due to the importance in a project financing of the assets of the borrower and the projected cashflows from the project, the representations and undertakings are extensive and seek to give the lenders substantial control over the borrower. The provisions are heavily negotiated, as the borrower seeks a sufficient level of flexibility to conduct its business and the ability to respond quickly to any issues which arise during the lifecycle of the project without the involvement of the lenders. That being said, the borrower also needs to consider the documents from a bankability perspective and maintain an appropriate balance between flexibility and ensuring it receives sufficient debt to develop the project.

The procedure for obtaining the prior approval of (or any waiver of breach of these provisions from) the lender group can take a significant period of time depending on the issue under consideration. A consent request may require the involvement of a number of project participants and their consultants and can be costly and time-consuming, particularly for the borrower. A material waiver of a breach of any borrower representations and undertakings, an amendment or a consent request is likely to require consequential amendments to the finance documents. Lenders may insist on delivery of additional documentation as a condition to agreeing such waiver, amendment, or consent and this can take some time to finalize, collate, and deliver. Lenders may also require internal sign-offs to approve material requests and ultimately may simply reject any request.

Project lenders customarily take security over all of the assets of the borrower. This security becomes enforceable on the occurrence and continuance of an event of default, unless remedied by the borrower. Relevant to the construction arrangements, the security package includes security over the facility or project owned by the borrower, fixed charges or mortgages over the project site, security over the borrower's bank accounts (and therefore ability to make payments) and security assignments of the borrower's rights and benefits under the project agreements (including the construction contracts), performance bonds and guarantees, licenses and consents, and insurances.[11] Lenders also take share security over the borrower and may ultimately "own" the borrower.

In addition to the specific provisions, lenders expect to see in construction contacts and other key project agreements described in Section "Project Risks, Risk Allocation, and Mitigation," lenders require that key project counterparties enter into direct agreements with the lenders or an agent appointed on their behalf. Direct agreements provide privity of contract between the lender group and those key counterparties, and seek to ensure that the project remains under continuous development or operational (as applicable) in the event of a default by the borrower or sponsors.[12] The ultimate aim is to ensure that the project can generate revenue and therefore service the debt, notwithstanding any default under the financing or project agreements. Direct agreements provide lenders with the right to "step in" and assume the borrower's position under those project agreements, in particular to permit the lenders (or their delegate or other appointee) to remedy any default by the borrower.[13] In addition to the step in right, direct agreements require that the counterparty acknowledges the lenders' security interest in the project agreement, suspends their right (for an agreed minimum time period, and subject to finding a replacement entity to fulfil the technical and financial obligations (as applicable) of the defaulting borrower) to terminate the project

---

11 Carmel de Nahlik and Chris Jackson, *Project Finance in Practice Case Studies* (London: Euromoney Institutional Investor PLC, 2021), 42.

12 Graham D. Vinter, *Project finance: a legal guide* (London: Sweet & Maxwell, 2006), 272 –273.

13 David Armstrong, and Gregory D. Howling, "Chapter 8 Lender Relationship With Project Counterparties" in *The Project Finance Law Review*, ed. by David F Asmus (London: Law Business Research, 2019), 71.

agreement in the event of default by the borrower, agrees to make any payment due to the borrower to a preidentified bank account of the borrower, and agrees not to amend or vary the terms of the project agreement without the prior written consent of the lenders.

Summary and Conclusion

Project finance plays a vital role in the development of major capital-intensive energy and infrastructure projects worldwide. Its importance in facilitating the transition to renewable and lower carbon energy sources cannot be overstated. While project finance offers significant advantages, including off-balance sheet treatment for sponsors and access to substantial long-term debt, it also presents unique challenges and complexities.

The success of a project finance transaction hinges on a carefully structured framework that allocates risks effectively among project participants. Construction contracts are particularly crucial in this structure, as they address one of the most significant risks: project completion. Contractors must understand and adapt to the stringent requirements of project finance lenders, demonstrating flexibility in contract negotiations to ensure "bankability."

For construction professionals, navigating the world of project finance requires a nuanced understanding of lender perspectives, risk allocation strategies and the intricate web of project agreements. By grasping these concepts, contractors can position themselves as valuable partners in project-financed ventures, contributing to the successful development of critical infrastructure and energy projects globally.

As the project finance landscape evolves, particularly in response to the growing focus on sustainable and renewable energy projects, construction professionals who can adeptly handle the unique demands of project-financed deals will find themselves at a distinct advantage. The ability to balance innovation with risk management, and to work within the constraints of project finance structures, will be key skills in shaping the future of global infrastructure development.

10

Subcontractor Claims

Virginie Colaiuta and Basil Thévignot

Introduction

Construction projects typically involve multiple stakeholders, including owners, main contractors and subcontractors, as well as interconnected contracts, including main contracts and subcontracts. While the terms of the main contract between the owner and the main contractor typically do not directly apply to the subcontract, disputes under one contract may trigger claims and conflicts under the other.[1]

A subcontract in a construction project is often shaped by the terms of the main contract. For example, subcontractors generally rely on designs, drawings, or specifications provided by the owner, either directly or via the main contractor. Delays in the performance of the subcontract can lead to contractor liabilities under the main contract. Similarly, if the owner delays site access or fails to timely provide the design, the subcontract's scope and timelines will likely be affected.

Nonetheless, the main contract and subcontract remain distinct agreements. Typically, no multiparty contractual relationship exists among the owner, contractor, and subcontractor.[2] The main contract and subcontract form a "contractual chain" where each party's rights, duties, and liabilities are confined to their direct contractual counterpart (subject to 'back-to-back' provisions, whereby

1 Alejandro López Ortiz and Patricia Ugalde-Revilla, "Chapter 4: Multiparty Construction Projects: An Arbitration to Bind Them All?", in Crina Baltag and Cosmin Vasile (eds), Construction Arbitration in Central and Eastern Europe: Contemporary Issues, pp. 47–62.

2 Stavros Brekoulakis and Ahmed El Far, "*Subcontracts and Multiparty Arbitration in Construction Disputes*," in Stavros Brekoulakis and David Brynmor Thomas, *Global Arbitration Review: The Guide to Construction Arbitration – Fifth Edition*, Law Business Research Ltd.

obligations under the subcontract may refer to obligations under the main contract). Consequently, the parties at the extremities of the chain – the owner and the subcontractor – lack a direct contractual relationship and corresponding liabilities or remedies.

The absence of a comprehensive contractual framework creates significant risks for subcontractors. When a subcontractor raises a claim against the main contractor, the contractual obligations between the main contractor and the owner can jeopardize the subcontractor's position. For instance, in cases of nonpayment by the main contractor, the subcontractor may file a claim. However, the main contractor may invoke a "pay-when-paid" or "pay-if-paid" clause in the subcontract, stipulating that payment to the subcontractor is contingent upon receiving payment from the owner.[3] Similarly, the main contractor might reject a subcontractor's request for additional compensation due to delays, arguing that the delays were caused by the owner. These provisions often place subcontractors at a disadvantage, limiting their ability to recover payments effectively.

These issues derive from the lack of direct liability between the owner and subcontractor, as such liability is uncommon unless expressly stipulated in the relevant agreement or mandated by applicable national laws.

In this regard, French Law No. 75-1334 of 31 December 1975 on subcontracting includes several provisions designed to address the issues outlined above.[4] This chapter will examine the provisions of Law No. 75-1334 and analyze how they assist subcontractors in overcoming common challenges when asserting claims against the main contractor and the owner.[5] In particular, this chapter will analyze how, in the context of international arbitration, the Law of 1975 may strengthen the subcontracts' claims against the main contractor and the owner.

3 Whether conditional payment provisions (including "pay-when-paid" or "pay-if-paid" clauses) are enforceable depends on the applicable law. While such clauses a usually prohibited in common law country, including in England and Wales pursuant to Section 113(1) of the Housing Grants, Construction and Regeneration Act 1996, these clauses are usually enforceable in civil law countries, subject to the application regulations ("Chapter IV: Clause-by-Clause Commentary," in Christopher R. Seppälä, The FIDIC Red Book Contract: An International Clause-by-Clause Commentary, pp. 561–562).

4 Loi n° 75-1334 du 31 décembre 1975 relative à la sous-traitance.

5 Law No. 75-1334 of December 31, 1975, on subcontracting distinguishes between contracts entered into by public companies (Section "The French Law No. 75-1334 of 31 December 1975 on Subcontracting") and those by private companies (Section "Subcontractors' Claims in Multiparty Construction Arbitration Under French Law"). The present analysis focuses exclusively on contracts concluded by private companies.

The French Law No. 75-1334 of 31 December 1975 on Subcontracting

Protections Granted to Subcontractors Under the Law of 1975

Article 1 of the Law of 1975 defines subcontracting as "the operation by which a contractor entrusts, through a subcontract and under their responsibility, another person called the subcontractor to perform all or part of the works or a part of a public contract concluded with the client."[6]

In this context, a subcontract involves a series of contracts sharing the same nature and purpose, at least in part. The second contract (the "subcontract") serves as a means to fulfill all or part of the first contract (the "main contract").[7]

Subcontract agreements must be clearly distinguished from other similar contracts, such as supply or sales agreements. The distinction is crucial for determining the legal obligations and protections afforded to the parties involved. To draw this distinction, the French Court of Cassation applies the criterion of "specificity." Under this approach, a supply or sales agreement may be classified as a subcontract if the goods or services provided are specifically tailored to meet the requirements of the requested works.

For instance, in a decision of 5 February 1985, the French Court of Cassation held that a company qualified as a subcontract, rather than as a mere supplier, as it was requested to supply "metal bars, delivered in the form of assembled 'cages,' with specific dimensions."[8] The Court of Cassation also held that "[the company] could not fulfil the order except after performing a specific task in accordance with particular instructions making it impossible to substitute to the ordered product another equivalent."[9]

Similarly, in a decision of 18 November 2009, the French Court of Cassation held that a company qualified as a subcontractor within the meaning of the Law of 1975 as "the multitude of data to be taken into account for preparing the calculation notes and manufacturing plans constituted a specific task requiring constant

---

6 *"Au sens de la présente loi, la sous-traitance est l'opération par laquelle un entrepreneur confie par un sous-traité, et sous sa responsabilité, à une autre personne appelée sous-traitant l'exécution de tout ou partie du contrat d'entreprise ou d'une partie du marché public conclu avec le maître de l'ouvrage."*

7 Jean-Pierre Karila and Laurent Karila, Fasc. 206: SOUS-TRAITANCE. – Loi n° 75-1334 du 31 décembre 1975. – Qualification et modalités de la sous-traitance. – Protection du sous-traitant – JurisClasseur Construction – Urbanisme.

8 Cass. Civ. 3, February 5, 1985, 83-16.675.

9 *Ibid.*

adaptation to the requirements of the manufacturing plans designed solely for the execution of the elements intended for the site."[10]

On the other hand, the sale of goods or the provision of services will not be considered subcontracted work under the Law of 1975 unless the goods or services being sold or provided are specifically tailored to meet the requirements of the requested works.

This is why in a decision of 17 March 2010, the French Court of Cassation upheld a decision of the Paris Court of Appeal that rejected the application of the Law of 1975 on the ground that "the evidence was not provided that [the work] included, in addition to supplying the carpentry, an installation service of these, nor that the carpentry manufactured was done so under particular specifications [...] nor, furthermore, was the evidence provided that these were not standard products."[11] In this case, the wood work services were provided on the basis of standard requirements and as such did not entail the qualification of subcontract.

In addition, as the series of contracts concluded within the context of construction projects may include several layers and sublayers, subcontractors, with respect to their own subcontractors, are considered the main contractor pursuant to Article 2 of the Law of 1975.

As explained in the Circular of 7 October 1976, regarding the reform of the subcontracting system in public contracts, the main objective of the 1975 law is twofold.[12] The first objective of the Law of 1975 is to ensure the protection of the subcontractor and guarantee their payment, notwithstanding the potential failure of the main contractor, by providing that the subcontractor can be paid directly by the owner, without relieving the main contractor of their contractual obligation to pay for the work.[13]

As for contracts entered by private owners, the relevant provisions of the Law of 1975 are included in Title III. First, Article 12 of the Law of 1975 provides that "the subcontractor has a direct action against the owner if the main contractor fails to pay, one month after being formally notified, the amounts due under the subcontract; a copy of this formal notice is sent to the owner." Paragraph 2 of Article 12 further specifies that "any waiver of direct action is deemed to be null and void." In addition, Article 13 of the Law of 1975 provides that "the direct action can only relate to the payment for the services provided under the subcontract and for which the owner is actually the beneficiary" and that "the obligations of the

10 Cass. Civ. 3, November 18, 2009, 08-19.355.

11 Cass. Civ. 3, March 17, 2010, 09-12.208.

12 Also see Jean-Pierre Karila and Laurent Karila, Fasc. 206: SOUS-TRAITANCE. – Loi n° 75-1334 du 31 décembre 1975. – Qualification et modalités de la sous-traitance. – Protection du sous-traitant – JurisClasseur Construction – Urbanisme.

13 Circular of October 7, 1976, regarding the reform of subcontracting in public contracts.

owner are limited to what is still owed to the main contractor as of the date of receipt of the copy of the formal notice."

The subcontractor's direct action against the owner provides significant protection to the subcontractor: if the subcontractor is not paid by the main contractor, they can, one month after the formal notice to the latter, claim to be paid directly by the owner. This mechanism ensures that the subcontractor is not solely dependent on the main contractor's financial stability or good faith.

However, there is an important limitation to the direct action: it can only be exercised in cases where the owner has accepted the subcontractor and agreed to their payment terms in accordance with Article 3 of the Law of 1975, as confirmed in an important ruling by the French Court of Cassation of 13 March 1981.[14]

Thus, even though the obligation of acceptance and approval lies with the main contractor, and not the subcontractor, it is very important for the latter to ensure that the formalities of the Law of 1975 have been respected in order to be able to benefit from the direct action against the owner.

Moreover, an important limitation to the effectiveness of the direct action is set by Article 13 of the Law of 1975, as it provides that the owner's obligations to the subcontractor are limited to what is still owed to the main contractor at the time of the payment request. In practice, the subcontractor may face a refusal to pay from the owner, who could argue that no amount is owed to the main contractor. As subcontractors often have limited visibility into the contractual relationship between the owner and the main contractor, this can create a blocking situation for them.

Second, Article 14 of the Law of 1975 provides that "under penalty of nullity of the subcontract, the payments of all amounts owed by the main contractor to the subcontractor, under the subcontract, are guaranteed by a personal and joint guarantee obtained by the contractor from a qualified institution."

In practice, this enables the unpaid subcontractor to request payment of the sums owed by the main contractor from the guarantor (usually a bank). Upon payment of the sums due, the guarantor will have the right to seek payment from the main contractor. Furthermore, the guarantor is subrogated to the rights and actions of the subcontractor, which also implies that the guarantor will have the ability to exercise the direct action against the owner to obtain payment. This was confirmed by the French Court of Cassation in a decision on 18 May 2017: "the bank, which had provided the guarantee as stipulated in Article 14 of the Law of 31 December 1975, was subrogated, after payment, to the rights and actions of

14 Cass. Mix, March 13, 1981, 80-12.125.

the subcontractor and entitled to exercise the direct action available to the latter against the client."[15]

However, the provision of a guarantee by the main contractor does not offer an absolute protection to the subcontractor. In particular, Article 2298 of the French Civil Code provides that "the guarantor may invoke against the creditor all exceptions, whether personal or inherent to the debt, that belong to the debtor." Thus, the guarantor may refuse payment to the subcontractor by invoking the same exceptions that the main contractor could have relied upon, for example, the guarantor may refuse payment by alleging poor performance of the subcontracted work.

In addition, Article 14 of the Law of 1975 provides that "the guarantee shall not be required if the contractor delegates the owner to the subcontractor under the terms of Article 1338 of the Civil Code, up to the amount of the services rendered by the subcontractor." "Delegation" refers to the situation where the owner undertakes to pay the subcontractor in place of the main contractor, with the latter remaining liable in the event the owner fails to make the payment.

Article 14 of the Law of 1975 stipulates that the penalty for noncompliance with the guarantee obligation is the nullity of the subcontract. This penalty, which benefits the subcontractor and can only be invoked by the subcontractor (the nullity could not be requested by the main contractor), is subject to strict interpretation by French case law. Thus, in a decision dated 11 October 1989, the French Court of Cassation confirmed that "the subcontractor could not waive the rights conferred upon him by Article 14 of the Law of 31 December 1975."[16]

However, this inability to waive the guarantee obligation only applies prior to the conclusion of the subcontract. Indeed, the subcontractor may, by voluntarily performing the subcontract despite the absence of a guarantee, waive the nullity of the contract, provided that the subcontractor is aware of the potential nullity. In this regard, the French Court of Cassation indicated in a decision dated 9 September 2020 that "the violation of the formalities in Article 14, paragraph 1, of the Law of 31 December 1975, which aim to protect the interests of the subcontractor, is sanctioned by relative nullity, which the subcontractor may waive through voluntary performance of their irregular commitment, with knowledge of the defect affecting it."[17]

In summary, Article 14 of the Law of 1975 provides additional protection to the subcontractor, but it also has weaknesses in that the guarantor may invoke against the subcontractor the exceptions arising from their contractual relationship with

15 Cass. Civ. 3, May 18, 2017, 16-10.719.

16 Cass. Civ. 3, October 11, 1989, 88-11.960.

17 Cass. com., September 9, 2020, 18-19.250.

the main contractor. Thus, the usefulness of Article 14 is particularly evident in situations of insolvency of the main contractor, as the guarantor remains liable for the amounts owed to the subcontractor.

The second objective of the Law of 1975 is to address certain unfair practices that could be imposed by main contractors on subcontractors.[18] To achieve this, the Law of 1975 grants the owner the right to scrutinize certain provisions contained in subcontract agreements. Thus, Article 3 of the Law of 1975 imposes on the main contractor who intends to subcontract part of the main contract to, "at the time of the conclusion and throughout the duration of the contract or project, have each subcontractor accepted and the payment terms of each subcontract approved by the owner."

Article 3 of the Law of 1975 further provides the sanctions for noncompliance with this provision: when the subcontractor has not been accepted or the payment terms have not been approved by the owner, the main contractor will remain liable to the subcontractor but cannot invoke the subcontract against the subcontractor. For example, the main contractor will not be able to invoke any contractual exceptions to oppose payment owed to the subcontractor.

Under Article 3 of the Law of 1975, the French Court of Cassation, in a decision dated 30 June 2015, ruled that a confidentiality clause included in a subcontract prohibiting the subcontractor from disclosing their involvement to the owner must be deemed null and void. This decision confirms the role of the owner in the relationship between the main contractor and the subcontractor: the owner is responsible for ensuring that the main contractor fulfills their obligations toward the subcontractor. In addition, this allows the owner to verify the quality of the services provided by the subcontractors selected by the main contractor.

Law of 1975, Public Order, and Mandatory Rules

The question arose whether the Law of 1975 qualifies as a matter of international public order ("*ordre public international*") and/or a mandatory rule ("*loi de police*") under French. Under French law, the notion of "international public order" refers to the most fundamental rules, values, or principles of the French legal system which must be preserved.[19] The French public order has both negative and positive effects.

On the one hand, the negative effect of international public order excludes the application of foreign law or the recognition of foreign judgments or awards

18 Circular of October 7, 1976, regarding the reform of subcontracting in public contracts.

19 Johanna Guillaumé Fasc. 40: ORDRE PUBLIC INTERNATIONAL. – Notion d'ordre public international, JurisClasseur Civil Code.

by French courts if they contradict the fundamental values of the French legal system. For instance, in a decision on 4 November 2010, the French Court of Cassation refused to recognize an American divorce judgment that granted full parental authority to the mother to the detriment of the father. Such a judgment was found to violate public order principles concerning the equality of parents in the exercise of parental authority and the respect for private and family life.[20]

On the other hand, the positive effect of international public order imposes the application of mandatory rules by the French Court, whether or not French law is applicable to the dispute and whether or not these rules were contractually excluded.[21] Mandatory rules are traditionally defined as the rules "necessary for the safeguarding of the political, social, or economic organization of the country."[22] In other words, mandatory rules are material rules of domestic law of such importance that they must be applied immediately without the need to resort to the conflict of laws rule.[23]

As for the Law of 1975, the French Court of Cassation has already had the opportunity to confirm its status as a mandatory rule, while introducing limitations to such a qualification. In a decision on 30 November 2007,[24] the Court of Cassation indicated that "concerning the construction of a building in France, the law of 31 December 1975 related to subcontracting, in its provisions protecting the subcontractor, is a mandatory rule within the meaning of the combined provisions of Article 3 of the Civil Code and Articles 3 and 7 of the Rome Convention of 19 June 1980 on the law applicable to contractual obligations." On this basis, the Court of Cassation considered that Articles 12 and 13 of the Law of 1975, relating to the subcontractor's direct action against the owner, were applicable even though the dispute between the parties was subject to German law.

However, this does not imply that the Law of 1975 applies whenever French courts are seized. Already, the decision of 30 November 2007 indicates that the solution applies to construction works in France. In addition, in a decision of 27 April 2011, the Court of Cassation clarified that the Law of 1975 should be considered a mandatory rule only in cases where there exists "a connection between the operation and France with regard to the objective of protecting

20 Cass. Civ. 1, November 4, 2010, 09-15.302.

21 Johanna Guillaumé Fasc. 40: ORDRE PUBLIC INTERNATIONAL. — Notion d'ordre public international, JurisClasseur Civil Code.

22 Ph. Francescakis, Y a-t-il du nouveau en matière d'ordre public?: Travaux comité fr. DIP 1966-1969, p. 149 ("*[les lois] dont l'observation est nécessaire pour la sauvegarde de l'organisation politique, sociale ou économique du pays*").

23 Cédric Latil, Fasc. 552-100: Lois de police, JurisClasseur Droit international.

24 Cass. Mixt. November 30, 2007, 06-14.006.

subcontractors."[25] In this case, the qualification of the Law of 1975 as a mandatory rule was not justified in relation to telecommunications equipment supplied by an Italian subcontractor to a French main contractor under a contract governed by Swiss law.

For the Law of 1975 to apply as a mandatory law, there must be consistent indications linking the operation to France, particularly the place of performance of the subcontract and/or the location of the subcontractor's registered office.[26] Accordingly, in a decision of 20 April 2017, the Court of Cassation refused to apply Articles 12 and 13 of the Law of 1975 concerning direct payment to the subcontractor by the owner to a situation where an Italian subcontractor had entered into a subcontract performed in Italy, even though the main contractor was a French company.[27]

As for decisions made by arbitral tribunals, few available arbitral awards, which predate the Court of Cassation's decision of 30 November 2007, have denied the qualification of the Law of 1975 as a mandatory rule.[28] For instance, in ICC Case No. 7528, a dispute arose in relation to a subcontract agreement governed by French law.[29] The arbitration was seated in Paris, France. The claimant argued that the Law of 1975 was a mandatory rule and should be applied by the arbitral tribunal notwithstanding any conflicting provisions included in the subcontract. In particular, the claimant argued that pursuant to Article 14 of the Law of 1975, the subcontract was null and void for lack of guarantee provided by the main contractor. In this case, the arbitral tribunal excluded the application of the Law of 1975:[30]

> It is difficult to accept that the parties meant to refer to French law while concluding a contract which would be void under that same law .... In order to apply Art. 14 as a loi de police, it should be proven first that such provision was intended to have such a character .... The tribunal therefore holds that the parties clearly intended to avoid the provisions of Law no. 75-1334 relating to the formation of the sub-contract, including Art. 14.

---

25 Cass. Com., April 27, 2011, 09-13.524.

26 Jean-Baptiste Racine, RDC Oct. 2011, p. 1294.

27 Cass. Com., April 20, 2017, 15-16.922.

28 Peter Rosher, "Forty years on: French law on sub-contracting," 32(1) Int'l Constr. L. Rev. 44 (2015).

29 "Sub-contractor v Contractor, Partial Award, ICC Case No. 7528, 1993," in Albert Jan van den Berg (ed), ICCA Yearbook Commercial Arbitration 1997 – Volume XXII, Yearbook Commercial Arbitration, Volume 22, pp. 125–131.

30 *Ibid.*

However, the arbitral tribunal explicitly noted that, in certain specific circumstances, including those related to subcontracts performed in France, the Law of 1975 could qualify as a mandatory rule:[31]

> No reported French decision has been put before the Tribunal where an international contract was avoided for non-compliance with Art. 14. This does not necessarily mean that it could not happen. But if Art. 15 were to be applied to this international contract, as claimant requests, such application would have to be warranted by a specific contact with France. In the present situation, the common nationality of the parties is important; but the place of performance of the contract is of equal significance. A number of commentators have convincingly pointed to the place of performance of the sub-contract as the relevant factor in order to determine the applicable law in contracts of soustraitance [subcontracts], absent any choice by the parties.

This appears consistent with the contemporaneous position of the French Court of Cassation, i.e., that the Law of 1975 should be applied by French courts and arbitral tribunals seated in France as a mandatory rule when the dispute has close connections to France (including in relation to construction works in France). For instance, this approach was followed by the Court of Appeal of Paris in a decision dated 19 November 2019.[32] In this case, the owner had entrusted the main contractor with the design, operation, and maintenance of a railway track in France. The main contractor had subcontracted part of the work to a subcontractor.

Following a dispute between the parties, the main contractor referred the matter to an arbitral tribunal acting as an *amiable compositeur* under the arbitration clause contained in the subcontract. In an award dated 12 September 2017, the arbitral tribunal ruled that the subcontract was null and void because the main contractor had not provided a guarantee, in violation of Article 14 of the Law of 1975. The main contractor applied for annulment of the award before the Court of Appeal of Paris, arguing that the arbitral tribunal had exceeded its role as an *amiable compositeur* by refusing to set aside the nullity provided for in Article 14 of the Law of 31 December 1975, based on equity. However, the Court of Appeal of Paris dismissed the application, holding that the arbitral tribunal was required to respect the mandatory rules of the Law of 1975, even when acting as an *amiable compositeur*:

> The arbitrator concluded that "under the risk of the award being annulled," his powers as an amiable compositeur did not allow him, in this case, to set aside the nullity provided for by the law of 31 December 1975 on the basis

31 *Ibid.*

32 Paris Court of Appeal, November 19, 2019, No. 17/20392.

> of equity. By acknowledging that he was still required to respect the provisions of public order, at the risk of the award being annulled, the arbitrator acting as an amiable compositeur fulfilled his mission.

This decision confirms that, with regard to construction projects located in France, arbitral tribunals are required to ensure the application of the provisions of the Law of 1975 as a mandatory rule.

Subcontractors' Claims in Multiparty Construction Arbitration Under French Law

Section A: Transmission and Extension of Arbitration Agreement Under French Law

Under French law, subcontractors are provided with legal remedies to handle disputes involving both the main contractor and the owner. However, in international arbitration, subcontractors often encounter a significant hurdle: when an arbitration clause is included in the subcontract, it typically does not extend to the owner, who is not a party to the subcontract.

Consequently, subcontractors may need to initiate two separate proceedings: arbitration against the main contractor based on the subcontract's arbitration clause, and judicial proceedings against the owner, potentially under the direct action provided by the Law of 1975.

This creates challenges for subcontractors: both the main contractor and the owner might raise defenses related to their main contract, in which the subcontractor is not involved. Thus, subcontractors may struggle to address these defenses effectively.

One approach to resolve this issue could be consolidating the proceedings, such as by involving the owner in the arbitration. However, this requires demonstrating that the owner is bound by the arbitration clause. Similarly, the main contractor may seek to involve the subcontractor in disputes with the owner, but he would need to prove that the subcontractor is bound by the arbitration clause contained in the main contract.

Under French law (and, in principle, under most legal systems), the principle of *pacta sunt servanda* dictates that the parties to a contract are bound by its terms,[33] and a contract typically does not bind third parties.[34] However, French case law has developed the theory of transmission and extension of the arbitration clause,

33 Article 1103 of the French Civil Code.

34 Article 1199 of the French Civil Code.

allowing third parties to be bound by the arbitration agreement even though they did not conclude the contract containing it.[35]

The transmission and extension of the arbitration clause must be distinguished: transmission involves successors, that is, legal subjects who receive pre-existing rights that are transferred to them, whereas extension involves persons who are connected to the original contract containing the arbitration clause, even if they are not the ones who signed the contract.[36]

The leading decision on this matter is that of the French Court of Cassation dated 27 March 2007.[37] In this case, the company ABS collaborated with another company from the same group, AME, for the manufacturing of electronic chips. The company AME entered a contract with Amkor for the supply of electronic components. This contract included an arbitration clause. Amkor itself ordered the electronic components from the company Anam. Anam delivered directly to AME, which, in turn, delivered to ABS.

A dispute arose between the parties, and ABS filed a claim against both Amkor and Anam before the French courts to seek damages. The court declared that it did not have jurisdiction due to the arbitration clause in the contract between AME and Amkor, to which ABS was not a party. ABS appealed to the Court of Cassation, which upheld the decision of the Court of Appeal on two grounds. First, the Court of Cassation stated that in a chain of contracts transferring ownership, the arbitration clause was transmitted from one contract to another:

> in a chain of contracts transferring ownership, the arbitration clause is automatically transferred as an accessory to the right of action ... the arbitration clause, contained in the contract between Amkor and AME, to which Anam had adhered, had binding force with respect to ABS, since this clause is transferred as an accessory to the right of action, which is itself an accessory to the substantive right.

Second, the Court of Cassation held that the arbitration clause contained in a contract extends to third parties directly involved in the execution of the contract:

> The effect of the international arbitration clause extends to the parties directly involved in the execution of the contract and the disputes that may arise from it; the Court of Appeal, having noted that the two French subsidiaries of Amkor were involved in the approval of the electronic

35 Jean-Baptiste Racine, Journal du droit international (Clunet) n° 1, Janvier-Février-Mars 2021, comm. 4.

36 Éric Loquin, Fasc. 1800-40: ARBITRAGE – Compromis et clause compromissoire, JurisClasseur Procédure civile.

37 Cass. Civ. 1, March 27, 2007, 04-20.842.

> microprocessors by AME, correctly concluded that these companies were entitled to invoke the arbitration clause stipulated in the contract between their parent company and AME, with respect to ABS and its subrogated insurer.

This solution is justified by the fact that the third party, by intervening in the execution of the contract, is presumed to be aware of its terms, and to some extent, to accept them.[38] It appears coherent that such behavior implies at least an implicit consent to be bound by the terms of the arbitration clause.

More recently, in a decision dated 23 June 2020, the Paris Court of Appeal confirmed this approach.[39] In this case, Kabab-Ji, a Lebanese company, granted a license to a Kuwaiti company, Al Homaizi, to operate a franchise in Kuwai pursuant to a franchise agreement. The franchise agreement was governed by English law and contained an ICC arbitration clause with a seat in Paris. The Al Homaizi Group went under a corporate restructuring leading to the establishment of a new holding company, Kout Food Group. Al Homaizi became a subsidiary of Kout Food Group.

A dispute subsequently arose under the franchise agreement. Kabab-Ji referred the dispute to arbitration seated in Paris under the ICC Rules. The arbitration was commenced against Kout Food Group alone and Al-Homaizi was not included as a respondent. Kout Food Group argued that it was not a party to the franchise agreement and was not bound by the arbitration clauses contained therein. The arbitral tribunal decided that that the arbitration clause was governed by French law, as the law of the seat, even though the franchise agreement were governed by English law, and that as a matter of French law, Kout Food Group was a party to the arbitration agreement. Accordingly, Kout Food Group commenced proceedings before the Paris Court of Appeal seeking to have the award annulled on the grounds that the arbitral tribunal had no jurisdiction over it.

On 23 June 2020, the Paris Court of Appeal dismissed the application to annul the award and confirmed that, as a matter of French law, the arbitration extended to Kout Food Group even though it was not a party to the franchise agreement containing the arbitration clause:

> The arbitration clause inserted in an international contract has its own validity and effectiveness, which necessitate extending its application to parties directly involved in the execution of the contract and in disputes

---

38 Jean-Baptiste Racine, Journal du droit international (Clunet) n° 1, Janvier-Février-Mars 2021, comm. 4.

39 Paris Court of Appeal, June 23, 2020, 17/22943.

> that may arise from it, provided it is established that their contractual situation and activities suggest they accepted the arbitration clause, of which they were aware, even though they were not signatories to the contract that included it .... Given the organizational structure of Kout Food Group, its involvement in the execution of the [franchise agreement] for several years, and in their termination and renegotiation, the arbitral tribunal rightly extended this clause to Kout Food Group, without needing to rule on the transmission of the arbitration clause from Al-Homaizi to Kout Food Group.

This confirms that the involvement of a third party in the performance of a contract may result in the extension of the arbitration clause to that party, provided that their actions or participation in the performance of the contract suggest an implicit acceptance of the terms, including the arbitration clause.

In addition, this decision confirms the requirement that the third party to whom the arbitration agreement is extended must not only have participated in the performance of the contract but also be aware of the existence of the arbitration agreement. This limitation is reasonable, given that arbitration is an exceptional method of dispute resolution that relies on consent – whether explicit or implicit – which cannot be assumed without knowledge of the arbitration clause.

Subcontractor's Claims Against the Owner in Arbitration Proceedings Under French Law

The theories of transmission and extension of arbitration clauses are particularly relevant in construction disputes involving subcontractors, main contractors, and owners. These theories may allow a subcontractor to rely on an arbitration clause included in the subcontract to bring claims against both the main contractor, for example, based on a breach of the main contract, and the owner, for example, under the provisions of the Law of 1975.

In a decision by the French Court of Cassation on 26 October 2011, the possibility of extending an arbitration clause in construction projects involving subcontractors was confirmed.[40] In this case, the French main contractor, CMN, had subcontracted part of the work involving the construction of two yachts to a Swedish subcontractor, FMS. The subcontract contained an arbitration clause. Subsequently, FMS subcontracted part of the work to another German subcontractor, PKC (as explained above, pursuant to Article 2 of the Law of 1975, the subcontractor is considered the main contractor with respect to their own

40 Cass. Civ. 1, October 26, 2011, 10-17.708.

subcontractors). After the first subcontract was terminated, PKC filed a claim against CMN and FMS before the French courts to seek compensation. CMN and FMS raised the issue of the court's lack of jurisdiction, arguing that the matter should be referred to an arbitral tribunal in accordance with the arbitration clause in their contract.

The court of appeal upheld the first-instance decision declaring that the French courts had jurisdiction, despite the arbitration clause in the first subcontract. As a result, CMN and FMS filed an appeal before the French Court of Cassation. On the one hand, the Court of Cassation rejected CMN's arguments, stating that CMN had, in the context of an arbitration initiated by FMS, expressly contested the jurisdiction of the arbitral tribunal with respect to PKC. By doing so, and in application of the principle of estoppel, CMN could no longer contest the jurisdiction of the French courts. On the other hand, the Court of Cassation accepted FMS's arguments, based on the theory of the extension of arbitration clauses, stating that the second-tier subcontractor was bound by the arbitration clause contained in the first subcontract:

> The effect of the international arbitration clause contained in the original contract extends to the subcontractor who was aware of it at the time of signing their contract and who is directly involved in the execution of the initial contract.

This solution appears appropriate, particularly because it allows for an efficient resolution of disputes in construction projects by avoiding the dispersion of disputes across multiple jurisdictions, while ensuring consent to arbitration through the dual condition that the third party was involved in the execution of the contract and was aware of the existence of the arbitration clause.

Authors have suggested that such a solution should also apply to the direct action brought by the subcontractor against the owner, allowing the subcontractor to rely on the arbitration clause contained in the main contract, even if the subcontractor's claims are not based on that contract.[41]

This approach was confirmed by the Paris Court of Appeal in a decision dated 14 October 2014.[42] In this case, the company Goro, as the owner, entrusted the construction of a pipeline and a pumping station to the company Lucas, as the main contractor. The company Lucas subcontracted part of the work to the company Cegelec. Both the main contract and the subcontract contained an arbitration clause. The subcontractor, facing payment issues from the main

41 Xavier Boucobza, Yves-Marie Serinet, RDC avr. 2012, p. 545.

42 Court of Appeal of Paris, October 14, 2014, 14/06434.

contractor, requested payment from the owner based on Article 12 of the 1975 Law. The owner refused to make the payment and the subcontractor brought the case before the Paris Commercial Court under the Law of 1975. The owner raised the issue of the lack of jurisdiction of the Paris Commercial Court, relying on the arbitration clauses contained in the subcontract and the main contract. By judgment dated 28 February 2014, the Paris Commercial Court declined jurisdiction.

The subcontractor appealed, arguing that the arbitration clause contained in the main contract was not applicable to them and that the owner could not invoke the arbitration clause in the subcontract, meaning the Paris Commercial Court had jurisdiction. In addition, the subcontractor argued that their claim against the owner was not based on the subcontract but on Article 12 of the 1975 Law, and therefore that the arbitration clauses contained in the main contract and subcontract were inapplicable. The Paris Court of Appeal rejected the subcontractor's arguments and confirmed that the French courts had no jurisdiction:

> Since [the subcontractor] had agreed under the subcontract to submit "all disputes arising from or related to this contract" to arbitration, the fact that [the subcontractor] intended to bring a direct action against the owner under Article 12 of Law No. 75-1334 of 31 December 1975, which is available to unpaid subcontractors, is irrelevant in this case, as the legal nature of this action does not have the effect of making the arbitration clause it invokes manifestly inapplicable, such an action being related to the subcontract.
>
> Equally irrelevant for the same reason is the argument that the clause contained in the subcontract does not include any provision in favor of the [owner], a third party to the contract, while the latter is being sought for payment of the services performed under the subcontract, by substitution of the defaulting main contractor.
>
> Therefore, the tribunal rightly declined jurisdiction, as the arbitration clause was not manifestly null or inapplicable, and referred the parties to seek further recourse.

This decision is particularly interesting in that it confirms the possibility for the subcontractor to initiate an action based on the Law of 1975 against the owner, relying on an arbitration clause contained in the subcontract. This decision also dispels any doubts regarding the feasibility of such an action by clearly stating that it is possible even though a claim based on the Law of 1975 is not contractual in nature and the owner is not a party to the subcontract.

However, it is important to consider the limitations of this decision, especially since, in this case, the Court of Appeal does not seem to have considered whether the owner was directly involved in the execution of the subcontract. In a decision dated 13 September 2017, the French Court of Cassation seemed to have, at least partially, validated this approach.[43] In this case, the owner had entrusted the main contractor with the construction of a railway line. The main contractor subcontracted part of the work to a subcontractor under a contract containing an arbitration clause. Following a dispute, the subcontractor initiated court proceedings against the main contractor and the owner for payment and the annulment of the subcontract. The subcontractor's claims were based on the subcontract and the Law of 1975.

The court decided that it had jurisdiction to hear the dispute, ruling that the subcontractor had no contractual relationship with the owner and that the action against the latter had no contractual basis. However, the Court of Cassation overturned the decision of the Court of Appeal, denying its jurisdiction to hear the dispute:

> By conducting a thorough examination of the contractual relationships between the parties to refuse the extension of the arbitration agreement to the relationship between the [subcontractor] and [the owner], the Court of Appeal [ruled] with reasons that were insufficient to establish the manifest inapplicability of the arbitration clause stipulated in the subcontract.

While this decision undeniably lacks clarity, it still emerges that the French Court of Cassation confirms, at least in principle, that the subcontractor would have been justified in bringing both his action against the main contractor based on the subcontract and his action against the owner based on the Law of 1975 before an arbitral tribunal under the arbitration clause contained in the subcontract.

Finally, it is worth mentioning a decision by the Court of Appeal of Lyon of 14 October 2021, which seems to contradict the previously cited decisions.[44] In this case, the company Aix Bords du Lac 4, as the owner, entrusted the construction of a real estate complex to the company Cabrero, as the main contractor. The company Battaglino intervened as the subcontractor. A dispute arose between the three parties regarding the payment of the subcontractor. The subcontractor initiated proceedings against the owner and the main contractor before the Tribunal of Lyon. The main contractor raised the lack of

---

43 Cass. Civ. 1, September 13, 2017, 16-22.326.

44 Court of Appeal of Lyon, October 14, 2021, 21/01459.

jurisdiction of the tribunal, relying on the arbitration clause contained in the subcontract.

The Court of Appeal of Lyon concluded that the arbitration clause was not applicable to the dispute because it was initiated against the main contractor and the owner, and the latter was not a party to the arbitration agreement. In particular, the Court of Appeal of Lyon concluded that

> the action brought jointly against the main contractor and the owner is inseparable.... The arbitrator's decision, whatever it may be, would have the effect of governing the relationship between the subcontractor and the main contractor while being unenforceable against the owner.... The application of the arbitration clause would therefore deprive the subcontractor of the effectiveness of his direct action against the owner [under the Law of 1975].

This solution, which has been subject to criticism,[45] seems to overlook the principles outlined above. Indeed, the Court of Appeal of Lyon appears to have considered that the arbitration initiated by the subcontractor would necessarily have been limited to a contractual action against the main contractor, excluding any action against the owner based on the Law of 1975. However, as discussed above, the subcontractor would actually have had the possibility to initiate arbitration against both the main contractor and the owner, even though the latter was not a party to the contract containing the arbitration clause. In addition, the Court of Appeal of Lyon strangely assumed that any decision rendered by an arbitral tribunal would be "*unenforceable against the owner*," which is legally incorrect: arbitral awards, like any judicial decisions, can have effects on third parties. In summary, this decision seems to deserve the criticism directed at it and does not call into question the principles outlined above.

45 Jérémy Jourdan-Marques, Chronique d'arbitrage: l'art de l'esquive en matière de corruption, *Dalloz Actualité*, 2021.

11

Testing and Certification Under Construction Contracts

Geoff White

Introduction

Testing and certification are fundamental to the administration of construction contracts, playing a crucial role in ensuring that projects meet technical, contractual, and legal requirements. The Engineer, as a key figure in construction projects, is typically responsible for overseeing testing procedures, certifying work completion, and verifying compliance with contractual obligations. These functions not only influence project quality and performance but also impact contractual entitlements such as payment, extensions of time, and defect rectification.

Testing and certification mechanisms vary across different standard form construction contracts, reflecting distinctions in risk allocation, procurement models, and project complexity. This chapter explores different testing regimes, the nature and effect of certificates and the Engineer's role in testing and certification. It also compares approaches across the following common contract forms: the New Zealand Standard 3910 Conditions of Contract for Building and Civil Engineering Construction (NZS 3910); the FIDIC Conditions of Contract for Construction (FIDIC); and the New Engineering Contract (NEC).

Testing

The Importance of Testing Under a Construction Contract

Put simply, the testing regime under a construction contract tells the Owner and Contractor when the works have been "completed." Construction contracts will typically have a detailed testing regime which is intended to provide certainty to both parties – for the Owner, that the works have been completed in accordance with the requirements of the contract and for the Contractor, that it

Major Issues in International Construction Law, First Edition. Mauro Rubino-Sammartano and Troy Harris.
Published 2026 by John Wiley & Sons, Inc.

has satisfied its obligations in this regard. Certainty as to testing requirements is critical, including in relation to the applicable performance standards, testing methodology and processes for conducting applicable tests and documenting compliance (typically through a certification process). Testing requirements under a construction contract can range from being complicated and prescriptive (for example, meeting guaranteed output requirements in a windfarm project or a processing plant constructed under an EPC arrangement) to the much more straightforward arrangements that would apply in a residential building contract, for example.

Testing and certification go hand in hand in a typical construction contract and both processes are usually administered by the Engineer. Further comments concerning the relationship between these two critical aspects of a construction contract are set out below.

Testing Procedures and Regimes – a Comparative Analysis

Testing regimes are project and contract specific and vary significantly in terms of complexity depending on the project characteristics and approach to procurement. The following section compares a typical testing regime under a traditional "build-only" construction contract with the more complex regimes that are a feature of "design and build" and EPC contracts.

Build-only Contracts

Traditional build-only contracts, exemplified by NZS 3910:2013 and the FIDIC Red Book (2017), incorporate relatively straightforward testing requirements. Under these forms, the Contractor's primary obligation is to demonstrate compliance with Owner-provided specifications.

The characteristics of a typical build-only contract testing regime include:

- A focus on workmanship requirements with relatively basic completion testing requirements;
- Specific testing and inspection requirements (if any) are agreed between the parties. Otherwise, the Engineer has significant control over directing testing;
- The Contractor has limited responsibility for testing; and
- Testing is primarily carried out during the construction phase of the project.

NZS 3910:2013 deals with testing primarily in Clause 6, requiring the Contractor to carry out tests specified in the contract documents. The Engineer maintains significant control, with authority to direct additional testing (Subclause 6.4.1). Completion of the applicable tests is an effective precondition for practical completion and/or final completion. The FIDIC Red Book similarly provides for "Tests on Completion" under Clause 9, but with limited scope reflecting the Contractor's restricted design responsibility.

The FIDIC Red Book (2017) contains detailed provisions dealing with the circumstances where tests are failed. The Contractor has obligations to remedy defects and undertake retesting pursuant to Subclauses 7.6 [Remedial Work] and 11.6 [Further Tests after Remedying Defects]. Under Subclause 7.5 [Defects and Rejection], the Engineer may reject failed elements of the Works and require retesting, with the ultimate sanction of complete rejection where the failure deprives the Owner of substantially the whole benefit of the Works. In such cases, the Owner may recover all sums paid plus financing charges and dismantling costs (see subclause 11.4 [Failure to Remedy Defects]).

By contrast, NZS 3910:2013 adopts a more generalized approach, primarily addressing test failures through its defective materials and work provisions under Clause 11. Under the 3910:2013 framework, the Engineer has broad powers to reject defective work and materials, require remediation, and withhold a Practical Completion Certificate for the Works. While less prescriptive than the FIDIC Red Book approach, these provisions should achieve similar practical outcomes if the Engineer properly utilizes its administrative powers.

Both forms ultimately allocate the cost of remediation and retesting to the Contractor, although FIDIC provides prescriptive cost recovery mechanisms. FIDIC's more structured approach provides greater certainty but less flexibility, while NZS 3910's broader framework allows more discretion in addressing test failures. The Engineer's role is ultimately central to both regimes.

Design and Build Contracts

Design and build contracts, such as the FIDIC Yellow Book (2017) and NZS 3916:2013, introduce additional testing complexity reflecting the Contractor's design responsibility. While maintaining similar basic provisions to build-only contracts, these forms incorporate expanded testing requirements to verify design compliance.

The key features of testing under design and build contracts, compared to a build only contract, include:

- Design verification;
- Increased documentation requirements;
- Extended testing duration; and
- Greater Contractor responsibility for test specifications.

NZS 3916:2013 builds upon the NZS 3910 framework but requires the Contractor to demonstrate that both design and construction meet specified requirements. The FIDIC Yellow Book introduces more comprehensive testing provisions, particularly regarding design verification.

The FIDIC Yellow Book largely mirrors the Red Book's test failure provisions, albeit reflecting the Contractor's design responsibility and recognizing that failures may stem from design deficiencies rather than just construction defects. The ultimate sanction of complete rejection remains available where failure deprives the Owner of substantially the whole benefit, allowing the recovery of all sums paid plus associated costs. Importantly, the Contractor's design responsibility means that remedial work under Subclause 7.6 [Remedial Works] may require design modifications in addition to physical rectification.

NZS 3916:2013, as a design and build form, maintains similar provisions to NZS 3910:2013 regarding defective work but includes design failures within its scope. The Engineer's powers to reject work and require remediation extend to design deficiencies, reflecting the Contractor's expanded obligations. The relationship between design defects and test failures is particularly significant, as unsuccessful testing may indicate underlying design inadequacies rather than just construction defects.

Both contracts reflect a fundamental principle that in design and build procurement, the Contractor assumes responsibility for both design and construction performance. This expanded liability means test failures may require more complex and costly remediation than under build-only contracts, potentially involving redesign as well as reconstruction. The corresponding cost implications are greater, although both forms maintain the basic position that the Contractor bears remediation costs.

EPC Contracts

Testing regimes under Engineering, Procurement, and Construction (EPC) contracts tend to be more comprehensive and complicated. This reflects the risk transfer under an EPC contract and the "turnkey" nature of the arrangement between the Owner and Contractor where the primary objective is to obtain an asset on time, on budget, and which performs to a required specification (for example, a power generating asset or a processing facility). A typical EPC testing regime will include the following categories of performance tests:

- Functional Tests: these tests evaluate specific components of the project or asset in isolation.
- Performance Guarantee Tests: these tests measure the plant's performance against contractual specifications, typically incorporating minimum and maximum performance thresholds. While meeting minimum guarantees is an absolute requirement, performance below specified levels usually triggers performance liquidated damages (these damages can be significant and are usually linked to guaranteed output levels). The timing of plant handover relative to guarantee testing has significant implications for Contractor liability.

- Posthandover Tests: these tests eliminate the contractor's exposure to delay liquidated damages, whereas testing prior to handover – particularly advantageous in project-financed ventures – maintains this liability until performance criteria are met or performance liquidated damages are paid.

By way of example, the FIDIC Silver Book (2017) EPC contract, establishes a hierarchical testing structure primarily through Subclauses 7.4 [Testing by the Contractor] and Clause 12 [Tests after Completion], incorporating the above categories of tests as follows:

- Factory acceptance tests;
- Construction and installation tests;
- Precommissioning of the plant;
- Commissioning of the plant; and
- Performance and reliability tests.

The scope for dispute concerning testing methodology, associated procedures, testing conditions, and tolerances (among other matters) is significant in the context of an EPC contract. The performance and delay liquidated damages provisions can be very onerous on the Contractor.

Most EPC contracts contain a structured approach to failed performance tests, including procedures for test repetition when performance guarantees are not met (see for example, Subclause 12.4 in the FIDIC Silver Book). Where minimum performance guarantees are achieved, an extended testing period typically allows the Contractor to optimize plant performance rather than immediately triggering termination rights or performance liquidated damages. This approach leverages the Contractor's technical expertise and knowledge of the project (the rationale being that the Contractor is best placed to fix its own mistakes) while protecting the Owner through delay liquidated damages during the extension period (which would typically apply).

Where repeated failures occur, under the FIDIC Silver Book, Owners can either order further repetition, reject the works if the failure substantially deprives them of the whole benefit, or issue a Taking-Over Certificate subject to performance damages (refer to Subclauses 11.4 [Failure to Remedy Defects] and 12.4 [Failure to Pass Tests after Completion]). This framework reflects the construction industry's preference for remediation over termination.

Courts have generally favored the application of performance damages over termination rights where such damages are expressly provided for in the contract, as demonstrated in *PBS Energo AS v. Bester Generacion UK Ltd* [2020].[1] This approach acknowledges both the commercial reality of complex engineering

1 *PBS Energo AS v. Bester Generacion UK Ltd* [2020] EWHC 223 (TCC).

projects and the parties' presumed intentions in including liquidated damages provisions, following the reasoning in *Photo Production Ltd v. Securicor Transport Ltd* [1980].[2] However, termination rights may still be exercised where failures fundamentally frustrate the contract's purpose or deprive the Owner of substantially the whole benefit of the works, as established in *Hong Kong Fir Shipping Co Ltd v. Kawasaki Kisen Kaisha Ltd* [1962].[3]

Comparative Analysis – Risk Allocation and Testing Scope

As set out above, testing obligations reflect the complexity and risk allocation under different contract forms. Understanding these distinctions is crucial for effective contract drafting and project execution. The progression from build-only through design and build to EPC contracts reflects both risk allocation and the increasing sophistication of project delivery methods. The key differences in these approaches are summarized in the table below.

Aspect of testing regime	Build-only contract	Design and Build contract	EPC contract
General risk allocation	The Owner retains design risk therefore, testing focuses on workmanship.	The Contractor assumes design risk therefore, testing encompasses design verification.	The Contractor accepts comprehensive risk therefore, testing must verify both design and performance.
The sophistication of the performance verification and testing regime increases across this spectrum of contracts	Basic functional testing.	Compliance with a specified performance criteria (level of detail subject to the complexity of the project).	Detailed performance criteria subject to performance guarantees secured by performance liquidated damages.
Documentation obligations increase with contract complexity	Basic test certificates and compliance records.	Design verification documentation (as built/operational documentation) and detailed test results.	Comprehensive procedures, comprehensive documentation demonstrating performance guarantees have been achieved and certified.

2 *Photo Production Ltd v. Securicor Transport Ltd* [1980] AC 827.

3 *Hong Kong Fir Shipping Co Ltd v. Kawasaki Kisen Kaisha Ltd* [1962] 2QB 26.

Aspect of testing regime	Build-only contract	Design and Build contract	EPC contract
The Engineer's involvement in testing varies significantly	Significant Engineer control.	Reduced Engineer involvement.	No Engineer role under the FIDIC Silver Book. The Owner or Owner's Representative fills primarily a monitoring role.
The testing duration reflects contract complexity	Testing integrated within construction program.	Extended testing phase for design verification.	Extended testing regime including reliability testing.
Defect identification and remedy provisions vary	The primary focus is on construction defects.	A broader focus on design (including associated documentation) and construction defects.	Comprehensive defect provisions relating to design and construction defects and mechanisms to deal with performance shortfalls.

Certification in a Construction Contract

What Are Certificates?

Hudson's Building and Engineering Contracts (2021) describes "certificates" in the following terms:

> the expression in a definite form of the exercise of opinion of the certifier in relation to some matter provided for by the terms of the contract.[4]

While this description is necessarily broad, it does hint at some of the common points of contention associated with certificates, and the role of the certifier in high value and complex construction projects. Certificates often reflect matters of opinion which may have significant consequences for the Owner (for example, progress of the works) and the Contractor (for example, an assessment of amounts payable).

4 Clay, N. and Dennys, N. (ed). *Hudson's Building and Engineering Contracts.* 14th ed. (UK: Sweet & Maxwell, 2021) at (4-001).

Common Types of Certificates

There are various common types of certificates under a construction contract which are a feature of different phases of the project lifecycle. These include:

- Interim and progress certificates – an assessment of progress based on work completed, these are often linked to payment under the contract.
- Certificates of practical or substantial completion – typically this category of certificate indicates provisional satisfaction but not the discharge of all obligations under a contract. In addition, certificates of practical or substantial completion usually signify:
  - the end of accrual of liquidated damages for delay;
  - the start of the maintenance/defect liability period; and
  - often, trigger the release of retention monies (or a portion of retention monies).
- Final certificates – indicating satisfaction with the works and the final balance of payment due to the Contractor following satisfaction of its obligations under the contract.
- Certificates indicating that the work should have been completed by a certain time – including, to take account of any extensions of time. This is required to enable the Owner to know when it is entitled to deduct liquidated damages under the contract.

The Effect of Certificates – Interim vs Final Certificates

As highlighted by the examples set out above, the purpose and effect of certificates can vary significantly. That being said, two broad categories of certification feature in nearly all construction contracts:

- administrative certificates – which apply on a largely interim basis; and
- final certificates – typically related to the completion of works and satisfaction of rights.

To summarize the difference between these categories of certificate in general terms, the first category of administrative or interim certificates form part of the mechanics of the contract, whereas the second category of final certificates define (or redefine) the rights of the parties.

Our current understanding of the distinction between these different categories of certificate in a construction contract context was well articulated in the case of *Beaufort Developments (NI) v. Gilbert Ash (NI) Ltd* (1999).[5] In particular,

5 *Beaufort Developments (NI) v. Gilbert Ash (NI) Ltd*, [1999] 1 AC 266 (HL).

Lord Hoffman identified that certain certificates will establish contractual rights on an interim or provisional basis and may simply be a matter of the contractual machinery rather than, absent clear terms to this effect, finally binding the parties (at 273 to 276).

The decision in *Beaufort* concerned the power of an arbitrator to revise certificates issued by the architect. In this case, it was decided that a broadly drafted arbitration clause, which expressly empowered the arbitrator to decide all disputes between the parties in connection with the contract, implicitly allowed the arbitrator to revise such certificates unless clear language was used to establish their finality. While a conclusion that certificates can be issued without excluding the rights of the parties appears to reflect a common-sense position – as does the distinction between the different categories of certificate described above – the *Beaufort* decision overturned a previous decision in *Northern Regional Health Authority v. Derek Crouch Construction Co. Limited* (1984),[6] which had determined that a court was not empowered to review such certificates.

The *Beaufort* position aligns with the earlier decision in *Modern Engineering v. Gilbert Ash Northern* (1974).[7] Both decisions establish the proposition that, absent express terms to the contrary, a certificate will create an interim position and not finally bind either the Contractor or Owner. This position (or at least the reasoning that underpins it) has generally been followed in most common law jurisdictions.

Statutory adjudication regimes applicable to construction contracts (increasingly common in a number of jurisdictions) and the proliferation of arbitration provisions in construction contracts has further undermined the notion that certificates are determinative of a final and binding position. In this regard:

- While the power of the adjudicator to review and revise certificates will depend on the statutory power granted to them and the terms of the contract in question, statutory adjudication regimes enable parties to quickly dispute a certificate in "real time" while the contract is underway. Parties can (and often will) use adjudication to challenge defective or controversial certificates contemporaneously, arguably reducing the importance of the certificate in this context.
- Furthermore, there is a current trend to include broadly drafted arbitration provisions in construction contracts, which will often provide an arbitrator with the express powers to review, revisit, or open up any previous decision

6 *Northern Regional Health Authority v. Derek Crouch Construction Co. Limited* [1984] QB 644; [1984] 2 WLR 676; [1984] 2 All ER 175, C.A.

7 *Modern Engineering v. Gilbert Ash Northern* [1974] A.C. 689, 1 B.L.R. 73.

or certificate made under the contract. As above, the presence of this type of arbitration provision supports the presumption that certificates will not be determined to be final and binding at the exclusion of the rights of the parties.

This can give rise to a question about which of these categories a particular certificate falls within. Whether a certificate is intended to be final and binding is primarily a question of the applicable law and the terms of a particular contract. That said, this can be an area where divergence occurs so careful consideration is necessary to ensure that the terms of the contract reflect the parties' intentions as to whether particular forms of certificate are to have a final and binding effect or shall be subject to review.

Certificates Under Related Contracts

Certificates issued under one contract may trigger rights under another related contract (such as sub contract arrangements) or ancillary arrangements (such as leases, funding arrangements, and purchase agreements). Linking certificates under one contract to the performance of obligations, or exercise of rights, under another contract, can be contentious.

Certain statutory regimes prohibit conditional payment mechanisms, which effectively link payment obligations under one contract to the certification and discharge of obligations under a related contract (also known as "pay when paid" provisions). For example, in the recent case of *Stevensons Structural Engineers v. McMillan Lockwood* (2024),[8] a retention release under a subcontract was linked to the issuance of a certificate of practical completion under the associated main contract. The Court held that this mechanism was prohibited by the New Zealand Construction Contracts Act 2002 on the basis that it was conditional on the performance of separate obligations under a different contract and the entire retention regime under the contract was determined to fall away. This New Zealand case adopted a similar position to the Australian courts in *Maxcon Constructions v. Vadasz* (2018)[9] where contractually interlinked retention release provisions contravened the prohibition on "pay when paid" provisions under Section 12 of the Building and Construction Industry Security of Payment Act 2009 (South Australia). A similar position would likely be held in the UK, where most forms of pay when paid clauses have been prohibited under the Housing

8 *Stevensons Structural Engineers 1978 Ltd (In Liq) v. McMillan & Lockwood (PN) Ltd.* [2024] NZHC 2415.

9 *Maxcon Constructions Pty Ltd v. Vadasz* [2018] HCA 5, (2018) 351 ALR 369.

Grants Construction and Regeneration Act 1996. By contrast, conditional payment provisions are valid in other jurisdictions, for example, the United Arab Emirates, and are routinely upheld by those courts.

The Relationship Between Testing and Certification

Overview

The quality of work is one of the most important factors in any construction contract. Failures in performance or the presence of defective work are some of the most common sources of delay, additional costs, and disputes in construction projects.

Testing demonstrates technical satisfaction of certain contractual obligations. As touched on above, certification serves a far broader contractual purpose, including in relation to payment entitlements and the completion of the works (including the passing of applicable tests).

That being said, both testing and certification are essential parts of most construction contracts and are inextricably linked. Different standard form contracts adopt different approaches to the interaction between testing and certification. Examples are set out below.

Certifying the Quality of the Works

While the individual or the entity that issues certificates under a construction contract (sometimes referred to as the "Certifier") does not have to be the Engineer, it is typical that the Engineer fills this role (or the equivalent individual or entity under forms that adopt different terminology). This is the approach that is adopted in a number of common standard form construction contracts.

The role of the Engineer has been discussed at length in other parts of this book. The Engineer's role in certifying the quality of the works is fundamental to ensuring the project meets its specified requirements and ultimately serves its intended purpose. This aspect of the Engineer's role is the focus of the following section. This role requires technical expertise, impartiality, and some discretion to effectively manage quality throughout project delivery. In this regard, some typical characteristics of the Engineer's role include:

- The responsibility for ongoing quality certification during the life of the project;
- Proactively inspecting and testing and reacting when issues arise;
- The Engineer's quality certifications often trigger contractual milestones and impact on entitlement to payment; and

- The Engineer must balance the interests of the Owner (who wants high-quality work) and the Contractor (who may face cost implications for additional testing or rework).

While contracts provide frameworks, the Engineer often needs to apply professional judgment and discretion in assessing quality, especially in complex or ambiguous situations. The Engineer's decisions on quality can be a source of disputes, highlighting the importance of clear performance requirements, testing processes, communication, and documentation.

Defining and certifying practical completion is a useful example of the discretion that the Engineer must apply when assessing the completion of the works. *Mears Ltd v. Costplan Services (South East) Ltd & Ors* (2019)[10] examined English and Commonwealth authorities in relation to practical completion. Among other matters, Coulson LJ noted that absent an express contractual definition, practical completion is initially a question for the certifier but that practical completion is easier to recognize than to define. A practical approach was adopted, describing the concept as a state of affairs in which the works have been completed free from patent defects, other than ones to be ignored as trifling. Whether or not a defect is trifling is a matter of fact and degree, to be measured against "the purpose of allowing the employer to take possession of the works and to use them as intended" (applying the rationale from *Mariner International Hotels Ltd v. Atlas Ltd* (2007)[11] and *Jarvis & Sons Limited v. Westminster Corporation & Another* (1969)[12] citing Salmon LJ at 1458 D – F).

This case emphasizes the importance of clarity concerning the assessment of practical completion and the Certifier's role in this regard. It also highlights the practical elements of the Engineer's role and the discretion that it must apply in its capacity as Certifier.

How Standard Form Construction Contracts Approach Testing and Certification

Overview

Testing and certification mechanisms are critical to construction contracts. Testing and certification directly impacts on fundamental aspects of a project such as the Contractor's entitlement to payment, the project's progress, and the

10 *Mears Ltd v. Costplan Services (South East) Ltd & Ors* [2019] EWCA Civ 502.

11 *Mariner International Hotels Ltd v. Atlas Ltd* [2007] 10 HKCFAR 1.

12 Jarvis & Sons Limited v. Westminster Corporation & Another [1969] 1 WLR 1448.

overall quality of the completed work. Decisions in relation to these matters can significantly impact on the success of a project and potentially lead to disputes if not handled properly.

Comparing NZS3910:2013 to the NEC4 Engineering and Construction Contract (ECC) and the FIDIC Red Book (2017) (Conditions of Contract for Construction), while there are universal concepts that apply in each, there is some divergence in how different standard form construction contracts deal with the relationship between testing and certification and the role of the certifier. Similarities include that the Engineer (or equivalent) fills the role of "certifier" under these forms and it has a broad authority to order tests, reject noncompliant work, payments, and certify the completion of the works. However, there are some notable differences, including:

- Differences in Terminology: FIDIC and NZS3910:2013 use the term "Engineer" who fills the dual role of a contract administrator and certifier, while NEC uses "Project Manager" and introduces a separate "Supervisor" role for some testing duties.
- Delineation of Duties: The typical approach under the FIDIC and NZS3910 standard form contracts is to combine the roles of project management and quality control within the Engineer's duties, leading to potential conflicts of interest. By contrast, NEC contracts delineate these responsibilities between the Project Manager and the Supervisor, promoting impartiality in testing and certification. While all three contract forms mandate rigorous testing and certification processes, FIDIC and NZS3910:2013 emphasize the Engineer's comprehensive oversight of all aspects of the contract, whereas NEC entrusts the Supervisor with quality assurance and the Project Manager with contractual compliance.
- The Importance of Impartiality: FIDIC and NZS3910:2013 explicitly require the Engineer to act impartially in certain matters, while NEC does not emphasize this as strongly.

These matters ultimately shape the role and responsibilities of the Engineer (as certifier) under these standard forms and the framework in which the Engineer (or equivalent) must discharge its duties. Further examples of the subtle but significant differences in the way that common standard form construction contracts deal with these obligations are set out below.

The FIDIC Approach

The Engineer

The FIDIC Red Book has a single Engineer that is appointed by the Owner. The Engineer does not have to be a natural person. FIDIC recognizes the commercial relationship that exists between the Engineer and the Owner. The Engineer is deemed to act for, and on behalf of, the Owner when carrying out his duties and exercising any authority under the contract (except where otherwise stated) (Clause 3). Except as expressly provided, the Engineer is not required to act "impartially," but they must act "neutrally" and come to a "fair determination" and "not be deemed to act for the Employer" in circumstances where they are empowered to assess and determine matters under the contract (Subclause 3.7). In discharging this obligation (and its other obligations), while the Engineer is not expressly required to act "impartially," it must act in an honest, even-handed manner, and exercise professional judgment.

Approach to Testing and Certification

The FIDIC suite of contracts establish a structured relationship between testing and certification that reflects the progressive nature of construction quality assurance. This framework operates across multiple stages of the project lifecycle, from interim works through to completion.

For example, under the FIDIC Red Book, testing serves as a prerequisite to certification at several key stages of the project. The Tests on Completion, detailed in Subclause 9.1 [Contractor's Obligations], must be successfully conducted before the Engineer can issue the Taking-Over Certificate under Subclause 10.1 [Taking Over of the Works and Sections]. These tests verify that the Works, or a Section, meet the contractual requirements and are ready for the Owner's use.

The certification process builds upon successful testing outcomes through a hierarchical structure governed by Clause 14 [Contract Price and Payment]:

- The Engineer reviews test results and supporting documentation (Subclause 7.4);
- Upon satisfaction, Interim Payment Certificates may be issued (Subclause 14.6);
- The Taking-Over Certificate follows successful completion of specified tests (Subclause 10.1); and
- The Performance Certificate represents the final validation after defects liability (Subclause 11.9).

Importantly, FIDIC maintains the distinction between testing for technical compliance (Clause 7 [Plant, Materials and Workmanship]) and certification for contractual purposes (Clauses 10 [Employer's Taking Over] and 14 [Contract Price and Payment]). This highlights a key distinction – while testing demonstrates technical achievement, certification serves broader contractual functions including payment entitlement and the satisfaction of completion obligations.

The Engineer's role bridges these functions by evaluating test results before issuing certificates under Subclause 3.7 [Agreement or Determination]. Following the principle established in *Beaufort Developments v. Gilbert-Ash*,[13] these certificates remain subject to later review through the dispute resolution mechanisms in Clause 21 unless the contract explicitly states otherwise.

NZS3910:2013

The Engineer

NZS3910:2013 requires that the singular Engineer be a natural person (Subclause 6.1.4 [Appointment of Engineer]). The Engineer is expressly required to make decisions, value work, issue certificates "fairly and impartially," and act "independently" of the parties to the construction contract (Subclause 6.2.1 (b) [Role of Engineer]), including when issuing practical completion certificates or defects liability certificates, and when verifying and certifying payments.

Approach to Testing and Certification

Under Subclause 6.4.1 [Inspection, recording, measuring, and testing], the Engineer has broad powers to require that any materials or work forming part of the defined contract works be inspected or tested (both at the place of manufacture or on site). The cost of any such tests or inspections is borne by the Principal unless either: (1) the test or inspection is required by the contract; or (2) the relevant Materials or work do not meet the requirements of the Contract (Subclause 6.4.2). The Engineer can also order tests and inspection of works or materials that have already been covered up, including requiring such works to be opened up for inspection (Subclauses 6.4.5 and 6.4.7).

Where any workmanship or materials are not in accordance with the contract, the Engineer may instruct the Contractor to remedy the work (Subclause 6.5.1 [Removal and making good]). The cost of any such remedial work is borne by the Contractor, whether that work is completed by the Contractor or, after sufficient written notice by the Engineer, another party (Subclause 6.5.1 to 6.5.4). The Engineer is also given discretionary power to accept work or Materials that do not meet the requirements of the Contract. If the Engineer exercises this discretion, they may decide the price of the noncompliant work or materials (Subclause 6.5.5).

In contrast to the Engineer's discretionary powers in relation to testing and remediation works, the Engineer's obligations in relation to certification of the practical completion certificate, final completion certificate and payment

13 *Beaufort Developments (NI) v. Gilbert Ash (NI) Ltd,* [1999] 1 AC 266, (HL).

schedules are mandatory. For example, the Engineer must issue the practical completion certificate if the contract works are completed to the standard required by Subclause 10.4.1 [Practical Completion Certificate], or provide written notice of any outstanding work (Subclause 10.4.3). The Engineer must issue the final completion certificate at the end of the defects notification period if all notified defects have been remedied (Subclause 11.3.1 [Final Completion Certificate]).

These certification obligations can have significant cost implications for the Contractor. Where the Engineer considers that works have not been completed in accordance with the contract, the Engineer can decline to pay any amounts submitted in payment claims for that work until the Engineer is satisfied (Subclause 12.2.1 [Payments]). If the Engineer declines issuance of the practical completion certificate under Subclause 10.4.3, the Contractor is not entitled to payment of retentions (Subclause 12.3.2(a)) and may become liable for liquidated damages (Subclause 10.5.1). The issuance of the final completion certificate by the Engineer is also a prerequisite to the release of the defects liability retention (Subclauses 12.3.2(a) and (b)).

The NEC Approach

The Project Manager and Supervisor

The NEC4 ECC contemplates two distinct contract administration roles: the Project Manager and the Supervisor. The separation of these roles is intentional, in an effort to promote a clear division of responsibilities between contract management and compliance monitoring.

The key differences between the roles are summarized as follows:

- Project Manager: this is primarily a commercial and contractual role but also includes some technical oversight. The Project Manager is responsible for managing contract execution, ensuring progress, and making decisions that affect time, cost, and scope. For example, the Project Manager has authority to change the Scope (Subclause 14.3) and certify completion and payment (Subclauses 30.2 and 51.1); and
- The Supervisor: this role is quality-focused and primarily involves ensuring compliance with the contract specifications (Scope). By way of example, the Supervisor is responsible for testing and inspections (Clause 40), notifying defects and issuing the defects certificate (Subclauses 43.2 and 44.3).

NEC4 ECC requires that the roles are performed applying a "spirit of mutual trust and cooperation" (Subclause 10.2). This arguably imposes a lesser obligation than under the other standard form contracts and a lesser standard than the common law but has given rise to arguments about whether such language creates an obligation of "good faith" (see the discussion in *Van Oord UK Ltd v. Dragados*

UK Ltd (2021))[14] and establishes a fiduciary relationship (the decisions in *Gold Group Properties Ltd v. BDW Trading Ltd* (2010)[15] and *Automasters Australia PTY Limited v. Bruness PTY Limited* (2002)[16] suggest that the answer is "no").

Theoretically, the separation of the Engineer's powers across distinct roles under the NEC4 ECC form appears intended to address any real or perceived tensions within the traditional "dual role" of the Engineer. That being said, it arguably creates an additional financial burden and administrative complexity as a result of multiple parties performing this split role (as is typical on larger projects).

The Approach to Testing and Certification

Under the NEC4 ECC, testing forms part of the quality assurance framework that leads to the completion certification. The contract requires the Contractor to provide works in accordance with the Scope (Subclause 20.1), with testing requirements typically detailed within the Scope document. The Project Manager and Supervisor have different but complementary responsibilities in this process. While the Project Manager has overall certification authority, the Supervisor's technical assessment and certification of tests is also a crucial aspect of this process. In summary, the testing regime operates as follows:

- Tests and inspections (Clause 41): the Supervisor has authority to carry out or witness these tests and review results, as required by the Scope. The Project Manager reviews the test results provided by the Supervisor and decides on contractual action if a test fails (e.g., corrective work) and considers test outcomes when assessing payment entitlements and completion;
- Failed Tests and Retesting (Subclause 41.4): the Supervisor notifies defects arising from failed tests. If a test fails, the Contractor must correct the issue and conduct a retest at their own cost unless the failure was due to an Owner's risk. The Supervisor witnesses retesting and confirms if compliance has been achieved. The Project Manager decides on compensation events if a failure was not due to the Contractor's fault and shall delay the certification of payments for work that has not passed required tests (Subclause 50.3);
- Certification (Completion and Defects):
  - The Project Manager is responsible for the issuing the following certificates:
    - Payment Certificates (Subclause 51.1): issued at regular intervals, factoring in test results and compliance;

14 *Van Oord UK Ltd v. Dragados UK Ltd*, [2021] CSIH 50.

15 *Gold Group Properties Ltd v. BDW Trading Ltd*, [2010] EWHC 1632.

16 Automasters Australia PTY Limited v. Bruness PTY Limited [2002] WASC 286.

- o Completion Certificate (Subclause 30.2)/Take Over Certificate (Subclause 35.3): issued when the Contractor has completed the works and corrected any major defects and/or the Owner has taken possession of the works.
- The Supervisor issues the Defects Certificate (Subclause 44.3) which is issued at the end of the defect correction period.

The relationship between testing and certification is particularly important for sectional completion, where the completion certificate for each section must be preceded by successful testing of that section as required by the Scope. Failed tests may lead to the notification of defects (Subclause 43.2), requiring correction before certification can occur and delaying payment until the issue is resolved.

Obligations and Liabilities of the Certifier

Overview

The "certifier" role under a construction contract is typically performed by the Engineer (or the equivalent entity). It is well established that the Engineer plays a "*dual role*" in administering a construction contract. In this regard, the Engineer's two roles are often described as being:

- The Agent of the Owner: in this role, the Engineer administers and supervises the works and monitors the performance of the contract on behalf of the Owner; and
- The Independent Certifier: in this role, the Engineer must independently issue certificates (including related to testing), resolve disputes and evaluate claims.

There is an obvious tension between these two roles, which provides an indication as to the potential liabilities that the Engineer can attract in performing its certifying and testing function.

The Engineer has a duty to comply with the terms of the applicable construction contract when discharging its certification duties. As outlined earlier in this chapter, different standard form construction contracts deal with these duties in different ways and impose different obligations on the Engineer. The Engineer must also comply with the terms of its contract of engagement with the Owner. In addition, the Engineer also has various duties to the parties to the construction contract which are established at law.

Common Law Position

The Engineer's Obligations

In addition to its primary contractual obligations, the Engineer's role is shaped by common law principles of impartiality and fairness – particularly when performing its role as certifier. The Engineer is expected to act independently, even when appointed by the Owner. Case law, such as *Sutcliffe v. Thackrah* (1974),[17] underscores the duty of care owed by the Engineer to all parties.

The limits of the Engineer's duty to act fairly can be illustrated by the following scenarios. First, despite the Contractor making a late claim for extension of time, the NSW Court of Appeal held that the Engineer (the term "Superintendent" is used in this case) had a duty to act "honestly and impartially" in exercising its power to grant an extension of time despite the claim being submitted late. The Court of Appeal described this duty in the following terms:

> "'This power is one capable of being exercised in the interests of both the owner and the builder, and in my opinion, the Superintendent is obliged to act honestly and impartially in deciding whether to exercise this power. Of course, if a timely claim has not been made, and the ground on which an extension is claimed is one which is difficult to decide because of the time that has elapsed since the time when the claim should have been made, that may be a ground on which the Superintendent can fairly refuse the extension, but there is no suggestion that that is the case here.[18]"

In a termination context, the court in *Brown v. Whangarei County Council* (1988)[19] determined that the Engineer's close involvement with preparing the Owner's termination documentation, failure to comply with procedural requirements in the contract, and failure to give the Contactor clear and unambiguous warning before recommending termination to the Owner, collectively constituted a breach of the Engineer's duty to act fairly in the circumstances.

Liability in Contract

While the Engineer has duties under a construction contract, typically, the Engineer is not a party to the contract and therefore has no direct contractual liability to either the Owner or the Contractor.

---

17 *Sutcliffe v. Thackrah*, [1974] AC 727, (HL).

18 *Peninsula Balmain Pty Ltd v. Abigroup Contractors Pty Ltd*, [2002] NSWCA 211.

19 *Brown & Doherty Ltd v. Whangarei County Council*, [1988] 1 NZLR 33, (HC).

The Owner and Engineer will have a separate engagement agreement. In broad terms, a negligent certifier who causes loss to the Owner, will be liable to the Owner for breach of the contract of engagement and for economic loss in tort (*Sutcliffe v. Thackrah* (1974);[20] *Hedley Byrne & Co Ltd v. Heller & Partners Ltd* (1963)).[21] Where the Owner wishes to advance a claim against the Engineer, it will most often be via that contract of engagement.

In relation to the Contractor, typically the Engineer will not owe a contractual duty and there will be no direct contractual remedies available. This is certainly the case in the standard forms examined throughout this chapter. In almost all cases, the Contractor will not be a party to the Engineer's separate engagement agreement. Therefore, the Engineer will usually not have any contractual liability to the Contractor.

Duty of Care in Tort (Negligence)

In circumstances where the Engineer has made a mistake, for example in certifying the amounts due to the Contractor or in awarding extensions of time, the Contractor generally will not have an entitlement to recover from the Engineer.

There may be tortious relief available for economic loss in circumstances where the Engineer has been negligent in the discharge of its duties (*e.g.*, issuing an incorrect certificate). That said, this is not a settled matter and the courts have historically been unwilling to recognize a duty of care between an Engineer and Contractor.

Pacific Associates v. Baxter (1990)[22] is the leading decision concerning whether an Engineer owes a duty of care to a Contractor. In the context of a strike out application, the English Court of Appeal determined that an Engineer did not owe a duty of care to a Contractor (p 179). Among other matters, the Court concluded that the Engineer did not have a direct responsibility in relation to the contract because the Contractor had recourse against the Engineer and, in that case, the Contractor could challenge the Engineer's certification in arbitration.

Whether *Pacific Associates*[23] establishes a complete bar on claims in negligence by a Contractor against an Engineer remains an open question – a more nuanced view is that without particular circumstances creating an assumption of responsibility and a direct duty between the Engineer and Contactor, no such duty shall exist as an ordinary matter of course.

20 *Sutcliffe v. Thackrah*, [1974] AC 727, (HL).

21 *Hedley Byrne & Co Ltd v. Heller & Partners Ltd*, [1964] AC 465, (HL).

22 *Pacific Associates Inc v. Baxter*, [1990] 1 QB 993, (CA).

23 Ibid.

The types of circumstances that might establish a duty, beyond the *Pacific Associates*[24] position, include:

- Where the Engineer provides assurances to the Contactor that it would be paid monies under the contract; or
- If the Engineer has assumed responsibility for certain specifications or technical aspects of a tender, upon which a Contractor relies in pricing its works (*Day v. Ost* (1973);[25] *RM Turton v. Kerslake & Partners* (2000)).[26]

Contractor's Remedies Against the Owner

As a starting point, the Owner is generally liable for the Engineer's performance of its obligations and any omissions and failures in this regard (both as a matter of contract and at common law).

At a basic level, if the Engineer issues an incorrect certificate or makes a mistaken determination with which the Contractor disagrees, it will have recourse through the contract. In particular, the Contractor can dispute the Engineer's decision or certificate under the applicable dispute resolution mechanism of the construction contract.

As a matter of common law, if the Owner has failed to ensure that the Engineer acts fairly and impartially when performing its duty as certifier, the Owner may be liable to a Contractor who suffers loss as a result. In *Perini Corporation v. The Commonwealth of Australia* (1969),[27] it was held that there was an implied term in the contract that the Principal/Owner would ensure that the Engineer will act in a fair, unbiased and competent manner when performing its duty as certifier (see also *Fredrick Leyland & Co v. Compania Panamena Europea Navegacio Lmitada* (1943)).[28]

Common standard form construction contracts typically go further than the common law position. They impose express obligations on the Owner in relation to the performance of the Engineer and provide remedies to the Contractor in this regard.

---

24 Ibid.

25 *Day v. Ost*, [1973] 2 NZLR 385, (SC).

26 *RM Turton & Co Ltd (in Liq) v. Kerslake*, [2000] 3 NZLR 406, (CA).

27 *Perini Corp v. Commonwealth (Redfern Mail Exchange Case)*, [1969] 2 NSWR 530.

28 Frederick Leyland & Co., Ltd. (J. Russell & Co.) v. Compania Panamena Europea Navegacion, Limitada, (1943) Ll L Rep 113 (CA).

Bibliography

Automasters Australia PTY Limited v. Bruness PTY Limited [2002] WASC 286.

Beaufort Developments (NI) v. Gilbert Ash (NI) Ltd, [1999] 1 AC 266, (HL).

Brown & Doherty Ltd v. Whangarei County Council, [1988] 1 NZLR 33, (HC).

Building and Construction Industry Security of Payment Act 2009. Available at Building and Construction Industry Security of Payment Act 2009 | South Australian Legislation (Accessed 24 January 2025).

Canterbury Pipe Lines Ltd v. Christchurch Drainage Board, [1979] 2 NZLR 347, (CA).

Clay, N. and Dennys, N. (ed). *Hudson's Building and Engineering Contracts.* 14th ed. (UK: Sweet & Maxwell, 2021).

Construction Contracts Act 2002. Available at Construction Contracts Act 2002 No 46 (as at 05 October 2023), Public Act Contents – New Zealand Legislation (Accessed 24 January 2025).

Day v. Ost, [1973] 2 NZLR 385, (SC).

Frederick Leyland & Co., Ltd. (J. Russell & Co.) v. Compania Panamena Europea Navegacion, Limitada, (1943) Ll L Rep 113 (CA).

Gold Group Properties Ltd v. BDW Trading Ltd, [2010] EWHC 1632.

Hedley Byrne & Co Ltd v. Heller & Partners Ltd, [1964] AC 465, (HL).

International Federation of Consulting Engineers ("FIDIC") (2017) *Second Edition Conditions of Contract for Plant & Design ("FIDIC Red Book")*. Switzerland: FIDIC.

International Federation of Consulting Engineers ("FIDIC") (2017) *Second Edition Conditions of Contract for Plant & Design ("FIDIC Silver Book")*. Switzerland: FIDIC.

International Federation of Consulting Engineers ("FIDIC") (2017) *Second Edition Conditions of Contract for Plant & Design Build ("FIDIC Yellow Book")*. Switzerland: FIDIC.

Jackson v. Barry Rail Co, [1893] 1 Ch 238, (CA).

Jarvis & Sons Limited v. Westminster Corporation & Another [1969] 1 WLR 1448.

Mariner International Hotels Ltd v. Atlas Ltd [2007] 10 HKCFAR 1.

Mears Ltd v. Costplan Services (South East) Ltd & Ors [2019] EWCA Civ 502.

Maxcon Constructions Pty Ltd v. Vadasz, [2018] HCA 5, (2018) 351 ALR 369.

Modern Engineering v. Gilbert Ash Northern, [1974] A.C. 689, 1 B.L.R. 73.

Nelson Carlton Construction Co (in liq) v. A C Hatrick (NZ) Ltd, [1965] NZLR 144, (CA).

New Engineering Contract ("NEC4") (June 2017) *Engineering and Construction Contract*. Glasgow: Bell & Bain Ltd.

New Engineering Contract ("NEC4") (June 2017) *Managing an Engineering Construction Contract Volume Four*. Glasgow: Bell & Bain Ltd.

Northern Regional Health Authority v. Derek Crouch Construction Co. Limited [1984] QB 644; [1984] 2 WLR 676; [1984] 2 All ER 175, C.A.

Pacific Associates Inc v. Baxter, [1990] 1 QB 993, (CA).

Panamena Europea Navigacion (Compania Limitada) v. Frederick Leyland & Company Limited (J. Russell & Company), [1947] AC 428, (HL).

PBS Energo AS v. Bester Generacion UK Ltd [2020] EWHC 223 (TCC).

Peninsula Balmain Pty Ltd v. Abigroup Contractors Pty Ltd, [2002] NSWCA 270.

Perini Corp v. Commonwealth (Redfern Mail Exchange Case), [1969] 2 NSWR.

Photo Production Ltd v. Securicor Transport Ltd [1980] AC 827530,(NSWSC).

RM Turton & Co Ltd (in Liq) v. Kerslake, [2000] 3 NZLR 406, (CA).

Standards New Zealand Paerewa Aotearoa (2013) NZS3910:2013 New Zealand Standard Conditions of Contract for Building and Civil Engineering Construction. New Zealand: Standards New Zealand.

Standards New Zealand Te Mana Tautikanga O Aotearoa (2023) NZS3910: 2013 New Zealand Standard Conditions of Contract for Building and Civil Engineering Construction. New Zealand: Standards New Zealand.

Stephen Furst KC and the Hon Sir Vivian Ramsey (eds), Keating on Construction Contracts, 10th ed., (Sweet & Maxwell, 2016).

Stevensons Structural Engineers 1978 Ltd (In Liq) v. McMillan & Lockwood (PN) Ltd, [2024] NZHC 2415.

Sutcliffe v. Thackrah, [1974] AC 727, (HL).

Van Oord UK Ltd v. Dragados UK Ltd, [2021] CSIH 50.

12

Insurance

Amy Lacey

> *Any risk is insurable at the right price.*
>
> – Cuthbert Heath

> *Anything that can go wrong, will ... if it is possible for several things to go wrong, the one that will cause the most damage will occur.*
>
> – Murphy's Law

Introduction

Construction is inherently hazardous and (re)insurance plays a critical role in managing the associated risks. Natural perils, including storm, flood, and landslip, pose major challenges for incomplete contract works, exposed to the elements; building sites face many intrinsic dangers of physical loss arising from human error or defective condition; and construction entails significant financial risk, particularly of delay to completion and increased costs.

Insurance contracts specify the terms on which an insurer agrees, in return for payment of the premium, to indemnify the policyholder against loss, or to restore a loss that may be suffered. Coverage may be provided for the benefit of one party or several, named or identified by description, in the construction contract or the insurance policy. There must be some uncertainty as to whether, or at least when, the event giving rise to a loss will happen, so that its occurrence is outside the parties' control.

Insurance is founded on the principle of utmost good faith, whereby the policyholder is obliged to provide the insurer with a fair presentation of the proposed risk, including disclosure of material circumstances likely to influence the judgment of a reasonable underwriter. Policy conditions may also include

Major Issues in International Construction Law, First Edition. Mauro Rubino-Sammartano and Troy Harris.
Published 2026 by John Wiley & Sons, Inc.

requirements to take reasonable precautions to avoid and mitigate loss, or to notify claims in a specified manner and time.

The policy limit is the maximum amount that the insurer will pay, for each and every claim, or alternatively as an aggregate limit – so that once the total of all claims under the policy in a given year reaches that level, the coverage is effectively exhausted, even if the period of insurance has not yet expired. The policy excess and limits of indemnity may be different for each type of property insured.

Governing Law and Jurisdiction

Policies providing cover for international construction projects should specify which country's laws govern the insurance and how any disputes arising between the policyholder and the insurer will be determined. Different jurisdictions may interpret policy terms differently, leading to varying outcomes. The absence of express provision on governing law and jurisdiction could lead to increased costs of preliminary disputes over the appropriate forum.

Insurance contracts typically provide for resolution of disputes by court proceedings or arbitration and may specify a muti-tiered process, requiring mediation initially, or expert determination. The option to have disputes heard in open court is usually advantageous for policyholders, especially where consistency is required for related claims arising under multiple policies or excess layers. Insurers tend to favor arbitration as a private means of settling disputes, without adverse publicity or fixed precedent in subsequent cases.

The composite nature of insurance contracts can give rise to conflicting provisions and careful consideration should be given to dispute resolution options prior to inception. Where alternative scenarios are dealt with in a single clause, the intention on triggering events and orders of application must be clearly expressed. Ambiguous language, imprecise punctuation, or typographical errors could lead to disputes over the meaning of words used and effective scope of coverage.[1]

Specific rules governing construction insurance depend on the relevant location in which contract works are undertaken. Local regulations may prescribe certain types of coverage, or this may be left to contractual arrangements between the various project participants. Many countries have strict regulations regarding foreign insurers, including mandatory reinsurance with domestic carriers or limitations on the use of nonadmitted insurance. If the required insurance is not placed with a locally approved provider, the coverage may be invalid or unenforceable in that jurisdiction.

1 For example, see *Al Mana Lifestyle Trading LLC & others v. United Fidelity Insurance Co PSC & others* [2023] EWCA Civ 6; *Project Angel Bidco Ltd v. Axis & others* [2024] EWCA Civ 446.

Construction Risks

Construction and engineering projects are subject to a multitude of risks that may be covered by a diverse range of insurance policies. The two main categories of construction insurance relate to property and liability risks:

- Property insurance covers the cost of remedying damage to physical property occurring during the period of insurance, and may extend to financial loss resulting from business interruption ("BI") caused by property damage, and any perils specified in nondamage extensions to the policy.
- Liability policies typically cover claims made against the policyholder, by: (a) third parties, in respect of injury or property damage; (b) employees, in respect of injury; and (c) clients or third parties, in respect of breaches of duty in the course of professional services.

The terminology used to describe risks arising in a construction context is not uniform. For example, public liability ("PL") policies are often expressed as third-party liability policies; and "delay in start-up" ("DSU") or "advanced loss of profits" ("ALOP") are variously adopted for business interruption. The label given to a policy or extension is not necessarily determinative: what matters is whether the policy wording engages with the events giving rise to the particular claim or loss.

Where multiple insurance policies are taken out in respect of a project, this may give rise to overlapping cover. For example, financial liability extensions to PL or product liability policies may overlap with coverage under professional indemnity ("PI") insurance; and defective design resulting in damage can lead to related claims under contract works and PI policies. If there is more than one policy available, the policyholder is entitled to indemnity under whichever it chooses but cannot claim a double contribution, in excess of its total loss. Policies may include "other insurance" clauses specifying an order of priority or contribution between insurers. If more than one policy is written on terms that it will not apply if there is other insurance available for the loss in question, the exclusions may be treated as cancelling each other out, so that the insurers remain liable.[2]

Insurance for construction operations may be provided by a policy covering a single type of risk, or expressed in a combined policy, with numerous categories of risk covered in a single insurance arrangement. Policies may be project-specific, with several parties participating jointly in the insurance and payment of the premium, and/or parties may take out separate policies covering all aspects of their own business operations, often renewed annually.

2 *Allianz Australia Ltd v. Certain Underwriters at Lloyd's* [2019] NSWCA 271.

Construction projects typically require insurance against many types of risk, including:

First-party Cover

Contract Works

Contract works insurance covers specified risks of physical loss of or damage to the works under construction, including temporary works. This type of coverage is also commonly referred to as construction all risks ("CAR"), erection all risks ("EAR"), builders' risk or contractors' indemnity. EAR policies are typically used for projects that are primarily concerned with the installation of electrical or mechanical plant and machinery. Although the name suggests otherwise, "all risks" insurance will not cover certain categories of risk, such as costs to remediate defective workmanship, and other perils expressly excluded.

Business Interruption

BI insurance responds to loss of revenue, or increased cost, caused by a delay to project completion due to insured property damage, or other specified perils (such as denial of access, terrorism, or infectious disease). The indemnity period and sums insured should be sufficient to cover financial loss during the time it would take to rebuild the works in consequence of the delay in completion arising from an insured event. Declaration-linked cover can help to ameliorate the risks of underinsurance.

Delays in construction projects often lead to substantial financial losses for the owners. Causation is key in proving BI claims and applying trends clauses, to evaluate the level of indemnity based on income projections, with allowance for factors that would have impacted the business in any event. Assessing whether delay is covered under the policy may be complex, particularly where several causes combine to bring about a loss.

DSU/ALOP is usually viewed as a "single shot" cover: if the policy is triggered by a claim during the period of insurance, the policyholder should look to reinstate the cover, if required for any further period of delay, rather than merely extending the policy period.

Latent Defects

Latent defects insurance covers costs to remediate structural problems or other safety concerns arising from inherent defects in design, materials, or workmanship, which only become discovered (or discoverable with reasonable diligence) after completion of the construction works.

In the UK, it is customary for builders to offer defects insurance, such as the Buildmark policy for private housing offered by the National House Building Council. Some countries, including France and Saudi Arabia, have compulsory inherent defects insurance requirements for construction projects in the private sector.

The UK Building Safety Act 2022 introduced mandatory requirements for new home warranties with a minimum term of 15 years from practical completion, as compared with the previous coverage of typically 10 years (i.e., "decennial" policies) which had been required by mortgage lenders. Minimum standards of cover are expected to be set in regulations following a period of consultation and a New Homes Ombudsman will be introduced to handle complaints arising in this sector.

Insolvency

Insurance cover may be available to a project participant against the risk of loss due to the insolvency of another party. This type of policy will usually protect the owner against economic loss suffered due to the contractor ceasing work because of its insolvency, but not for defects in its works arising prior to the insolvency. Subcontractor default insurance is widely used in the US to protect main (or "general") contractors against the financial exposures that they can face if their subcontractors become insolvent and are unable to complete their work. Subcontractor default insurance remains rare in the London market but is provided by a small number of insurers.

Third-party Liability Cover

Public Liability

PL insurance covers the risk of liability for bodily injury, or damage to third-party property. In the absence of clear wording to the contrary, which is sometimes provided, a PL policy does not provide coverage for physical damage to property of the insured, or economic loss suffered by third parties arising from contractual duties.

In *Tesco Stores Ltd v. Constable*, the English Court of Appeal held that a financial loss extension to a PL policy would not cover sums which Tesco became liable to pay to Chiltern Railways under the terms of a deed of covenant, as this was pure economic loss arising from contractual liability:

> A public liability policy provides cover against liability to the public at large. By contrast, private liability arises from contracts entered into between individuals. Public liability in this sense arises in tort; it does not and

> cannot arise solely in contract. As a general rule, a claim in tort cannot be founded upon pure economic loss. So the judge was right to say that the fact that this was public liability insurance was important and that such policies do not generally cover liability in contract for pure economic loss.[3]

Employers' Liability

Employers' liability insurance covers risks associated with claims by employees for harm or injury arising from their employment. This type of insurance is compulsory in many jurisdictions including the UK.[4] Similar requirements apply in Australia, Hong Kong, and Singapore, among other places.

Product Liability

Product liability insurance covers risks arising from the use of defective materials or goods, causing bodily injury or physical damage to third parties. Loss associated with damage to the defective product itself is usually excluded, although some policy wordings cover the cost of removing defective materials attached to a structure.[5]

In *Pilkington (UK) Ltd v. CGU Insurance plc*,[6] a supplier of glass panels for the Eurostar terminal at Waterloo station was unable to recover under its product liability insurance for costs incurred in defending a claim by the main contractor, where an inherent defect was identified that made a small number of panels prone to shatter. The policy wording made clear that damage or deterioration confined to the product itself was excluded, i.e., the policy would only answer in respect of physical damage to other property, to which the product had been introduced or juxtaposed. The Court of Appeal held that no physical damage to third-party property had occurred, and cover under the policy was not engaged where the owner did not remove the affected panels (so there was no access damage to replace defective parts), and chose instead to implement safety measures, to avoid any shattered glass falling.

Professional Indemnity

PI insurance covers legal liability for breaches of duty committed in the course of professional duties, and usually operates on a claims-made basis, so that the policy in force at the time of notification usually provides cover for claims which subsequently arise out of circumstances notified.

---

3 [2008] EWCA Civ 1242, at paragraph 14 per Tuckey LJ.

4 Employers' Liability (Compulsory Insurance) Act 1969.

5 For example, see *Austral Plywoods Pty Ltd v. FAI General Insurance Co Ltd* [1992] QCA 004.

6 [2004] EWCA Civ 23.

PI policies do not respond to claims resulting exclusively from manual labor and physical operations. Claims arising from contract terms imposing performance standards beyond reasonable skill and care, or requiring payment of fixed penalties such as liquidated damages which include a noncompensatory element, are typically excluded. Contracting parties should endeavor to agree back-to-back terms throughout the contractual chain and avoid onerous provisions such as warranties or guarantees of fitness for purpose or design life, which could lead to uninsured claims.

Escalating concerns over building safety defects have led to many PI claims, including block notification of projects involving the installation of external cladding systems. Disagreements with insurers frequently arise over the threshold for notification of claims or circumstances, the scope of notification, "prior knowledge" exclusion clauses, and aggregation issues, where losses could stem from many projects, with several blocks or units on each site. Insurers have responded by adding new policy terms limiting or excluding cover for these types of claims and imposing retroactive dates.

Directors and Officers

Directors' and officers' insurance covers management liability for wrongful acts, and particularly for breach of statutory or fiduciary duties owed to companies and shareholders, including costs of defending claims made against directors and officers. The global rise in climate litigation and increasing focus on environmental, social, and governance responsibilities highlights the importance of adequate coverage. Implementation of decarbonization and climate strategies is subject to scrutiny across many industries, with claims proceeding in multiple jurisdictions seeking compensation for environmental harm, or strategic influence over regulatory, corporate, and investment decisions.

Allocation of Insurance Responsibility

Statutory laws may impose mandatory insurance requirements. For example, most states of Australia require builders to be insured in relation to residential projects; and developers of buildings in Nigeria which are over two floors in height must take out compulsory all risks and liability insurance.

Construction and engineering contracts specify how risks are allocated, and usually require one or more parties to arrange insurance cover in favor of identified persons, and to maintain that cover throughout the construction works. Owners, contractors, and consultants must ensure that their insurance policies align with the agreed allocation of risks in the underlying construction contracts. The optimal approach varies, depending on specific project risks, and preferences in terms of the owner or contractor exercising control over the scope of cover and the conduct of any claims.

Often the principal contractor is obliged to take out and maintain contract works insurance in the joint names of the owner, contractor, nominated subcontractors, and possibly consultants, and to produce evidence of the policy to the owner, or contract administrator. This corresponds with the standard contractual risk allocation, where the contractor takes on risk for the physical site once handed over by the owner, up until practical completion.

The Fédération International des Ingénieur Conseils (FIDIC) standard forms of major works contracts include general conditions requiring the contractor to provide insurance, subject to any express variation agreed between the parties. The scope of the obligation extends to cover for the works, bodily injury to persons including employees, damage to property, and all other insurances required by local laws in the countries where the works are being carried out. The contract data may also specify that the contractor's liability in respect of any breach of its fitness for purpose obligation to the owner, for design of the works, shall be covered by PI insurance.

In the Joint Contracts Tribunal (JCT) standard form contracts, there are three alternative options for the insurance of works and existing structures:

- Option A – where the parties agree that the contractor will take out and maintain all risks insurance in joint names for the contract works.
- Option B – where the parties agree that the owner will take out and maintain all risks insurance in joint names for the contract works.
- Option C – where the parties agree that the owner will take out and maintain all risks insurance in joint names in respect of the existing structures and their contents, and the contract works.

Options A or B are intended to be used where the works comprise the construction of a new building. Option C is suitable for refurbishment projects to an existing building.

The 2024 edition of the JCT Design & Build Contract introduced some minor amendments to the insurance provisions: (i) the contractor's obligation to maintain PI insurance is now subject to the cover being available on reasonable terms (and at reasonable cost); (ii) the contract particulars have been changed to include a placeholder for any insurance sub-limits; and (iii) in relation to Option A, a new clause has been added stating that any reinstatement work will be treated as a "Change" with the contractor entitled to an extension of time.

Other standard forms of contract, such as those published by the Institution of Civil Engineers (ICE), Institution of Chemical Engineers (IChemE), and New Engineering Contract (NEC), include slightly different provisions on insurance allocation.

Co-insurance

Multi-party insurance is especially common in the context of construction. The commercial rationale for placement of project insurance policies was explained in *Petrofina (UK) Ltd v. Magnaload Ltd:*[7]

> where numerous different subcontractors may be engaged, there can be no doubt about the convenience from everybody's point of view, including the insurers, of allowing the head contractor to take out a single policy covering the whole risk, that is to say covering all contractors and subcontractors in respect of loss of or damage to the entire contract works. Otherwise each subcontractor would be compelled to take out his own separate policy. This would mean, at the very least, extra paperwork; at worst it would lead to overlapping claims and cross-claims. Furthermore ... in the case of a small subcontractor, may be uneconomic. The premium might be out of all proportion to the value of the subcontract. If the subcontractor had to insure his liability in respect of the entire works, he might well have to decline the contract.

Composite policies insure multiple parties for their separate interests in the insured subject matter, such as owners and contractors, landlords and tenants, or lenders and borrowers. A breach of policy terms by one policyholder does not usually affect the rights of other insured parties under a composite policy.

To determine the extent of cover for subcontractors and consultants under a project insurance policy, it is necessary to evaluate the intention of the parties to the underlying building contract, based on agreed terms for the allocation of insurance responsibility. This is crucial in the context of subrogation, whereby insurers, having paid a policy claim, are entitled to "step into the shoes" of the policyholder and pursue recovery action against any third parties with responsibility for the loss.

Insurers are prevented from bringing subrogated claims against co-insured parties under a composite policy, if the building contract between the co-insured parties shows that this was the intention.

In *Cooperative Retail Services Ltd v. Taylor Young Partnership Ltd,*[8] the owner arranged insurance in respect of fire risk, during construction of a new office block, in the joint names of all contractors and subcontractors, while the architects and engineers were not within the insurance cover and made their own

7 [1984] QB 127 at 136, per Lloyd J.

8 [2002] UKHL 17.

arrangements. A fire occurred on site and the owner sought to recover damages from the architects and engineers, who, in turn, sought contribution from the main contractor and electrical subcontractor, on the basis that they were liable for the same damage and it would be inequitable, if the owner's claim succeeded, that they should be required to meet the claim in full, when the loss had arisen from alleged failings by the contractors.

The UK House of Lords held that where a construction contract stipulates that parties are to be named in an insurance policy, the risk is also to be shared equally between the parties. It was clear that the effect of the main contract and the subcontract was to exclude the contractors from liability to the owner for loss and damage resulting from fire, and therefore no contribution could be claimed by the architects and engineers from any of the contractors for insured losses.

In *Haberdashers Aske's Federation Trust Ltd v. Lakehouse Contracts Ltd*,[9] a project-wide insurance policy taken out by the main contractor in respect of construction works at a school provided that coverage would extend to all subcontractors. However, the subcontract in respect of roofing works provided that the roofing subcontractor was required to take out its own insurance cover. Following a fire on site, the owner claimed damages against the main contractor and roofing subcontractor, the main contractor sought a contribution or indemnity from the subcontractor, and the subcontractor sought a declaration that it could rely on the project insurance. The UK Technology & Construction Court held that the project policy insurer, having indemnified the main contractor for loss resulting from fire damage, was entitled to exercise subrogation rights against the contractor, because the arrangements between the parties were contrary to any possible intention by the main contractor to insure on behalf of the subcontractor. Viewed objectively, the subcontractor had intended that it would not rely on the project-wide policy to the extent of its own policy.

The relationship between an obligation on party A to insure on behalf of parties A and B, and an obligation on B to indemnify A for any loss, was considered in *Gard Marine & Energy Ltd v. China National Chartering Co Ltd.*[10] This case concerned a ship that sank in the port of Kashima in Japan. The contract provided for the ship owner and the charterer to have joint names hull insurance, and a dispute arose as to whether this co-insurance precluded the insurer from bringing subrogated claims in the name of one policyholder against the other. By a 3:2 majority, the UK Supreme Court held that where there was co-insurance in place, there was an implied term in the arrangements between

9 [2018] EWHC 588 (TCC).
10 [2017] UKSC 35.

the parties that they would look to the insurance rather than to each other in the event of a loss.

Lord Toulson, for the majority, said:

> It has become a common practice in various industries for the parties to provide for specified loss or damage to be covered by insurance for their mutual benefit, whether caused by one party's fault or not, thus avoiding potential litigation between them. The question in each case is whether the parties are to be taken to have intended to create an insurance fund which would be the sole avenue for making good the relevant loss or damage, or whether the existence of the fund co-exists with an independent right of action for breach of a term of the contract which has caused that loss. Like all questions of construction, it depends on the provisions of the particular contract.[11]

In *Rugby Football Union v. Clark Smith Partnership*,[12] the owner appointed a contractor to install ductwork to carry electric cables at a sports stadium. The ductwork was allegedly defective due to the negligence of the contractor and the cables were damaged as a result. The agreed contract terms provided for the owner to take out joint names insurance of existing structures and works in respect of specified perils; and the question arose whether the insurer, having indemnified the owner, was entitled to bring a subrogated claim against the contractor.

At first instance, the Judge decided that the contractor was not insured under the project policy for damage to existing structures caused by its own defective work. Cover was instead restricted to specified perils in accordance with the (unamended) JCT Option C insurance provision. This was upheld on appeal, where Lord Justice Coulson said:

> The mere fact that A and B are insured under the same policy does not, by itself, mean that A and B are covered for the same loss or cannot make claims against one another ... In circumstances where it is alleged that A has procured insurance for B, it will usually be necessary to consider issues such as authority, intention (and the related issue of scope of cover) ... where there is an underlying contract then, in most cases, it will be much the best place to find evidence of authority, intention and scope.[13]

---

11 At paragraph 139.

12 [2022] EWHC 956 (TCC).

13 [2023] EWCA Civ 418, at paragraph 53.

In *Sky UK Ltd & Mace Ltd v. Riverstone Managing Agency Ltd & others*,[14] the main contractor specifically named in the project policy was found to be a co-insured, but only up until practical completion of the works, and not to expiry of the maintenance liability period, in view of the building contract terms. Only the owner had the benefit of cover for that extended period, which meant that the main contractor could be liable to indemnify the owner in respect of any damage to the roof occurring after practical completion.

There has been significant debate historically over how a subcontractor, not specifically named in a project policy and appointed after inception of the policy period, comes to be a party to the insurance contract and entitled to its benefit. Recent authorities[15] confirm that an insurer writing the policy makes an open offer to insure anyone who subsequently comes within the class of "insureds" as defined in the policy. Subcontractors appointed to the project can accept the offer by signing their subcontracts. But if a subcontractor is engaged under a contract that makes clear that the subcontractor was not to be covered by the project policy, then the subcontractor should arrange separate insurance, and may be exposed to potential subrogated claims by the project insurer.

For parties not specifically named in a project insurance policy and belonging to a class of potential insureds identifiable by description (e.g., "subcontractors of any tier"), it is important to ensure that the contractual arrangements governing their involvement make clear whether, and to what extent, they are intended to benefit from coverage under the project policy and, where necessary, that any standard forms are appropriately amended.

Causation

Insurers are only liable for losses caused by an insured peril.

In English law, a test of "proximate causation" applies, i.e., the dominant, direct or efficient cause, from which the loss arises as a natural consequence, whether or not this is the last to occur in a sequence of events. "But for" causation (*i.e.*, but for the existence of X, would Y have occurred?) is often applied as a threshold test, although not always necessary or sufficient. In *FCA v. Arch & others*,[16] a test case by the UK financial services regulator on BI insurance following the Covid-19 pandemic, the Supreme Court held that "but for" causation should not be applied where multiple proximate causes bring about a loss, and

14 [2023] EWHC 1207 (Comm); [2024] EWCA Civ 1567.

15 Including *Haberdashers* and *Gard Marine*.

16 [2021] UKSC 1.

both the insured peril and uninsured peril operating concurrently arise from the same underlying fortuity – overturning the decision in *Orient-Express Hotels v. Generali.*[17]

Construction losses often result from a combination of causes, such as design flaws and poor workmanship, or adverse weather. Where two or more causes of approximately equal efficacy operate together to bring about the loss, they are regarded as concurrent proximate causes. If one proximate cause is an insured peril and the other is uninsured (but not excluded), the loss will be covered under the policy.[18] If one of the proximate causes is excluded, the loss will not be covered.[19]

The degree of causal connection required between losses and insured or excluded perils depends on the policy wording. Phrases such as "caused by," "resulting from," or "due to" are understood to mean proximate causation. Broader cover could be provided for loss "directly or indirectly caused by," "arising out of," "in connection with," or "in any way involving" specified perils; as compared with more restrictive language on losses "solely" or "exclusively" resulting from a particular cause.

Aggregation

The purpose of aggregation wording in insurance policies is to enable two or more separate losses covered by the policy to be treated as a single loss, when linked by a particular unifying factor or number of factors. This may benefit the policyholder or the insurer, depending on the facts of any particular loss or claim, and whether the losses are being aggregated for the purpose of a policy deductible or limit. Aggregation is pivotal in many insurance claims.

Relevant language may be contained in a general definition of "loss" or "event," in the policy schedule of limits for items of insured property, or in an excess or limit of liability clause. Aggregation based on a common "originating cause" is a broad concept that can include a state of affairs or omission to act. Aggregation by reference to an event is a narrower concept, with reference to something happening in a particular manner, time, and place.[20]

---

17 [2010] [2010] EWHC 1186 (Comm).

18 *Zagora Management Ltd v. Zurich Insurance plc* [2019] EWHC 140; *JJ Lloyd Instruments v. Northern Star Insurance Co (The Miss Jay Jay)* [1987] 1 Lloyd's Rep. 32.

19 *Wayne Tank v. Employers' Liability Assurance* [1974] QB 57.

20 *Axa Reinsurance (UK) plc v. Field* [1996] 3 All ER 517; see also *Marketform Managing Agency Ltd v. Amashaw Pty Ltd* [2018] NSWCA 70.

Where a policyholder's loss arises out of a single problem, such as a design defect, the consequences could arise in various ways, and lead to disagreement over whether a single or many deductibles apply. Each case will turn upon the underlying factual matrix and policy wording. In *Mitsubishi Electric UK Ltd v. Royal London Insurance (UK) Ltd,*[21] it was held that a single deductible applied to the entire claim in respect of 94 modular toilets in a building, as the cementitious board used in the design was the defective component, giving rise to a single composite claim.

In *Sky v. Riverstone*, it was held that a decision not to install a temporary roof to protect the various sections of a new permanent roof, pending the completion of waterproofing work, was an "event" giving rise to the damage to individual sections. Only one deductible applied to the loss, rather than a separate deductible for each of the modular cassettes making up the roof construction. The trial judge's conclusion on the aggregation of loss was upheld on appeal.

Damage

In most common law jurisdictions, "damage" is understood to mean an adverse change in physical condition, impairing the value or usefulness of insured property.[22] The adverse change in physical condition does not have to be permanent;[23] and could include microscopic changes, indiscernible to the naked eye.[24]

The precise dividing line between defects and consequential damage is often contentious. In *Promet Engineering (Singapore) Pte Ltd v. Sturge (The Nukila),*[25] Ward LJ said:

> If one is considering whether there is damage to the hull and whether such damage is caused by a latent defect in the hull, it follows that the damage must be something different from ... and incrementally greater than the latent defect itself. Where the line is to be drawn is a matter of fact and degree.

---

21 (1994) 74 BLR 87.

22 *Ranicar v. Frigmobile Pty Ltd* [1983] Tas. R. 113; *Hunter v. Canary Wharf Ltd* [1996] 2 W.L.R 348.

23 *Hunter v. Canary Wharf Ltd* [1996] 2 W.L.R 348; *Losinjska Plovidba v. Transco Overseas Ltd (The Orjula)* [1995] 2 Lloyd's Rep. 395.

24 In *Quorum A/S v. Schramm* [2001] EWHC 494 (Comm), sub molecular change to a Degas pastel painting sustained by close proximity to a fire was held to constitute damage, as this increased the risk of future deterioration and shortened the anticipated life span.

25 [1997] 2 Lloyd's Rep. 146 at 157.

In *The Nukila*, the legs of an offshore platform were susceptible to excessive metal fatigue stress as a result of defective welding in their construction, and the evidence showed that this defect had initiated tiny undetectable cracks at the time the owners took out the policy, which developed into larger and visible fractures during the period of cover, necessitating repairs. This was held to constitute damage to the legs of the platform, notwithstanding the pre-existing defects within the relevant insured property.

In *Sky v Riverstone*, the Court of Appeal held that insurers were liable to indemnify all damage that occurred during the period of insurance but deteriorated or developed thereafter, overturning the trial judge's decision that the claimants were only entitled under the policy to indemnity in respect of the cost of repair of such damage as existed at the end of the insured period. Further, the trial judge had erred in classifying investigation costs as "speculative opening up works," concluding that investigation costs reasonably incurred to remediate damage were covered, whether or not damage was revealed.

Recent authority in Australia suggests that affixation of combustible cladding panels may constitute damage for purposes of a products liability policy, given the physical alteration involved in attaching a harmful external wall system to the building, necessitating remedial works.[26] That decision was made in an interlocutory application to join the insurer to proceedings, not a final determination, and the Court was understandably influenced by public policy considerations.

Defects

A contract of insurance is not a guarantee of good condition or proper working order, and defects alone are not indemnifiable under contract works policies. However, insurers are often willing to provide cover for damage to defective property, or damage caused by defects in the design or construction of the insured property.

Policies providing all risks cover for contract works usually contain a standard clause based on the Defect Exclusion ("DE") or London Engineering Group ("LEG") suite of options, offering progressive levels of cover for costs of repair associated with defective design, workmanship or materials. The DE clauses were originally drafted with a focus on building works, while the LEG clauses are primarily intended for engineering risks. The LEG clauses, or derivatives, are used in

26 *Owners – Strata Plan No 91806 v. Fairview Architectural* [2023] FCA 814; see also *Capral v. Insurance Australia* [2024] FCA 775, and *Austral Plywoods Pty Ltd v. FAI General Insurance Co Ltd* (1992) 7 ANZ Insurance cases 61–110.

most global engineering and construction projects. The relevant defect exclusion wording is often crucial in determining the scope of cover.

DE Clauses

- DE1 excludes damage caused by defects.
- DE2 excludes damage to insured property which was in a defective condition before the damage occurred, structures that rely for support on insured property in a defective condition, and insured property lost or damaged to enable the replacement repair or rectification of defective insured property, while providing cover for other parts of the insured property which are damaged as a consequence of the defect.
- DE3 excludes damage to insured property in a defective condition, and insured property lost or damaged to enable the replacement repair or rectification of defective insured property, while providing cover for damage to other nondefective insured property.
- DE4 excludes damage to insured property in a defective condition, and insured property lost or damaged to enable the replacement repair or rectification of defective component parts of insured property, while providing cover for damage to other nondefective parts of the insured property.
- DE5 excludes only the additional costs (if any) of improving the original design plan specification materials or workmanship.

As above, the DE3 and DE4 clauses provide cover for damage to "other" insured property, caused by defects, but not the defective part itself. Disputes often arise over divisibility of defective and nondefective insured property or parts, and the courts will look to the commercial reality of whether a particular aspect of construction would be viewed by the contracting parties as a separate package or phase of the works.[27]

LEG Clauses

- Like DE1, LEG1 excludes damage caused by defects.
- LEG2 is engaged where insured property was in defective condition before the damage occurred. In that situation, LEG2 excludes the hypothetical cost that would have been incurred if the defect had been discovered, and fixed, immediately before the damage occurred.

---

27 *Seele Austria GmbH & Co KG* [2008] EWCA Civ 441; see also *CA Blackwell (Contractors) Ltd* [2007] EWCA Civ 1450.

- Like DE5, LEG3 excludes only the costs (if any) incurred to improve the original material workmanship design plan or specification.

DE5 or LEG3 are intended to provide the broadest coverage, including costs incurred to gain access to property damaged as a result of defects, and excluding only the cost of improvements. Subject to any impact on the policy premium or limits/deductible, policies may provide a choice exercisable at the policyholder's option as to application of DE5 or LEG3, in place of one of the less generous defect exclusions, e.g., DE3 or LEG2, allowing flexibility at the claims stage to determine which alternative produces the best outcome for the policyholder.

Concrete Cases

Recent decisions from the US courts on application of LEG3 have caused controversy in the insurance market.

Firstly, in *South Capitol Bridgebuilders v. Lexington Insurance Co*,[28] a court in the District of Columbia, applying Illinois law, considered a claim for losses arising from sub-standard concrete used in construction of the Frederick Douglas Memorial Bridge in Washington DC.

The design involved consecutive steel arches on either side of the bridge, supported by concrete abutments on either side, and concrete piers toward the center of the river. The concrete was poured separately into each of the abutments and piers, with workers standing inside the formwork and vibrating the concrete in order to achieve even placement. Due to the vibration being carried out inadequately on a third pour, this section of dried concrete contained "honeycombing" voids, diminishing its structural strength. Substantial repairs were required to remove and repair the affected concrete.

The policyholder was insured under a Builder's Risk (CAR) policy containing a LEG3 defects exclusion. The insurer refused indemnity on grounds that the honeycombed concrete components were in a defective condition from the outset, no damage had occurred and, even if the concrete had been damaged, the LEG3 clause excluded coverage for all the remedial works, because fixing something broken inevitably involves improvement of the relevant property.

Damage was not a defined term in the policy and the Judge relied upon a general meaning of "injury" or "any bad effect," taken from Black's Law Dictionary, to conclude that the bridge and its support structures had suffered damage by the presence of honeycombed concrete. The Judge described the wording of LEG3 as "ambiguous ... egregiously so" and construed the clause contra proferentem, i.e., against the insurer. The court rejected the insurer's argument that simply

28 2023 US Dist. LEXIS 176573 (D.D.C. Sep. 29, 2023).

replacing defective property amounts to an improvement and held that only costs incurred to make something better than it would have been, but for the defective work, are excluded.

This was followed by *Archer Western – De Moya JV v. Ace American Insurance Co*,[29] in the District Court of Florida, where a design and build contractor undertaking highway reconstruction works used batches of concrete that were contaminated with excessive amounts of fly ash, due to malfunction of a pressurized silo used on site in the mixing process, before being incorporated at various pour sites throughout the project. Fly ash, together with cement and aggregates, is a necessary component of concrete, but the wrong proportions were combined, reducing the concrete's compressive strength. This meant that elements of the supporting structures had to be rectified.

The contractor was insured under an all-risks policy containing a LEG3 exclusion. Proceedings were issued on the policy claim, after coverage was declined, and the insurer applied to strike out. The court noted that previous cases had established that the test for damage "*requires a tangible alteration to the covered property*" and, based on the high bar required to obtain summary judgment under local laws (i.e., where there is no genuine issue as to any material fact), the insurer's argument that no damage had occurred was not accepted. This issue was therefore due to be determined at trial, had the case not subsequently settled on commercial terms.

Faulty concrete mixes are usually viewed by insurers as merely defective, but the precise dividing line with physical damage may be unclear. Difficult issues of interpretation can arise, depending on how the question of coverage is framed. On the facts of *Archer* and *South Capitol*, for example: the relevant insured property could be viewed as the cement, which becomes damaged by being overlaid with excessive quantities of fly ash,[30] or the concrete mix, which was under-strength as soon as it came into existence;[31] and a third pour, using defective concrete, changes the physical condition of the (previously nondefective) sections of concrete completed in earlier stages.

While these decisions are not binding on courts in other jurisdictions, they are significant decisions globally and likely to influence insurers' underwriting on international construction risks. The outcome in *Archer* confirms that *South Capitol* is not just an outlier case, and these developments may lead to review and possible amendment to the standard defect exclusion clauses.

---

29 Docket No: CASE NO. 1:22-CV-21160-GOODMAN.

30 The deposit on to insured property of excessive quantities of a benign substance is capable of constituting damage, as in *Hunter v. Canary Wharf*.

31 Incorporation of a defective component into a finished product was held not to amount to damage in *Bacardi-Martini Beverages Ltd v. Thomas Hardy Packaging Ltd* [2002] EWCA Civ 549.

13

Variations

Scott Chalmers

Introduction

Construction contracts have a distinctive characteristic of being designed from the outset to be contemplated and accommodate unknown changes. The owner's[1] right to vary or change the scope of the contactor's works during the course of performing the work is a longstanding and fundamental element of construction contracts globally.

Inevitably, writing precisely about legal aspects of variations will necessitate a close examination of the construction laws applicable in the relevant jurisdiction. This chapter does not endeavor to do so – rather, its express intent is to provides a general international comparison and discussion of the themes common throughout most construction contracts, with a focus on the common law jurisdictions of England, Canada, Australia, the United States, and Pakistan, and on some of the more widely known standard forms of contract used in those jurisdictions and the Middle East.

Nomenclature: Variations and Changes

"Variations," or "variation orders" as they are called in much of the common law world,[2] and "changes" or "change orders" as they are called in the United States and Canada, are the same thing. Similarly, a "change directive," as it is called in the United States, is simply a "unilateral variation" or "owner-directed variation."

---

1 We use owner here as the counterparty to the contractor – though it can be variously called other names – e.g., "Employer" in the JCT and FIDIC forms, and "Principal" in Australian Standard forms. Similarly, we have used contractor to refer to the constructing counterparty.

2 And as we will refer to them in this chapter.

Major Issues in International Construction Law, First Edition. Mauro Rubino-Sammartano and Troy Harris.
Published 2026 by John Wiley & Sons, Inc.

A "variation" is a change in the scope of work to be performed by a contractor, typically as expressly directed by an owner or on an owner's behalf by its representatives, whether project managers, contract managers, or consultants. Variations may add to the scope of work, omit work, require the deletion or demolition of work, or may change the nature, character, sequence, completion date or manner of performing of the work, all as the relevant contract form may expressly provide.

In perhaps the best-known global suite of standard form contracts, the FIDIC suite of contracts, the Red Book states that a variation may include: changes to quantities of any item of work, changes to quality or characteristics of work, changes to levels, positions and/or dimensions of any part of the work, omission of any work, additional work, plant, materials or services necessary for permanent works including any associated tests on completion, boreholes and other testing and exploratory work, or changes to sequence or timing of the execution of the work.[3]

The reasons for, or circumstances in which a contractor may be entitled to, a variation are diverse. The right to a variation is not typically limited to an owner only, and a right to a variation may arise from a contractor's contractual entitlement for any number of reasons, discussed in more detail below and Chapter 18 – Contractors' Claims.

The Right to a Variation

Common and Civil Law

In English common law,[4] it has been settled since the mid-nineteenth century that an owner may not issue a variation to the agreed scope of work unless there is express authority to do so under the contract.[5] As an obvious corollary to this, a contractor equally is not entitled to unilaterally depart from the scope of work that it was hired to construct, and accordingly will have no entitlement to payment for those unauthorized departures – indeed, the contractor will be in breach of the contract.

Absent a right to vary the work, under common law principles, parties would be forced to amend the contract for each variation and reach full agreement on all changes to the contract.

---

3 *FIDIC Red Book (2017 edition) Subclause 13.1.*

4 And, we submit, almost all common law jurisdictions.

5 See *Cooper v. Langdon* (1841), 9 M&W 60 at 67. On the other hand, in *Watts v. Moore* [1988] HHJ Fox-Andrews QC apparently held that it was an implied term of a construction contract that the owner could order extra work for which the contractor would be entitled to a reasonable remuneration.

Instead, construction contracts commonly include an owner's right to change the scope of the work by ordering variations, and principles and a process by which the parties determine a contractor's entitlement to relief, typically in the form of payment for the varied work, an extension to the time to the current date for completion of the work, and where applicable, relief in relation to the performance criteria that may be altered arising out of the change in scope.

In contrast, the approach in Europe is still inconclusive in regard to this For example, under French and German law, the right to a variation under construction contracts remains the subject of on-going discussion.[6] Additionally, the Draft Common Frame of Reference (DCFR) of 2009, which attempts to codify model contract rules of European private law, includes a general right for either party, by notice to change the service to be provided if such change is reasonable and takes into account the following factors: (i) the result to be achieved; (ii) the client's interest; (iii) the service provider's interests; and (iv) the circumstances at the time of the change. The DCFR also restricts application of the provision by qualifying that a change is reasonable only if it meets specified criteria which include consideration of whether, among other things, the change is necessary in order to enable the service provider to act in accordance with its specified obligations or is required by a change of circumstances which would justify a variation of the service provider's obligations following a unilateral juridical act by the court to vary the contractor obligations.[7]

Express Power to Instruct Variations

Almost all construction contracts contain express provisions by which the parties may vary the scope of work under the contract. The owner or the owner's representative may instruct a variation by giving a notice that describes the required change and sets out any requirements for the recording of contractor's costs.[8] The contractor typically proceeds with preparing the variation and submitting detailed particulars to the owner, within agreed timelines, relating to the description of the varied work performed or to be performed, including details of resources and methods to be adopted by contractor, program, and modification to the existing project schedule and time for completion, and the contractor's proposal for adjustment to the contract price with supporting particulars such as the estimated quantities or additional cost incurred. Where

6 "Unilateral variations in construction contracts," *Construction Law Journal* 29, no. 3 (2013): 211–233.

7 https://sakig.pl/wp-content/uploads/2019/01/dfcr.pdf.

8 *FIDIC Red Book (2017 edition) Subclause 13.3.1.*

variation instruction relates to omission of work, the contractor's proposal may also illustrate the adjustment for any loss of expectation profit or other losses and damages to the contractor as a result of the variation. The owner then proceeds to agree or determine the extension of time and/or adjustment to contract price. If the parties cannot reach agreement on any, or on a material portion of the variation, the owner may abandon the agreed variation process and resort to issuing a unilateral variation/change directive.

The Nature of Varied Work

At the outset, it is important to distinguish variations from additional work that would require, or be the subject of, a separate contract. The principle of variations is to carry out work that was either implied to be within the general scope of work contemplated by the parties, or a change to that generally contemplated work as may be necessary or otherwise approved by an owner. In either case the variation procedure in the contract is followed to provide notice and determine the compensation payable to the contractor for such variation. If a Contractor does carry out work that is, or is determined to be, work outside the contract, that work will be valued on a *quantum meruit* basis.

However, not every change to the contractor's intended scope of work either requires or necessitates the instruction of a variation.

Implied Work

In the oft-quoted English 1858 case, *Williams v. Fitzmaurice*,[9] a contractor was to build a house "dry and fit for occupation." The contract documents specifically note floor joists, but failed to mention flooring. The contractor claimed that the provision of flooring constituted a variation. The court clarified that where a contractor agrees to do work for a lump sum, the courts will infer a promise on the contractor's part to provide everything that is "indispensably necessary" to complete a project. Even where work has not been specified, the contractor is required to perform implied work as part of the overall scope of work and not as variation or an "extra."

Whether an owner's instruction is in fact a variation or a direction to carry out work that already forms part of the contractor's scope of work is a contentious area for dispute between the parties. An owner may issue an instruction or provide a written comment on a drawing that an owner considers is merely an instruction or clarification in relation to work that the contractor was already obliged to

9 *Williams v. Fitzmaurice* (1858) 157 ER 709.

undertake, while the contractor may consider this instruction or comment to be a direction to carry out an "extra." Whether the work is a variation or an "extra" is determined by referencing to the original scope of work and whether the subject work is impliedly or expressly included within the contract price payable. In most cases,[10] it will be critical to differentiate whether the work was that which the contactor was initially contracted for, or whether that work is beyond what the contactor agreed to construct.

Chittick v. Taylor, an internationally known Canadian case, provided four guideline points for determining whether an item or work is or is not a variation or extra.[11] Those points were:

> "**Rule 1**. An item specifically provided for in the contract is not an "extra."
> **Rule 2**. When the plaintiff [contractor] supplied material of a better quality than the minimum quality necessary for the fulfilment of the contract, without any instructions, express or implied, from the defendant [owner] to do so, he is not entitled to charge the extra cost as an "extra."
> **Rule 3**. When the plaintiff [contractor] does work or supplied materials not called for by the contract (plans or specifications) without instructions, express or implied, from the defendant [owner], or the consent of the defendant, he is not entitled to charge this additional work or materials as an "extra."
> **Rule 4**. When the plaintiff [contractor] did work or supplied materials not called for by the contract on the instructions, express or implied, of the defendant [owner], he is entitled to charge for additional work or materials as an "extra.""

Similarly, in a US mail services contract case, the court considered that in asking whether a variation instruction is within the scope of a variations power, it may be relevant to consider whether, if the variation were to be permitted, the contractor would be adequately compensated or remunerated for the additional work under the original contract.[12] If not, the court would be likely to find that the work instructed would not be of the nature of varied work.

The English, as well, have wrestled with what is and is not a variation. As an illustration of what was considered not to be a variation, in *Strachan and Henshaw v. Stein Industrie (UK) Ltd*,[13] an instruction requiring a subcontractor to relocate

10 Setting aside common law claims for *quantum meruit*, where work has been performed on a noncontractual basis, and in respect of which a contractor has a right to payment regardless of whether the work originally formed part of its base scope.

11 (1954) 12 WWR 653.

12 *United States v. Uth, Nevada and California Stage Co.* 199 US. 41 (1905).

13 [1997] EWCA Civ 2940, 87 BLR 52.

its "tea cabins" [i.e., site sheds] some half mile from the work site. The court held that the instruction was *not* a variation, despite the fact that the instruction cost the contractor an estimated £1.6m in productivity loss for "walking time." The instruction did not constitute a change in the "work to be done by the Contractor under the Contract" so as to trigger a variation under the general conditions of Contract and the contractor was not entitled to any additional costs.

Contractual Risk Allocation

Notwithstanding a general right to vary the scope of a contract, it is also not uncommon for parties to allocate what might become variation risks, and limit the instances in which variations will be permitted at the time of contract signing. That is, except where an owner initiates a variation to the scope of the work entitling the contractor to an adjustment in cost and time, a contractor is entitled to variation only in the circumstances expressly listed in the contract. These risk allocations may well legislate that certain events or risks are in the contractor's control and accordingly, expressly or impliedly, not the subject of a variation. The classic example of this is the allocation of certain concealed or subsurface conditions – in the parlance of the FIDIC 2017 contracts, "unforeseeable physical conditions."[14] While contracts may allocate this risk to an owner, it remains that where the contract allocates the risk to the contractor, the contractor would not be entitled to a variation, even where the work to be done was significantly different. This same allocation of risk principle can be variously applied, in whole or in part, to numerous other provisions in construction contracts – including with respect to force majeure events, changes in applicable law, delays caused by third parties, and in the "no damages for delay" provisions commonly seen in the United States, even to breach of the contract by the owner.[15]

In construction contracts for complex infrastructure projects, particularly projects being implemented in the public-private partnership mode, it is common for the project agreement to outline delay and relief events upon the occurrence of which the contractor will be entitled to a variation. Additionally, contractors are in many such contracts permitted or encouraged contractually to submit innovation and value engineering proposals that may reduce the project cost or the service fees payable to the contractor.

---

14 For other examples: the American A201 and Canadian CCDC 2 "concealed or unknown conditions," the Australian AS4300 "unforeseen conditions," the English NEC4 Contract "unforeseen physical conditions."

15 To be clear, "no damages for delay" clauses do not deprive a contractor of a remedy entirely, but rather limit their remedy by agreement to time, but not money, as has been the case for force majeure events in many common law contracts.

No Oral Variations Clauses

Many contracts contain a provision prohibiting any express oral variations. The purpose of the provision is, on the face of it, commendable and benign, as it no doubt intended to ensure that variations are clear for both parties, and backed by the appropriate written documentation. In practice, however, this provision can be problematic. In England, this fact scenario was at issue in the 2018 Supreme court case, *Rock Advertising Limited (Respondent) v MWB Business Exchange Centres Limited [2018] UKSC 24*[16] in which the Supreme Court determined that a "no oral modifications" clause needed to be given legal effect, and to ignore such a clause would override the parties' intentions.

However, the English position is not ubiquitous, even in the common law world. In Australia, a "no oral variations/no oral modification" clause is generally not considered legally binding.[17] In Canada, while there is considerable lower court authority, there are Ontario appellate court decisions holding that a contract may be varied or amended orally or by conduct even if the contract contains a no-oral variation clause, on the basis that under the principles of contract law, the parties always have the autonomy to enter into a new agreement, although the onus of proving that such a variation or amendment meets the requirements of contract law may be heavy in light of that clause.[18] In the United States, there too is a patchwork of case law on the common "no oral change orders" clauses, wherein in at least some states, courts have found that oral change orders may generally be enforceable.[19]

Alternative Materials

A variation or change order may be required if there is a change in the materials being used for the project or new materials are required. A change order may be required to account for the increased cost of procuring new materials, price difference payable following change of materials (if the original specified material was cheaper) and the lead time and cost for the materials. Lead time is the time from when the material is ordered until it is delivered to site. This may vary based on material availability, distance from location of origin to delivery site or time for building custom material for the project. The construction of the Malir Expressway in Pakistan was an illustrative example of this. A court order

---

16 [2018] UKSC 214.

17 *Sara Stockham Pty Ltd v. WLD Practice Holdings Pty Ltd* [2021] NSWCA 51 at [15].

18 *Shelanu Inc. v. Print Three Franchising Corp.*, 2003 CarswellOnt 2038, 64 O.R. (3d) 533 (Ont. C.A.) at [5], discussed in Heintzman, West and Goldsmith on Canadian Building Contracts, 5th Ed.

19 For example, in Pennsylvania, see *STI Oilfield Services v. The Williams Companies, Inc. f/k/a Access Midstream Partners*, No. 2018-1003 C.P. (Pa. Com. Pl., Susquehanna County, March 16, 2020 Opinion).

prohibited the planned excavation of sand from the Malir riverbed for the use in construction of the expressway.[20] In anticipation of the court proceedings not being resolved in time for commencement of construction, the parties negotiated that the contractor would be entitled to the lead time and cost of procuring of sand from an alternative site.

Instruction to Correct Defective Works

Many standard form contracts give an owner or its representatives the power to direct an instruction or no-cost variation in respect of overcoming defective work. Where a breach by the contractor results in defective work, the instruction by the owner (e.g., to adjust the remainder of a design to accommodate a wrongly located concrete wall) will not amount to a variation entitling the contractor to additional payment.[21] Similarly, a variation directed to overcome the consequences of a defect shall also not entitle a contractor to a variation.[22] As one example, the FIDIC Red Book[23] states that any defects in the work that are attributable to the contractor's design, breach, or improper operation and maintenance shall be remedied at its cost and risk.

Instruction to Change Work Methods

In the *Joint Contracts Tribunal* (JCT) standard form contracts, the definition of "Changes" includes, in addition to the change of contractors scope of work, the imposition by the owner of obligations or restrictions that relate to: (1) access to site or specific parts of the site, (2) limitations of working space, (3) limitations of working hours, or (4) execution of completion of work in any specific order.[24]

These types of changes to the contractor's work methods are particularly relevant on construction sites that may be partially occupied, such as an addition to an existing building. In construction projects where the site is under complete control of the contractor, the contractor has more flexibility as to its work methods that relate to access to site or working hours. However, where an owner instructs or changes the work methods, it follows that the contractor is entitled to a change. For example, if a contractor has to perform work out of hours, or with restricted access, that may impact the project schedule and result in an increase in overhead

20 See discussion and background at https://www.dawn.com/news/1678926 published March 8, 2022.

21 See, in England, *Dixon Kerly Ltd v. Robinson* [1965] 2 Lloyd's Rep 404.

22 See, as one example, *Howard de Walden Estates v. Costain Management Design Ltd* (1991) 55 BLR 124.

23 *FIDIC Red Book (2017 edition) Subclause 11.2.*

24 See, for example, clause 5.1.2 of JCT Design and Build Contract, 2024 Edition (DB 2024).

costs and time required for completion. Where the contactor is required to work within prescribed methods that were not contemplated by the contractor at the time of execution of the contract, the contractor is entitled to a change.

Extensions of Time, Acceleration, and Disruption

The power to vary the contract scope does not, without an express agreement to the contrary, allow an owner to reduce the time for completion of the work or any part of it. Typically, while a contractor has the power[25] to accelerate the progress of its work as it may see fit, the same entitlement is not afforded to any owner. Under common law, the contractor is entitled to the time allowed to perform the works as at the date it enters into the contract.[26]

Notwithstanding the general position, there are circumstances in which an owner may well wish to accelerate, and many contracts, including standard form contracts, do contain provisions allowing for acceleration to be directed as a variation, though generally only by agreement.[27] Other standard form contracts provide for acceleration of work when the contractor is in inexcusable delay, such as Clause 8.7 of the FIDIC 2017 suite and § 8.3.4 of the American AIA Document A-201 (2007).[28] These contractual provisions effectively shift the state of play at common law, as absent the Owner's express right, it is a well-established principle that a contractor (being then "a party in breach of a contract") would typically have discretion as to whether to rectify the breach or elect instead to accept the consequences – typically damages – for its failure.

Constructive Acceleration

Constructive acceleration is a concept that emerged in the United States from the Federal Government's Boards of Contracts Appeals in the late 1970s.[29] In its simplest summary, constructive acceleration occurs when a contractor claims an

25 Generally, and absent a provision to the contrary preventing the Contractor from accelerating.

26 In the UK, see *Ascon Contracting Ltd v. Alfred McAlpine Construction Isle of Man Ltd* (1999) 66 Con LR 119.

27 See, for example the JCT Standard Building Contract 2024 (acceleration by agreement following an Acceleration Quotation, or the FIDIC Red Book (2017) clause 13.1(vi) "changes to the sequence or timing of the execution of the Works."

28 Section 8.3.4 provides that "In the event of Inexcusable delay by Contractor, Owner may direct that the Work be accelerated by means of overtime, additional crews or additional shifts or re sequencing of the Work. All such acceleration shall be at no cost to Owner."

29 See *Constructive Acceleration – A Global Tour*, James G. Zack, Jr, Navigant Construction Forum, ©Navigant Consulting Inc., 2011, and *Constructive Acceleration: Exploring the Delay/ Accelerate Quandary, Robert D'Onofrio, Professor Doug Jones AO, Dr Hamish Lal And Kim Rosenberg, [2024] ICLR 1–147 Volume 41, Part 1, January 2024.*

extension of time in a proper and timely manner, but its claim was wrongly denied or not timely granted, the effect of which was to force the contractor to accelerate so as to avoid being in delay under the contract. Constructive acceleration is now a well-established legal doctrine in the United States, and the concept has made inroads into various other common law countries to various extents, including Australia and Canada.[30] Under the doctrine, where a contractor incurs cost in accelerating its work to try to achieve substantial completion by the unadjusted date, the contractor may be entitled to claim damages for its acceleration costs from the owner. The fundamental legal basis for the compensation would be that the owner, having committed a breach of contract in refusing to grant a proper extension of time, will be liable for the consequences of its breach. However, the concept of "constructive acceleration" does not seem to have found the same purchase in all common law, or any civil law jurisdictions.[31]

The Variation Process

Generally

Construction contracts internationally follow generally consistent formalities from inception of the variation through to its conclusion. Broadly speaking the process involves, at its simplest: (i) an owner's request in respect of a proposed variation; (ii) the contractor's proposal in respect of an owner's request, setting out the time, cost, schedule and quality (or performance) impacts of the proposed variation on the work; (iii) an owner or its representatives enacting the variation by way of a formal variation order; and (iv) the contractor counter signing or acknowledging the variation. While the process is dependent entirely on the terms of the contract, it seems appropriate to make some general observations.

Equally, most contracts provide for a unilateral variation (or, in the United States and Canada, a "change directive") to be issued by an owner on the basis of, typically, a time unit rates and materials cost payment regime with predetermined percentages added for overhead and profit, and with schedule impact to be

30 In Canada, see, for example, *Morrison-Knudsen Co. Inc. v. British Columbia Hydro and Power Authority* (1978), D.L.R. (3d) 186 (B.C.C.A.). In Australia, while courts have not used the term "constructive acceleration," in *Perini Pacific v. Commonwealth of Australia* [1969] [1969] 2 NSWLR 530 and *in V601 Developments Pty Ltd v. Probuild Constructions (Aust) Pty Ltd* [2021] VSC 849, the courts recognised similar claims, although more commonly, Australian courts have looked to the English law "prevention principle."

31 Zack, n30.

determined later, based on a contractual algorithm. Those unilateral variations/change directives are subject to the same limitations as are variations generally.

Where an owner is contemplating a variation, the typical first step in the process is for the owner to make a written request for a variation to the contractor. The owner may provide details of the variation contemplated,[32] or may make a general request and leave the detail to the contractor.[33] Most contract provisions require the request for the proposal to be in writing and may also set out minimum criteria for the detail to be covered in the contractor's proposal – for example, the detailed description of the proposed variation, impact on the project schedule, project operations or performance guarantees, expected usage of utilities and contract price and whether any additional permits and approvals are required to be obtained. If the contractor considers that its request falls within the limitations of the variation regime (in respect of which, see Section "Limits of the Power to Direct Variations"), the contractor will typically submit a written proposal to the owner for it to consider, setting out details of the proposed variation and its impact on the time, money and quality aspects of the work under the contract.

Some contracts provide negotiated, contractual limitations on an owner's right to vary the work, which is discussed in Section "The Power to Direct Variations."

Occasionally, given the formality of the process, the timelines built in contracts for preparing, reviewing and approving the proposals, and the cost of variation process itself, parties sometimes negotiate a provision that the variation procedure will not be followed for very small works. Such changes to the scope of work are negotiated and implemented by the parties in the ordinary course of business, though they are usually ultimately documented through the variation regime either individually or on a periodic basis.

The cost of preparing variation proposals is occasionally a contentious point as those costs may not be recoverable as a cost of the work itself, particularly where proposals are abandoned. A typical owner's position is that all costs for preparing a variation proposal are "free" – that is, to be borne by the contractor as a cost of doing business, covered in the contractor's overhead, or the cost of the variations that do proceed. However, where the contractor is required to carry out detailed design or extensive preparatory work, or where owners have a reputation for abandoning many variation proposals, it is not uncommon for contractors to seek the cost recovery of a variation as a separate cost, to limit their burden of stranded costs of variation proposals that do not proceed.

32 As might be the case in a construct-only contract.

33 As might be the case in a design-build or an engineering, procurement, and construction contract.

Upon receiving the variation proposal, an owner might typically be required to revert within a specified time period (or "a reasonable period"), to approve, reject, or provide comments and questions on the proposal. The response period and the need for a quick response is critical for two reasons: Firstly, to ensure that the proposal is not out-of-date upon approval, as the contractor must proceed with unaffected work while the variation process proceeds. As well, material and subcontractor rates may fluctuate between the time they are quoted and approved, or the validity period for subcontractor and supplier quotes may be limited. Secondly, the proposed variation may have a disruptive overall impact on the progress of the works – including, at the extreme, that the contractor may be required to demolish newly constructed work to accommodate the variation.

If a variation proposal is approved, the owner issues or countersigns a formal written variation order, recording the agreed terms details of the order. In its purest form, a variation order is anticipated to deal with all of the consequences of the variation. However, it is unfortunately not uncommon for some aspects of the variation to be deferred to a later date.

Contractor Originating Variations

The entitlement to variations is not limited to an owner only, and under most forms of construction contract, a contractor has various explicit entitlements to make a claim for a variation – but not, for perhaps obvious reasons, to issue the variation itself. Typical entitlements of this nature may include where relief is granted by the contract for changes in law, encountering unforeseeable physical conditions, the discovery of archaeological artifacts or hazardous substances, breaches of contract or acts of prevention by an owner, or delays caused by an owner or third parties. All of these entitlements will be specific to, and must be expressly listed as relief events in, the contract, and depending on the language of the contract, may entitle the contractor to a claim to be granted a variation. For example, subclauses 2.24 to 2.26 of the JCT Design and Build Contract, 2024 Edition permit adjustment to the completion date, by way of the change [variation] regime, in the event of specified "Relevant Events." Those "Relevant Events" include delay in handing over possession of site, suspension of work by contractor in the event of delay/nonpayment by owner, owner's instructions relating to postponement of work or opening up works for inspection or testing (unless the work, materials or goods are not as per contract requirements in which case there may be no entitlement for the contractor), force majeure events or delay in obtaining relevant statutory approvals.[34]

---

34 Section 5.1(2) of JCT Design and Build Contract, 2024 Edition (DB 2024).

Contracts[35] often contain provisions relating to variations "for the convenience of the contractor."[36] These types of variations generally do not entitle a contractor to time or money under a contract, so contractors often use this clause only when they can demonstrate a benefit to the owner.

Value Engineering Proposal Process

Many construction contracts include value engineering proposals, by which a contractor is permitted or obliged to submit proposals that propose to accelerate completion, reduce the contract price, or to reduce the long-term cost of maintaining or operating the works, or otherwise benefit the owner by improving the project efficiency.[37] Value engineering proposals are customarily prepared by the contractor at its cost, on the premise that the contractor will benefit from a contractually allocated share of savings realized by the owner.

Under the FIDIC Red Book regime, for example, following an engineer's consent to the value engineering proposal (with or without comments), the engineer is to instruct a variation. The Red Book additionally clarifies that where a value engineering proposal consented to by the engineer (on behalf of the owner) and the proposal includes a change in the design of the permanent works, then unless agreed by both parties, the contractor will design this part at its own cost and be bound by its other obligations that relate to warranting the adequacy, stability, and safety of all the contractor's operations and activities, of all methods of construction.

Value engineering for long-term operational savings can be both beneficial and hotly contested, and much depends on the precise drafting of the clause. The author is aware of at least one disputed proposal where a contractor value engineered a reduction in size of numerous pumps that were genuinely overdesigned, but then sought, in the value engineering gain share, not only the reduced capital cost of the pumps, but the exponentially larger operational savings over the life of the relevant plant to which the engineering related.

Variation Entitlements in Civil Law

The German Civil Code (the Bürgerliches Gesetzbuch (BGB)[38] codifies the owner's right to issue a variation. Section 650b states that the owner may seek a modification to obtain (i) the agreed result to be obtained by the work for example

35 Largely "construct-only" contracts, where the detail of the work is not left to the broader discretion of the contractor.

36 See, for example, Clause 36.3 of the Australian AS 4902 Contract, under which a contractor can request the superintendent to direct a variation.

37 *FIDIC Red Book (2017 edition) Subclause 13.2.*

38 https://www.gesetze-im-internet.de/englisch_bgb/englisch_bgb.html

if an essential item of work has not been expressly stated in the scope such as sealing which would be in any case implied work; and (ii) the agreed result to be obtained by the work and a modification is necessary to achieve the same. Thereafter, contractual parties will strive to reach mutual agreement on the modification and remuneration payable as a result. If the modification is necessary for the successful performance and for the work to achieve the agreed results under part (ii), the contractor must implement the modification. However, a contractor is only obligated to proceed with a modification under part (i), which may be categorized as a mere owner's preference and not necessary from a technical standpoint then the burden of proof is on the contractor to establish that the modification cannot be reasonably expected of it.[39]

The BGB entitles a contractor to be remunerated for additional works for their actual costs plus a reasonable surcharge for general administrative costs, business risk, and profits, and in calculating the remuneration the contractor may take recourse to the calculation approaches used in the initial costing calculation.[40]

Variations to construction contracts are permitted in the UAE and the right has also been codified. The UAE Civil Code under Article 887 and Article 888[41] states that if a variation or addition is made to a plan with the owner's consent, the existing agreement with the contractor must be observed and if consideration for work is not specified in a contract, the contractor shall be entitled to fair remuneration. Variations are also expressly permitted in context of procurement of federal government public works contracts governed by the Federal Finance Order No. 16 of 1975. The order regulates the tendering process and permits variations provided the work is carried out and delivered as per the contract except for any stoppage caused by a variation.[42]

It is worth a minor note that there are some stylistic differences in the use of change orders and variations between the North American market and the rest of the common law world. Most notably, in North America, claims involving only changes in time are almost invariably administered through the change order. In many other jurisdictions, contracts[43] make a distinction between extensions of time orders and variations, which are typically separate documents. In the author's experience, however, this is a matter of mere form and not substance.

---

39 The New German Construction Law, 12 No. 4 Construction L. Int'l 46.

40 The New German Construction Law, 12 No. 4 Construction L. Int'l 46.

41 https://uaelegislation.gov.ae/en/legislations/1025/download.

42 Tony Pennington, "Variations (Change Orders) In Construction Contracts For U.A.E. Turnkey Projects" (1997) 20 No. 3 Middle E. Executive Rep. 9.

43 See, for example, JCT Design and Build Contract, 2024 Edition clause 2.23; FIDIC Red Book clause 8.5.

Limits of the Power to Direct Variations

Introduction

As between the owner and contractor, there is arguably no limits at law on the extent which the parties may agree a variation. There may be limits outside the contract that prevent this – particularly where public procurement and fairness to others may be involved, and there may be practical limits relating to a contractor's capabilities or restrictions on the efficacy of the contract language to address variations that involve a fundamental change in the nature of the parties' bargain.

Common Law Limits

In most common law jurisdictions, there are no limits on the extent or quantum of the variations permitted under a contract. That said, however, a variation clause cannot be used to require the contractor to perform work that is drastically different from the original scope of work to be performed at the original contract rates.[44] That work, if directed, would be categorized as additional work, falling outside the original scope of work, and one that a contractor could refuse to perform.[45]

For those familiar with American law, this is closely related to the American "cardinal change doctrine." Under that doctrine, a cardinal change is "a drastic modification beyond the scope of the contract" and constitutes a breach of contract. It is worth highlighting that the cardinal change doctrine developed in context of federal government contracts and constituted a breach of contract in instances where the federal government of the United States altered the contractor duties to be materially different from the duties originally bargained for. However, the doctrine is deemed to be applicable to all variations under contracts. The requirement of the change being "materially different" is an imprecise subjective criterion that must be assessed on a case to case basis on the facts and circumstances of the change. A key contrast in the American and English law approach is the legal effect of a variation. The cardinal change doctrine is breach-based and provides that a variation resulting in a material difference in the scope of contract constitutes a breach entitling the contractor to damages for breach of contract. Under English law, the contractor has the option to decline to implement the out-of-scope work, resulting in a dispute, or in extreme circumstances, termination, or alternatively to perform the additional work for increased compensation.

---

44 See, for example, *Blue Circle Industries PLC v. Holland Dredging Company* [1987] 37 BLR 40.

45 See section "The Nature of Varied Work."

The Power to Direct Variations

Because the power to direct a variation under a contract is not unlimited, whether an instruction to vary the work is permissible will depend up on the exact terms of the contract when viewed in the context of the commercial deal between the parties. Absent an express provision to the contrary, variation powers will not be interpreted as allowing an owner to give variation instructions that have the effect of changing the work entirely. Similarly, an owner will not be entitled to direct a contractor to perform work that is beyond the contractor's capability or to extend to work that the parties had expressly excluded from the scope.

The English case *Blue Circle Industries plc v Holland Dredging Company (UK) Ltd. (1987) 37 BLR 40*[46] illustrates one limitation on variations powers. In that case, the court held that the construction of an island using disposed materials from dredging works was deemed to be work wholly outside the scope of the original dredging contract, unexpected and so different from anything contemplated by the contract. Therefore, it is subject to a separate agreement. This type of limitation has two noteworthy aspects. First, an owner does not have the right to insist that an unlimited scope of work be performed under the existing contract terms, or even at all. Second, if the contract provides that varied work will be performed under agreed contract rates, the parties will not be bound by those rates for work not falling under the scope of the contract. Where the work is additional work (as was the case here), the parties will be free to negotiate new rates that may differ from those quoted in the original contract.

Express Contract Limits on Variations

In most standard form contracts, the contract itself imposes limits, typically based on either practicality, custom, or the legal regime of the relevant jurisdiction. For instance, in the American A201 form, the Owner may in a Construction Change Directive, "order changes in the Work within the general scope of the Contract consisting of additions, deletions or other revisions." In the Australian AS 2124, the Contractor is "only bound to execute a variation which is within the general scope of the contract" and provides that the Contractor may advise that the variation cannot be effected.[47] The FIDIC Red Book expressly states a variation does not include omission of any work that has to be carried out by the owner or other unless agreed between the parties.[48]

46 At the time of contract signing.

47 AS 2124-1992, clauses 40.1 and 40.2. We note that the 2124 remains in use in 2025.

48 *FIDIC Red Book (2017 edition) Subclause 13.1.*

The FIDIC Red Book.[49] as another example, states that a contractor is bound by each variation unless it gives notice (with supporting particulars) refusing to implement a variation on the basis that: the varied work is unforeseeable having regard to the scope and nature of work specified in the contract, the contractor cannot readily obtain goods required to implement a variation, or it will adversely affect contractor's ability to comply with health and safety and/or environmental protection provisions and requirements under the contract.

At the more contractor-friendly end of the construction contract spectrum, some contracts[50] provide the contractor with discretion as to whether the contractor even wishes to perform a variation at all, regardless of whether the contractor is capable of doing so. Other forms may seek to constrain the contractor's discretion in various ways, such as limiting the contractor's right to reject a variation price request only for specific reasons, such as an assertion that performing the variation is beyond the contractor's technological or manpower capability, or that the extension of time even if granted will extend the contractor's duration on site beyond a specified calendar date.[51]

Similarly, in public-private-partnership (P3) contracts globally, it is common practice to specify grounds on which a contractor and/or the project company owner can refuse or object to providing a proposal and proceeding with a variation. The limitations may be those such as where implementing a variation will infringe applicable laws, cause existing permits or approvals to be revoked or require new permits and approvals, have a material adverse impact on health and safety or performance on the project operations or performance guarantees, if time specified for completion of the works cannot be met despite commercially reasonable practices, be contrary to good industry practices, or where an owner has not provided adequate information to develop a variation proposal.

In this context, multiple variations, each of which is individually within the overall scope of work, can cumulatively result in a change of scope of the contractor duties. This is the "cumulative change doctrine" under American law, and in which the collective, or cumulative, impact of the variations on the construction process and costs to the contractor fundamentally change the scope that was originally agreed.[52]

---

49 *FIDIC Red Book (2017 edition) Subclause 13.1.*

50 In the author's experience, this is typically limited to those proffered and drafted by the contractors themselves.

51 E.g., in a dredging contract, where a barge may be booked only for a specific chronological window.

52 See part "*Limits of the Power to Direct Variations: Express Contract Limits on Variations.*" Note that a "cardinal change" refers to a single, drastic alteration to a contract that fundamentally changes the scope of work, while the "cumulative change doctrine" relates to a series of smaller changes that, *when taken together*, create a substantial change to the contract, even in the absence of any cardinal changes.

Increases and Decreases in the Volume of Work

Changes to quantity of the work under the contract will only be categorized as a variation if it is additional or extra work the contractor is required to carry out and is distinguishable from the overall works required to be performed by a contractor in any case to complete the contract.

Construction contracts do not typically contain an express limitation on an owner's right to omit work and give that work to other contractors. However, in common law jurisdictions, that power is not unlimited and cannot be such that the variation fundamentally changes the scope of the original contract, or requires work that is not within the general scope contemplated by the parties at the time of entering into the contract. There are common law limits on an owner's ability to omit work by variation, and without an express contract provision allowing it, an owner may not remove work from a contractor to give it to another contractor, or carry out the work itself. Absent an express provision entitling an owner to do so, courts have held it "self-evident that the building owner must permit the contractor to carry out the whole of the work... It is implicit in most contracts that an owner who exercises a power to omit work must genuinely require the work not to be done at all."[53]

The Australian case *Carr v J.A. Berriman Pty Ltd (1953) 89 CLR 327*[54] is the leading case in many common law countries on an owner's right to remove work from the contractor's scope. In that case, after the contractor had placed its steel order, the owner's architect informed the contractor that the owner had (before informing the contractor) unilaterally removed the steel fabrication work from the contractor scope of work and given it to another party – at a lower price. As a result of such work being carved out of` its scope the contractor suffered a combined loss of profit on the project and incurred damages that were payable to its steel fabricator. The High Court of Australia determined that in so doing, the owner had "given... Berriman the right to believe that the contract would not be performed according to its true construction" and held that the contract had been repudiated by owner's actions. The court held that the owner did not have the power under the building contract to omit or reduce the contractor's scope of work to give to another party.

---

53 *Tancred Arrol v. The Steel Company of Scotland* (1890) 15 App. Cas 125.

54 (1953) 89 CLR 327. American law has taken a similar position in *Gallagher v. Hirsch (1899) 45 AD 467* where the court held that the word "omission" contemplates only things to be left out of the contract altogether, not taken out and given to others.

Similarly, absent an express contractual provision to the contrary, an owner's variation power is not sufficient to omit or delete the entirety of the contractor's work or a substantial portion of it.[55]

Instruction beyond Scope of Variations Power

While an owner has some discretion to change the time and sequence of the works as necessary, this power is not unlimited. In addition to the acceleration limitations discussed in parts 14.4(g) and (h) above, an owner is limited in its discretion to vary the contractors methods for execution of the works, both at law and in most standard form contracts.[56] As one example, in Canada, the Supreme Court has opined:

> "It has long been established that, barring specific arrangements to the contrary, the method by which a contractor chooses to execute the work falls within its sphere of responsibility, and that neither the owner nor the design professionals employed by the owner have a duty to advise the contractor as to what method to choose, or how to go about accomplishing the work by whatever method the contractor in fact chooses."[57]

A unilateral variation to the scope of work by the owner may also impact the ultimate performance of the work[58] – and in engineering, procurement, and construction, contracts may prejudice the contactor's ability to achieve the performance guarantees it has agreed to achieve.

No Power to Instruct Variations After Practical Completion

Most standard form contracts restrict not only the scope of the power to instruct variations but also the timing. The FIDIC Red Book states that variations may be initiated by the owner at any time before the issuing of the Taking-Over

55 In Australia – see *Chadmax Plastics Pty Limited v. Hansen and Yuncken (SA) Pty Ltd* (1984) 1 BCL 52. In that case, a head contract variation resulted in a 98% deletion of Chadmax's subcontract works. Relevantly, the court held that the contract gave power to the principal to vary the subcontract works, but not the power to virtually cancel them.

56 See, for example, FIDIC Red Book 4.1, AIA A201 § 3.3.1, CCDC 2 at GC3.1.2.

57 *Auto Concrete Curb Ltd. v. South Nation River Conservation Authority*, 1993 CarswellOnt 806, [1993] 3 S.C.R. 201 (S.C.C.) at p. 901.

58 Design- build contracts also often give a contractor the ability to refuse a variation where the revised design leads to unacceptable risk – e.g., the FIDIC Silver Book 1999, cl 13.1.

Certificate for the works.[59] Once the contractor has achieved practical completion[60] of the works as required by the contract and the relevant completion certificate has been issued, it is not uncommon for contracts to provide that the owner cannot issue a variation. Traditional wisdom in respect of instructing variations after the scheduled date for practical completion was that doing so would prejudice an owner's claim to delay liquidated damages as the scheduled time for completion would be correspondingly adjusted to account for increased time required to perform the varied work, making the variation at the least a bad idea. Moreover, following completion a contractor has presumably demobilized from site and may be unable or unwilling to retrospectively carry out additional work. However, as Julian Bailey has observed: "There is, however, no reason in principle why a construction contract may not empower the owner or its agent to instruct the contractor to perform variation work at any time, including after the achievement of practical completion or even final completion."[61]

Variations Without a Variation Order

Introduction

Any change to the scope of the work requires the owner's approval, whether implicit or explicit. A variation order is, of course, one example of that approval, but there are other ways by which a contractor may be entitled to a variation.[62]

Recovery in the Absence of Written Variation Instruction

At the simplest, an owner may have either directed a variation or consented to a contractor's variation request without issuing a formal variation order. Where the

59 *FIDIC Red Book (2017 edition) Subclause 13.1.*

60 Or the contractual equivalent however described – e.g., taking over, handover, turnover, completion, and substantial completion are just some typical names for this point in time. See, for example, the FIDIC Red Book clause 13.1 (2017 ed), AS 4000-1997 at 36.1, which address this explicitly. Other contract forms (CCDC 2 in Canada, AIS A201 in the USA making reference to adjusting the "Contract Time," JCT in the UK revising the "Completion Date") deal with this only indirectly.

61 Julian Bailey, Construction Law, 4th Ed, 2024, London Publishing Partnership. II.7.37.

62 This chapter deals with variations narrowly, and does not discuss other claims or collateral contracts, such as where a contractor has undertaken work at the owner's request outside the contract and may be entitled to restitution, or where the parties agree to enter into a separate agreement (orally or in writing) outside the contract for carrying out additional work. Nor does it deal with owner directed suspension or force majeure, and the consequent variations that may arise out of those.

owner does this, the lack of a formal order is no bar to the contractor's right to a claim for variation.[63] In a similar manner, an owner that has issued an instruction may refuse to issue a variation because it considers that the relevant work in discussion does not constitute or require a variation e.g., on the basis that the instruction relates to work that the contractor is already required to undertake, or is required to make the work comply with the contract, or is correct contractor's work the owner considers to be defective.

The American doctrine of "constructive change" has developed to deal with this problem, where a written instruction has not been issued under the contract, but the contractor is under pressure to implement the change anyway, because otherwise the project cannot proceed. This doctrine allows the court to find that an implied or constructive change order has been issued.

The courts of England and Wales and other common law jurisdictions have not adopted the same approach, notwithstanding that the same issue is ubiquitous in construction. Under English law, where a contract specifies that a written instruction is a condition precedent to a contractor's entitlement to payment, courts have upheld that provision and the owner will not be obliged to pay in the absence of a written variation.[64] However, there are also cases dating back to the nineteenth century in which the principles of waiver and estoppel were applied to defeat an owner's argument that the lack of a formal variation instruction deprived the contractor of its right to payment for changes, though most commonly where the owner had directly informed the contractor that it required the additional work to be performed but did not issue the formal instruction.[65]

Directions Changing Work Methodology or Sequence

English cases further demonstrate that whether a direction to alter the sequence or method of work constitutes a variation will depend on the terms of the contract. Where an owner exercises and expresses power in a contract to change the sequence of construction or the methodology by which the contractor performs the work, the exercise of that power will entitle the contractor to a claim for variation.

Cumulative Impact

The US "cumulative impact" or "cumulative change" doctrine, familiar to US construction practitioners, is a relative unique development that has not found

---

63 See further discussion in Chapter 18 – Contractors' Claims.

64 Bailey, II 7.43 refers to a litany of nineteenth-century cases.

65 See the more thorough discussion of waiver and estoppel in relation to variation claims in *Construction Contract Variations*, Sergeant and Wieliczko, Informa 2014, at Chapter 8D.

much purchase in other jurisdictions to date. The authoritative American construction law text *Waiver of impact costs, 2A Bruner & O'Connor on Construction Law § 7:245* describes the cumulative impact thus:

> "Where there are numerous changes made during construction it is possible that the changed work can affect the sequence and timing of the unchanged portions of the work. These costs are sometimes referred to as impact or cumulative impact costs. Because change orders often contain broad release or waiver language, owners frequently adopt a position that by executing the change order the contractor has waived or released any right to impact costs. Contractors, on the other hand, argue that the change orders should not be construed to cover these costs, as they were unknown at the time the change order was executed... It can be difficult for a contractor, at least with respect to the earlier change orders, to realize or appreciate the impact costs associated with the changed work."[66]

The cumulative impact doctrine is often countered in US contracts with broad waiver and "full accord and satisfaction of claims" language, inserted with the core purpose of closing off such claims for cumulative impact.

In other jurisdictions, this philosophy or its relation in the form of "global claims" has had lesser success. Bailey has written, "A claim will be unobjectionable if it can be established, to a level of persuasive satisfaction, that the events in question, being events for which the owner bears contractual responsibility, had the cumulative effect of causing the contractor to separate delay and/or a loss."[67] Conversely, in the leading English case on this issue *McAlpine Humberoak v McDermott International(1992) 58 BLR 1*,[68] the Court of Appeal held that each variation must be assessed individually and the contractor is entitled to be paid for each variation as per the contract prescribed mechanism, and the contractor's entitlements assessed by reference to the variations clause in the contract. The contractor could not abandon the regime and seek to claim for extra work on a different basis.

The UK Society for Construction Law (SCL) Delay and Disruption Protocol update in 2017 says that, "There is recognition of an apparent trend for the construction legal industry and the courts to take a more lenient approach towards global claims, albeit the risks in proceeding on this basis remain" and that "the

66 §7:245. Waiver of impact costs, 2A Bruner & O'Connor Construction Law §7 : 245. (citations omitted).

67 Bailey, II 11.70., citing, inter alia, *London Borough of Merton v. Leach* (1985) 32 BLR 51 at 102.

68 *McAlpine Humberoak v. McDermott* (1992) 58 BLR 1.

practice of contractors making composite or global claims without attempting to substantiate cause and effect is discouraged by the Protocol, despite an apparent trend for the courts to take a more lenient approach when considering global claims".

However, the landscape in England and Wales may be shifting, and as recently as December 2024, the Court of Appeal provided guidance on a plaintiff's ability to plead global claims, saying, "No judge or court is entitled to require a party to establish their case by a particular method. A party should be permitted to formulate their claims as they wish, not forced into a straitjacket (or corner or cul-de-sac) of the judge's or their opponent's choosing."

Valuation of Variations

Generally

While it is not tautologically the case that all variations will result in a change in time and money, in most cases, variations will result in changes in time and/or money.

The process of determining the value of a variation is fundamentally determined by the terms of the contract. In common law jurisdictions, there are no fixed legal principles that determine how variations must be valued. Of course, the process of determining the price of variation may vary depending on whether the contract price is a lump sum fixed price or a "measure and value contracts." In rare cases, a contract may simply not contain a variation pricing protocol. In those circumstances, the contractor's entitlement will need to be determined under the appropriate common law (or civil law) principles, such as the law of restitution.

A contract with a lump sum fixed price is where the contract price is fixed and the fixed price is adjusted upwards or downwards to account for the variation. There is no basis in the contracts to account for the revised scope of work in its totality and to renegotiate a fixed price for the revised scope of work. The contractor remains bound by the lump sum contract price at the rates quoted on the scope of the work and the only adjustment to the contract price is to account for cost of the change in scope of work.

Nonetheless, most forms of contract generally contain a familiar process regime for dealing with the time and cost consequences of variations. Loosely, that process often involves: (i) seeking to value the variation by agreement; followed by (ii) determination of the variation costs not agreed, using a cost determination algorithm.

Generally, for lump sum fixed price contracts, where a contemplated variation[69] is initiated by the owner or the contractor, the contractor typically prepares and submits a proposal outlining the impact of the variation on its work, timelines and other consequences to the work for the owner's review and approval. The procedure may include a consultation period during which the parties may review and discuss the variation impacts set out in the proposal. Once the consequences of the proposed variation are settled, the parties proceed with a formal variation (or, in North American parlance where price and timing are agreed, a Change Order). Once issued – and in North America, generally only upon it being signed by both parties – the variation becomes part of the contract work and is implemented by the contractor.

Almost all construction contracts contain a procedure for determining the cost and time consequences of a variation, failing an initial agreement between the parties. Those regimes vary significantly in detail and scope. At the simpler end, there may be a general obligation on the owner (or a third-party contract administrator) to determine a "fair and reasonable" valuation of time and money, which is binding in the interim, and in respect of which the contractor has a right of dispute. More thorough regimes contain a process of seeking to calculate all or part of the price and time consequences by application of a predetermined process. For cost, this is typically a reference to preagreed contract rates for pricing, and for timing, by reference to a demonstrated impact on the contract schedule for timing – often by reference to the critical path for completing the work by the scheduled date.

In contrast to lump sum contracts, in what are often called "measure and value contracts," the contractor provides the agreed rates for items of work in its bill of quantities included in the contract, and those rates form the basis for determining the contract price payable to the contractor for actual work performed. The valuation methodology in such contracts varies depending on the nature of the work being performed as a result of variation but is generally thus: Firstly, if the varied work is identical or similar to the work stated in the bill of quantities then then rates and quantities set out in the bill of quantities will be used to value the additional cost payable to the contractor. Secondly, if the varied work is of similar character but carried out under significantly different conditions or there are significant changes to the quantity, then rates and quantities set out in the bill of quantities may form the basis of interpolating or extrapolating the additional cost payable to the contractor. Lastly, if the varied work is not similar, then the varied work will be valued at fair rates and prices, determined by the owner or the contract administrator.

The JCT DB 2024[70] expressly clarifies that there is no allowance for any effect upon the regular progress of the works, in whole or in part, for any other direct

---

69 Terminology for this request varies, but might be called a proposed or contemplated change notice, or a variation price request or notice.

70 Section 5.7. 2 of JCT Design and Build Contract, 2024 Edition (DB 2024).

loss and/or expense for which the contractor would be reimbursed under any other provision of the contract.

JCT and FIDIC standard construction contracts apply a staged valuation methodology according to the type of variation in question. For example, in JCT DB 2024:[71]

- 1st stage (contract rates and prices): Where the varied work is of similar character to the contract work and is carried out in similar conditions without significantly changing the quantity of the work to be carried out, the rates and quantities set out in the contract will apply;
- 2nd stage (allowance to contract rates): Where the varied work is of similar character but is carried out under significantly different conditions or significantly changes the quantity, then the rates and quantities set out in the contract will be the "basis" for determination and include a fair allowance for the change in conditions or quantity. Additionally, valuation includes allowance for addition/reduction of the provision of site administration, site facilities and temporary works; and
- 3rd stage (fair rates and prices): Where the varied work is not of a similar character to the work set out in the contract, it will be valued at fair rates and prices.

The FIDIC Red Book[72] conditions provide an analogous framework for the valuing variations:

- 1st stage (contract rates and prices): The appropriate rate or price for the item shall be the rate or price specified for such item or similar item in the bill of quantities or other schedules.
- 2nd stage (adjustment to contract rates): A new rate or price shall be appropriate where: (i) the item is not identified in the bill of quantities or other schedules; (ii) no rate or price for the item is specified in the bill of quantities or other schedules; or (iii) the item of works is not of similar character or is not executed under similar conditions. Each new rate or price shall be derived from any relevant rates and prices specified in the bill of quantities or other schedules with reasonable adjustments.
- 3rd stage (cost plus profit): If no specified rates or prices are relevant for the derivation of a new rate or price, it shall be derived from the cost of executing the work, together with the applicable percentage for profit stated in the contract (if not stated, 5%), taking account of any other relevant matters.

---

71 Section 5.4.3 of JCT Design and Build Contract, 2024 Edition (DB 2024).

72 *FIDIC Red Book (2017 edition) Subclause 12.3.*

JCT DB 2024[73] also states that valuation of the omission of work in the contract price is based on the rates and quantities set out in the contract and such valuation will include allowance for addition/reduction of the provision of site administration, site facilities and temporary works.

The Application of Contract Rates and Prices

The general practice in international construction project and the UK is similar to the third stage under JCT DB 2024 which provides that where varied work is not of a similar character to the work set out in the contract bills of quantities, then such varied work will be valued at fair rates and prices and not the contract rates and prices.

Fair and Reasonable Rates

Once it has been determined the contract rates are not applicable, then the pricing mechanism may default to a "fair and reasonable price" payable for performance of such varied work. In English law and most related common law jurisdictions, where the obligation to make a valuation as to what is fair and reasonable is left to a contract administrator, "the valuation made by the contract administrator under the contract will be unassailable unless it was made in bad faith, or the contract otherwise contemplates the valuation being reviewed or revised."[74]

Cost or Cost Plus

Variation clauses often provided a contractor will be entitled to payment for variation based on its "actual" or "reasonably incurred" costs of performing the varied work, plus a margin for its overhead and/or profit – the latter two often being applied as an agreed percentage on the cost of the work. There may be a significant distinction, particularly in the administration of changes, between what the actual cost may be, and what was considered reasonable by the courts (and which may be less than a contractor's actual costs). In Australia, the courts have held:

> The reference to "actual cost" in the contract should, in any event, be moderated to mean reasonable cost. The respondent could not recover for unnecessary or extremely inefficient labour, even if it did comprise his "actual cost."[75]

73 Section 5.4 *(Measurable Work)* of JCT Design and Build Contract, 2024 Edition (DB 2024).

74 Bailey, II.7.53.

75 See, for example the Australian case *Skinner & Anor v. Harnass* [2007] SASC 122 at 21.

Omissions and Negative Variations

An omission, or negative variation, is a variation under which an owner omits or removes work from the contractor's original scope under a contract. As a general principle, upon entering into a contract, a contractor and owner agree not only to the contractor's duty to perform the work, but also to a corresponding right to carry out and complete such work and earn its expected profit.[76]

Generally, under common law and standard forms of contract globally,[77] where an owner omits work, performs the works itself, or gives the works to a third party, the contractor is entitled to damages (or under relevant contract provisions to its expected profit) arising out of the lost opportunity to carry out the work. By way of example in the Canadian CCDC suite of contracts, the contract specifies that for the purpose of valuing changes, changes in the work that are not related to each other shall not be grouped together in the same change directive – ostensibly to preserve a contractor's entitlement to its lost profit, which may be prejudiced where positive (additions) and negative (omissions) variations are netted against each other.

The English Court of Appeal case of *MT Højgaard A/S (Respondent) v E.On Climate & Renewables UK Robin Rigg East Limited and another (Appellants) [2017] UKSC 59*[78] related to valuation of omitted works and the question arose whether omitted works are valued the same way as additional work. The contractor argued that a component of the original contract price should be deducted for the omitted works and claimed €12.9 million based on a contract sum analysis applying the quoted rates, whereas the owner alleged the value of the omitted works was €34.65 million applying the schedule of rates appended to the contract to the time the alleged omitted works would be taken to be constructed if the contract work had proceeded without any omission. The court found for the contractor by interpreting the contract as whole including the variation provision to make a reasonable and fair assessment of the omitted work. The case illustrates the significance of the valuation methodology applied in determining the cost of varied or omitted work.

76 In the UK, *Abbey Developments Ltd v. PP Brickwork Ltd* [2003] EWHC 1987 (TCC); (3002) CILL 2033 at para 49.

77 As just one example, the Canadian CCDC 2 Stipulated Sum Contract (2020) at GC 6.3.6.2, says, "If the change results in a net decrease in the Contractor's cost, the Contract Price shall be decreased by the amount of the net decrease in the Contractor's cost, *without adjustment for the Contractor's fee.*" For clarity, note that the Contractor's cost is the cost incurred in the performance of the works and the contractor's fee is the fee payable in consideration of the contractor's performance of the works. The contractor's fee is generally a percentage of the overall cost of the works.

78 *MT Højgaard A/S v. E. On Climate & Renewables UK Robin Rigg East Ltd* [2013] EWHC 967 (TCC) (23 April 2013).

Recovery for Ancillary Financial Impacts

Contractors may also claim, in addition to the direct cost for varied work, other ancillary costs for financial impacts attributable to a change. Such financial impacts may include: (i) preliminary costs consequential to the variation; (ii) delay damages; and/or (iii) disruption costs of unvaried works.

Overhead and Preliminary Costs

Preliminary costs or site overheads are costs related to administering a project such as costs of staff, site-based services, site management, temporary facility costs, welfare expenses, bonds, insurance, guarantees, rental fees, and travel and transport expenses. In the event, contractor is required to deploy additional resources (as may be necessary for varied works) for an extended period that is referred to as thickening of the preliminary costs (as set out in the bill of quantities). In addition to the direct cost of varied works, a contractor will also claim overhead costs proportionate to the direct costs payable for the varied works. Where the valuation of varied works is based on the rates and prices set out in the contract, then the overhead cost or thickening of the preliminary costs is also determinable in reference to the preliminary cost set out in the bill of quantities. In the event, if there are no preliminary costs set out in the bill of quantities, then the overhead cost claimed by the contractor may be based on the actual costs incurred by the contractor.

Delay Costs

Ascertainment of costs arising from adverse effects of variation is set out separately from other costs under the JCT DB 2024.[79] In that regime, the contractor is to notify the owner of its claim accompanied with an initial assessment of the loss expense incurred and any further amounts likely to be incurred along with information reasonably necessary for the owner to ascertain the loss/expense incurred. Such ascertained amounts, to the extent accepted and not disputed by the owner, are added to the contract sum without double payment with any sums payable for variable works.

Cost of Disruption to Unvaried Work

A consequence of the varied works may be to disrupt, hinder or interrupt the unvaried works and the contractor may claim any additional costs incurred in the performance of such unvaried work. For example, such costs may be claimed as loss and/or expense under Subclause 4.20 in JCT DB 2024.

79 JCT SBC/Q 2024, Subclause 4.20.1.

14

Prevention of Claims: The Various Paths

Aarta Alkarimi

Introduction

With the claims and disputes – minor or major – being near ubiquitous on construction projects, proactive avoidance and mitigation of conflicts are crucial for the successful delivery of projects. Conflict avoidance and mitigation not only ensure smoother project execution but also help maintain positive working relationships, enhance project outcomes, and protect the financial interests of all parties involved.

Effective conflict avoidance strategies can prevent minor disagreements from escalating into formal claims or legal battles, thereby saving significant time and resources. Much like the multitude factors to consider on the battlefield, the execution of construction projects revolves around strategic planning, coordination, and the ability to tackle unforeseen challenges; all in an environment under the considerable pressures of time and money.

Several, often overlapping, stages and activities during the execution of a project (permitting, engineering, material procurement, managing deliveries, performing foundation and structural works, fabricating and installation of cladding, MEP works, finishes, inspections, testing and commissioning, etc.) are akin to battlefronts of their own; each with its own set of issues and objectives and – ultimately – interdependent and affecting the outcome of the greater mission.

And, like war, building and infrastructure projects involve multiple parties (including peripheral stakeholders) with common and competing interests, complex alliances, and the constant risk of disruption due to external factors, which all illustrate the intricacies and high stakes involved in managing projects, particularly large-scale ones.

In such circumstances, claims are almost inevitable (often enough, perhaps, to be considered the norm) and they can be costly, time-consuming, and damaging to relationships; often leading to delays, cost overruns, and legal disputes.

The battlefield analogy was aptly and succinctly expressed in 1981 by the judges presiding over the *Blake Construction Co. vs. C.J. Coakley Co.* case in Washington DC where the battleground extended from the site to the courtroom over project delays, miscommunication, and the allocation of blame.[1]

Over four decades later, despite several technological advancements to aid the management of construction projects (more rapid means of document production and transfer, sophisticated document control systems and scheduling software, Building Information Modelling (BIM) to aid design, laser-based scanning and surveying, long-scope inspection cameras, sensors, and drones, to name a few), there has been little fundamental change to suggest any diminishment of the battlefield analogy.

In fact, as projects have become more complex, one is faced with new and unfamiliar terrain. As such, reliance on past practices is often misplaced and the lack of adaptable strategies can lead to mission failure.

Understanding such complexities is crucial for managing risk, mitigating disputes, and improving outcomes at both the project-level and across the construction industry.

With large construction contracts commonly running into the hundreds of millions of dollars, the stakes are high and the tolerance for errors is low. Mistakes can not only jeopardize projects but prove fatal to businesses.

In this aspect, the judges hearing *Blake Construction Co. vs. C. J. Coakley Co.* were acutely aware that construction endeavors are inherently risky and litigating them can be very complicated. Therefore, proactive prevention and timely treatment are key to mitigating claims and disputes. Otherwise, relatively small claims and disputes can metastasize to pose existential threats to projects and businesses.

Examples of this are rife and even the giants of the industry (contractors and developers), with their ample resources and skills, have not been spared the consequences. For example:

---

1 Blake Const. Co., Inc. v. C. J. Coakley Co., Inc., 431 A.2d 569, 675 (D.C. 1981) ("Except in the middle of a battlefield, nowhere must men coordinate the movement of other men and all materials in the midst of such chaos and with such limited certainty of present facts and future occurrences as in a huge construction project [...] Even the most painstaking planning frequently turns out to be mere conjecture and accommodation to changes must necessarily be of the rough, quick and ad hoc sort, analogous to ever-changing commands on the battlefield.").

Tutor Perini Corporation (Perini Building Company)

The Contractor was involved in the CityCenter project in Las Vegas, one of the largest privately funded construction projects in the United States. Fraught with disputes and legal claims, particularly over structural defects and cost overruns, the matter was litigated for years.

Tutor Perini filed a $492 m lawsuit against the several entities controlled by MGM Mirage and Infinity World Development Corp. for unpaid works alleging that the contract value had grown from $3.5 billion to $6.3 billion due to "numerous design changes and substantial project modifications."[2,3]

MGM Resorts International (coowners of the CityCenter development) filed a lawsuit in 2010 alleging construction defects. The Harmon Tower, which was a part of the $8.5 billion development, was ultimately demolished before completion despite Tutor Perini maintaining that the building was structurally sound.

Although no party admitted wrongdoing, Tutor Perini reached a settlement with MGM in 2014. The confidential settlement (eventually disclosed in regulatory filings) stated that MGM was paying the contractor $153 million to satisfy construction lien claims unrelated to the Harmon Building (later Perini filed a clarification that it was receiving $189.5 million to settle its CityCenter lien claims).

Several other lawsuits resulted, including a claim by a subcontractor against the structural engineers.[4]

Fluor Corporation

Texas-based Fluor Corporation, a major engineering and construction firm, has faced significant difficulties in recent years due to claims and disputes largely stemming from project delays, cost overruns, and contractual disagreements.

Several of its large projects – particularly in oil, gas, and petrochemical sectors – experienced disputes over compensation, project scope, and delays resulting in significant financial losses and expensive legal battles. As a result, Fluor needed to improve its risk management practices, restructure its operations, and adjust its project execution strategies to mitigate future risks associated with claims and disputes.

---

2 *Perini Sues MGM Mirage Over CityCenter Hotel-Casino Project in Las Vegas*, ENGINEERING NEWS-RECORD (Mar. 31, 2010), https://www.enr.com/articles/6083-perini-sues-mgm-mirage-over-citycenter-hotel-casino-project-in-las-vegas.

3 Deena Beasley, *MGM Mirage says CityCenter contractor to file suit*, REUTERS (Mar. 15, 2010, 10:35 AM), https://www.reuters.com/article/business/mgm-mirage-says-citycenter-contractor-to-file-suit-idUSLDE62C04W/.

4 *Tutor Perini Settles Vegas Case*, LOS ANGELES BUSINESS JOURNAL (Dec. 17, 2014), https://labusinessjournal.com/real-estate/construction/tutor-perini-settles-vegas-case/.

For example, changes in the design and construction of a chemical plant (the "Chocolate Bayou") in Alvin, Texas, led to dispute with Solutia Inc. While Fluor successfully settled this dispute in 2002, the case provided another example of challenges construction companies face in managing complex projects and maintaining profitability amidst contractual disputes.[5]

Another Fluor case, a dispute lasting approximately nine years, demonstrated similar challenges and further underscored the financial and human-capital toll disputes can take on the contesting parties. General Electric (GE) had sub-contracted Fluor to build the Rabigh power plant for a Saudi Arabian state-owned entity. The project encountered significant delays, technical issues, and cost escalations that led to legal and financial disputes, with both parties seeking compensation and damages. Fluor was accused of mismanaging the project, while Fluor claimed that GE's misrepresentation of the project's complexity, changes in scope, and local factors were responsible for the delays. The case, initiated in 1998, was eventually moved to arbitration, concluding in 2007, and highlighted the added challenges of operating in foreign markets with different regulatory and operational environments.[6]

Carillion

Carillion, a very large and prominent UK-based contractor with several domestic and international mega-projects, went into liquidation due to insolvency in 2018.

While the lack of accurate financial reporting and aggressive pricing were cited as key reasons for the company's failure, several major disputes and unrealized claims on five high-value projects exasperated Carillion's problems and assured its demise.[7]

---

5 *Fluor Wins $20-million Settlement In Solutia Contract Dispute*, ENGINEERING NEWS-RECORD, https://www.enr.com/articles/30306-fluor-wins-20-million-settlement-in-solutia-contract-dispute(last visited Oct. 25, 2024); *Flour, Solutia Settle Project Dispute*, FLUOR (Oct. 11, 2002, 08:08 AM), https://newsroom.fluor.com/news-releases/news-details/2002/Fluor-Solutia-Settle-Project-Dispute/default.aspx.

6 John O'Dell, *Fluor Sues G.E. Over Costs to Build Plant*, LOS ANGELES TIMES (Oct. 14, 1998, 12:00 AM PT), https://www.latimes.com/archives/la-xpm-1998-oct-14-fi-32217-story.html; *Fluor Daniel Intercontinental, Inc v. General Elec. Co*, No. 06 Civ. 3294 (GEL), 2007 WL 766290 (S.D.N.Y. Mar. 13, 2007).

7 David Price, *Five jobs that killed Carillion: the £1.3bn lawsuit in detail*, CONSTRUCTION NEWS (Feb. 7, 2022), https://www.constructionnews.co.uk/contractors/carillion/five-jobs-that-killed-carillion-the-1-3bn-lawsuit-in-detail-07-02-2022/; Simon Goodley, *The four contracts that finished Carillion*, THE GUARDIAN (Jan. 15, 2018, 19:55 GMT), https://www.theguardian.com/business/2018/jan/15/the-four-contracts-that-finished-carillion-public-private-partnership.

For example, Carillion's GBP 500 m ($653 m) contract to construct significant portions of Qatar's Msheireb Downtown Doha was saddled by claims of approximately GBP 200 m ($261 m) for purported unpaid work, significant number of design changes, delays, and other disagreements (including the owner's appointment of third-party contractors to complete works).[8]

Common Sources of Claims and Disputes in Construction Projects

Claims and disputes can arise from various sources, often due to misunderstandings, miscommunication, or unmet expectations and can range from minor disagreements that are easily resolved through effective communication to larger, more complex issues that may escalate into formal disputes or legal action if not managed effectively.

Key factors that contribute to disputes and claims in construction include:

- Contractual Disputes: ambiguous contract terms, scope changes, or differing interpretations of contract clauses can lead to disagreements between parties.
- Design Errors, Omissions, and Changes: incomplete or inaccurate design documents and late design changes can result in conflicts over the work required and associated costs.
- Delays: delays caused by unforeseen circumstances, such as weather, labor shortages, or supply chain issues, can lead to claims for additional time and costs and disputes over the responsibilities for associated costs.
- Resource Allocations: conflicts over the adequacy and timely availability of labor, materials, and equipment are common both at owner-contractor level and at contractor-subcontractor level.
- Payment Issues: disputes over payment terms, delayed payments, or disagreements over the valuation of work completed can strain relationships and lead to claims.
- Quality and Workmanship Disputes: disagreements over the quality of materials or workmanship can lead to conflicts, especially when standards are not clearly defined or met.

8 *Qatari developer disputes swathes of Carillion evidence*, UK PARLIAMENT (Feb. 28, 2018), https://committees.parliament.uk/work/5425/carillion-joint-inquiry/news/98025/qatari-developer-disputes-swathes-of-carillion-evidence/; Noor Nanji, *Carillion chiefs blame Qataris for company's collapse*, THE NATIONAL (Feb. 6, 2018), https://www.thenationalnews.com/business/carillion-chiefs-blame-qataris-for-company-s-collapse-1.702242.

- Communication Breakdowns: lack of clear and consistent communication among stakeholders can lead to misunderstandings that create conditions for conflicts and claims that otherwise may not arise.

Where one or more stakeholders have created an adversarial rather than collaborative management approach, the communication breakdowns can further exasperate claims and disputes; even create them.

Common Types of Claims

A claim in construction is a formal request made by one party, typically a contractor or subcontractor, seeking compensation for additional costs, time, or resources due to an event or condition not originally accounted for in the contract.

Common claims in construction include:

Delay Claims and Prolongation Claims

Delay Claims

Delay claims are to compensate the contractor for additional time resulting from project delays which may be caused by Owner instructions, delayed access to site, design changes, or external factors like weather, force majeure events, etc.

Delay claims may also be raised by the Owner during the execution of the project or upon completion where a contractor exhibits poor management (e.g., poor scheduling, lack of coordination, and resource misallocation), fails to mobilize in time, causes significant rework due to quality issues and noncompliance with the specifications, provides inadequate manpower or materials, and other avoidable errors that result in key project milestones being missed or exceed the project's agreed completion date without valid justification.

In such circumstances, the Owner may demand that the contractor take measures to speed up the work at its own cost or, once the project is completed, apply penalties that are usually stipulated in the contract (liquidated damages).

Prolongation Claims

Prolongation claims are to compensate the contractor for additional overhead costs and expenses incurred because of an extension of the project duration beyond its agreed completion date for reasons not attributable to the contractor's actions.

While similar to delay claims in their causes, prolongation claims should not be confused with the other; a delay claim may be satisfied through the granting of an extension of time to the contractor while the latter is addressed through monetary compensation (in addition to an extension of time).

The following table is provided for further clarity:

	Delay claim	Prolongation claim
Issue	Time impact	Financial impact
Relief sought	Extension of time (without penalties)	Compensation for additional costs
Outcome sought	Adjusted schedule/revised completion date	Monetary claim allowed

Disruption Claims

Claims to compensate the contractor for additional time and related costs related to disruptions in the contractor's planned workflow that will lead to inefficiencies/loss of productivity, and their associated costs.

Payment Delay Claims

Claims to compensate the contractor for financial burdens such as interest charges or penalties to creditors and suppliers.

Where an Owner's delays in payment are significant or frequent, a contractor's inability to pay its workers, subcontractors, and suppliers may detrimentally affect project activities and timelines due to interim demobilizations of labor and late delivery of material and equipment, giving rise to disruption claims, delay claims, and/or prolongation claims.

Acceleration Claims

Claims to compensate the contractor for additional labor and equipment, and overtime costs based on requests to complete the work or part of the work earlier than the originally planned dates.

Variation Claims

Claims to compensate the contractor for additional costs (and, in many cases, additional time) resulting from design changes or changes to the project scope that require additional work, materials, or durations beyond the original agreed scope.

Variation claims may also include those to compensate the contractor for costs incurred and loss of profit arising from significant reductions to the originally contracted scope of work such that it affects the contractor's anticipated revenue, and profitability. Further, excessive descoping – depending on its nature and timing – can also give rise to disruption claims.[9]

Defective Works Claims

Claims by the Owner against the contractor (or by a contractor against a sub-contractor or supplier) to remedy or otherwise compensate for work that has not been executed to the required specifications, quality standards, or industry norms including patent defects (visible/easily discoverable during inspection) and latent defects (hidden flaws that may emerge later; often after project completion).

Differing Site Condition Claims

Claims to compensate the contractor for unforeseen complications and costs arising from actual site conditions differing from those described in the contract.

Termination Claims

Claims to compensate the contractor for costs incurred to date, demobilization costs, and loss of profit arising from the premature termination of a contract without just cause.

Benefits of Proactive Claims and Disputes Management

Claims are usually governed by the terms of the construction contract, and their resolution can involve negotiation, mediation, arbitration, or even litigation, depending on the severity and complexity of the issue.

Where the parties may have taken proactive measures to structure the process by which disputes and claims are raised and managed, many disputes may be mitigated or avoided altogether.

The benefits of proactive conflict management can lead to:

- Improved Project Performance: projects are more likely to be completed nearer to the original time for completion and closer to the budget and required quality standards when conflicts are minimized and resources are not diverted to time-consuming disputes and legal action.

9 Claire King, *Can I omit that?*, FENWICK ELLIOTT, https://www.fenwickelliott.com/research-insight/newsletters/insight/84#footnote3_qufrrpc (last visited Oct. 25, 2024).

- Enhanced Stakeholder Relationships: collaborative approaches and conflict avoidance foster trust and respect among project participants, leading to more effective teamwork and communication.
- Cost Savings: by preventing disputes and claims, projects can avoid costly arbitration and litigation legal fees, and the indirect costs associated with delays and rework.
- Reputation Management: successful conflict management enhances the reputation of the involved parties, increasing the likelihood of future business opportunities between the parties with the resultant benefits of "lessons-learned" that may be applied to future projects for improved outcomes.

Key Strategies for Disputes and Claims Avoidance

Determine the Suitable Project Delivery Model

Selecting the right project delivery model is critical for the success of a construction project as it defines how responsibilities, risks, communication channels, and processes will be managed throughout the project lifecycle (from design through completion and beyond).

The delivery model outlines the roles and relationships among key stakeholders, including the Owner, contractor(s), designer(s), and subcontractors, influencing the project's schedule, cost, and quality. An appropriately selected delivery model aligns with the project's goals, type, complexity, and budget.

Further, suitable delivery models help address uncertainties at an early stage, ensuring that risk is distributed appropriately among parties, setting up better project controls, and facilitating smoother collaboration and more predictable outcomes.

The right delivery model – whether Design-Bid-Build (DBB), Design and Build (D&B), Construction Management (CM) "at risk" and "not at risk," or Integrated Project Delivery (IPD) – provides clarity on responsibilities, which help with better project set-up, reduce misunderstandings, and preempt conflicts.

The appendix to this chapter provides the key characteristics of each delivery model including contractual and team structures, prerequisites, types of projects to which models are commonly applied, and their advantages and disadvantages.

As a first step, a selected project delivery model provides the Owner with a road map for the type of consultants required, the extents of their scopes of services, and potential for staged design deliverables (e.g., if an advance "early works"/ foundations package is to be prepared for a fast-track project and whether this package may be separately procured in advance of the main works).

By facilitating better alignment of expectations across all stakeholders, a suitable delivery model can also help reduce disputes and claims. With clear risk apportionment and efficient communication protocols in place, potential issues can be identified and resolved proactively, preventing escalation into costly legal battles.

DBB offers clear separation between design and construction, while D&B consolidates both functions (at an early design stage) under one contract. While the D&B approach may reduce the potential for blame-shifting between designers and contractors, it is only appropriate for some project types and depends highly on the readiness of an Owner to pursue the project using this approach (feasibilities in place, clear design guidelines, and performance standards established with earlier and greater financial commitments).

In cases of complex or fast-track projects where a traditional DBB approach will not meet the completion date and the D&B approach is not suitable because the Owner prefers greater control of the design, an Owner may directly procure several "prime package" contractors separately while the design is completed (*e.g.*, prime packages for foundations, superstructure, building envelope, or interior fit-out). In such cases, the selection of the right project model helps structure the early consideration of procurement routes and timely appointment of consultants and contractors for these packages with their appropriate scopes, which – in turn – help mitigate certain types of disputes. However, with the risks associated with untimely procurement of consultants and package contractors or scope gaps between packages resting with the Owner, this approach is best employed by very seasoned Owners or pursued through a CM-At Risk delivery model.

Integrated models like IPD promote collaboration by aligning incentives across all stakeholders, fostering greater transparency, and a cooperative approach to problem-solving. However, this model may only be appropriate in highly mature markets with stakeholders possessing the right skill sets, and the legal structures of the jurisdiction being equally prepared.

Ultimately, selecting an appropriate delivery method not only provides clarity and enhances project efficiency but also contributes to a more positive working environment, helping reduce the likelihood of adversarial relationships.

Clear Contractual Agreements and Detailed Scopes

While it is the most evident and elementary of efforts to reduce the potentials for disputes and claims, it is not uncommon to see contracts assembled in a patchwork manner by several people and departments within the Owner's organization or outsourced parties who are often working under tight deadlines.

The conditions of contract, general requirements and general conditions documents, drawings and specifications, durational schedules, schedule of values (SOVs) (known as a bill of quantities (BOQs) in many jurisdictions outside the United States) add layers of additional information to a construction contract from a variety of separate parties (lawyers, project managers, designers, engineers, cost consultants, etc.), with a greater potential of discrepancies between documents and large sums of money at risk.[10] As such, the need to reiterate the need for clear contractual agreements that are well-vetted and prepared in a collaborative manner is warranted.

A well-prepared contract is the foundation for preventing conflicts in construction projects. Contracts should clearly define each party's roles, responsibilities, and expectations to avoid misunderstandings.

Common Forms of Construction Contracts

Much like the selection of the suitable project delivery model may reduce disputes and claims, the use of right form of contract is crucial to ensure that it is jurisdictionally enforceable, familiar to the market it is being applied to (local industry practice) and tried and tested; the first step to help eliminate disputes that may arise over some of the most fundamental issues.

Some of the most widely recognized forms of construction contracts are listed here:

Contract Type	Description	Regions Used
AIA (American Institute of Architects)	Several AIA forms of contract widely used for building projects via various delivery models.[11]	United States
ABIC (Australian Building Industry Contracts)	Widely used for building projects	Australia

(Continued)

10 General requirements and general conditions documents (often referred to as "Division 01" documents in the United States and "Employer's Requirements" (ER) in the United Kingdom and some other countries) supplement the drawings and specifications to further define aspects of a contractor's overall scope of work that are not typically covered by drawings and specifications alone. These include descriptions of the scope of work, project goals and objectives, team/stakeholder structure, responsibilities matrices, the project schedule, logistics expectations/constraints/challenges, permits and other regulatory approvals, communications systems and protocols, security, health and safety, quality assurance and control, environmental and sustainability, security, insurances, temporary facilities (owner and consultant site office requirements, washrooms, break-out rooms/canteen, storage, sever rooms, etc.), waste management, pest control, snagging/de-snagging, testing and commissioning strategy, project completion and handover, etc.

11 Various AIA forms of contract are available. For example, the "Conventional A201 Family" of contracts are used for design-bid-build projects, while AIA's "Design-Build Family" of contracts cater to the design and build approach, and AIA Document C191–2009 caters for multi-party agreements including the Integrated Project Delivery (IPD) model.

(Continued)

Contract Type	Description	Regions Used
ConcensusDocs	For projects using collaborative/ integrated projects	United States
CCDC (Canadian Construction Documents Committee)	Suite of standard construction contracts	Canada
EJCDC (Engineers Joint Contract Documents Committee)	Suite of standard contracts including engineering, infrastructure, and public works	United States
FIDIC (International Federation of Consulting Engineers)	Suite of internationally recognized contracts used for buildings, engineering, and infrastructure	Africa, Asia, Europe, Latin America, Middle East
GC21 (General Conditions of Contract 21)	Collaborative government-specific contracts	Australia
HKIA/HKIS (Hong Kong Inst. of Architects/Hong Kong Inst. of Surveyors)	Standard contracts used for building projects	Hong Kong
ICE (Institution of Civil Engineers)	Standard contracts for civil engineering and infrastructure	United Kingdom, Hong Kong, some Commonwealth countries
JCT (Joint Contracts Tribunal)	Standard contracts used for buildings	United Kingdom
NEC (New Engineering Contract)	Used for construction and maintenance projects	United Kingdon, New Zealand, South Africa, Hong Kong
PSSCOS (Public Sector Standard Conditions of Contract)	Standard conditions for public sector projects	Singapore
SAICE (South African Institution of Civil Engineering)	Contracts for civil engineering projects	South Africa

Having selected the appropriate form of contract, the following key elements should be carefully considered and clearly described with minimum ambiguity.

Hierarchy of Documents/Contract Precedence

As it is fairly common to find instances where contract documents conflict with another (*e.g.*, drawings not matching specifications, or the quantities in the SOV (BOQ) not aligned with what is shown on the drawings), a hierarchy clause listing the precedence of documents (general conditions, particular conditions, drawings, specifications, schedule of values/bill of quantities, etc.) is essential to preclude several potential areas of dispute.

Comprehensive Payment Terms

Clearly defined payment terms and criteria for progress payments reduce the chances of payment disputes. To avoid ambiguity about when and how payments will be made, contracts should provide details pertaining to:

- Advance payments;
- Required documents for substantiation of payment applications;
- Established timelines for review of payment applications, certification period, and remittance;
- Payment for materials on site or in factory;
- Payment retentions and conditions for retention releases; and
- Milestone/bonus payments and the criteria for their realization.

Precise Scope of Work

Clearly outline the scope of work, including sufficiently detailed and vetted drawings and specifications, descriptions of materials and methods. A thorough and well-defined scope minimizes the risk of scope creep and disputes over what is or is not included in the project.

Complete Drawings and Specifications List

The contract must include a complete list of drawings and specifications (with drawing/document numbers, accurate titles, revision numbers, *etc.*) as it will form the basis upon which the scope shall be carried out and administered. This will help eliminate disputes over whether a certain drawing or speciation (and which version) was included or not.

Exclusions

A comprehensive list of exclusions with sufficient clarity and detail help enhance the understanding of all parties and reduce ambiguities. If an excluded items is to be provided by another party, they should also be identified to minimize future disputes between stakeholders.

"By Others"

Consultants' and contractors' labelling of materials, components, and details on their drawings must avoid the use of words like "by others." This often leads to scope misunderstandings and unintended exclusions which may not be easily picked up and corrected before the execution of contracts between Owners and contractors or contractors and their subcontractors.[12] Instead, items or components that are shown on a drawing but are excluded from a particular contract/package should be identified on the drawings using the name/number of the contract/package in which it is intended to be included or clearly listed in a sufficiently detailed list of exclusions for the respective package.

Construction Interface Documents

Where two or more contractors are directly engaged by an Owner on the same project (whether working sequentially or overlapping in activities) or in cases such as a tenant fit out that commences before project completion, formalized interface documents should be included in all relevant contracts.[13]

An interface document is used to delineate scope boundaries, align systems interfaces, identify and address potential issues at interface points, and coordinate schedules and sequencing of works by independent contractors on a project. Interface documents are particularly useful in the case of fast-track projects where a project is procured through several distinct works packages. The following are three practical examples where interface documents are useful and – when considered properly – will help mitigate disputes:

- An Owner/Architect mandated use of the same foundation waterproofing system and applicator used by their predecessor where a main works contractor is building upon a high-rise sub-structure/foundations constructed by a preceding contractor. This helps ensure that waterproofing products are compatible, there is one point of responsibility if there are leaks, and one comprehensive warranty.

12 For example, fire and smoke seals ("firestopping") are essential components required around penetrations where mechanical, electrical, and plumbing (MEP) systems pass through fire-rated walls and floors. However, when these seals are labeled as "by others" in the architectural or MEP drawings, it often leads to disputes over responsibility. While a mechanical subcontractor might assume the firestopping is the responsibility of the general contractor or a specialist team, the general contractor may have assumed the MEP trades would install firestopping around their respective pipes, ducts, and cables. Such ambiguity can result in delays while teams on site dispute responsibility for the missing seals.

13 A "tenant fit-out" refers to design and construction related to the completion of a leased or purchased space within a building to meet the specific needs of the occupant (e.g., an office space, retail outlet, restaurant, or condominium). While many tenant fit-outs occur after a building project is completed and handed over to the owner, large-scale projects such as corporate offices or retail centers often have such activities occur early (when the building owner and contractor have established the "base building" (also known as "core and shell") comprised of its primary elements such as the structure, building envelope, primary MEP lines) and have the building's main contractor concurrently completing the remainder of their work.

A tenant typically manages their own fit-out, often hiring their own designer and contractor (the "fit-out contractor") to carry out the works within their spaces.

- The accommodation of anchors embedded into the concrete structure to support curtain wall panels that will be installed by a cladding contractor in due course (after the panels are fabricated and delivered to site and once the tower's structure is sufficiently advanced). In such a case, an interface document will establish the timeline for engineering details for the anchors to be provided by the cladding contractor (including setting out dimensions and tolerances), and the timely furnishing of the anchors for embedment into scheduled concrete pours.
- The accommodation of a major tenant occupying many floors of a high-rise office tower who has a lease agreement that allows them to start their interior fit-out works while the main contractor is still constructing the building (so that the tenant may move in soon after the building is ready for occupancy). In such a case, an interface document is very useful in establishing the rules of engagement between two parties that are not contractually bound to each other (the building contractor and a tenant's fit-out contractor) such as:
  - the frequency and timing of the use of the building crane and construction hoists by the tenant fit-out contractor to move material and labor;
  - designated areas within a project where the tenant fit-out contractor may set up their site offices to manage their operations (which may need to be relocated one or more times as required by the building contractor to complete areas of their work); and
  - project working hours and working week, and observed holidays when the site will be shut.

Disputes between contractors that are directly engaged by an Owner would require the Owner's intervention to resolve, often at its own cost and its accommodation of the impact. With well-considered interface documents incorporated into the contractors' contracts, many such disputes can be avoided altogether and others that may arise can be more efficiently mitigated.

Mock-ups and "First-in-place" Constructions

Consider mock-ups and "first-in-place" constructions for major project elements and clearly list these requirements in project scopes.[14] Ensure that sufficient time

14 "First-in-place" constructions are similar to mock-ups but are distinguished by the fact that they are built in their permanent locations and, if approved, remain a part of the final works. Mock-ups may be built on-site or off-site and may be temporary. Mock-ups are also used to assist an Owner evaluate a design and make changes (e.g., building a mock-up room (with its furnishings) for a 100-key hotel before committing to the build of a 100 of the same). In contrast, "first-in-place" constructions are not meant to be used for the purpose of re-evaluating designs and making changes; they are predominantly meant to establish agreed benchmarks for fit and finish (quality) and finalizing some of the detailing before the work on similar conditions on a project proceed.

is built into project schedules to allow for the construction, review, and adjustments prior to the continuation of the respective works across a project. Mock-ups and "first-in-place" constructions are very useful for the review of tangible project efforts by identifying unresolved details at an early stage and establish agreed references for quality and durability before the respective works progress further.

Building Information Models (BIM)

Require the architects/engineers to prepare their designs using BIM (to the appropriate Level of Development (LOD)) for enhanced cross-discipline coordination and better visualization of the project by Owners.[15]

Require the contractors to further develop the BIM to the appropriate LOD to prepare engineering/shop drawings and allow better coordination across trades (e.g., use BIM's clash detection features to identify conflict areas), plan resources, and tie it to the project schedule.

Effectively using BIM during the design, construction, and testing and commissioning stages of a project provides several layers of checks to remedy errors, clashes, unresolved details, and unexpected conditions that may otherwise lead to changes in design or scope, delays, and disputes.

Through its ability to provide interactive simulations of the construction process and allow virtual walk-throughs of project spaces, BIM reduces the risk of abortive works and costly disputes.

Provisional Sums

Provisional sums are specifically allocated budgets included in a contract to cover works that are anticipated in a contractor's scope but cannot be sufficiently defined at the time the contract is executed because the designs or selections for these works have not been completed. The budgeted amounts are placeholders for the future works which will be defined and priced as the project progresses. Common provisional sum examples include:

- Unspecified design elements that require selection (decorative lighting, door hardware, sanitary ware, security systems, artwork, etc.).
- Interior fit-outs, landscaping, specialist installations (media walls, water features, signage, etc.) for which designs and scopes are yet to be defined.

15 BIM Levels of Development (LOD) – introduced by the American Institute of Architects (AIA) and now adopted across the world – refer to the progressive level of information/detail provided at different design stages (LOD 100, 200, 300 and 350 from pre-Concept/Concept, Schematic, and Design Development/Detailed Design during the design stages, and LOD 400 and 500 by the contractor during shop drawings and construction stages). *Level of Development (LOD) Specification*, BIM FORUM (Dec. 28, 2023), https://bimforum.org/resource/lod-level-of-development-lod-specification/.

- Allowances for existing utilities that may need to be relocated (once they are adequately revealed).
- Allowance for potential remedial/improvement works required to address site conditions that are not apparent or the extent of which cannot not be measured until works have progressed (e.g., sub-surface ground conditions, abatement measures for asbestos discovered during demolition works).

While provisional sums secure budgets in advance and offer flexibility, they also present risks that can impact project costs, timelines, and quality, which can lead to disputes and claims. Risks include:

- Cost overruns due to potentially insufficient budget allocations for a provisional sum. Once the designs and details are available, actual costs can significantly exceed the initial sum allocated, straining the project budget and potentially leading to disputes over additional payments.
- Project delays incurred due to late release of designs for provisional sum items or protracted pricing exercises can affect progress, disrupt planned sequence of the works, and cause delays.
- Quality and scope disputes due to lack of properly worded and sufficiently detailed definition of what was intended to be covered by the provisional sum.

To mitigate the risks associated with provisional sums, the following strategies should be employed:

- Clearly define provisional sums. Include detailed assumptions, scope outlines with inclusions and exclusions (with reference projects of similar nature executed by the same contractor, where available), and key criteria for how the work will be defined and valued later in order to avoid disputes over interpretations.
- Implement a robust and systematic process as early as possible to develop the scope, agree how they will be approved and priced (e.g., who will select the subcontractors/suppliers that will price the work, whether open-book tendering will be expected), and establish clear timelines for the process.
- Agree on the dates by when selections or designs of provisional sum items must be furnished to the contractor to avoid unnecessary delays and provide regular updates to the parties to keep everyone aligned so as to allow mitigative adjustments to the project schedule and sequence of work, if necessary. These dates should ideally be established in writing before the awarding of construction contracts or included in the baseline schedule developed by the contractor (typically, within the first 30 days of their appointment) and agreed upon by the parties.
- Avoid over-reliance on provisional sums by ensuring that substantial portions of the scope are already designed and priced into the contract;

specially avoid provisional sums for elements of the work that will require early coordination.

- Include added contingencies in the overall project budget to accommodate potential cost increases for provisional sum items to buffer their financial impact and reduce the risk of disputes triggered by budget overruns.
- Include contractual safeguards to limit the financial exposure presented by provisional sums. Clearly define the process for procuring and valuing work under provisional sums to avoid disputes.

Problem-solving, Decision-making, Dispute Resolution Forums

Timely problem-solving and decision-making are key to keeping a project on schedule and within budget. Indecision, ambiguity, and concerns about fairness in dealing with disputes can plague a project and lead to the compounding of issues that can significantly compromise all three primary Owner objectives (quality, time, and costs) and a contractor's key concerns of disruption, abortive works, cashflow, profitability, and liabilities.

To address these, processes that foster collaboration and facilitate rapid problem-solving and decision-making help keep the project on schedule, prevent disagreements from becoming project-debilitating disputes, and provide early warnings to appropriate levels of management.[16] Some of the most effective processes in this regard are:

WAR Rooms

Contrary to what the acronym may imply, a project's Work, Action, Resolution (WAR) Room is the nerve center for up-to-date project information and on-site coordination efforts. Key WAR Room features include:

- A dedicated room within the project site offices.
- Wall space for drawings, photographs, issues and submittal logs, dashboards, project schedules, digital screens for the project's Building Information Model (BIM) and other pertinent project information for immediate reference.

16 The WAR Room implemented at the Burj Khalifa project in Dubai had its frequency increased from weekly to daily and – at peak periods of activity – maintained a 24-hour/day "service desk" to cater to issues, facilitate rapid problem-solving/decision-making, monitor progress, and mitigate disputes and claims on the project. With a daily labor force reaching 12,000 workers at its peak activity period, the WAR Room was pivotal to mitigating costs, reducing instances of abortive works, increasing productivity, maintaining safety, and reducing delays. (Based on the author's interview with Mr. Basel Tanbouz, Program Director (PCS Manager during the Burj Khalifa project), Turner Construction – International/Turner International Middle East.)

- Regularly scheduled sessions (bi-weekly, weekly, daily), as required.
- Better overall representation of the respective issue stakeholders (sessions attended by representatives from the Owner, project manager, architects/engineers, cost consultants, and contractors and their subcontractors who are involved at the day-to-day level such as managers, superintendents, site architects, and inspectors).

Key WAR Room objectives include:

- Providing a structured platform for cross-disciplinary involvement in project matters.
- Fostering a collaborative and proactive problem-solving culture.
- Data-driven decision-making.
- Real-time issue spotting and resolution.
- Regular monitoring and control of progress and budgets.
- Regular alignment and re-aliment of project priorities.
- Documentation and reporting of decisions and action plans.
- Serving as a ready forum for emergencies and crisis management.

Steering Committees

Large/complex projects and deadline driven projects can greatly benefit from a steering committee comprised of executive level representatives from the Owner, project manager, architects/engineers, cost consultants, and contractors. In this forum, issues that cannot be solved at the WAR Room level are escalated or matters that require a higher Delegation of Authority (DOA) can be raised and resolved collaboratively. With the interests of all affected parties represented, the committee can serve as a well-informed body with a disposition for consensus building. If held regularly (e.g., monthly), the committee also serves as an ideal opportunity for progress reporting to the leadership, the exchange and setting of priorities, anticipate and address issues that may be at their fruition stage.

Key steering committee features include:

- High-level engagement: provides strategic oversight to ensure alignment with organizational goals.
- Regularly scheduled sessions: bi-weekly or monthly sessions as required to keep parties sufficiently informed and updated.
- Empowered senior representation: comprises senior representation from the Owner, project manager, architects/engineers, cost consultants, and contractors.
- Primary decision-making authority: resolves major project matters and handles escalated issues to ensure streamlined progress.

Key steering committee objectives include:

- Ensuring project success through strategic alignment and prioritization of key objectives (quality, time, and cost).
- Facilitating better communication and transparency between stakeholders to achieve greater clarity and avoid misaligned expectations.
- Monitoring key performance indicators (KPIs) and agreeing on actions required where they may not be met.
- Monitoring and allocating/re-allocating resources as changing project conditions require.
- High-level oversight and control of project budgets.
- Managing and approving significant changes to project scope, budget, and schedule.
- Resolving major issues and matters escalated by project team (e.g., from WAR Room).
- Addressing and settling disputes through negotiation and consensus-building by adequately empowered representatives.

Dispute Review Boards

Larger or more contentious disputes may not be resolved at WAR Room or steering committee level. Further, some disputes may require review by subject-matter/construction experts and – therefore – may not be suitable for steering committees to resolve. In anticipation that such cases are likely to arise, particularly on large or complex projects with high-stakes and high value, a standing Dispute Review Board (DRB) may be formed. For greatest efficacy, provisions for a DRB (including agreement on its structure and individual members) should be included in contracts between the project parties (particularly between Owners and contractors); with the costs to retain the board – from project commencement through completion – being shared by the parties. Further, if a DRB's duties include early review of project documents (drawings, specifications, project schedules, interface documents, the form and language of Owner instructions/scope changes, etc.), there are greater opportunities to preclude disputes, and claims.

Key DRB features include:

- Independent/neutral panel (typically comprised of three members) agreed upon by the Owner and contractor.
- Comprised of professionals with expertise in construction, engineering, and law.
- Meets regularly/periodically to monitor project progress for proactive dispute prevention.
- Maintains continuous involvement throughout the project.
- Potentially, more cost effective than arbitration and litigation.

- Proceedings are confidential (with recommendations not admissible in future legal actions unless agreed by the parties).
- Recommendations of the DRB are nonbinding.

Key DRB objectives include:

- Early identification of potential disputes and resolution before they escalate.
- Facilitating better collaboration and communication between parties to reduce misunderstandings.
- Providing expert guidance and insight on technical and contractual issues.
- Mitigating project delays by swiftly addressing conflicts that may interfere with progress.
- Reducing the need for parties to resort to arbitration or litigation.
- Preserving working relationships between parties through nonadversarial means.

An alternative mechanism to address disputes is the Dispute Adjudication Board (DAB).[17] Given that this chapter's primary focus is on the avoidance and mitigation of disputes and claims, the table below provides a summary-level comparison between DRBs (which are largely collaborative) and DABs (which cater to disputes after they have risen to a level requiring adjudication).

	DRB	DAB
Role	Proactive dispute prevention	Adjudication of disputes
Nature of Decision	Nonbinding recommendation	Binding decision
Involvement	Continuous during project execution	Ad-hoc or standing
Focus	Prevention and collaborative resolution	Formal adjudication (noncollaborative)
Legal Standing	Advisory	Enforceable (may be challenged via arbitration or litigation)
Use	United States, Canada	FIDIC contracts, other international projects

Design Errors, Omissions, and Changes

Incomplete or inaccurate design documents and late design changes can result in conflicts over the work required and associated costs. Peer reviews of critical and high-value project components such as the structure, main MEP systems,

17 The provisions for DABs have been incorporated into FIDC contracts.

or specialist components are essential to curtail errors, omissions, and untimely changes to realign the project with its intended purpose.

Defined Risk Allocation

To reduce disagreements that are colored by the stresses of a risk event, contracts should anticipate and explicitly define how risks are shared or allocated. For example, the risks and impacts resulting from delays caused by unusual weather, unforeseen site conditions, or material shortages should be clearly assigned to the relevant parties.

The supply chain disruptions, lockdown restrictions, and significant cost increases (particularly relevant to lump-sum contracts) caused by the COVID-19 pandemic were stark reminders that force majeure clauses were susceptible to disputes and led to legal challenges in the courts.

Learning from the pandemic and its ramifications, parties should supplement force majeure clauses with explicit and robust provisions that address pandemics. Early identification of potential disruptions, transparent and collaborative efforts to mitigate impacts, and sharing risks can help manage expectations and reduce conflicts.

Dispute Resolution Mechanisms

Contracts should include clauses for dispute resolution (such as mediation or arbitration) to facilitate a quicker, more amicable settlement of conflicts and discourage resorting to lengthy and costly litigation.

Effective Communication

Clear, consistent communication is key to avoiding conflicts in construction. Miscommunication or lack of communication between project participants can lead to misunderstandings, delays, or mistakes. To avoid this, it is advisable to develop and formally distribute comprehensive communications protocols at the project's "kick-off" meeting and routinely review and disseminate updated protocols as personnel change and new entities join or leave the project.

Clear Documentation of Decisions

Over-reliance on informal communications (e.g., phone calls, "water-cooler" conversations, and casual e-mail language) can lead to misinterpretations or missed details. To mitigate such issues, detailed records of all decisions, agreements, and changes should be maintained and disseminated to the appropriate parties in a timely manner (via letter, instruction forms, software-based project communications platforms, and the like). Documenting key decisions and changes during the course of a project (including the nature of the direction given and by whom, its impact, and the parties involved) reduces the likelihood of disputes and provides clear references if there are disagreements later.

Regular Progress Meetings

Regular progress meetings and trade-specific meetings (for major components such as building cladding, MEP systems, interior fit-outs, specialist constructions) with all key stakeholders should be held to review progress, discuss and resolve issues, align expectations and priorities, and identify expected actions and by whom. These meetings should be minuted with sufficient detail and with clear identification of parties in attendance and those who have follow-up actions. While many will consider this self-evident, we see this disciple often over-looked on projects during their most frenetic periods when people are dealing with urgent issues. However, these periods are the very ones that are most prone to misunderstandings and the seeds of disputes.

Project Planning

Detailed and well-coordinated project plans can help identify potential issues before they arise and provide timely opportunities to develop roadmaps for stakeholders to mitigate conflicts related to delays, scope changes, and unforeseen problems. Due to the dynamic nature of construction projects, project plans need to be flexible, realistic, and regularly revisited to ensure the remain proactive rather than reactive tools.

Detailed Scheduling

A detailed project schedule that outlines all tasks, deadlines, and dependencies helps ensure all parties understand the timelines and critical milestones so that an agreed baseline schedule is put into place.

This schedule should be updated regularly to reflect actual progress against the baseline and disseminated to all parties so that they are informed through a common document for planning, coordination, and taking mitigative measures where required.

Risk Management and Contingency Planning

Conduct regular meetings dedicated to risk management involving all key stakeholders and prepare and maintain an active and detailed risk register. Proactive identification of potential risks, their likelihood, and the severity of their impact allows for timely development of agreed contingency plans to address them and reduce the instances of risk or their consequences, if risks materialize.

Regular Inspections and Audits

Institute a structure for regular site inspections and audits (including periodic site walks by senior stakeholders) to ensure work is being carried out according to plan and expectations. The early addressing of any deviations helps prevent more significant disputes over misunderstood design intents and quality expectations at a later stage.

Change Management Procedures

Establish clear procedures for managing changes to the project scope, schedule, and budget. Ensure changes are documented and approved before they are implemented so that all parties are aligned on the nature and impact of these changes and disputes are avoided.

Conflict Resolution Training

Facilitate training in conflict resolution techniques, such as mediation or negotiation within the respective stakeholders' teams (at all levels including foreman, job captains, superintendents, managers, and more senior personnel). Conflict resolution skills are crucial even when minor disputes arise, as they can often de-escalate conflicts amicably and before they become more serious and affect project progress and team morale.

Early Contractor Involvement

Depending on the type and complexity of a project, early engagement of contractors during design and planning stages (whether a general contractor or specialized contractors such as an elevator manufacturer/installer, structural steel specialist, curtain wall cladding contractor, etc.) may prove beneficial for proper sizing of components, feasibility of envisioned construction methods, early identification of problems and scope gaps, and assistance in the completion of well-coordinated and sufficiently detailed design documents.

Conclusion

Construction projects are multi-faceted, highly complex, and dynamic long-term endeavors involving significant technical aptitude, resilience, and seasoned interpersonal skills. Disputes and claims, if not mitigated and managed with great skill and care, can derail projects and have the potential to be existential threats to Owners, contractors, and subcontractors. Managed well through several strategies applied at the appropriate stages of a project's lifecycle – from initial set-up to the drafting of contracts, administration of the works, and project close-out – disputes can be more effectively resolved, and claims may be prevented to ensure smoother project completion, maintain relationships, and preserve the financial health of all its stakeholders.

Appendix: Project Delivery Models

Design-Bid-Build (DBB)

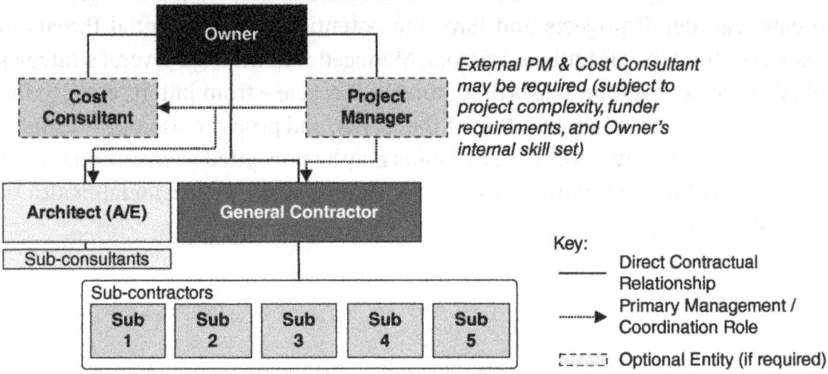

Prerequisites

- Regulatory Site/Zoning Plan (aka Affection Plan)
- Basic Development Plan/Project Brief
- Project Feasibility (high level, progressively updated)

Suitable Use

- Uncomplicated projects of small/medium size
- Comfortable timeframe to complete design then construct
- Construction innovation is not required
- Site conditions are well known
- Owner wishes control of design and construction quality
- Owner wishes to retain overall control through construction
- High degree of cost certainty required at Contractor award
- Owner team has project appropriate skills and experience (particularly if no PM is used)

Organizational Framework

- Owner hires Architect (A/E) first
- Owner and A/E develop the program

Advantages

- Relatively simple communication protocols
- Owner has overall control of the project
- Owner selects Architect on qualifications/ability
- Architect completes design before Contractor award
- Architect's role as Owner advocate
- Architect's supervision role preserves design intent
- Well defined roles, responsibilities and liabilities
- Allows for a large pool of potential bidders
- Bid competitively on complete/defined scope
- Relatively uncomplicated bidding process
- Relatively low tendering costs
- Bid evaluations are simpler than other delivery methods
- Risk retention by Owner encourages competitive bids
- Maximized built product for price (quality/value)
- High level of cost certainty at time of Contractor award
- Delivery model is very common and well understood

- A/E completes the documents (drawings/specifications)
- A/E responsible for design and constructability
- Documents are tendered
- Contractor is selected and contracted directly by the Owner
- A/E supervises the work
- Communication is directed through A/E to Owner (if no PM) or through PM (if one is appointed)
- Complex/large projects may require Project Manager
- Complex/large projects may require Cost Consultant (QS)
- Type of contract options: Lump Sum, GMP, Cost Plus, etc.

Common Uses

- Single Family and Multi-family Dwellings
- Community Centers/Small Retail Malls/Strip Malls
- Small/Medium Mixed-use Developments
- Small/Medium Business Parks
- Mid-rise and High-rise Towers (Single/Dual-use)
- Small/Medium Architecturally Significant Developments
- Schools
- Clinics
- Assembly Buildings (Multi-purpose Halls, Religious, etc.)

Disadvantages

- "Fast-tracking" option not available
- Design to budget may take several rounds*
- Owner control of Contractor is limited*
- Limited Owner control of subcontractor selections*
- Adversarial set-up (particularly without experienced PM)*
- Cost certainty relies on accuracy/completeness of Owner docs.
- If PM used/required, it is added cost (do cost benefit analysis)
- Late/no Contractor input on constructability*
- Owner retains design and constructability risks
- Can be prone to numerous claims/variations
- Level of Owner resources required to administer may be large
- Limited teamwork incentives*
- Limited incentives for innovation during construction
- Lifecycle considerations may not be properly considered

Notes

* Project Manager may help mitigate issues

Design & Build (D&B)

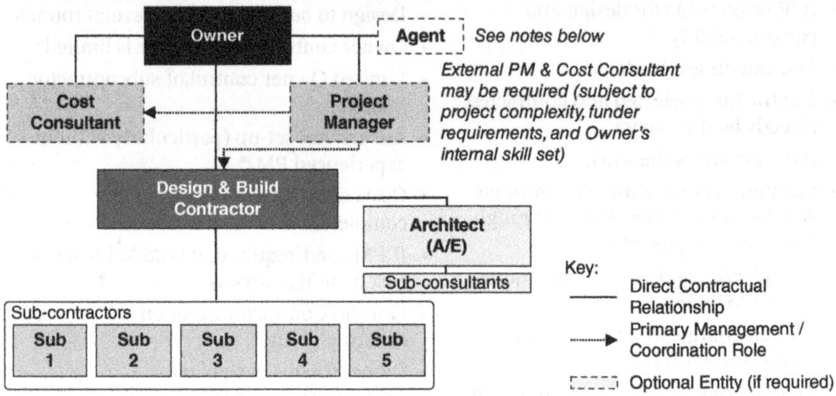

Prerequisites

- Regulatory Site/Zoning Plan (aka Affection Plan)
- Clear and Detailed Development/ Project Brief
- Robust Design Standards & Performance Specifications
- Project Feasibility (fairly robust – large early commitment)

Suitable Use

- For uncomplicated projects of small/ medium size
- For specialized projects (see Common Uses below)
- Robust design, quality, finishes standards set before award
- Building program (uses) firmly set
- Little/no change expected after Contractor appointment
- Single point of accountability is highly important to Owner
- Ideally, high level of trust/good prior working relationship
- Functionality is priority over design quality/character

Advantages

- Relatively simple communication protocols
- Owner administrative resources reduced
- "Fast-tracking" is an option
- Single contract to administer for design and construction
- Fairly early cost certainty
- Owner risk controlled
- Change orders largely limited to Owner changes
- Issues between Contractor and A/E not Owner's responsibility to resolve
- Contractor solves constructability issues early
- Contractor incentivized to seek economical solutions
- Contractor warrants design to be "fit-for-purpose"

Disadvantages

- Limited Owner involvement/control (design and construction)
- Establishing robust Owner criteria for a D&B is challenging
- Low tolerance for design changes (high costs)

Organizational Framework

- Owner contracts with Contractor (D&B Team)
- D&B Team consists of A/E under Contractor
- Design is managed by the Contractor
- D&B Team fully responsible to deliver the project
- GMP may be established in early construction phase
- D&B Team supervises own work
- Communication is direct between D&B Team and Owner
- Complex/large projects may require Project Manager
- Complex/large projects may require Cost Consultant (QS)
- Type of contract options: a D&B contract, potentially GMP

Common Uses

- Warehouses, "Big-Box" Retail
- Franchise Retail and Fast-Food Outlets
- Strip Malls
- Parking Structures
- Factories
- Plants (Power, Desalination, District Cooling, etc.)
- Single-use Repetitive Bldgs. (Public Housing, Motels)
- Suburban Residential Developments (Developer/Builder)
- Dormitories, Barracks and Labor Housing
- Correctional Facilities
- Business Parks
- Data Centers
- Some Stadiums and Airports
- Roads and Bridges

- Owner uncertainties in requirements can be exploited
- May not match best A/E and Contractor for Owner
- Commercially less competitive that other delivery models
- Contractor focus on maximizing own profit at all stages
- Primary A/E duty is to the Contractor, not the Owner
- Reduced built product for price (value/quality)*
- Postaward savings will typically not be passed on to Owner
- Contractor's savings focus outweighs lifecycle costs
- Disagreements about quality and design intent are common
- Conflict of interest acting as both Designer and Contractor
- Contractor supervises and approves own work*
- Smaller pool of tenderers than Design-Bid-Build model
- Owner takes over in breach/default more complicated
- "Buyer remorse" if delivery model is misapplied
- Can turn into Cost Plus job if model is mis-applied

Notes

* An agent/ advisor (Arch. or Eng.) may be hired by Owner to oversee key project components; however, their influence will be limited by the robustness of initial Owner criteria (Project Brief, Standards and Specifications). Appointment of an agent should be known to Contractor before their appointment.

Construction Management – Agency (Not at Risk)

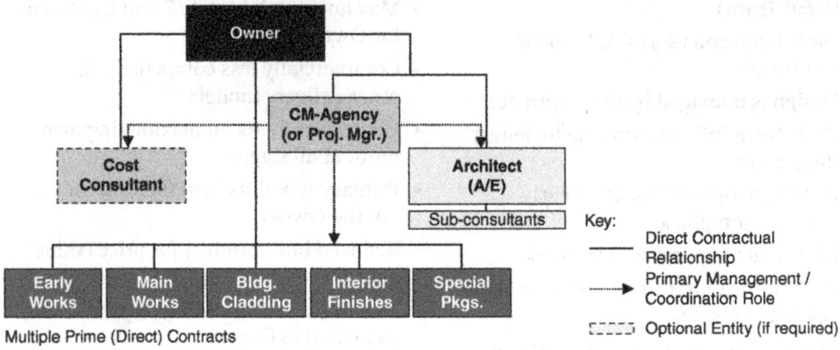

Prerequisites

- Regulatory Site/Zoning Plan (aka Affection Plan)
- Basic Development Plan
- Project Feasibility (high level, progressively updated)

Suitable Use

- For complex/various project types (see Common Uses below)
- Tight timeframes to complete the project
- Owner wishes control of design and construction quality
- Owner wishes to retain overall control through construction
- Project complexities warranting expert advice of CM/PM
- Project works can be broken down into parts for early start

Organizational Framework

- Owner hires Architect (A/E)
- Owner hires Construction/Project Manager (CM/PM)
- Owner hires Cost Consultant (Q/S)

Advantages

- "Fast-tracking" of project by experts (CM/PM)
- Owner selects CM/PM based on qualifications/experience
- Owner retains control with expert advice from CM/PM
- Owner selects Architect on qualifications/ability
- CM/PM advises on design priorities to suit packaging of works
- Owner commitments can be more gradual/incremental
- Owner can select prime/package Contractors with CM/PM
- CM/PM identifies constructability issues during design
- Expert (CM/PM) tasked with schedule and budget goals
- Level of day-to-day Owner management/oversight reduced
- Contract administration undertaken by expert CM/PM
- Respected/influential CM/PM has pull with Contractors

- Owner contracts with multiple Contractors (work packages)
- CM/PM may be hired first for expertise to set-up project
- CM/PM may assist Owner selecting appropriate A/E
- CM/PM will package works contracts
- Type of contract options: Lump Sum, GMP, Cost Plus, etc.
- Some work packages may get novated under others

Common Uses

- Mixed-use Developments
- Master-planned Developments and Re-developments
- Medium/Large Hotels, Resorts and Theme Parks
- Medium/Large Educational Campuses
- Super-tall/Mega-Tall Towers (particularly mixed-use)
- Regional and Super-regional Retail Malls
- Complex Buildings (large Hospitals, Museums, etc.)
- Major Performance Venues
- Large Stadiums and Airports
- Large Exhibition Centers/Halls
- Major Ground Transport/Transit Hubs
- Major Road and Bridge Infrastructure

- Packaging allows A/E focus and more time to complete design
- CM/PM "bridge" between int'l designer and local considerations
- A/E engages with some prime Contractors to improve design
- Promotes team-oriented approach to project delivery
- CM/PM viewed favorably by funders (as experts and third-party)
- Owner benefits from CM/PM industry relationships
- Owner benefits from CM/PM knowledge of market/territory

Disadvantages

- CM/PM has no direct contractual control of Contractors
- Owner holds multiple works contracts
- No single point of project accountability
- Administratively more complex than other delivery models
- Weak CM/PM may have limited ability to resolve disputes
- Cost certainty gradually attained (some works underway)
- Owner holds and administers multiple works contracts
- Higher Owner risk with multiple contracts*
- CM/PM is added cost

Notes

-

Construction Management – at Risk

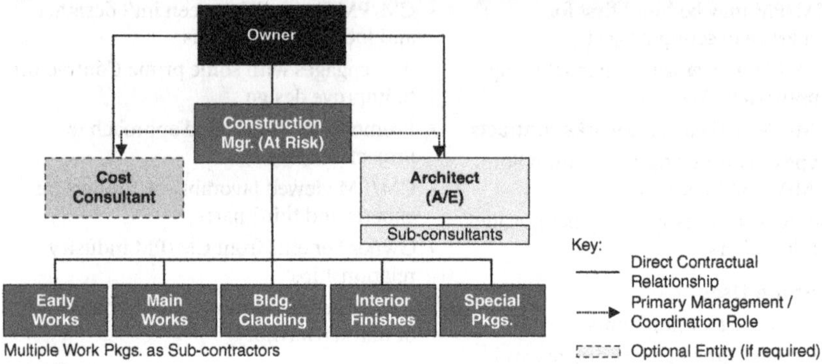

Prerequisites
- Regulatory Site/Zoning Plan (aka Affection Plan)
- Fairly Robust Development Brief
- Project Feasibility (fairly robust large early commitment)

Suitable Use
- For complex/various project types (see Common Uses below)
- Tight timeframes to complete the project
- Where project scope is uncertain/ many risks unidentified
- Where stakeholder interfaces are complex
- Where Owner resources/expertise is limited or unsuitable
- Project works can be broken down into parts for early start

Organizational Framework
- Owner hires Architect (A/E)
- Owner hires Construction Manager (CM)
- Cost Consultant (Q/S) may be under CM, AE or both
- CM contracts directly with multiple Contractors

Advantages
- "Fast-tracking" of project by experts (CM)
- CM and Architect (A/E) selected on qualifications/experience
- Potential to bring together best suited project team
- Reduced oversight/management required from Owner
- Promotes team-oriented approach to project delivery
- Owner/key stakeholders provide input with less cost and risk
- Fewer works contracts held by Owner
- Owner involved in selecting sub/prime Contractors with CM
- Early CM involvement allows better cost/ schedule control
- Greater efficiencies in design with early CM involvement
- Early constructability/logistics input with CM involvement
- Early CM involvement in design could preclude some claims
- Packaging allows A/E focus and more time to complete design
- Cost/GMP is established relatively early
- GMP helps cap construction cost risk for Owner

- CM hired first for expertise (or after A/E selected)
- CM will package works contracts
- Type of contract options: Lump Sum, GMP, Cost Plus, etc.

Common Uses

- Mixed-use Developments
- Master-planned Developments and Re-developments
- Medium/Large Hotels, Resorts and Theme Parks
- Medium/Large Educational Campuses
- Super-tall/Mega-Tall Towers (particularly Mixed-use)
- Regional and Super-regional Retail Malls
- Complex Buildings (large Hospitals, Museums, etc.)
- Major Performance Venues
- Large Stadiums and Airports
- Large Exhibition Centers/Halls
- Major Ground Transport/Transit Hubs
- Major Road and Bridge Infrastructure

- Provision to share savings can incentivize CM to reduce costs
- "Open book" process with good transparency
- Maintains higher quality at lower costs than other models

Disadvantages

- Smaller pool of capable/specialized Contractors
- Requires early commitment by Owner to select right CM
- Some reduced Owner control of project
- CM's management role is added cost
- GMP established after award/ involvement of CM
- Protracted negotiations to establish GMP could delay project
- Conflict of interest being both manager and contractor*

Notes

* A Project Management team to oversee CM and protect Owner's interests can be beneficial (particularly for very complex/ large projects or for Owner who is not in real estate/development sector

Integrated Project Delivery (IPD)

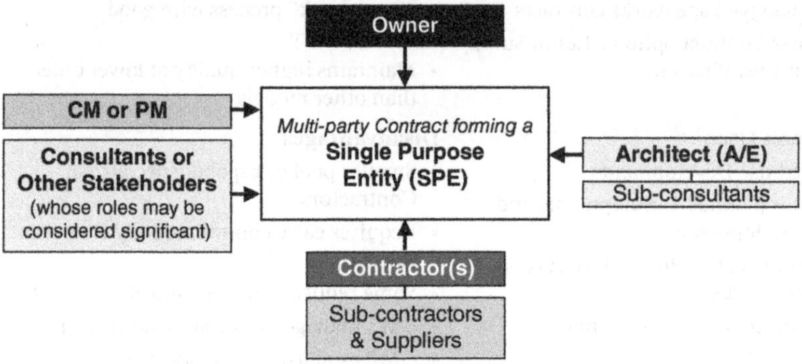

Prerequisites

- Regulatory Site/Zoning Plan (aka Affection Plan)
- Basic Development Brief (refined in first stage)
- Project Feasibility (high level, progressively updated)

Suitable Use

- For complex/various project types (see Common Uses below)
- Tight timeframes to complete the project
- Where project scope is uncertain/ many risks unidentified
- Where stakeholder interfaces are complex
- Where Owner resources/expertise is limited or unsuitable
- Where there is identifiable benefit in collaborative design
- Where there is requirement to obtain cost certainty early
- Where there is requirement to demonstrate transparency

Advantages

- "Fast-tracking" of project by experts (CM)
- Potentially high efficiency delivery model
- Optimum team-oriented approach and project delivery focus
- Early involvement of CM, A/E and some prime Contractors
- Early integration of design and constructability
- Owner risk mitigated by team approach/ joint contract
- Owner can be more involved in selecting sub/prime Contractors with CM
- Team member successes tied to project success
- Shared cost savings
- Higher incentive to achieve project within budget and on time
- Easier novation of A/E to CM for Stage 2 (construction)
- "Open book" process with higher level of transparency
- Team structure has the potential to greatly reduce disputes and claims

Organizational Framework*

- Owner selects CM and Architect (A/E) on qualifications/suitability
- Owner, CM and Architect (A/E) (and other parties, as required) enter joint contract (e.g., forming a "Single Purpose Entity" (SPE) limited liability company) with each party's roles, responsibilities, obligations clearly outlined.
- The joint team establishes project goals and objectives
- May be developed in two stages: Stage 1 – Design/Tender, Stage 2 – Construct

Common Uses

See **Design-Build** and **Construction Management** delivery models

Disadvantages

- Relatively new delivery method (not fully tested and not widely used/known); adoption may be challenging
- Requires high degree of Owner involvement
- Contractually complex to set up; some jurisdictions' courts may not be sufficiently experienced to deal with disputes
- Owner may lose some independence in dealing with A/E they typically have DBB and CM delivery models
- Owner's level of investment in Stage 1 may make Stage 2 change of a problematic CM more onerous
- Requires early Owner selection of right CM and A/E for team
- Departure of key staff or relationship breakdowns may have greater impact on project that other delivery models

Notes

*For example, in its guidance on SPEs for IPDs, the AIA states:

a. *"The SPE is managed by its members through a Governance Board. The Governance Board consists of representatives from each member. Except for decisions that the members specifically identified in the agreement, the Governance Board makes all decisions by unanimous vote."*

b. *"The Governance Board consists of representatives from each of the members. Each non-owner member appoints one representative to the Governance Board. The Owner appoints a number of representatives to the Governance Board so that the total number of Owner representatives on the Governance Board will be one more than the total number of non-owner representatives."*

15

Claims: Their Administration

Thiago Fernandes Moreira, Aline Piteres Porto, and Vitor Parreiras Almeida

Introduction

Regardless of how well-written a contract may be, parties involved in construction projects will likely encounter numerous claims and challenges during the project's development. Given the inherent complexity and technical nature of construction contracts, controversies are bound to arise, requiring revision of a long-term agreement structure, which will evolve over time due to external and internal factors.

It is not only possible, but natural, to face price adjustment and time extension claims during construction projects, which arise from various reasons, such as: (i) delays in the project development – and most likely concurrent delays, giving rise to a discussion on the parties' responsibilities; (ii) quality issues, that can prevent the safe and/or reliable operation of the project; (iii) quantity issues, that might be related to a wrong prediction of the materials necessary to perform certain scope or even a disconnection between design projects and reality; (iv) change orders, which are issued by the owner altering the original scope of works; or (v) hardship events, which entitle the aggrieved party to request renegotiation with a view to adapting the contract or terminating it. It is not always possible to address the inherent risks in advance.

An emblematic example is the COVID-19 pandemic. Parties are aware that contracts may be affected by external events, such as acts of God, strikes, labor disturbances, changes in economic or financial conditions, or even adverse weather. Consequently, most agreements will allocate such risks, assuring a certain level of predictability in the project. Notwithstanding, the global economy was significantly affected by the COVID-19 pandemic, leading to multiple lockdowns, importation and supply chain issues, project delays, additional costs, and workforce losses. These impacts were difficult to predict and lasted over two years,

Major Issues in International Construction Law, First Edition. Mauro Rubino-Sammartano and Troy Harris.
Published 2026 by John Wiley & Sons, Inc.

with some still impacting the industry, such as fluctuations in logistics prices. As a consequence, several construction contracts were impacted by these unforeseeable events beyond the parties' control, prompting a review of the contracts.

Is there something that could have been done differently? To avoid the claims, not much. However, the consequences arising from each contract varied in significance depending on the parties' ability to properly administrate the presented claims – which goes beyond merely choosing an appropriate method of conflict resolution.

It is difficult to deny that the chosen method of conflict resolution will significantly impact the time and costs required to achieve a decision or a settlement. On the one hand, if a dispute is brought before state courts, it will probably take longer to obtain an award. On the other hand, if a dispute is solved through arbitration, it can be faster, but is likely to be more costly. Whenever the claims escalate to a dispute, the parties inevitably lose something: time, money, political capital, and commercial relationships. Accordingly, infrastructure projects might benefit from mechanisms that allow for better management of claims and encourage settlements – such as formalizing occurrences and executing amendments and change orders – which has been a growing tendency in market practice.

Specialized organizations have developed international construction contract forms with prominent examples including those from the International Federation of Consulting Engineers (FIDIC), the New Engineering Contract (NEC), and the Alliance Contracts developed by the Association of Consultant Architects and King's College Centre of Construction Law (TAC-1). These standardized forms help define contractual relationships and establish essential criteria for their users, and there has been a growing emphasis on minimizing disputes and promoting the amicable resolution of claims.

Notably, the FIDIC Red Book, which initially included a single clause covering "claims, dispute, and arbitration," underwent a significant change in its 2017 edition. The revised version now divides the clause into two distinct sections: (1) "employer's and contractor's claims" and (2) "disputes and arbitration."

This modification intends to clarify that a claim does not necessarily equate to a full-blown dispute. "A claim is a request under the contract and the dispute arises when this claim is rejected."[1] Accordingly, there is a need for a distinct approach that prioritizes effective communication between parties and facilitates conflict resolution when addressing claims. This has been done through provisions regarding duty to notify, time management clauses that provide a period in which the parties can exercise a certain right, and mechanisms for claim presentation.

Provisions of dispute review or adjudication proceedings have also proven valuable, and their use has increased in the market practice, either as a mechanism

1 BEGRICH, M., KERBER, C., 2022.

of claim management or as a means of ensuring proper collection of evidence for future disputes that may arise. Since the implementation of large and complex infrastructure projects typically extend over years, most controversies often arise after a significant portion of the work has been completed. This is when evidence collection becomes a challenge. As far as the construction industry is concerned, "project and contract management are by far not as formally structured and not yet as process driven as in other industry sectors."[2]

This lack of evidence and claim administration creates a situation where the substantive complexity of the dispute is inversely proportional to the level of fact-finding.[3] Meanwhile, proper management of claims may provide a possibility for addressing conflict effectively and with minimal financial impact. Accordingly, this chapter aims to explore critical challenges encountered during contract administration, specifically focusing on claim management alternatives across various jurisdictions.

Major Issues Regarding Infrastructure Projects

"There are several reasons why parties to a construction contract may find themselves amid a claiming procedure."[4] Infrastructure projects may be affected by a series of external events, including economic and political circumstances that can drastically alter the terms of the agreement, as well as *force majeure* events, hardship situations, and even supervening impracticability.

Nevertheless, the need for effective claim management in infrastructure projects arises especially from their inherent spectrum of risks. "The construction service industry is a sensitive and a complex industry and has a fairly high construction uncertainty, related to the development of processes, methods, construction techniques, and the natural conditions factor in construction activities."[5] Martin Begrich and Cornel L. Kerber highlight some of the risks and regulatory burden involved in infrastructure projects:[6]

"From an English perspective, the specific risks of a construction project include the following:

- Execution of the works: subsoil conditions; unnatural obstructions; defective materials or labor; tests and samples; weather; site preparation; inadequate staff, labor, materials, time and finance.

2 KUNZLE, J., 2020.

3 KUNZLE, J., 2020.

4 BEGRICH, M., KERBER, C., 2022.

5 HAYATI, K., 2019.

6 BEGRICH, M., KERBER, C., 2022.

- Dispute and delay: ground ownership; late provision of information; inefficient execution of work; delays due to circumstances beyond a party's control; disagreements over execution.
- Instruction and supervision: greed; incompetence; inefficiency; unreasonableness; partiality; poor communication; errors in documents; faulty design; compliance with requirements; unclear requirements; unsuitable consultants and contractors; changes in requirements.
- Damage and loss to persons and property: negligence or breach of representations; uninsurable matters; accidents; uninsurable risks; consequential loss; exclusions (disclaimers); gaps in coverage and time limitations on insurance policies.
- External factors: government policies relating to taxation, labor, safety and other laws; building permits; financial embarrassment; energy or payment matters; war costs or civil commotion; malicious infliction of damage; intimidation; industrial disputes.
- Payment: delay in invoicing and invoice approval; delay in payment; legal restrictions on claiming interest; insolvency; shortage of funds; inadequacies in the measurement and valuation process; exchange rate, inflation.
- Law and jurisdiction: delay in adjudicating disputes; injustice; uncertainty due to lack of records or ambiguous contracts; costs of adjudication and litigation; enforcement decisions; changes in the law; changes in case law."

A technical study[7] also listed similar reasons for claims within infrastructure projects:

Reasons for claims	Mean score	Rank
Changes in owner's requirements	4.67	1
Changes of design from owner during the posttender award stage	4.38	2
Project being implemented in short time period with inadequate site investigation, design works, tender, and contract documents	4.29	3
Inadequate definition and/or specification of the precise scope of contract	4.29	3
Incomplete and/or uncoordinated design	4.05	4
Lack of clarity in the owner's requirements and/or inadequacy of the design brief	3.81	5

7 HAYATI, K., 2019.

Reasons for claims	Mean score	Rank
Parties generally more aware of their rights and relatively more litigious	3.67	6
Changes arising out of statutory, local authority sources	3.67	6
Employer's/contract administrator's failure and/or neglect to meet relevant contract obligations	3.38	7
Sudden swings in economic and market conditions	3.24	8
Negative effects of political factors	2.95	9
Contracting environment relatively more competitive with a larger number of players and narrow profit margins	2.90	10

Accordingly, most construction contracts will allocate risks between the parties and provide for specific scenarios in which they will be able to require an extension of time, price adjustment, or compensation. For example, the contract may provide that the contractor has the right to an extension of time proportional to the impact of delays caused by the suspension of works attributable to the owner, delay in clearance areas that were under owner's responsibilities, or even *force majeure*. Similarly, a contract might stipulate that owner's suspension of the work, changes in law, or delay in the clearance of areas, will also allow for the reimbursement of approved and evidenced costs that are reasonable and compatible with the market practice.

Nevertheless, the parties often cannot address all the risks associated with the contract, nor can they always agree on the characterization of certain contractual possibilities, such as extensions of time or price reviews. This can lead to the development of claims.

Construction Claim Management

Kongkoon Tochaiwat defined that "'construction claim management' can be construed as the process of dealing with or controlling the seeking of consideration or change by one of the parties involved in the construction process."[8] To achieve this, there are seven basic procedures that can be followed: (i) recognition and identification of a change of circumstances, which involves timely and accurate detection of a construction claim; (ii) notifications of change, alerting the other party of a potential situation; (iii) systematic and accurate documentation of

8 TOCHAIWAT, K., 2004.

change; (iv) proper analysis of time and costs impacts of change; (v) detailed pricing of change, so the parties can have a substantive description and details of the extra costs incurred or to be incurred due to the claim or contractual modification; (vi) negotiation of claim; and (vii) dispute resolution and settlement.[9] The incorporation of these standards is increasingly being observed in construction contracts as well as in international forms.

Time-barred Claims

In international practice, construction contracts commonly provide for time-bar clauses, which stipulate that the noncompliance with specific timeframes prevents the claiming party from presenting an extension of time or additional payment claim. For example, the FIDIC Red Book 2017 edition provides that the claiming party shall give notice to the client describing the event or circumstance giving rise to the cost, loss, or delay which the claim is based as soon as practicable, and no later than 28 days after the claiming party became aware, or should have become aware, of the event or circumstance. If the claiming party fails to comply with this timeframe, section 20.2.1 of FIDIC Red Book provide that:

> "20.2.1. (...) the claiming Party shall not be entitled to any additional payment, the Contract Price shall not be reduced (in the case of the Employer as the claiming Party), the Time for Completion (in the case of the Contractor as the claiming Party) or the [Defect Notification Period] DNP (in the case of the Employer as the claiming Party) shall not be extended, and the other Party shall be discharged from any liability in connection with the event or circumstance giving rise to the Claim."

In other words, if not timely presented, the right to submit the claim is lost, and the claiming party shall accept the burden arising from the event or circumstance that gave rise to the claim.

The FIDIC Red Book also provides that, within either 84 days after the claiming party became aware, or should have become aware, of the event or circumstance giving rise to the claim,[10] or within any other period agreed between the parties, the claiming party shall also submit a fully detailed claim, appointing (i) a detailed description of the event or circumstance giving rise to the claim; (ii) a statement of

9 TOCHAIWAT, K., 2004.

10 In the 2017 edition, the timeframe was extended. The 1999 edition and the 2005 MDB edition provided a period of 42 days after the claiming party became aware, or should have become aware, of the event or circumstance giving rise to the claim.

the contractual and/or other legal basis of the claim; (iii) all contemporary records on which the claiming party relies; and (iv) a detailed account of the supporting particulars of the amount of additional payment claimed, and/or the extension of time claimed.

If the claiming party fails to comply with this timeframe, section 20.2.4 of FIDIC Red Book provides that "the Notice of Claim shall be deemed to have lapsed, it shall no longer be considered as a valid Notice, and the Engineer shall, within 14 days after this time limit has expired, give a Notice to the claiming Party accordingly." In this sense, even if the claiming party meets the first timeframe to present a claim, its validity is subject to the compliance with the second timeframe.

In this scenario, the claiming party will only recover its right to present the claim if the other party fails to notice the course of time, as per sections 20.2.2 and 20.2.4 of the FIDIC Red Book.

The latest version of the NEC suite also includes time-bar clauses, specifically in the chapter on compensation events. The NEC suite has evolved over three decades to reflect industry developments, management of information, and supply chain engagement, resulting in the draft of NEC4. One of the main characteristics of this standard suite of contracts is its modularity, which allows greater adaptability to the specificities of each project and encourages its use in different markets and jurisdictions. Nonetheless, it provides for a set of core clauses that leaves no doubt regarding the importance of communication records and document management.

Analyzing NEC 4 model for engineering and construction contract as an example, it expressly states that "If the Contractor does not notify a compensation event within eight weeks of becoming aware that the event has happened, the Prices, the Completion Date or a Key Date are not changed unless the event arises from the Project Manager or the Supervisor giving an instruction or notification, issuing a certificate or changing an early decision," as per section 61.3 of the core clauses.

These time-bar clauses, which are being increasingly used, reflect the first step in effective claim management: the timely and accurate detection of a construction claim. During the development of a project, the parties are focused on completing the work and, in most cases, commencing commercial operation within the project schedule. Consequently, discussions are often postponed, accumulating at the end of the project when evidence has already been lost, the relationship between the parties may be strained, and the project's cash flow could be impaired. The inclusion of time-bar clauses appears as a way of avoiding this kind of situation, inviting and encouraging the parties to address their claims before they escalate into disputes.

Common-law tribunals have accepted that a party who fails to provide timely notice of its claim might be subject to losing its right:

> "At present the attitude of the courts is that a party who has failed to comply with the time-bar clause will have to persuade a court that something akin to frustration or force majeure has prevented it from complying with the time bar or that the conduct of the other party makes it unjust to uphold the time-bar provision. This will never be easy and case law indicates that only in exceptional circumstances have the courts permitted an extension of time."[11]

Under some civil law jurisdictions – such as Brazil – there are grounds for challenging and circumventing contractual time-bar provisions. As a general note, tribunals tend to decide based on the substance of the controversy (*i.e.*, analyze the right to a certain extension of time and/or reimbursement of cost), favoring the reality of the facts during the entire performance of the work and, possibly, overlooking the more restrictive and formalistic provisions. Tribunals have accepted informal communications, email exchanges, and testimonies as compliance with the time-bar provisions, if made without unreasonable delay.

Additionally, under some civil law systems, the soundness and effectiveness of those clauses could be understood and argued under different legal concepts. For example, Article 211 of the Brazilian Civil Code allows the parties to establish that one party's right ceases to exist if not exercised within a certain timeframe specified in the contract (*decadência convencional*), which would allow time-bar provisions. However, this legal concept of *decadência convencional* is quite controversial under Brazilian law and is generally understood to apply only to the extinction of unilateral rights. Considering that a contractor would not have a unilateral right, but rather a proper claim to demand an extension of time and/or reimbursement of costs, it is debatable if the time-bar clause is legally acceptable. In this sense, proper claims would not be subject to *decadência convencional*, but rather to legal statute of limitations (*prescrição*), which is considered a public policy issue in Brazil and thus, is not subject to contractual arrangements.

In any case, market practice has shown that the provision of time-bar clauses under construction agreements has been growing. One of the few obstacles that have been broadly accepted to the enforcement of time-bar clauses relates to (1) the absence of an express consequence for the noncompliance with the time-bar and (2) the principle of "good-faith" or the "prevention principle," depending on the jurisdiction.

For example, previous editions of the FIDIC Red Book did not specify major consequences for noncompliance with the stipulated timeframe for presenting a

---

11 TWEEDDALE, 2009.

detailed notice of the claim. As a result, an arbitral tribunal may accept a claim as valid even if it was not detailed by the claiming party, provided the contract is governed by the FIDIC Red Book's earlier provisions on this matter. The party's noncompliance shall only account for the impacts over the claim's investigation, as provided in the last paragraph of such clause.

Regarding the second obstacle, several civil law jurisdictions require contracts to be performed in "good faith." One of the many consequences of the good faith principle is that the parties to a contract are generally obliged to cooperate with each other (within reasonable boundaries considering the commercial nature of the relationship between the parties) and cannot benefit from their own wrongdoing. In that sense, in a hypothesis where the owner is in breach of its contractual obligations, an arbitral tribunal will likely find that the legal time-bar argument is not fair.

The owner cannot, on one hand, default on the contract, causing delays and extra costs and expenses to contractor, and, on the other hand, allege that noncompliance with the agreed timeframe prevents contractor from submitting a claim arising precisely from owner's breach of contract. This time-bar argument would run counter to the rule that no one can benefit from their own wrongdoing and would hardly be accepted in an arbitration in some jurisdictions.

The same goes for the "prevention principle." Under this principle, "in case any delay is attributable to the employer, the time bar provision may not result in loss of entitlement to claim, as one party cannot insist upon the performance of a contractual obligation by the other if the former has caused the non-performance."[12]

Apart from these exceptions, time-bar provisions have been more and more accepted as an adequate mechanism of claim management.

Proper Registration of the Matters

Systematic and accurate documentation of any supervening situations is also crucial in terms of claim management. Accordingly, legal practitioners have increasingly been developing agreements that outline rules for parties' communications and establish a duty to cooperate.

FIDIC Red Book, for example, provides in Section 1.3 for the criteria to be met for a valid notice and/or formal communication:

> **1.3** Notices and Other Communications
>
> Wherever these Conditions provide for the giving of a Notice (including a Notice of Dissatisfaction) or the issuing, providing, sending, submitting or transmitting of another type of communication (including acceptance,

12 ÇETINEL, Y., 2014.

acknowledgement, advising, agreement, approval, certificate, Claim, consent, decision, determination, discharge, instruction, No-objection, record(s) of meeting, permission, proposal, record, reply, report, request, Review, Statement, statement, submission or any other similar type of communication), the Notice or other communication shall be in writing and:

a. shall be:

   i. a paper-original signed by the Contractor's Representative, the Engineer, or the authorized representative of the Employer (as the case may be); or

   ii. an electronic original generated from any of the systems of electronic transmission stated in the Contract Data (if not stated, system(s) acceptable to the Engineer), where the electronic original is transmitted by the electronic address uniquely assigned to each of such authorized representatives, or both, as stated in these Conditions;

b. if it is a Notice, it shall be identified as a Notice. If it is another form of communication, it shall be identified as such and include reference to the provision(s) of the Contract under which it is issued where appropriate;

c. delivered by hand (against receipt), or sent by mail or courier (against receipt), or transmitted using any of the systems of electronic transmission under sub-paragraph (a)(ii) above; and

d. delivered, sent, or transmitted to the address for the recipient's communications as stated in the Contract Data. However, if the recipient gives a Notice of another address, all Notices and other communications shall be delivered accordingly after the sender receives such Notice."

The 1999 edition had already established criteria to be met for a valid notice and/or formal communication in Section 1.3, which were reproduced in the 2005 MDB version with only a few adjustments. However, FIDIC expanded the list of communications subject to those criteria, detailed the conditions to be met, and introduced the concept of Notice. Accordingly, it ensured that relevant communications between the parties are formalized, thereby avoiding further disputes.

Under Sections 13.3 and 18.2, FIDIC Red Book outlines the procedures for "variations and adjustments" and "exceptional events," respectively, seeking to establish an appropriate procedure for potential contractual changes and to avoid future claims. The same logic can be found under section 20, which establishes the claims procedure, as detailed in the topic above. Through the establishment of these procedures, FIDIC Red Book not only reinforces the importance of proper communication between the parties but also allows for the proper collection of evidence in case of claims and/or disputes.

Similarly, the Engineering and Construction Contract from NEC4 suite provides for clauses governing communications between the parties, such as sections 13 (Communications) and 15 (Early Warning), ensuring that issues arising during the construction project are addressed throughout the performance of the contract. Additionally, section 10.2 encourages the parties to act in the spirit of mutual trust and cooperation, supported by various processes within the contract. As stated in the NEC suite's official page, "anyone with experience of NEC4 ECC will tell you that talking and listening to each other, such as reviewing draft submissions so the project manager will be able to accept them, is the best way to act,"[13] clearly demonstrating the goal of improving claim resolution before they escalate into disputes.

The importance of proper documentation, however, goes beyond evidence collection and the possibility of solving claims throughout the performance of the contract. The mechanisms established by the international forms of construction contracts, currently applied by legal practitioners, reinforce the commercial conditions agreed upon by the parties, preventing misunderstandings based on their behavior.

Some legal systems stipulate in their rules on contract interpretation that the general course of the parties' behavior, including actions taken after the conclusion of the contract, shall be taken into account when interpretating the contract. In this regard, the legal systems aim to prevent a result contrary to the parties' intentions in the execution of the agreement.[14] This is the case of article 1,362 of the Italian Civil Code and article 113 of the Brazilian Civil Code, for example.

Therefore, it is important to formalize the parties' supervening understandings regarding the agreement, such as an understanding related to a time-barred claim, or the conditions for the performance of an additional scope. Otherwise, the parties' intention can be misinterpreted based on emails exchanges, conversations, or other informal means of communications that were not intended to be binding. A mere negotiation regarding a time-barred claim – accepted for business purposes – might be interpreted as a waiver of the time limitation provided in the contract for claims presentation if the parties do not expressly document that the negotiations are occurring out of mere liberality, and it is understood that the claim is time-barred and the receiving party has no obligation to evaluate it.

In addition, Joshua Karton developed research on the role of arbitrators in adjudicating the merits of international commercial disputes. He aimed to evaluate if arbitrators were more like judges, bound to apply the governing law according to strictly defined rules, or more like "commercially minded problem-solvers."

13 NEC, 2023.

14 YILDIRIM, 2019, p. 46.

As a result, Karton identified a widespread tendency to interpret contracts based on the parties' intent and trade usages, rather than strictly applying the governing law's rules of interpretation.[15] Karton's research indicated that arbitrators preferred an interpretative method characterized by "subjective interpretation supported by liberal consideration of extrinsic evidence," even though they acknowledged the priority given to the plain meaning of clearly drafted terms. This underscores the importance of proper documentation.

Lucy Greenwood presents an interesting perspective on the rule of parties' behavior on contract interpretation:

> "Even in situations where a tribunal is, for whatever reason, applying the principles of the governing law perhaps more strictly than is usual in international arbitration, the way in which arbitrations are pleaded and the way evidence is introduced in international arbitration may affect the tribunal's ruling, whether the tribunal is aware of this or not. In interpreting contracts under English law, a tribunal should not consider (absent particular circumstances which are beyond the scope of this article) the subjective views of the parties, negotiations between the parties, and postcontractual behavior. Yet, how easy is it in fact to ignore these key parts of a commercial relationship once they have been introduced in evidence by the parties?
>
> Although there are reasonably clear rules under English law on what should not be taken into account, research shows how difficult it is to exclude evidence once it has been introduced."[16]

Considering the above, proper documentation of claims, change orders, variations, formal notifications, and formalizing agreements by means of executing amendments, have proven to be an essential part of the contract and claims administration.

Conflict Resolution

In addition to the contractual frameworks aiming to improve the management of projects – allowing for a more organized approach to issues presented throughout the performance of the agreement – alternative dispute resolution techniques have proven to be a valuable option for claims resolution and preventing major disputes.

---

15 KARTON, 2015.

16 GREENWOOD, 2019, p. 26.

The term alternative dispute resolution ("ADR") can be applied to define a wide range of processes which aim to provide solutions other than the traditional litigation in national courts. ADR methods have gained relevance in the last decades as they can provide a more personalized assistance, as well as render more adequate and accurate decisions to the matters in discussion.

The construction industry is no stranger to ADR mechanisms. Given the high technical complexity involved in construction claims, several methods have evolved, with a growing preference for negotiation processes, expert determination, and dispute boards.

Negotiation and Mediation

It has become an increasingly common practice among construction contracts' negotiations to include an escalation clause as for the dispute resolution and claim management process. In this sense, the parties are required to negotiate on the claim raised or the matter discussed before entering an arbitral proceeding or litigation in national court.

The adoption of this matter can largely be attributed to the attempt to preserve an amicable relationship between the parties involved in a project, which can quickly wear very thin once claims are raised and the dispute has escalated into a more costly and serious proceeding. With this strategy in mind, many issues can be resolved without the need of further discussions, – better preserving future and current commercial partners.

Considering the importance of promoting a negotiation environment before the discussions evolve to litigation/arbitration, some courts have ruled on the matter. For example, the Swiss Federal Tribunal, pursuant to Article 190(2)(b) of the Swiss Private International Law Act, allows arbitral awards to be set aside if there is a lack of compliance with prearbitral procedures.[17]

In any case, regardless of national jurisprudence on the matter, given that the parties may encounter themselves in a very different spirit once disputes arise, it is advisable that some aspects are determined in the contract, aiming to avoid further discussions regarding the interpretation of contractual terms and procedures. A common strategy is to provide contractual clauses on the matter that are as specific as possible and establish important concepts beforehand.

In this context, for a negotiation process to be considered valid, it is usual that meetings between the parties' representatives are held during a determined period of time. For example, the parties may agree to hold a minimum number of meetings over a specific number of days so their respective representatives can discuss and negotiate the subjects in question. Additionally, the negotiation clause should

---

17 SWISS FEDERAL TRIBUNAL, 2014, 2016.

specify the executive level of the parties involved to ensure decision-making authority and the effectiveness of the procedure.

If the parties fail to reach an independent agreement, or if they have previously agreed that discussions should occur in the presence of a third party, there has been a noticeable increase in the preference for mediation – an approach less risky and less expensive than arbitration. Many legal systems have incorporated mediation as a viable mechanism for conflict resolution, aiming to foster consensual settlements and prevent major disputes. For instance, the Dispute Resolution Statistics of the International Chamber of Commerce recorded that in 2023, "newly-registered mediation cases involved 90 parties from 28 countries," with the United States (21), France (10), and Italy (8) accounting for the highest number of parties. This marks another step toward effective claims management and the amicable resolution of controversies.

Expert Determination

Expert determination has emerged as an interesting alternative to arbitration, being more frequently used to resolve issues to, among others, infrastructure, construction, and energy.[18] It involves hiring a reliable third-party expert to provide a decision on controversial technical matters,[19] and it can be very useful for preventing disputes or settling the existing ones amicably.

Given the wide variety and technicality of issues arising from construction contracts, such as the quality of the works and the compliance with performance tests, expert determination can be a quicker and more cost-effective way to resolve such claims. The decisions given by the specialist are usually binding, except if specifically provided otherwise in the corresponding contract. Therefore, as a rule, it can only be challenged on serious and duly justified grounds.[20]

The main advantages of expert determination lie in its expedited nature, as it offers one of the quickest resolutions to disputed matters without the need for more complex proceedings. In this context, while the specialist is typically required to analyze up to two main points, they can be central and critical to resolving many related issues, thereby preventing the need for arbitration or litigation.

Such practicality also reflects on the costs of the procedure, as the parties are not required to submit many written submissions or other important information, such as evidence, witness statements, and expert reports. It is meant to be as straightforward as possible.

18 YESILIRMAK, 2015.

19 YESILIRMAK, 2015.

20 CAPPS, 2022.

Moreover, expert determination can provide further flexibility, as the parties and the specialists are free to determine their own tasks, documents, and deadlines, given that they are not under any specific rules. It can also be less confrontational in comparison to other ADR methods, which can provide greater protection for the parties' relationship.

All in all, although expert determination may not be suitable for every category of dispute, it is typically ideal for construction and infrastructure disputes, as they require the analysis of deeply technical matters.

Dispute Boards

Dispute boards have increasingly grown in popularity over the last 30 years, especially in relation to long-term construction and infrastructure contracts.[21] They serve as a dispute settlement mechanism that can be established at the beginning of a project and followed through the entire process. Typically, an independent and impartial person or panel of three individuals provides recommendations, solutions, and decisions to controversial matters during the period of execution of the contract.

Although the dispute board members can be lawyers, they are typically specialists on technical issues regarding the contract's subject matter. In construction and infrastructure cases, for example, the dispute board members can be engineers or architects specialized in that specific type of project. As to provide better suitability to the disputes at hand, whether contractually or technically, the board can include both lawyers and specialists.

Traditionally, dispute boards can be divided in two categories: Dispute Review Boards ("DRB") and Dispute Adjudication Boards ("DAB"). In the first type, the board issues nonbinding "recommendations" in the form of a written report that includes the reasoning that led to each recommendation. In this sense, if none of the parties express dissatisfaction with the issued recommendation, compliance is required. However, if one party expresses dissatisfactions within a specified timeframe, they may resort to arbitration/litigation, and compliance is not required while the matter remains unresolved. DRBs can minimize animosity between the parties and, as a result, further negotiations can provide better economical and strategical solutions. However, this system relies heavily on the parties' trust in the member's expertise, impartiality, and judgment, amplified by the possibility of subsequent arbitration/litigation.

In the second type, the board issues "decisions," which are, consequently, binding where the parties must comply immediately with the decision. If either party expresses dissatisfaction with the decision, they can submit the matter to

---

21 DISPUTE RESOLUTION BOARD FOUNDATION, 2007.

arbitration/litigation. However, they must comply with the decision until the arbitral tribunal/courts decide otherwise.

It is also important to note that there is a third type of dispute board, which is less common by its nature: Combined Dispute Board ("CDB"). With CDBs, the board issues recommendations; however, if requested, it can issue a decision, which must be fully complied with and does not leave room for objections.

Preferably, dispute boards should be established at the beginning of the project, before the work begins, or shortly thereafter, and remain in place until the project is completed. Regular visits to the site where the works are being performed will maximize their benefit and value. This approach allows members of the dispute board to oversee the entire project development, understand its specifics, and monitor the decisions made by the parties throughout. Consequently, any claims can be resolved more quickly and efficiently, as the dispute board members will already have in-depth knowledge of the project.

However, there is also the possibility of appointing an *ad hoc* dispute board. In other words, the dispute board can be organized and formed once a dispute has arisen. Despite being less costly in comparison to a permanent dispute board established from the beginning, the process can be more challenging and less efficient due to the parties' relationship at that stage and a potential loss of evidence during the performance of the work and the onset of the dispute.

Over the years, experience has shown that dispute boards have been highly successful in preventing disputes.[22] Accordingly, even after a dispute board procedure, if an issue reaches arbitration, the discussion of the dispute board is, at some extent, refined and studied, which can prove to be efficient not only in terms of time, but also with costs and evidence collection.

The results of such popularity are noticeable: DABs and DRBs are present in different bodies of rules, such as the standard FIDIC Contracts, the Dispute Board Federation, the International Chamber of Commerce ("ICC"), the Center for Arbitration and Mediation of the Chamber of Commerce Brazil-Canada ("CAM-CCBC") and the World Bank.

Nevertheless, dispute boards have an important drawback: they must be constantly updated. Delays in rendering documents, copies of the construction progress reports, and minutes of weekly meetings reduce their effectiveness. Therefore, since it is an ongoing process, it is important to keep the board duly informed so when a matter arises, it is prepared to address it and issue a decision or recommendation.

---

22 PETKUTE-GURIENE, 2019.

Conclusion

International practice has shown a growing preference for amicable dispute resolution, which has been possible through proper contract management mechanisms. Whether through the inclusion of time-bar provisions in construction contracts – in order to induce the parties to resolve any issues in a timely manner – or through clauses that encourage a documentary record of the parties' relationship, more effective means of preserving the relationship between the parties and developing an efficient project have been achieved.

Alternative dispute resolution methods have also been increasingly adopted in search of extrajudicial solutions for technical and more complex disputes. This approach allows for quicker and more effective resolutions while preserving commercial relations.

Finally, even if a dispute cannot be avoided, effective claim management has proven to be an excellent method for evidence collection, enabling the proper resolution of disputes.

Bibliography

Begrich, M. and Kerber, C.L., "Chapter 1: Early Claims Resolution," in *International Construction Disputes: A Practitioner's Guide*, eds. Roquette, A.J. and Pröstler, T.C (Munich: Verlag, 2022), 1–34.

Brazil (2002) Código Civil Brasileiro. Lei No. 10.406, de 10 January 2002. Brasília: Presidência da República.

Brunner, C., "Chapter 5: Hardship (Change of Circumstances): Fundamental Change of the Equilibrium of the Contract, Section 13: Legal Effects of Hardship," in *Force Majeure and Hardship under General Contract Principles: Exemption for Non-performance in International Arbitration. International Arbitration Law Library* (Kluwer Law International, 2008), 18: 479–512.

Capps, D. and Sheddon, L. (2022) "Expert Determination," Ashurst. Available at: https://www.ashurst.com/en/news-and-insights/legal-updates/quickguide---expert-determination/(Accessed: 14 June 2022).

Çetinel, Y. (2014) "Losing Entitlement to Claim and Resort to Dispute Resolution in Construction: Time Bar Provisions and the Turkish Approach," Kluwer Arbitration Blog, 19 December. Available at: https://arbitrationblog.kluwerarbitration.com/2014/12/19/losing-entitlement-to-claim-and-resort-to dispute-resolution-in-construction-time-bar-provisions-and-the-turkish-approach/(Accessed: 09 February 2024).

Dispute Resolution Board Foundation, *DRBF Manual*. 1st ed. (Seattle: Dispute Resolution Board Foundation, 2007).

FIDIC, *Conditions of Contract for Construction: For Building and Engineering Works Designed by the Employer*. 2nd ed. (Geneva: Fédération Internationale des Ingénieurs-Conseils, 2017).

Greenwood, L. "Principles of interpretation of contracts under English law and their application in international arbitration," in *Arbitration International*, ed. Park, W.W. (Oxford: Oxford University Press, 2019), 35(1): 21–27.

Hayati, K., Latief, Y. and Jaka, A.S. "Risk-based Contract Management on the Design and Build Construction to Minimize Disputes in Infrastructure Projects," *IOP Conference Series: Materials Science and Engineering*, 469, (2019): 012082.

Italy, Codice Civile. (Rome: Istituto Poligrafico e Zecca dello Stato, 1942).

International Chamber of Commerce (ICC) 2024, 2023 Statistics: ICC Dispute Resolution, ICC, viewed 9 September 2024, https://iccwbo.org/wp-content/uploads/sites/3/2024/06/2023-Statistics_ICC_Dispute-Resolution_991.pdf.

Karton, J. "The Arbitral Role in Contractual Interpretation," in *Journal of International Dispute Settlement*, ed. Schultz, T, (Oxford: Oxford University Press, 2015), 6(1): 4–41.

Künzle, J. "Chapter 5: Dispute Resolution," in *Multiple Contracts and Coordination in International Construction Projects: A Swiss Law Analysis*. International Arbitration Law Library, (Alphen aan den Rijn: Kluwer Law International, 2020), 57: 183–204.

NEC, *NEC4: Engineering and Construction Contract*, 1st ed. (London: NEC Contracts, 2017).

NEC Contracts (2024) "Do what NEC4 ECC says or face the 'enforcement clauses,'" NEC Contracts, 15 August. Available at: https://www.neccontract.com/news/do-what-nec4-ecc-says-or-face-the-'enforcement-clauses' (Accessed: 31 August 2024).

Swiss Federal Tribunal (2016) Judgment of 16 March 2016, Case No. 4A_628/2015.

Swiss Federal Tribunal (2014) Judgment of 7 July 2014, Case No. 4A_124/2014.

TAC-1 (2016) Term Alliance Contract. 1st ed. United Kingdom: The Association of Consultant Architects (ACA).

Tweeddale, A. and Tweeddale, K. "Commencement of Arbitration and Time-Bar Clauses," *Arbitration: The International Journal of Arbitration, Mediation and Dispute Management*, (London: Institute of Arbitrators (CIArb); Kluwer Law International), 75 no. 4(2009): 480–486.

Tochaiwat, K. and Chovichien, V. "Contractors' Construction Claims and Claim Management Process," *Research and Development Journal*, 15, no. 4(2004): 66–73.

Yeşilırmak, A. "Chapter 2: ADR Mechanisms: Negotiation, Mediation and Expert Determination," in *Arbitration in Turkey*, eds. Yeşilırmak, A. and Esin, I.G. (Alphen aan den Rijn: Kluwer Law International, 2015), 15–36.

Yildirim, A.C. "Interpretation of Contracts in Comparative and Uniform Law," *International Arbitration Law Library*, (Alphen aan den Rijn: Kluwer Law International, 2019), 47: 25–124.

16

Owners' Claims*

Maude Lebois and Nicolas Pralica

Introduction on Owners' Typical Claims

In the context of construction disputes, it is common for owners to raise claims against contractors in relation to delays to completion and failure to carry out the works in accordance with the contract's specifications. Although less frequent, owners may also exercise their right to terminate the construction contract, either pursuant to its terms or under the applicable law. Owners' claims typically fall into three broad categories: (i) claims for delay liquidated damages, (ii) claims for defective or non-compliant work, and (iii) claims arising from termination (or replacement).[1]

First, owners may seek damages for delay if the contractor fails to complete the works by the agreed milestones or final completion date. Delays can occur due to various reasons such as inefficiencies, poor planning, or issues with subcontractors, materials, or labor. Most construction contracts contain liquidated damages clause in favor of the owner. Although this may vary depending on the applicable law, the owner will generally be entitled to claim liquidated damages without having to prove that the contractor was at fault, but only that the works were delayed beyond the agreed deadlines. In practice, however, most contractors will often challenge the owner's entitlement to delay liquidated damages and claim an extension of time, arguing that the delay was caused by owner-related events. The owner must therefore be prepared to present evidence to refute such counterclaims, particularly by demonstrating that the contractor is not entitled to

---

* The views expressed in this article are those of the individual co-authors and do not necessarily reflect a unanimous position; each co-author may hold differing opinions depending on the topics discussed.

1 As each construction contract is unique, the solutions discussed in this article may not apply in all cases and should be assessed in light of the specific contractual terms.

relief under the terms of the contract. The rate of liquidated damages may vary depending on whether the delay affects specific milestones or the overall completion of the project.

Secondly, claims for defective work may arise at various stages, including before completion, at the time of handover, during, or even after, the defects liability period. Prior to completion, many contracts allow owners to inspect the works and require the contractor to rectify defects before issuing interim payments. At completion, the owner typically conducts inspections before issuing a certificate of completion, and the defects identified at this stage must be remedied as a condition for the issuance of the provisional acceptance certificate (PAC). In many construction contracts, after the issuance of the PAC, the contractor remains responsible for fixing any defects that arise during the defects liability period. This period usually starts after provisional acceptance is granted and typically lasts from 6 months to 2 or 3 years, depending on the contract. The contractor is required to remedy any defects that appear during this period.

If the contractor fails to do so, the owner has the right to withhold payments or recover the costs necessary to rectify these defects. Furthermore, the owner may withhold the final acceptance certificate until all defects are resolved, as the purpose of this certificate is generally to confirm that the project has been fully completed, including that any defects identified during the defects liability period have been satisfactorily remedied.

This mechanism ensures that the contractor is still incentivized to deliver a full project that meets the contract's specifications, even after the provisional acceptance is issued. If additional defects are later discovered (including latent defects) affecting the structure, the owner may claim damages assessed either on the basis of the cost of rectification or the resulting diminution in value of the works, depending on the applicable law and the contract's provisions.

The owner may also claim that the contractor has failed to comply with the detailed specifications or design requirements stipulated in the contract, or has failed to carry out works required under the contract. This may involve issues such as defective designs, missing or non-conforming materials, or failure to comply with applicable technical standards. Construction contracts often include guaranteed performance levels or other contractual performance requirements for the completed works, such as efficiency levels or output targets. If the contractor fails to achieve these performance levels, the owner may require the contractor to carry out remedial works, or claim liquidated or unliquidated damages. In addition, the owner may be entitled to call on any performance bond or bank guarantee provided by the contractor.

Finally, the owner may elect to terminate the contract, especially in circumstances where the contractor is unable or unwilling to complete the works. Termination is a remedy that requires careful consideration, as it may cause disruption to the project and increase costs. If termination is validly exercised due to the contractor's default, the owner may be entitled to recover the additional costs of completing the project with a replacement contractor. These costs can be substantial, particularly since a replacement contractor may charge a premium for taking over partially completed works and assuming liability for defects. Subject to the terms of the contract and the applicable law, the owner may also retain the right to claim any accrued liquidated damages.

In less extreme cases, rather than terminating the contract altogether, the owner may choose to remove certain aspects of the work from the contractor's scope – often referred to as "descoping" – and reassign these works to another contractor. While descoping may help mitigate delays or ensure critical elements are completed to the required standard, any additional costs incurred due to this approach are generally not recoverable from the original contractor unless such recovery is provided for under the contract or subsequently agreed by the parties. However, whether through termination or descoping, the owner bears the burden of proving that the action was justified under the contract, making it crucial to ensure compliance with contractual provisions.

All the above claims will ultimately give rise to claims for monetary damages – whether by way of general damages or, for instance, liquidated damages for delay – and will therefore be subject to any liability cap set out in the construction contract. For this reason, although each of these heads of loss will raise different issues of entitlement, causation, and quantum, one major issue arising in relation to most claims is whether the owner may recover the full extent of its losses when the aggregate value of its claims exceeds the contractual limitation on the contractor's liability.

Two types of provisions commonly found in construction contracts may restrict the owner's ability to recover its full losses:

- Exemption clauses, which cover both limitation-of-liability clauses (*i.e.*, clauses setting out a "ceiling" on the contractor's liability, typically as a percentage of the contract price) and exclusion-of-liability clauses (excluding liability for certain types of breach or losses, especially consequential damages); and
- Liquidated damages clauses, which, although not limitation clauses in the strict sense, have a similar effect because they generally[2] operate both as a "floor" and a "ceiling" on the contractor's liability for specific breaches (e.g., by fixing the amount payable for delay or failure to meet specified performance requirements).

2 As discussed below, as an exception to this general statement, the rule under Swiss law is that, in the absence of clear language to the contrary, liquidated damages clauses will be interpreted as providing only for a "*floor*" on damages without precluding recovery of additional losses by way of general damages. In a contract governed by French or Belgian laws, the parties would be required to include some express language to that effect to allow recovery of additional damages beyond the agreed sum.

In practice, the owner's ability to recover fully for its losses will depend on two main factors: first, the specific wording of the relevant clause as interpreted by the court or tribunal, and second, any substantive rules under the contract's governing law that may, in certain circumstances, preclude the contractor from relying on such contractual limitations of liability.

While the modern approach under English law – following the decisions in *Photo Production* and *ICS* – focuses primarily on the proper interpretation of the clause, leaving limited scope for the application of substantive rules,[3] civil law systems, particularly French law, have traditionally allowed judges and arbitrators to review the substance of these clauses more closely, in light of their potentially drastic consequences.

This chapter examines the interpretative principles and substantive rules governing exemption clauses and liquidated damages provisions under French law. It also highlights key differences between French law and other civil law systems, and contrasts these with the position under English law.

Exemption Clauses

In this section, we discuss the substantive controls applicable to exemption clauses under French law relating to their enforceability and scope which may have in practice a significant impact on the owner's ability to be made whole. Importantly, even if the owner elects to terminate the construction contract, the clause may under certain circumstances continue to apply and survive after termination, unless otherwise agreed. In construction contracts governed by the laws of other civil law jurisdictions, owners should be wary of the fact that, although the basic legal concepts are substantially the same, the legal position and potential outcomes may vary in significant ways. Conversely, if the owner agrees to a choice of English law to govern the construction contract, particular care should be taken to set out express exceptions to the exemption for gross negligence and willful misconduct and other serious breaches, as the owner will not be able to rely on mandatory exclusions provided for by the general law.

---

3 The rule against limiting or excluding liability for fraud is arguably one of the only surviving substantive rules under the common law. To some extent, these common law rules have been replaced by the statutory controls introduced by the Unfair Contract Terms Act 1977 (**"UCTA"**). These statutory controls govern the substance of exemption clauses, but, in practice in international construction projects, they rarely find any meaningful application (s. 3 UCTA only applies to clauses inserted as part of the defendant's "*written standard terms of business*"), especially in relation to projects taking place outside England and Wales (s. 27 UCTA).

Substantive Controls on the Enforceability of Exemption Clauses

The main exception to the principle that exemption clauses are valid and enforceable under French law is when such clause deprives of any substance an essential obligation of the party invoking it.

Further, the 2016 reform introduced another exception in case of "significant imbalance" between the rights and obligations of the parties under the contract. Despite these new developments, the French legislature did not go as far as to introduce a statutory power to revise contractual limitations analogous to that of the judge or arbitrator in relation to liquidated damages.

Exemption Clauses May Not Negate One of the Contract's Essential Obligations

It is well-established under French law that a contractual term will not be enforced if its effect would be to negate an "essential obligation" of the contract. This rule focuses on the effect of the clause itself, and for this reason the assessment is generally made by reference to the time of the conclusion of the contract.[4] This rule is now codified in the new Article 1170 of the French Civil Code pursuant to which "any clause that deprives the promisor's[5] essential obligation of its substance is unenforceable."

Although the rule applies generally to all contract terms, it is particularly relevant in respect of exclusion of liability clauses. Such a clause will be unenforceable if it purports to exclude the defendant's liability for breach of one of its essential obligations under the contract.[6] By contrast, limitations of liability are less likely to be struck down under this rule, but any clause that sets a "derisory" limit on the defendant's liability may not be enforced.[7] In these cases, the claimant is entitled to recover uncapped damages for breach of contract.[8]

Many construction contracts generally limit the contractor's liability to the contract price or a percentage of the contract price and exclude liability for consequential losses. These clauses are unlikely to be struck down under Article 1170

---

4 P. le Tourneau, C. Bloch, A. Giudicelli, C. Guettier, J. Julien, D. Krajeski, M. Poumarède, *Dalloz Action – Droit de la responsabilité et des contrats 2023/2024 – Régime d'indemnisation* "Clauses limitant la responsabilité," 13e éd. (Paris: Dalloz, 2023, 1556, § 3225.13.

5 In our view, it is preferable to translate the general concepts of "*créancier*" and "*débiteur*" under French law (which apply in respect of any obligation and are not restricted to the payment of a sum of money) by using the terms "*promisor*" and "*promisee*" (as they relate to any contractual promise under English law) rather than "*creditor*" and "*debtor*." Depending on the context, we also refer more simply to "*claimant*" and "*defendant*."

6 Malinvaud, Ph., Mekki, M., Seubé, J.-B. *Droit des obligations*, 17e éd. (Paris: LexisNexis, 2023), 547, § 594.

7 Malinvaud, Ph., Mekki, M., Seubé, J.-B. *Droit des obligations*, 17e éd. (Paris: LexisNexis, 2023, p. 547, § 594

8 French Civil code, New Art. 1231-3.

of the French Civil Code. However, more specific clauses relating to particular obligations – for example excluding liability for deploying an integrated software across newly built facilities – may give rise to difficult questions of fact and degree. In this example, the court or tribunal will have to determine whether the obligation to complete this software integration constitutes an essential obligation of the contractor under the construction contract.

Exemption Clauses Inserted as Part of Nonnegotiable, Standard Terms May Not Create a "Significant Imbalance" in Contracts Concluded After 1 October 2016

For all contracts concluded after 1 October 2016, the new Article 1171 also provides a further substantive control on the enforceability of contract terms. This provision calls for the courts to make a holistic assessment as to the imbalance, if any, created by a term dictated by one party:

> *When contracting on a party's standard terms,*[9] *any non-negotiable term, set out in advance by that party, which creates a significant imbalance between the rights and obligations of the parties to the contract is unenforceable.*

While this type of test is more commonly found in relation to consumer contracts, the French courts have recently clarified that this provision also applies to contracts between commercial parties.[10] The scope of this provision is similar to that of the new Article 1170 of the French Civil Code in that it is not restricted to exemption clauses but extends to any contract terms. However, this substantive control will seldom apply, as construction contracts are typically negotiated – or at least open to negotiation.

Statutory Power to Adjust the Amount of a "Derisory" Limitation of Liability?

As discussed in the second part of this chapter, a key feature of the French law on liquidated damages provisions is that the judge or arbitrator enjoys a broad discretion to adjust the amount provided for under the liquidated damages provisions if it is manifestly inadequate in comparison to the actual loss.

There is some debate under French law whether a judge or arbitrator would have the power, instead of striking out an unreasonably low limitation, of simply substituting a higher one. Some authors argue that in cases in which a limitation of liability is found to be "derisory" (to the extent that it negates an essential

---

9 Like other authors, we use the expression "standard terms" to translate the French concept of "*contrat d'adhésion.*" See e.g., Chitty *on Contracts* (35th edn, Sweet & Maxwell 2022) ch 2, § 2-005, 103–104.

10 Cass. com. 26 janv. 2022, n°20-16.782.

obligation of the contract under Article 1170 or creates a significant imbalance within the meaning of Article 1171), the courts should be entitled to exercise the same power that they have in relation to liquidated damages provisions and have discretion to modify the contractual limit, rather than disregarding it altogether.[11]Other authors reject this analogy on the ground that the two types of provisions are generally assessed at different times. Whether the contractual limitation is "derisory" and unenforceable is generally assessed by reference to the time of the conclusion of the contract, whereas the review of liquidated damages is made with the benefit of hindsight in light of the actual loss suffered, as explained further below.[12]

Substantive Controls on the Scope of Exemption Clauses

As a matter of public policy, an exclusion or limitation of liability in a contract governed by French law does not apply if the breach in question is either deliberate (and constitutes "*dol*" or "*faute dolosive*")[13] or amounts to gross negligence ("*faute lourde*").[14] The application of this rule cannot be excluded even by express terms. This rule is reflected in Article 1231-3 of the French Civil Code which states as follows:

> *The promisor is only liable for damages [for breach of contract] to the extent provided for in the contract or otherwise foreseeable at the time of its formation, except if its failure to perform arises from a deliberate breach or gross negligence.*

---

11 Malinvaud, Ph., Mekki, M., Seubé, J.-B. *Droit des obligations*, 17e éd. (Paris: LexisNexis, 2023), 544, § 591 (citing in favor of this position Y. Gontier, *Plaidoyer pour une révision judiciaire des clauses limitatives de responsabilité*, préf. J. Mestre, éd. Y. Gontier, 2005).

12 Delebecque, Ph. "Régime de la réparation – Conventions relatives à la responsabilité," *JurisClasseur Civil Code*, (LexisNexis, 2022), 14, § 76.

13 Throughout this chapter, the term "*deliberate breach*" is used to translate the concept of "*dol*" (as a breach of contract), also referred to as "*faute dolosive*." The term "*dol*" in French encompasses two different concepts, one which relates to the contract formation and is a ground for rescission of the contract and the other that refers to a deliberate breach of contract. While "*dol*" in the former sense could be translated as "fraudulent misrepresentation," it would not be accurate to translate "*dol*" in the latter sense (also called "*faute dolosive*") with the word "*fraud*" because, contrary to the English law concept of "*fraud*," deceit or some form of dishonesty is not required to make out a claim for "*faute dolosive*." It is sufficient to show that the promisor intentionally breached the contract.

14 In English law, the term "negligence" covers "*both the separate tort of failing to use due care and also breach of contractual provision to exercise skill and care.*" (*Triple Point Technology Inc v. PTT Public Company Ltd* [2021] UKSC 29, [52]). Here, the term "gross negligence" refers to a contractual claim for breach of contract where the breach results from a gross lack of care and skill in the performance of the contract, as explained further below.

Establishing a deliberate breach or gross negligence within the meaning of Article 1231-3 has much wider ramifications than simply rendering any exemption clause in the contract unenforceable. Such a finding not only precludes the promisor from relying on any contractual provision limiting damages, but also displaces the ordinary rule that damages for breach of contract may only be awarded to compensate losses of a kind that was foreseeable at the time of contracting.

The latter aspect of this rule may be particularly relevant in the context of large construction projects, which typically span several years. Over the course of the project, circumstances may change, and the owner may incur significant losses that were entirely unforeseeable at the time of contracting. Provided the owner can establish gross negligence or a deliberate breach, it may nevertheless be entitled to recover those losses.

Deliberate Breach of Contract

The concept of "*faute dolosive*" or deliberate breach is defined under French law as a deliberate refusal to perform a contract without lawful excuse.[15] In this context, it is sufficient to prove that the breaching party intentionally failed to perform the contract, even though it may not have specifically intended to cause harm to the innocent party.[16] In other words, to establish a "deliberate breach" within the meaning of Article 1231-3 of the French Civil Code, the innocent party is required to prove the guilty party's general intent to breach the contract, without generally having also to prove a specific intent to cause harm.[17]

According to the rule laid down in Article 1231-3, where the defaulting party has deliberately breached its obligations under the contract (i.e., where the breach amounts to a "*faute dolosive*"), no contractual term may exclude or limit its liability for that breach. If the exemption clause is expressed in general terms without any specific reference to "*dol*," "*faute dolosive*" or similar terms, the clause will be generally construed to apply to unintentional breaches. On the other hand, if the clause expressly refers to "*faute dolosive*" or a similar concept, it will, to that extent, be void as contrary to public policy.

15 Terré, F., Simler, Ph., Lequette, Y., Chénedé, F. *Droit civil – Les obligations*, 12ᵉ éd. (Paris: Dalloz, 2018), 901, § 843.

16 Malinvaud, Ph., Mekki, M., Seubé, J.-B. *Droit des obligations*, 17ᵉ éd. (Paris: LexisNexis, 2023), 636, § 680.

17 It is not entirely free from doubt whether the breaching party must also have been aware of the risk of harm to the other party and decided to breach the contract regardless, i.e., whether recklessness is required in addition to general intent.

Gross Negligence

The innocent party may also seek to establish gross negligence (*i.e.*, "*faute lourde*"). According to the classic definition of "gross negligence," it involves "conduct of extreme gravity tantamount to a deliberate breach and evidencing an inability of the promisor to fulfil the contractual mission it accepted."[18] A prominent, eighteenth-century French jurist had defined this type of breach as "[the failure to] take the same care of other people's affairs as the most careless and stupid people would of their own."[19]

If the breach involves such gross negligence, it will generally be treated as tantamount to a deliberate breach and the breaching party will be precluded from relying on any contract term that purports to exclude or limit its liability. This principle is encapsulated in the legal maxim "*culpa lata aequiparatur dolo*" (gross negligence is equivalent to a deliberate breach). The basic idea is that, in the absence of direct evidence of the defaulting party's intent, the circumstances and manner of the breach may nonetheless be so severe as to cast doubt on whether the breach – although alleged to be merely negligent – was not in fact deliberate. In such circumstances, it is said that the party in breach should not be allowed to "hide behind the mask of ineptitude."[20]

The Relevant Factors That May Indicate Gross Negligence Under French Law

Courts assess gross negligence based on several factors, including the expertise and specialist knowledge of the defendant, the foreseeability and seriousness of the harm caused, and the recurring nature of the defendant's defaults.

First, the fact that the defendant held himself out as a professional with specific skills and expertise in the subject matter of the contract will weigh in favor of a finding of gross negligence. There is a significant body of French case law where professionals were held to have committed gross negligence because their conduct involved a significant failure in their specialized area of activity.

For example, the Paris Court of Appeal found gross negligence in the case of a professional who failed to distinguish between two materials – titanium and stainless steel. The task was deemed "simple" and could have been performed through basic visual inspection by a specialist. The professional also failed to

---

18 Terré, F., Simler, Ph., Lequette, Y., Chénedé, F. *Droit civil – Les obligations*, 12e éd. (Paris: Dalloz, 2018), 947, § 881. Cass. com. 3 avr. 1990, *Bull. civ.* IV, n° 108; rappr. Civ. 3e, 21 janv. 2009, CCC 2009, n° 94, note Leveneur.

19 le Tourneau P., Bloch, C., Giudicelli, A., Guettier, C., Julien, J., Krajeski, D. Poumarède, M. *Dalloz Action – Droit de la responsabilité et des contrats 2023/2024 – Régime d'indemnisation*, "L'inexécution d'une obligation contractuelle," (Paris: Dalloz, 2023), 1306, § 3121.152.

20 Mazeaud, L. "L'assimilation de la faute lourde au dol," *DH* 1933, chron. 49. cited in Ph. Malinvaud, M. Mekki, J.-B. Seubé, *Droit des obligations*, 17e éd. (Paris: LexisNexis, 2023, 544, § 591.

inform a worker about the material's nature, resulting in the destruction of the titanium plates. The court concluded:

> *The destruction of the titanium plates resulted from a gross error that a professional of average diligence would not commit.*
>
> *The stripping operation, which caused the destruction of the plates, thus demonstrates a blatant carelessness, lack of professionalism, and incompetence of this company in carrying out a simple task, amounting to gross negligence without requiring intent to harm, contrary to Polimet's allegations. Consequently, the company cannot rely on the limitation of liability clause.*[21]

Another factor that will lead the courts to lean in favor of gross negligence is whether the defendant could foresee the harmful consequences of their behavior. Indeed, a defendant who has voluntarily accepted an obligation to perform certain activities may reasonably be expected to have sufficient knowledge and skill to anticipate any risks associated with these activities. For instance, in the following passage, the French Cour de cassation approved the lower courts' conclusion that the defendant had committed gross negligence where the circumstances showed that the defendant's actions were reckless because it was fully aware that its actions might result in harm:

> *The judgment found that Mr. Di Z... was fully aware that his manoeuvre, conducted with the assistance of the employee of Planzer (acting as a temporary employee of Ectra), would likely result in damage. His conduct not only involved a repeated lack of care but also demonstrated recklessness.*[22]

The seriousness of the risk involved is also a key factor. Where the defendant's lack of care could lead to severe consequences, especially involving health and safety, gross negligence is more likely to be found. For instance, the French Cour de cassation confirmed the lower courts' finding of gross negligence in a case in which the defendant shipbuilder's defective works jeopardized the ship's seaworthiness.[23]

Another factor often referred to by the courts is the recurrence of defaults on the part of the defendant. A recurring lack of due care and skill may lead the courts to find that the defendant has shown itself incapable of fulfilling its contractual obligations. This was the view taken by the Paris Court of Appeal in a recent case involving repeated instances of poor workmanship on the part of the contractor, where it concluded: "*The numerous defaults of ERDF, as enumerated above,*

21 CA Paris, 30 juin 2009, n°07/09960.

22 Cass. com. 16 janv. 2007, n°05-11.507.

23 Cass. com. 22 mai 1968, n°67-10971.

constitute gross negligence."[24] Similarly, the French Cour de cassation upheld a finding of gross negligence against a company on the basis of repeated shortcomings throughout the project, including inadequate planning, delays, noncompliance with specifications and safety standards.[25]

Liability in Case of Vicarious Performance

Unless expressly or impliedly excluded under the contract, a promisor may rely on third parties such as employees, subcontractors, and other auxiliaries, to perform its contractual obligations.[26] However, in such case, vicarious performance (as opposed to novation or assignment) does not transfer liability; the promisor remains liable for the proper performance of his obligations under the contract. Any failure by these third parties is attributed directly to the promisor, as if the breach were his own.[27] Crucially, this means that a claimant need not prove a specific breach on the part of the defendant itself (i.e., failure to exercise reasonable care and skill in selecting, instructing or supervising such third parties). The defendant will be liable for any acts carried out by its auxiliaries that are in breach of contract. The underlying principle is that the promisor should not be in a better position by delegating the performance of his obligations than if he performs personally.[28] For this reason, while clauses purporting to limit or exclude liability for the actions of such third parties are generally enforceable, they are subject however to the same limitations in respect of deliberate breaches and gross negligence as noted above.[29]

Accordingly, the courts have generally held that this type of exemption clause expressed in general terms must be construed restrictively so as not to relieve or limit the promisor's liability in respect of his auxiliaries' gross negligence or deliberate breaches. Even if expressly provided for, a clause purporting to exonerate, even partially, the promisor for his auxiliaries' deliberate breaches would be void as against public policy. Considering the principle stated above according to which the promisor's liability should be the same irrespective of whether he performs personally or vicariously, some authors have argued that the same result should follow for a clause expressly excluding or limiting liability for an auxiliary's gross negligence. This will need to be assessed on a case-by-case basis. The policy

---

24 CA Paris, 9 déc. 2015, n°13/16066.

25 Cass. com. 29 avr. 2014, n°12-29546.

26 F. Terré, Ph. Simler, Y. Lequette, F. Chénedé, *Droit Civil – Les obligations*, 12e éd., (Paris: Dalloz, 2018), 910, § 851.

27 Boucard, H. "Responsabilité contractuelle," *Répertoire de droit civil*, (Paris: Dalloz, 2018), § 334.

28 Terré, F., Simler, Ph., Lequette, Y. Chénedé, F. *Droit civil – Les obligations*, 12e éd. (Paris: Dalloz, 2018), 910–912, § 851.

29 Delebecque, Ph. "Régime de la réparation – Conventions relatives à la responsabilité," *JurisClasseur Civil Code*, (Paris: LexisNexis, 2022), 12, §§ 61–62.

considerations against such a clause are arguably less compelling than those concerning deliberate breaches, especially where the parties have expressly agreed to this exemption in their contract. To the best of our knowledge, the French courts have never directly considered this issue and, in light of recent reforms in other civil law jurisdictions, it is not clear whether such a clause would be upheld.[30]

Liability for Personal Injury and Death

Another area of uncertainty under French law concerns whether liability for death and personal injury can be excluded or otherwise limited by contract. Most scholars argue that such clauses are void as against public policy[31] and the current draft for the upcoming reform of the French law on civil liability contemplates a general prohibition on such clauses for both contractual and tortious liability.[32] This majority view, however, is not free from dissent,[33] and both the legislature and the courts have already recognized several exceptions to this principle in certain specialized areas of law, such as contracts of carriage.[34]

In light of these uncertainties, it is best to include an express carve-out for liability for death or personal injury in any exemption clause in a construction contract governed by French law.

The Survival of Exemption Clauses upon Termination or Rescission of the Contract

Another issue is whether the claimant's decision to put an end to the contract would bring the entire contract to an end, including the exemption clause.[35] The debate has persisted under French law for some time notably in light of the differences between "termination" ("*résiliation*" – that only operates prospectively)

30 Delebecque, Ph. "Régime de la réparation – Conventions relatives à la responsabilité," *JurisClasseur Civil Code*, (Paris: LexisNexis, 2022), 12, § 63.

31 Terré, F., Simler, Ph., Lequette Y., Chénedé, F. *Droit civil – Les obligations*, 12e éd. (Paris: Dalloz, 2018), 949, § 881 b; Malinvaud, Ph. Mekki, M. Seubé, J.-B. *Droit des obligations*, 17e éd. (Paris: LexisNexis, 2023), 548–549, § 595; Bénabent, A. *Droit des obligations*, 20e éd. (LGDJ), 363, § 438 b; Delebecque, Ph. "Régime de la réparation – Conventions relatives à la responsabilité," *JurisClasseur Civil Code*, (Paris: LexisNexis, 2022), 16–17, §§ 91–95.

32 Terré, F., Simler, Ph., Lequette, Y., Chénedé, F. *Droit Civil – Les obligations*, 12e éd. (Dalloz, 2018), 949, § 881 b); Malinvaud, Ph., Mekki, M., Seubé, J.-B. *Droit des obligations*, 17e éd. (Paris: LexisNexis, 2023), 548–549, § 595.

33 Terré, F., Simler, Ph., Lequette, Y., Chénedé, F. *Droit Civil – Les obligations*, 12e éd. (Paris: Dalloz, 2018), 949, § 881 b); Malinvaud, Ph., Mekki, M., Seubé, J.-B. *Droit des obligations*, 17e éd. (Paris: LexisNexis, 2023), 548–549, §§ 595–596.

34 Malinvaud, Ph., Mekki, M., Seubé, J.-B. *Droit des obligations*, 17e éd. (Paris: LexisNexis, 2023), 549, § 595.

35 *Photo Production Ltd v. Securicor Transport Ltd* [1980] AC 827 (HL).

and "rescission" of the contract for breach.[36] The general rule is that the order of the court or tribunal granting rescission will seek to restore both parties to their original position as if the contract had never been entered into.[37] Although this rule is subject to several exceptions, the purpose of this remedy is to deprive the party that failed to perform its side of the bargain of any benefits it may have obtained from the other party's performance. Because this remedy is generally understood to wipe out the contract altogether with retrospective effect, the argument could be made that this remedy would also deprive the defendant of the benefit of any exemption clause in the contract. In a series of decisions handed down before the 2016 reform, some chambers of the French Cour de cassation seemed inclined to adopt this position.[38]

These decisions meant in practice that if the claimant had the right to rescind for breach under the contract's terms, the claimant's election to rescind the contract for breach would entitle it to claim full reparation for its loss, regardless of any exclusion or limitation agreed to in the contract. This development significantly enhanced the prospects of recovery of potential claimants for two reasons: first, because a breach may well be regarded as sufficiently serious to give the claimant the right to rescind for breach even though it does not rise to the level of gross negligence; and second, because the contract may expressly grant the claimant a right to rescind in specific situations, irrespective of the seriousness of the breach.

These decisions created, however, a distinction between two types of contractual provisions which seemed hardly justifiable as noted by some authors. On the one hand, liquidated damages clauses were generally held to survive rescission

---

36 We chose in this chapter to translate the French concept of "*résolution*" with the expression "*rescission for breach,*" or simply "*rescission,*" because it has <u>similar effects</u> in that "*résolution*" operates in principle retroactively (to the extent possible). It should be borne that this translation may obscure a key distinction between the two concepts: the French "*résolution*" is a <u>remedy for breach</u> whereas rescission in English law is a <u>remedy for misrepresentation</u> (which makes the contract voidable and is akin to the French concept of "*nullité relative*").

37 Terré, F., Simler, Ph., Lequette, Y., Chénedé, F. *Droit civil – Les obligations*, 12e éd. (Dalloz: 2018), 853, § 795.

38 Cass. com. 5 oct. 2010, n°08-11.630; Cass. com. 3 mai 2012, n°11-17.779; Cass. 3e civ. 20 juin 2012, n°11-16.197. See also D. Mazeaud, "Le sort des clauses limitatives de réparation en cas de résolution du contrat," *Recueil Dalloz*, 2018, p. 2, § 7. As noted above, there is a distinction in French law between "*rescission for misrepresentation*" or "*nullité relative*" according to which the contract is void ab initio and, on the other hand, "*rescission for breach*" or "*résolution*" in which case the court may make an order for restitution to put the parties in their original position *to the extent possible*. It is only in the former case, the exemption clause would inevitably fail as result of the contract being void.

for breach[39] and, on the other, exemption clauses were held not to.[40] It was hard to see any reason why the latter would be applicable in case of termination (*i.e.*, operating only prospectively) and not rescission (*i.e.*, with retroactive effect), while the former would apply in any case. The only possible explanation was that an exclusion or limitation of liability operates for the sole benefit of the defendant, whereas the provision for the payment of an agreed sum upon breach is also intended to benefit the claimant.[41]

Following the adoption of the 2016 reform on French contract law, the French Civil Code includes a new provision regarding the survival of certain clauses upon rescission for breach set out in the new Article 1230 as follows: "Rescission does not affect clauses relating to the settlement of disputes, nor those intended to be effective even in the event of rescission, such as confidentiality and non-competition clauses."

Although the provision does not include any express reference to exemption clauses, some authors have considered that the rule extends to such clauses. According to these authors, absent any clear language to the contrary, these clauses are clearly "*intended to be effective*" in the event of (any) breach, and, if the breach is sufficiently serious, "*even in the event of rescission.*"[42] This interpretation is supported by the parliamentary debate during the reading of the bill before the French senate where the senators considered that there was "*no dispute*" that the new provision would be applied equally to exemption clauses.[43] Accordingly, any exemption clause included in a contract entered into after 1 October 2016 will probably be held to be binding between the parties, even in the event of rescission on the basis of the new Article 1230 of the French Civil Code. This interpretation is even more likely considering the recent change in the French Cour de Cassation's jurisprudence under the previous legal regime. Two years after the enactment of the new French law on obligations, the French Cour de Cassation reversed its earlier position regarding contracts concluded before 1 October 2016, holding that

39 See discussion below.

40 Cass. com. 5 oct. 2010, n°08-11.630; Cass. com. 3 mai 2012, n°11-17.779; Cass. 3e civ. 20 juin 2012, n°11-16.197. See also D. Mazeaud, "Le sort des clauses limitatives de réparation en cas de résolution du contrat," *Recueil Dalloz*, 2018, p. 2, § 7.

41 D. Mazeaud, "Le sort des clauses limitatives de réparation en cas de résolution du contrat," (*Recueil Dalloz*, 2018), 3, § 8; p. 4, § 14.

42 D. Mazeaud, "Le sort des clauses limitatives de réparation en cas de résolution du contrat," (*Recueil Dalloz*, 2018), 4, § 14.

43 Terré, F., Simler, Ph., Lequette, Y., Chénedé, F. *Droit Civil – Les obligations*, 12e éd. (Paris: Dalloz, 2018) 886 citing to Pillet, F., *Rapport Sénat*, 1re lecture, 74.

a claimant cannot "evade" the effect of an exemption clause by rescinding the contract.[44]

Despite these recent developments, it is always open to the parties to deviate from this rule in their contract. Further, while these recent developments appear to put an end to this debate under French law, the French Cour de cassation's vacillation over this issue suggests that, in other jurisdictions where the matter has not been finally settled, there would be room to argue against the survival of these clauses. As such, where the law governing the construction contract does not provide for a clear rule on this issue, the owner could very well make the argument that its claims are not subject to any exclusion or limitation of liability in case of rescission or some other form of contract termination.

The Approach Followed in Other Civil Law Jurisdictions

Exemption clauses are in principle enforceable under Swiss and Belgian laws. Similarly to French law, both systems provide that such clauses are deemed unenforceable if their effect is to render an essential obligation meaningless.

By contrast, the rules developed under Belgian and Swiss laws deviate more significantly from the approach followed under French law in so far as the permissible scope of exemption clauses is concerned. First, the Belgian Civil Code contains an express prohibition against liability for death and personal injury[45] and most Swiss commentators consider that such rule follows from general principles of Swiss contract law.[46] Further, while the French, Swiss, and Belgian systems share the broad principle that a clause purporting to exclude liability for deliberate breach is void as against public policy,[47] its legal ramifications are markedly different. Both legal systems are generally more liberal and allow for greater party autonomy than the rules under French law.

In respect of vicarious performance, Swiss law is the most liberal in that a contractor may exclude part or all of its liability in respect of any person involved on its behalf in the performance of the contract without any restriction. An exemption

44 Cass. com. 7 févr. 2018, n°16-20.352; F. Terré, Ph. Simler, Y. Lequette, F. Chénedé, *Droit Civil – Les obligations*, 12e éd., Paris, Dalloz, 2018, p. 885-886, § 823 b); D. Mazeaud, "Le sort des clauses limitatives de réparation en cas de résolution du contrat," *Recueil Dalloz*, 2018, p. 2, §§ 3–5.

45 Belgian Civil code, Art. 5.89, § 1, 2°.

46 G. Geissbühler, *Le droit des obligations – Volume 1: partie générale*, (Zurich: Schulthess, 2020), § 952.

47 Belgian Civil code, Art. 5.89, § 1, 1; Swiss Code of obligations, Art. 100, § 1.

clause governed by Swiss law may therefore exclude the contractor's liability for the deliberate breach or gross negligence of auxiliaries or subcontractors.[48]

On the other hand, Swiss law still upholds the maxim "*culpa lata dolo aequiparatur*" and thus treats the contractor's own gross negligence (referred to as "*faute grave*") as tantamount to a deliberate breach of contract. As such, under Swiss law, the contractor may not exclude or restrict liability for its own gross negligence.[49] In this respect, depending on the contract terms, a contractor performing the contract vicariously will be in a better position under Swiss law than if he had performed the contract personally. The position is the same under Algerian law which draws the distinction between clauses in which the promisor seeks to exclude its own liability for deliberate breach or gross negligence (void as against public policy) and those seeking to exclude that of its auxiliaries (in principle valid).[50]

By contrast, the Belgian legal system broadens the scope for freedom of contract by rejecting the maxim "*culpa lata dolo aequiparatur.*"[51] Accordingly, clauses excluding liability for gross negligence – that of the contractor itself as well as its auxiliaries – are in principle valid under Belgian law.[52] In this respect, the Belgian Civil Code simply provides that "*such exclusion cannot be presumed*" and thus requires clear and unambiguous language to that effect.[53] In other words, the Belgian Civil Code substitutes a restrictive, yet flexible, rule of interpretation for the rigid substantive rule historically inherited from French law.

In respect of deliberate breach, however, Belgian law still holds that clauses purporting to exclude or limit liability for deliberate breaches – that of the contractor itself as well as its auxiliaries – are void as matter of public policy.

Although both systems recognize greater party autonomy in respect of exemption clauses, the focus of their control is different: under Swiss law the key distinction is whether the promisor intends to perform personally or vicariously while the decisive factor under Belgian law is whether the breach is deliberate or simply grossly negligent. At the other end of the spectrum, some jurisdictions such as Morocco adopt the same position as in French law and reject any distinction between deliberate breach and gross negligence[54] and between personal and vicarious performance.[55]

---

48 Swiss Code of obligations, Art. 101, § 2; Geissbühler Grégoire, *Le droit des obligations – Volume 1: partie générale*, (Zurich, Schulthess, 2020), § 954.

49 Swiss code of obligations, Art. 100, § 1.

50 Algerian Civil code, Art. 178, § 2.

51 B. Kohl, P. Wéry (dir.), *Le nouveau droit des obligations* (Liège: Anthémis, 2022), 204.

52 Belgian Civil code, Art. 5.89, § 1.

53 Belgian Civil code, Art. 5.89, § 1.

54 Moroccan Dahir on obligations, Art. 232.

55 Moroccan Dahir on obligations, Art. 233.

The Emphasis on the Parties' Bargain and the Limited Role of Substantive Rules Under English Law

In the last 50 years, the rules governing the interpretation and enforceability of exemption clauses in English law have undergone significant reform. As a result, the principles of interpretation applied in English law now largely align with those recognized under French law and other civil legal systems. On the other hand, the substantive rules – which once provided similar protections to those existing in French law – have been significantly rolled back and are now largely confined to the statutory controls in the Unfair Contract Terms Act 1975. Therefore, even in cases of willful misconduct, gross negligence, or similarly culpable conduct, the injured party's ability to claim compensation will typically be restricted by the exemption clause, unless the clause expressly carves out these types of breaches from its scope.

The Principles for the Interpretation of Exemption Clauses Under English Law

The first notable change has been to discard most of the special rules of contractual interpretation that had developed historically in respect of exemption clauses. These rules were a reflection of the courts' traditional hostility toward exemption clauses, which were perceived as depriving parties of valuable common law rights.

The orthodox view is now that exemption clauses should be interpreted like any other contract term in accordance with the approach set out by Lord Hoffmann in *Investors Compensation Scheme Ltd v. West Bromwich Building Society* in which he announced that "*almost all the old intellectual baggage of 'legal' interpretation has been discarded.*"[56] Under that modern approach, "rules" of interpretation have been replaced by "principles" or guidelines, allowing greater flexibility in considering the natural and ordinary meaning of the words used in the context of the contract as a whole and in light of the surrounding circumstances, while placing less emphasis on precedent interpreting similarly worded provisions.[57] The aim is to interpret and give effect to the clause in a way which is consistent with business common sense, rather than adopting a legalistic, artificial meaning at odds with the expectations of ordinary business people.[58]

---

56 *Investors Compensation Scheme Ltd v. West Bromwich Building Society* [1998] 1W.L.R. 896; [1997] C.L.C. 1243 at 912.

57 Hugh Beale (ed), *Chitty on Contracts*, 35th ed. (Sweet & Maxwell, 2024) ch 16, §§ 16-048–16-050, 1220–22.

58 Hugh Beale (ed), *Chitty on Contracts*, 35th ed. (Sweet & Maxwell, 2024), ch 16, §§ 16-094–16-110, 1250–61.

However, the English courts still consider that it is reasonable to assume that contracting parties do not intend to give up valuable rights and remedies provided for by the general law, unless they use clear words to that effect in their contract.[59] For this reason, exemption clauses are still construed narrowly under English law and the burden of proving that the clause applies to a particular breach is on the party invoking it. If the wording of the clause is unclear or ambiguous, the English courts may also resort to the principle of *contra proferentem* to resolve any doubt against the party that introduced the clause, particularly where the clause was not the subject of negotiation between the parties.[60]

The Demise of the Substantive Rules at Common Law and the Enactment of Statutory Controls Under the UCTA

The second major change to the approach to exemption clauses in English law came about around the time of the enactment of the Unfair Contract Terms Act ("**UCTA**") in 1977. At that time, consumer protection was still in its infancy in England and Wales and, as noted above, the common law had developed substantive rules to curb the potential for abuse, especially in case businesses inserted sweeping exemption clauses in their standard form contracts with customers with often little to no bargaining power.[61] Many of these substantive rules have striking similarities to the rules that have developed under French law and other civil law systems. As matter of substantive law, the English courts refused to enforce an exemption clause, however worded, in circumstances where the exemption clause would have relieved the breaching party of liability for a "fundamental breach" of the contract[62] or would otherwise have excused the breaching party from breaching a "fundamental term" of the contract.[63] A fundamental breach was deemed to have occurred if the party in default had refused or failed to carry out his contract in some essential respect, such that the breach could be said to "*go [...] to the root of the contract.*"[64]

---

59 Lewison, *The Interpretation of Contracts*, 8th ed. (Sweet & Maxwell, 2023), ch 12 §§ 12.24–12.41, 728–37.

60 Peel, *Treitel on The Law of Contract*, 15th ed. (Sweet & Maxwell, 2020), §§ 7-009–7-010, 280–81.

61 Peel, E. *Treitel on The Law of Contract*, 15th ed. (Sweet & Maxwell, 2020), § 7-018, 286.

62 See e.g., *Karsales (Harrow) Ltd v. Wallis* [1956] 1 W.L.R. 936; *Yeoman Credit Ltd v. Apps* [1962] 2 Q.B. 508; *Charterhouse Credit Co Ltd v. Tolly* [1963] 2 Q.B. 683.

63 *Smeaton Hanscomb & Co Ltd v. Sassoon I Setty & Co (No.1)* [1953] 1 W.L.R. 1468 at 1470; *RW Green Ltd v. Cade Bros Farms* [1978] 1 Lloyd's Rep. 602.

64 See, e.g., *Karsales (Harrow) Ltd v.* Wallis [1956] 1 WLR 936, 940 (CA) (Denning LJ): "it is now settled that exempting clauses of this kind, no matter how widely they are expressed, only avail the party when he is carrying out his contract in its essential respects. He is not allowed to use them as a cover for misconduct or indifference or to enable him to turn a blind eye to his obligations. They do not avail him when he is guilty of a breach which goes to the root of the contract."

In such circumstances, the party in default was precluded from relying on the exemption clause, regardless of the wording of the clause. In practice, this meant that the courts could refuse to enforce an exemption clause, regardless of the parties' intention, in light of the seriousness and manner of the breach in question. To that extent, this rule echoes the public policy found in French law and other civil law systems against any exclusion of liability for deliberate breach or gross negligence.

A variant of this substantive rule was to consider that some of the obligations undertaken by the parties in their contract should be treated as "fundamental terms" in respect of which an exemption clause could never apply. A "fundamental term" was defined as a term that "underlies the whole contract so that, if it is not complied with, the performance becomes totally different from that which the contract contemplates."[65] Under this variant, the focus was not on the seriousness and manner of the breach, but rather on the importance of the underlying obligation breached. In this way, this second rule is not too dissimilar from the rule found in French law against enforcing an exemption clause in respect of the "essential obligations" of the contract.[66] In both cases, the purpose of the rule is to prevent a party from relying on an exemption clause to evade some of its key obligations under the contract and deprive the innocent party of its bargain.

The English courts have, however, gradually abandoned this form of substantive control, considering that in many circumstances, such interference with the parties' bargain was not justified. The decisions of the House of Lords in *Suisse Atlantique*,[67] and subsequently in *Photo Production*,[68] concluded that these substantive rules – in that they applied irrespective of the parties' clear intention – constituted an excessive encroachment on the principle of freedom of contract and therefore ought to be abandoned.

The House of Lords' rejection of the doctrine of "fundamental breach" does not mean, however, that the seriousness and manner of the breach cannot be taken into account by the courts. Under the new approach chartered by the House of Lords, these considerations are still relevant but only to the extent that the courts seek to determine whether, as a matter of interpretation, the parties intended the clause to apply to the situation at hand. For example, in *Suisse Atlantique*, Lord Wilberforce confirmed that the seriousness of the breach may inform the courts' interpretation of the clause: "[It is] a question of contractual intention whether

---

65 *Smeaton Hanscomb & Co Ltd v. Sassoon I. Setty, Son & Co* [1953] 1W.L.R. 1468, 1470.

66 See above in Section 1.1.1.

67 *Suisse Atlantique Societe D'Armement Maritime SA v. NV Rotterdamsche Kolen Centrale* [1967] 1 A.C. 361.

68 *Photo Production Ltd v. Securicor Ltd* [1980] A.C. 827.

a particular breach is covered or not and the courts are entitled to insist, as they do, that the more radical the breach the clearer must the language be if it is to be covered."[69]

Moreover, even though English law no longer recognizes a category of "fundamental" terms, the breach of which can never be excused by an exemption clause, the UCTA provides a new set of statutory controls according to which some exemption clauses may be found to be ineffective.

These controls, however, only apply to the extent that the construction contract falls within the scope of the UCTA. Indeed, pursuant to s. 27(1) UCTA, the mere fact that English law has been chosen by the parties to govern the construction contract will not necessarily entail the application of the UCTA. According to s. 27(1) UCTA, the provisions of the UCTA do not apply if, in the absence of such choice-of-law clause, the contract would not have been governed by English law. In practice, this rule will exclude the application of the UCTA in many international construction contracts relating to works performed outside the UK where the parties involved are not necessarily all based in the UK but have still chosen English law for their contract.

On the contrary, if the construction contract has a sufficient connection with some part of the UK, there are two main provisions of the UCTA that may restrict the effect of any exemption clause found in the contract. First, s.2 UCTA restricts the ability of a party to exclude or limit liability for negligence. This is particularly relevant in construction disputes because, under English contract law, there is no general rule against concurrent liability in tort and contract, unlike in many civil law systems, which apply the so-called *règle de non-cumul.* As a result, a party may be liable both for the tort of negligence and for breach of contract (e.g., for failing to exercise such skill and care as required in the contract) and any clause seeking to exclude or limit "liability for negligence" – whether tortious, contractual or both – would be subject to the controls of the UCTA. Importantly, Section 2(1) of the UCTA provides that any term seeking to exclude or restrict liability for death or personal injury resulting from negligence is automatically ineffective. Meanwhile, s. 2(2) states that any clause attempting to limit or exclude liability for other types of loss or damage caused by negligence will only be enforceable if it meets the requirement of reasonableness under Section 11 of the Act. The burden of proving reasonableness is on the party seeking to rely on the clause.[70] The assessment is made at the time the contract was formed and such as the parties'

---

69 *Suisse Atlantique Societe D'Armement Maritime SA v. NV Rotterdamsche Kolen Centrale* [1967] 1 A.C. 361, 432 C.

70 E. Peel, *Treitel: The Law of Contract* (16th edn, Sweet & Maxwell 2025) para 7-062.

relative bargaining power, whether the term was clearly communicated, and whether the party seeking to enforce the term could have obtained insurance.[71]

Second, s.3 UCTA imposes restrictions on exemption clauses in contracts where one party is dealing on the other's written standard terms of business. However, in most complex construction projects, the contracting parties are not dealing on the contractor's standard terms. Typically, the contract will have been drafted by the owner's representatives during the tendering process, and even where the contractor initially proposes a standard form contract, the owner will generally have had ample opportunity to negotiate its terms. As a result, s. 3 UCTA is unlikely to apply to many construction contracts, since the requirement that one party is contracting on the other's standard terms is often not met.

These statutory restrictions mean that, in respect of construction contracts sufficiently connected to the UK, parties seeking to rely on exemption clauses – particularly those limiting liability for negligence or contractual breaches – must ensure that such clauses can withstand scrutiny under the UCTA's reasonableness test.

Residual Common Law Exceptions to the Enforceability of Exemption Clauses in English Law

Despite the extensive statutory regulation introduced by the UCTA and subsequent consumer protection legislation, certain residual common law exceptions to the enforceability of exemption clauses persist. The most significant remaining substantive rule concerns fraud or deceit. A party cannot rely on an exemption clause to exclude or limit liability for fraudulent misrepresentation or deceit, as such conduct fundamentally undermines the contract itself.[72] This principle reflects a long-standing commitment in English law to prevent contractual provisions from being used to shield fraudulent behavior.

The justification for these residual common law exceptions has diminished in light of modern statutory protections, particularly in contracts between commercial parties with equal bargaining power and access to legal advice. UCTA and the Consumer Rights Act 2015 provide comprehensive mechanisms for controlling unfair terms, making further common law intervention largely redundant. Nevertheless, the fraud exception remains a necessary safeguard, ensuring that no party can contractually exclude liability for conduct that goes to the root of the agreement and distorts the very basis of contractual relations.[73]

71 Peel, E., *Treitel: The Law of Contract*, 16th ed. (Sweet & Maxwell, 2025), paras 7-063–7-068.

72 Peel, E., *Treitel: The Law of Contract*, 16th ed. (Sweet & Maxwell, 2025), paras 7-033–7-034.

73 Peel, E. *Treitel: The Law of Contract*, 16th ed. (Sweet & Maxwell, 2025), paras 7-027.

Liquidated Damages Clauses

As noted above, while liquidated damages provisions are generally stipulated for the benefit of the owner[74] – allowing it to recover the agreed sum without having to establish the extent of its loss – these provisions may prove prejudicial to the owner if its actual loss greatly exceeds the agreed sum. In such circumstances, the question arises as to whether the owner can recover additional damages beyond the stipulated sum. Under a construction contract governed by French law, the owner is able to do so in three scenarios, which we consider further below: first, if the clause expressly or impliedly allows such recovery; second, if the owner demonstrates that the actual loss is out of proportion with the agreed sum; and third, if the breach is deliberate or amounts to gross negligence, as discussed above in relation to exemption clauses. More generally, we will consider the extent to which liquidated damages provisions may be subject to some of the controls that apply in respect of exemption clauses and whether such clauses survive rescission for breach or other forms of termination of contract under French law. Finally, we will contrast the approach adopted under French law with that of other civil law jurisdictions and with the position in English law.

The Clause Expressly or Impliedly Allows for Additional Damages to Be Recovered

In French law, liquidated damages provisions are commonly referred to as "*clauses pénales*" and this concept is defined as a contract term providing for (i) the payment of an agreed sum (ii) by way of compensation (iii) in the event of a breach (or specific type of breach, such as delays). Unless otherwise agreed by the parties, the default rule under French law is that the defendant is relieved from all liability for the breach upon payment of the agreed sum. In other words, absent any contrary indication in the contract, the agreed sum is the sole monetary remedy available to the aggrieved party (that is, to say, without prejudice to the availability of other nonmonetary remedies such as specific performance).[75]

This principle is codified in the new Article 1231-5 of the French Civil Code in the following terms:

---

74 The FIDIC Green Book 2021 has extended the use of liquidated damages (usually stipulated only in respect of the owner's delay damages) to the contractor's prolongation costs.

75 Malinvaud, Ph., Mekki, M., Seubé, J.-B. *Droit des obligations*, 17ᵉ éd. (Paris: LexisNexis, 2023), 554, § 598; Terré, F., Simler, Ph., Lequette, Y., Chénedé, F. *Droit Civil – Les obligations*, 12ᵉ éd. (Paris: Dalloz, 2018), 961, § 890 d; Malaurie, Ph., Aynès, L., Stoffel-Munck, Ph., *Droit des obligations*, (Paris: LDGJ, 2022), 578, § 634, 2.

> *When a contract provides that in the event of a breach, the breaching party shall pay a certain sum by way of damages,* ***the injured party may not be awarded a greater or lesser sum*** *(emphasis added).*

If the injured party's overall losses exceed the agreed sum, the liquidated damages clause in the contract may operate to the defendant's benefit so as to limit its overall liability.

In these situations, the first issue to consider for the claimant will be whether, as a matter of interpretation of the clause, all the losses it suffered can be properly said to be within its scope or whether certain types of loss or breach fall outside its scope and can be fully recovered by way of unliquidated damages.[76] For example, in some cases, the owner may seek to argue that liquidated damages were intended under the contract to compensate the owner for all losses arising from late completion, but not for all time-related loss or damage and seek to claim general damages for losses arising from the contractor's failure to prosecute certain parts of the works diligently, without any resulting delay to completion.[77]

The second related issue is to determine whether the language used in the clause indicates that the parties intended, expressly or impliedly, to depart from the default rule laid down in Article 1231-5 of the French Civil Code, thereby allowing the injury party to claim for any loss exceeding the agreed sum by way of general damages.[78] For instance, any indication that the agreed sum shall not to be treated as a fixed, lump-sum compensation ("*indemnité forfaitaire*") or does not amount to full and final reparation ("*réparation intégrale et définitive*") may be sufficient to rebut the presumption established under Article 1231-5, depending on the particular terms of the contract and the parties' intention. To avoid uncertainty, it is preferable to state in terms that the payment of the agreed sum shall not be treated as a release from liability for further losses or that the payment of liquidated damages is without prejudice to any claims for further losses that the injured party may have.

---

76 See e.g., Cass. Com. 20 mai 1997 cited in Terré, F., Simler, Ph., Lequette, Y., Chénedé, F. *Droit Civil – Les obligations*, 12e éd. (Paris: Dalloz, 2018), 962, § 891, footnote 6.

77 See e.g., Bailey, J., *Construction Law*, 4th ed. (London Publishing Parntership, 2023) paras 13.145-13.146.

78 Malaurie, Ph., Aynès, L., Stoffel-Munck, Ph. *Droit des obligations*, (Paris: LDGJ, 2022), 578, § 634, 2) ("*La règle n'est pas d'ordre public ; les parties peuvent convenir que le juge pourra allouer une somme autre que le montant fixé par la clause (arg., Art. 1231-5, al. 4). Cela conforte la pratique dite des clauses de minimum, par lesquelles la somme énoncée par la clause constitue un plancher auquel s'ajoutera la réparation intégrale du préjudice*").

The Owner's Actual Loss Greatly Exceeds the Agreed Sum

While liquidated damages clauses are in principle valid and enforceable, the French legislature introduced a significant qualification on the binding force of such terms in 1975 when it conferred on the courts (and arbitrators deciding on the basis of French law) an express power to adjust the fixed sum agreed to by the parties having regard to the actual loss suffered by the claimant.[79] As a result of the 2016 reform, this statutory power is now enshrined (in substantially identical terms) in the new Article 1231-5 of the French Civil Code which reads as follows:

> *Where the contract stipulates that the party failing to perform will pay a certain sum by way of damages, the other party cannot be awarded a greater or lesser sum.*
>
> *The judge may, <u>even on its own motion</u>, **reduce** or **increase** the agreed sum provided it is **manifestly excessive** or **insufficient**.*
>
> *When the obligation has been partially performed, the court may, even on its own motion, reduce the agreed sum in proportion to the benefit that the promisee has obtained from partial performance, without prejudice to the application to the preceding paragraph. (emphasis added)*

This provision fails to specify, however, the standard for assessing whether the agreed sum is "manifestly excessive or insufficient." It is now well-settled, as a matter of French law, that this assessment must be made at the time of the judge's decision by reference to the actual loss suffered by the claimant.[80] As such, it differs in two material respects from the approach followed under English law: first, the judge must consider the actual loss suffered and not a genuine pre-estimate of loss as at the time of contracting, and, secondly, the judge cannot in principle take into account any other factor, such as the claimant's legitimate interests in performance.

The courts may exercise their discretion "even on [their] own motion" based on the available evidence, but, importantly, the French Cour de cassation imposes an obligation to give reasons if the power to adjust the agreed sum is exercised[81] (whereas there is no such obligation if the courts refuse to alter the

---

79 See e.g., Terré, F., Simler, Ph., Lequette, Y., Chénedé, F. *Droit Civil – Les obligations*, 12e éd. (Paris: Dalloz, 2018), 963–965, §§ 892–893.

80 Malinvaud, Ph. Mekki, M., Seubé, J.-B. *Droit des obligations*, 17e éd. (Paris: LexisNexis, 2023), 552, § 598; Terré, F., Simler, Ph., Lequette, Y., Chénedé, F. *Droit Civil – Les obligations*, 12e éd. (Paris: Dalloz, 2018), 965–966, § 893.

81 Bénabent, A. *Droit des obligations*, 20e éd. (Paris, LGDJ, 2023), 367, § 443; F. Terré, Ph. Simler, Y. Lequette, F. Chénedé, *Droit Civil – Les obligations*, Paris, Dalloz, 12e éd., 2018, p. 966, § 893.

parties' bargain).[82] Moreover, the French Cour de cassation systematically overturns decisions of lower courts in which the reasons given are based on improper factors, such as the breaching party's conduct[83] or the financial position of the parties.[84]

Where the agreed sum is held to be manifestly disproportionate to the actual loss, the courts have broad discretion to fix a new amount, subject to two limitations[85] If the agreed sum is manifestly excessive in comparison to the actual loss, the courts may reduce the amount awarded to the claimant, provided the sum awarded is not lower than the loss actually incurred.[86] Conversely, if the agreed sum is manifestly insufficient, the courts may increase the amount due, but only up to the actual loss suffered, not beyond.[87] This is consistent with the broader principle against punitive damages under French law.

In practice, Article 1231-5 of the French Civil Code is more often relied on by contractors than owners as a way to reduce their liability under the clause. In such a case, the burden should in principle lie with the contractor to prove that the agreed sum is manifestly excessive in comparison to the actual loss suffered.[88] This means, however, that the contractor is placed in the unusual position of having to prove – at least estimate – the owner's losses, which will often be based on facts more readily available to the owner. The contractor will often be reluctant to put forward any figure that may be interpreted as an admission, albeit partial, of liability. Instead, the contractor may be tempted to argue that the owner's greater access to relevant evidence is sufficient to justify shifting the burden of proof onto the owner to establish these losses. Facing such arguments, the owner will probably seek to argue that the very essence of liquidated damages clause is to provide certainty to both parties in respect of a type of loss which is typically difficult to quantify. If the burden was so easily reversed, the owner would virtually lose all the benefits of the clause while the contractor would be in a better position than what it had bargained for in the contract.

The owner may also argue that the fact that the owner may be in possession of the relevant evidence does not create an insurmountable obstacle that would prevent the contractor from discharging its burden of proof. In most proceedings, the relevant rules of court (or arbitration rules as the case may be) will set out various procedures to ensure the disclosure of documents between the

82 Bénabent, A. *Droit des obligations*, 20e éd. (Paris, LGDJ, 2023), 367, § 443.

83 Malaurie, Ph., Aynès, L., Stoffel-Munck, Ph. *Droit des obligations*, (Paris, LDGJ, 2022), 579-580, § 635.

84 Bénabent, A. *Droit des obligations*, 20e éd. (Paris, LGDJ, 2023), 367, § 443.

85 Malinvaud, Ph., Mekki, M., Seubé, J.-B. *Droit des obligations*, 17e éd. (Paris, LexisNexis, 2023), 552–553, § 598.

86 Bénabent, A. *Droit des obligations*, 20e éd. (Paris, LGDJ, 2023), 367, § 443.

87 Terré, F., Simler, Ph., Lequette, Y., Chénedé, F. *Droit Civil – Les obligations*, 12e éd. (Paris: Dalloz, 2018), 966, § 893.

88 Pimont, S. "Clause pénale," *Répertoire de droit civil*, (Dalloz, 2010), § 33.

parties to the dispute. To the extent that the contractor considers that some documents in the owner's possession are relevant and necessary to support its argument that the agreed sum is manifestly excessive within the meaning of Article 1231-5 of the French Civil Code, it should apply for disclosure of such documents in accordance with the applicable rules.

The Contractor's Breach of Contract Was Deliberate or Otherwise Amounts to Gross Negligence

As discussed above, Article 1231-3 of the French Civil Code lays down a key principle of French contract law according to which any contract term that purports to prevent the injured party from claiming full compensation for losses caused by deliberate breach of contract or gross negligence is void as against public policy.[89] To recall, the provision is expressed in the following general terms:

> *The promisor is only liable for damages [for breach of contract] to the extent provided for in the contract or foreseeable at the time of its formation, unless its failure to perform arises from a deliberate breach or gross negligence.*

For this reason, the same principles discussed above in relation to exclusion and limitation of liability clauses are equally applicable to liquidated damages provisions.

Whether Liquidated Damages Clauses Are Subject to the Same Legislative Controls Governing Exemption Clauses

Contrary to the position under English law, all the legislative controls applicable in relation to exclusion and limitation clauses under French law apply equally to liquidated damages provisions. As we have seen above, the provisions of Article 1170 (contract terms may not deprive the essential obligation of its substance), Article 1171 (a party may not introduce standard terms that create a significant imbalance between the parties), and Article 1231-3 of the French Civil Code (exclusions of liability for deliberate breach of contract or gross negligence are void) apply generally to all contracts terms.[90] As such, these provisions also apply to liquidated damages provisions.[91] The claimant could therefore rely on these

89 Terré, F., Simler, Ph., Lequette, Y., Chénedé, F. *Droit Civil – Les obligations*, 12e éd. (Paris: Dalloz, 2018), 963, § 891; Malinvaud, Ph., Mekki, M., Seubé, J.-B. *Droit des obligations*, 17e éd. (Paris: LexisNexis, 2023), 550, § 597.

90 See sections 1.2.1., 1.2.2. and 1.3. above.

91 Malinvaud, Ph., Mekki, M. Seubé, J.-B. *Droit des obligations*, 17e éd. (Paris: LexisNexis, 2023), 550, § 597.

provisions to argue that the liquidated damages clause is unenforceable where the agreed sum grossly understates the actual loss.

Survival After Termination

Although liquidated damages clauses are not specifically listed in Article 1230 of the French Civil Code, some authors consider that, in keeping with several decisions of the French Cour de cassation issued before the 2016 reform, these clauses should also be deemed to have been "intended [by the parties] to be effective in the event of rescission" or any other form of termination, unless otherwise clearly stated in the clause.[92]

Main Differences with Other Civil Law Jurisdictions

All the civil law jurisdictions considered in this chapter impose some limitations to the enforceability of exemption clauses in case of deliberate breach or gross negligence.

Most of these legal systems also consider that a liquidated damages clause worded in general terms is akin to a limitation of liability clause, *i.e.*, the agreed sum provided for under the clause is in effect a cap on the defendant's liability. For example, the Belgian Civil Code contains the same rule as set out in Article 1231-5 of the French Civil Code according to which, where the parties have provided for liquidated damages in their contract, "[the injured party] *cannot be awarded a higher or lower compensation*."[93] As discussed above, this rule establishes a presumption that the parties intended for the agreed sum to be the exclusive monetary which the claimant is entitled to claim. Put differently, the presumption is that the parties impliedly intended the agreed sum to operate as a cap on the defendant's liability.

On the other hand, the injured party will retain the ability to claim full compensation for its losses arising from deliberate breach or gross negligence under Belgian, and Swiss laws as is the case under French law.

These three legal systems also grant the judge or arbitrator a statutory power to adjust the agreed sum provided under the liquidated damages provision. The extent of this power and the factors to be considered vary, however, across these three systems.

For instance, the factors that are relevant to the exercise of this discretion are not the same in all these jurisdictions. Under Algerian and Moroccan laws, similar

---

92 Cass. Civ. 3e, 15 févr. 2005, n°04-11.223; Cass. Com. 22 mars 2011, n°09-16.660. See also D. Mazeaud, "Le sort des clauses limitatives de réparation en cas de résolution du contrat," *Recueil Dalloz*, 2018, p. 2, § 7; Ph. Malinvaud, M. Mekki, J.-B.Seubé, *Droit des obligations*, 17e éd. (Paris: LexisNexis), 2023, p. 553, § 598.

93 Belgian Civil code, Art. 5.88, § 1.

to French law, the judge or arbitrator carries out an ex-post assessment by reference to the actual loss.[94] By contrast, Belgian law had initially favored an approach similar to that of English law where an ex-ante assessment is made by reference to the loss likely to be incurred as at the time of contracting.[95] Recently, the Belgian legislature has broadened the factors that may be taken into account to allow the judge or arbitrator to consider generally all the relevant circumstances of the case, and "in particular the legitimate interests of the claimant."[96]

In contrast to all the other legal systems discussed above, the approach under Swiss law proceeds from the opposite assumption. In a contract governed by Swiss law, a liquidated damages provision is not taken to imply, without more, a limitation of liability for the benefit of the defendant. In other words, in the absence of any indication to the contrary, the payment of the agreed sum is not presumed to have been intended by the parties to be the exclusive monetary remedy for the claimant.[97] Article 161 of the Swiss Code of obligations expressly provides that the claimant may recover any additional damages in accordance with the normal rules for establishing liability for breach of contract. In the absence of any limitation of liability implied by the clause, the Swiss Code of obligations only confers the power to the judge or arbitrator to reduce the amount of liquidated damages if it is found to be "excessive."[98] Since the clause does not operate as a cap on liability, the judge or arbitrator may always award damages in addition to the agreed sum.[99]

Liquidated Damages Provisions and the Rule against Penalties in English Law

At first sight, the rules on liquidated damages clauses under English law may appear to take a radically different approach from the rules described above in various civil law systems.

As we have seen, the common premise across these legal systems is that liquidated damages provisions are valid and enforceable, but, in certain circumstances,

94 Moroccan Dahir on obligations, Art. 264 § 3 (as to which it should be noted that it does not expressly require any manifest or gross disproportion between the actual loss and the agreed sum); Algerian Civil code, Art. 184.

95 Belgian Civil code, former Art. 1226 et 1231 (as in force before 1 January 2023).

96 Belgian Civil code, Art. 5.88. This expansion of the factors to be considered bears some similarities with the restatement of the rule against penalties under English law in *Cavendish Square Holding BV v. El Makdessi* [2015] UKSC 67.

97 Geissbühler, G. *Le droit des obligations – Volume 1: partie générale*, (Zurich: Schulthess, 2020), §§ 1315–1316.

98 Swiss code of obligations, Art. 163, § 3.

99 Geissbühler, G. *Le droit des obligations – Volume 1: partie générale*, (Zurich: Schulthess, 2020), § 1316.

the judge or arbitrator may vary the amount payable under the clause. In such a case, the judge or arbitrator will ordinarily revise the amount to be paid under the clause in order to reflect more closely the claimant's actual loss. By contrast, liquidated damages provisions are not necessarily valid and enforceable under English law and may be struck out if they amount to a penalty.[100] If the clause is held to be penal, the claimant is not left without remedy, however, and will be entitled to recover unliquidated damages for its actual loss in the ordinary way.[101]

Even though, under French law, the judge or arbitrator ostensibly gives effect to the clause (albeit in a modified amount), whereas the clause is held to be void as a penalty under English law, the outcome is in substance the same: in both instances, the claimant is entitled to recover damages commensurate with its actual loss. As such, whether the emphasis is placed on the validity of the clause under English law or the adequacy of the amount under French law is largely a distinction without a difference.

Apart from the different remedies available in the two systems, the rule against penalties under English law has a similar scope and purpose to the rules described above under French law. Its application is limited to liquidated damages clauses defined as clauses that (i) provide for the payment of a sum of money and (ii) are triggered by a breach of contract (for instance, failing to complete the works before the contractual completion date).[102] The second element (ii above) is essential: if the clause requires payment upon the occurrence of events other than breach, the rule against penalties does not apply.[103] By contrast, the first element (i above) is understood more broadly, and the rule also extends to clauses that, although not liquidated damages clauses in the strict sense, require the breaching party to forfeit sums to which it would otherwise be entitled or to transfer property to the innocent party for less than its full value.[104] Where both conditions are satisfied, the rule against penalties will apply to determine whether the clause in question is enforceable.

As in most legal systems discussed above, a valid liquidated damages clause is presumed to be the exclusive monetary remedy available to the innocent party,

100 Beale, H. (ed), *Chitty on Contracts*, 35th ed. (Sweet & Maxwell, 2023), para 30-205.

101 Beale, H. (ed), *Chitty on Contracts*, 35th ed. (Sweet & Maxwell, 2023), para 30-206.

102 *Cavendish Square Holdings BV v. Makdessi, ParkingEye Ltd v. Beavis* [2015] UKSC 67, [2016] A.C. 1172 at [14] and [32]. See also Hugh Beale (ed), *Chitty on Contracts*, 35th ed. (Sweet & Maxwell, 2023), paras 30-246, 30-255.

103 *Cavendish Square Holdings BV v. Makdessi, ParkingEye Ltd v. Beavis* [2015] UKSC 67, [2016] A.C. 1172. See also ed. Hugh Beale, *Chitty on Contracts*, 35th ed. Sweet & Maxwell, 2023), paras 30-250–30-251.

104 *Cavendish Square Holdings BV v. Makdessi, ParkingEye Ltd v. Beavis* [2015] UKSC 67, [2016] A.C. 1172 at [170] (Lord Mance), [226–227] (Lord Hodge). ed. Hugh Beale, *Chitty on Contracts*, 35th ed. (Sweet & Maxwell, 2023), paras 30-250–30-251.

who is entitled to the payment of no less and no more than the agreed sum.[105] The claimant may, therefore, decide to seek an injunction instead of payment of liquidated damages under the clause.[106]

In reality, the distinctive features of the English law approach to the enforceability of liquidated damages lies along the same, two main dividing lines outlined above: (i) whether the assessment is carried out ex ante – *i.e.*, before the breach occurs by reference to the estimated loss as of the time of contracting – or ex post – *i.e.*, once the breach has occurred, considering the actual loss suffered; (ii) whether other factors may be taken into consideration in deciding whether to uphold the clause. As already mentioned above, it is firmly established under English law that the court or arbitral tribunal must assess the enforceability of a liquidated damages provision ex ante as of the time of the making of the contract. The position on the second issue – whether the court or arbitral tribunal should solely consider the estimated loss as of the time of contracting to the exclusion of any other relevant factors – has recently evolved as a result of the UK Supreme Court's decision in the conjoined appeals of *Cavendish Square Holdings BV v. Talal El Makdessi* and *ParkingEye Ltd v. Beavis.*[107]

Before the Supreme Court's intervention, the classic rule, as laid down by Lord Dunedin in *Dunlop Pneumatic Tyre Ltd v. New Garage and Motor Co Ltd*, was that the clause was enforceable if the agreed sum was a "*genuine pre-estimate*" of the claimant's loss from breach as assessed at the time of contracting.[108] Although the term "genuine" may suggest a subjective test – whether, at the time of contracting, the parties genuinely or honestly believed that the sum fixed in the clause was a pre-estimate of the likely loss – this traditional test is an objective one. The courts simply consider, independently from the parties' intention, whether "*the sum stipulated for is extravagant and unconscionable in <u>amount</u> in comparison with the greatest loss that could conceivably be proved*" based on the information available to both parties at the time of formation of the contract.[109] Under this test, compensation was the only legitimate interest that could justify providing for liquidated damages in a contract and, as such, the enforceability of the clause turned exclusively on the comparison between the agreed sum and the likely loss.

In 2015, in *Cavendish* and *ParkingEye*, the Supreme Court reconsidered the law on penalties and decided to broaden the test so as to ensure that a clause intended

105 *Clydebank Engineering and Shipbuilding Co Ltd v. Don Jose Ramos Yzquierdo y Castaneda* [1905] A.C. 6. See also Hugh Beale (ed), *Chitty on Contracts*, 35th ed. (Sweet & Maxwell, 2023), paras 30-206 and 30-266.

106 *Sainter v. Ferguson* (1849) 1 Mac. & G. 286; *Carnes v. Nesbitt* (1862) 7 H. & N. 778; *General Accident Insurance Co v. Noel* [1902] 1 K.B. 377. See also ed. Hugh Beale, *Chitty on Contracts*, 35th ed. (Sweet & Maxwell, 2023), paras 30-009.

107 *Cavendish Square Holdings BV v. Makdessi, ParkingEye Ltd v. Beavis* [2015] UKSC 67, [2016] AC 1172.

108 *Dunlop Pneumatic Tyre Co Ltd v. New Garage & Motor Co Ltd* [1915] AC 79 (HL), 86–87.

109 *Dunlop Pneumatic Tyre Co Ltd v. New Garage & Motor Co Ltd* [1915] AC 79 (HL), 87.

to protect the legitimate interests of the innocent party may be upheld, even in circumstances where its amount might not be a genuine pre-estimate of loss. As recast by Lords Neuberger and Sumption, the modern test can be stated as follows: "*the true test is whether the impugned provision is a secondary obligation which imposes a detriment on the contract-breaker out of all proportion to any legitimate interest of the innocent party in the enforcement of the primary obligation.*"[110] For example, in *ParkingEye*, the claimant managed a car park and charged a £85 fee to any customer exceeding the free two-hour parking limit. Although the sum did not represent a genuine pre-estimate of the loss likely to be suffered by the claimant, the Supreme Court still found that the fee was not a penalty and was justified, among others, by the claimant's legitimate interest in ensuring a prompt turnover of parking spaces for its customers. Under this new approach, the Supreme Court allows the courts to take into account other legitimate interests that may justify upholding the clause, instead of focusing exclusively on comparing the agreed sum to the likely loss. As noted above, the Belgian legislature adopted a similar reform in 2023 to broaden the factors that may be taken into account in deciding whether to enforce a liquidated damages provision.

Under this widened approach, compensation is but one of the legitimate interests that may justify upholding the clause. While the courts will continue to enforce clauses providing for the payment of a sum which is a genuine pre-estimate of the likely loss, the inquiry no longer ends there and the courts may uphold a clause even if its predominant purpose is to achieve deterrence (which was traditionally not recognized as a valid commercial justification) as long as the claimant has an interest in securing performance rather than damages.[111] Such an interest may be present if the innocent party would face difficulties in proving the extent, if any, of its loss, or in identifying each individual breaches by the defendant and the correspondent losses, or where there is no substitute available for the defendant's performance and damages would not be an adequate remedy.[112] Assuming that such an interest exists, the clause will only be valid if the agreed sum is not extravagant or unconscionable compared to that interest. In other words, the sum stipulated under the clause must not be grossly disproportionate to what is required to achieve deterrence.[113] The decision in *Cavendish* suggests that, if the clause is included in a commercial contract between parties of comparable bargaining power, the courts will be reluctant to second-guess whether the amount agreed to between the parties is proportionate to the innocent party's legitimate interest.[114]

---

110 *Cavendish Square Holdings BV v. Talal El Makdessi, ParkingEye Ltd v. Beavis* [2015] UKSC 67, [2016] AC 1172, [32].

111 See also Beale, H. (ed), *Chitty on Contracts*, 35th ed. (Sweet & Maxwell, 2023), paras 30-224.

112 Beale, H. (ed), *Chitty on Contracts*, 35th ed. (Sweet & Maxwell, 2023), paras 30-230–30-232.

113 Beale, H. (ed), *Chitty on Contracts*, 35th ed. (Sweet & Maxwell, 2023), paras 30-242–30-245.

114 Beale, H. (ed), *Chitty on Contracts*, 35th ed. (Sweet & Maxwell, 2023), para 30-245.

In contrast to French law, however, English law does not generally apply the same principles that govern exemption clauses to liquidated damages provisions,[115] which are therefore considered to fall outside the scope of the UCTA.[116]

As under French law, it is well accepted that delay liquidated damages serve a legitimate purpose in light of the inherent difficulties in assessing losses resulting from delays and are therefore generally considered to be valid.[117] Under English law, it should be noted that the question whether the clause is a penalty or liquidated damages is a pure question of law which can be resolved on the basis of limited evidence.[118] More importantly, the court or tribunal must only consider the facts available to both parties objectively as of the date the contract was entered into, thereby preventing the procedural issues outlined above in relation to the burden of proof under Article 1231-5 from arising.

Conclusion

In conclusion, the enforceability of exemption and liquidated damages clauses under French law and other civil law jurisdictions reflects a nuanced balance between party autonomy and the protection of fundamental contractual values, including fairness and good faith. In particular, French law imposes substantive controls to prevent clauses that may negate essential obligations.

This approach stands in contrast to the position under English law. While English law once relied on similar substantive doctrines – including the concept of fundamental breach – these have largely been abandoned in favour of a more interpretation-driven approach, with limited statutory controls under the Unfair Contract Terms Act 1977. The English courts now focus primarily on the parties' intentions as expressed in the contract, subject to a narrow set of exceptions rooted in public policy, such as fraud.

As this comparative analysis shows, the legal framework applicable to exemption and liquidated damages clauses can vary significantly depending on the governing law. This has important implications in the context of international construction projects, where the risks associated with liability limitations must be carefully evaluated. Ultimately, clear drafting and properly tailored contractual language remain essential tools to mitigate these risks and safeguard the parties' respective interests.

---

115 Peel, E., *Treitel: The Law of Contract*, 16th ed. (Sweet & Maxwell, 2025), para 7-013.

116 Peel, E. *Treitel: The Law of Contract*, 16th ed. (Sweet & Maxwell, 2025), para 7-044.

117 *Clydebank Engineering Co v. Don Jose Ramos Yzquierdo y Castaneda* [1905] A.C. 6; *Philips Hong Kong Ltd v. Att-Gen of Hong Kong* (1993) 61 B.L.R. 41, 60, PC; *Alfred McAlpine Capital Projects Ltd v. Tilebox Ltd* [2005] Build. L.R. 271 (TCC). See also *Law v. Redditch Local Board* [1892] 1 Q.B. 127; *Cellulose Acetate Silk Co Ltd v. Widnes Foundry (1925) Ltd* [1933] A.C. 20.

118 Beale, H. (ed), *Chitty on Contracts*, 35th ed. (Sweet & Maxwell, 2023), para 30-211.

17

Delay Damages

Garrett J. Brown

Introduction

It is typical for construction contracts between owners and contractors to state what actions should be taken if a project experiences delay. Depending on who is responsible for the delay, an owner or contractor may be entitled to delay damages, and a contractor may be entitled to a time extension. Construction contracts typically include a clause or clauses for delay damages, including (1) liquidated damages (LDs); (2) actual delay damages;[1] and (3) consequential damages. This chapter discusses types of delay that may lead to damages, the three types of damages, and methods for calculating costs related to these damages.[2]

Considerations of Project Delay

If a delay event affects activities on what was the project critical path[3] at the time that the delay occurred, the event will delay project completion unless delay mitigation occurs. A critical delay caused by an owner-responsible event may entitle a contractor to a time extension and delay damages. If the delay is noncritical, meaning the delayed work is not part of the critical path, then in most cases, the

1 Actual damages are also called "direct damages."

2 The content in this chapter has been slightly modified from Garrett Brown's article titled "Delay Damages," originally published by Long International (https://www.long-intl.com/articles/delay-damages/).

3 The "critical path" is the longest sequence of activities from start to finish that must be completed to ensure a project is finished by a certain date.

Major Issues in International Construction Law, First Edition. Mauro Rubino-Sammartano and Troy Harris.
Published 2026 by John Wiley & Sons, Inc.

contractor cannot recover increased time-related costs.[4] A critical, *force majeure* delay may entitle the contractor to a time extension, but not delay damages, unless the contract terms and conditions state that such damages are compensable.[5]

If the objective of the delay analysis is to assess all delays during a window of time or the entire project duration, then a comparative analysis of planned and actual activity start dates and durations may be the appropriate first step before determining criticality. If the analysis is being performed on each monthly schedule update, then the analyst compares activity start dates and durations at the beginning of the month to the activity start dates and durations at the end of the month and quantifies the delays. This initial analysis also includes quantifying extended lag durations.

In order to determine delay criticality properly, the analyst must first examine the quality of the schedule to ensure that the schedule contains no open ends (missing predecessors or successors)[6] and that noncontractual mandatory constraints do not drive the schedule. Open-ended activities should be linked, and inappropriate mandatory constraints should be changed.[7]

After remedying schedule logic and inappropriate mandatory constraints, the analyst identifies critical path delays either on a prospective basis, such as when using a time impact analysis,[8] or on an as-built basis[9] when using other modeled schedule analysis methodologies such as an as-built but-for schedule delay analysis method[10] or an observational method using an as-planned vs. as-built comparison.[11]

---

4 If an owner delays a contractor or subcontractor on noncritical path work and causes the contractor or subcontractor to incur increased time-related costs, the contractor or subcontractor may be entitled to delay damages, in some cases.

5 If project delay is compensable, it means that the party affected by the delay should be given compensation for the delay damages or an extension of time on the project schedule.

6 Predecessors are schedule activities that must be completed or at least started prior to another activity. Successors are schedule activities that follow another activity. Successor activities follow predecessor activities.

7 Other schedule quality checks should also be performed. See Ronald J. Rider, "Forensic Schedule Assurance Review: A Vital Step Before Starting a Detailed Schedule Analysis," https://www.long-intl.com/articles/forensic-schedule-assurance.

8 See Richard J. Long, Ronald J. Rider, and Rod C. Carter, "Implementing Time Impact Analyses and Update Impact Analyses on Large, Complex EPC Projects," https://www.long-intl.com/articles/time-impact-analyses.

9 See Andrew Avalon, "Calculating the As-Built Critical Path," https://www.long-intl.com/articles/as-built-critical-path.

10 See Richard J. Long, "As-Built But-For Schedule Delay Analysis," https://www.long-intl.com/articles/as-built-but-for.

11 See AACE International Recommended Practice No. 29R-03, "Forensic Schedule Analysis," April 25, 2011, MIP 3.1 through 3.5, pp. 41–72 of the pdf.

If the analyst determines a delay to be critical, the next step is to determine if the delay is excusable or nonexcusable. The causes of a contractor's excusable delays are beyond the contractor's control. If an act of God (*e.g.*, extremely unusual weather) or other *force majeure* events cause an excusable, critical delay, the contractor may be entitled to a time extension, but usually not its delay costs. If an owner-responsible event causes an excusable, critical delay, then the contractor may be entitled to a time extension. If the contract does not include a "No Damages for Delay" clause, and the contractor has not caused concurrent delay, the contractor may also be entitled to delay damages.

Nonexcusable delays are generally within the contractor's control and include poor quality of work requiring rework, lower-than-planned labour productivity that is not caused by the contractor's underestimate of planned productivity or owner-responsible events, problems with construction equipment, or other events directly tied to the contractor's responsibility and performance. If contractor-caused, nonexcusable delay is critical, the owner may be entitled to LDs or actual delay damages if the contractor does not take action to make up the lost time.

AACE International[12] provides general rules related to excusable and compensable delay:

> *Excusability exists where there is contractual or equitable justification in a claimant's request for a contract time extension for relief from potential claims for liquidated/stipulated or actual delay damages. The showing of excusability does not necessarily mean that the claimant is also entitled to compensation for the delay. Conversely, delay is non-excusable when such justification does not exist.*
>
> *Compensability or compensable delay exists where the claimant is entitled to recover not only a time extension but compensation for expenses associated with the extension of completion date or the prolongation of the duration of the work. Excusability is a prerequisite to compensability. Therefore, where compensability can be established, excusability is assumed.*[13]

Before determining whether a contractor or owner can recover delay damages, the analyst should determine if the delay event is concurrent with any other project delays. A concurrent delay occurs when two or more individual delay events, one caused by the owner and the other caused by the contractor, occur

12 AACE International is the Association for the Advancement of Cost Engineering which is a nonprofit professional association that provides resources to cost management professionals.

13 See AACE International Recommended Practice No. 29R-03, "Forensic Schedule Analysis," April 25, 2011, Section 4.1.A, p. 100 of 136.

during the same timeframe, and both events would delay the project completion date. A core principle from the Society of Construction Law states:

> Where Employer Delay to Completion and Contractor Delay to Completion are concurrent and, as a result of that delay the Contractor incurs additional costs, then the Contractor should only recover compensation if it is able to separate the additional costs caused by the Employer Delay from those caused by the Contractor Delay. If it would have incurred the additional costs in any event as a result of Contractor Delay, the Contractor will not be entitled to recover those additional costs.[14]

Thus, neither the owner nor the contractor may be able to recover its delay damages for concurrent delay, although the contractor may still be entitled to a time extension.[15]

If the contractor does not complete work on time, the owner may be entitled to liquidated damages per day of delay. Similarly, a contractor may be entitled to LDs if its subcontractor delays the project, provided that the contract between the contractor and subcontractor includes LD terms. Parties typically establish the value of LDs and how they are to be applied during the contracting phase, before work begins.

The owner and/or the contractor may incur "actual" delay damages, but an owner may not be entitled to its actual delay damages if the contract contains an LD clause. Actual delay damages can be calculated and tabulated by different methods, which are discussed in Section "Calculating Delay Damages" of this chapter.

Determining Entitlement to Delay Damages

After determining criticality and concurrency, the analyst should evaluate a delay and any concurrent delays for causation and potential compensability. The analyst determines whether the delay is excusable or nonexcusable, and, if it is excusable, whether it is compensable to the contractor for its entire duration or only part, due to a concurrent, contractor-caused portion of the delay.

14 See SCL Delay and Disruption Protocol, 2nd Edition, February 2017, p. 7.

15 For a more complete discussion of concurrent delay, see Richard J. Long, "Analysis of Concurrent Delay on Construction Claims," https://www.long-intl.com/articles/concurrent-delay.

Noncritical Delay

Noncritical delays may be excusable to the contractor if the owner-caused noncritical delay causes the contractor's labour, equipment, or subcontractors to work longer than it reasonably planned. If an owner causes a noncritical delay, then delay to noncritical work activities may be compensable. In these cases, the contractor must prove that others (*i.e.*, owners and other owner-responsible contractors) caused its actual damages, and that it calculated its damages based on the costs specifically incurred for the noncritical work and not time-related costs associated with the completion of the overall project. Delay costs would include such things as the costs of indirect labour, overhead, and equipment working on the affected noncritical activity, but not for the entire project. In these situations, contractors should seek recovery via contract change orders for noncritical, unplanned delay caused by others. Figure 17.1 shows a five-day, owner-caused delay to a contractor's noncritical Activity C.

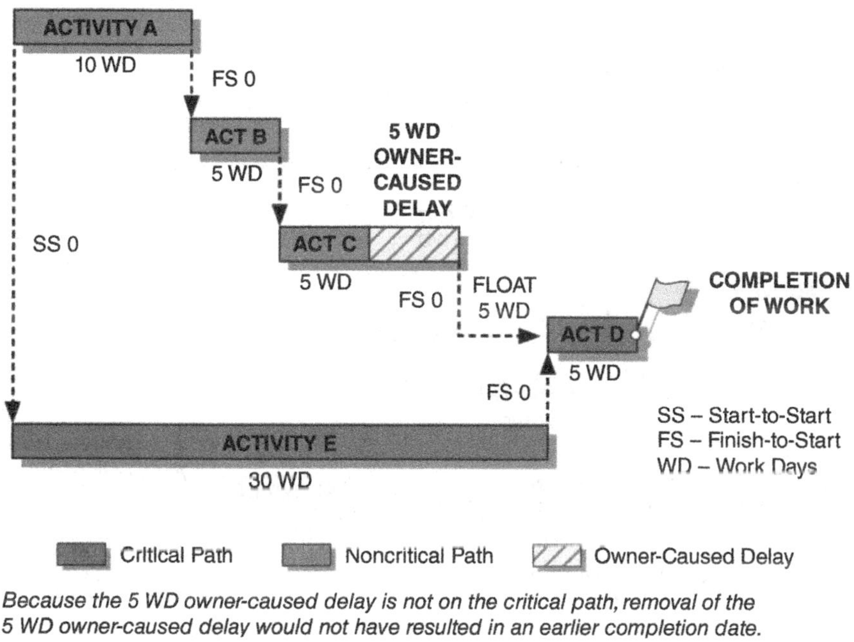

Figure 17.1 Noncritical, owner-caused delays.[16] Long International.

16 See Richard J. Long, "As-Built But-For Schedule Delay Analysis," https://www.long-intl.com/articles/as-built-but-for.

The figure shows that although the owner-caused delay to the contractor's Activity C does not affect the start of the contractor's Activity D and therefore delay completion of work, the delay may have caused unforeseen delay costs to the work on Activity C. This owner-caused, excusable delay and any associated delay costs may be compensable if the contractor can prove that its unplanned costs were a direct result of the owner-caused delay. On the contrary, if a contractor causes noncritical delay for which the owner incurs delay costs, those costs may be compensable for the owner.

Critical Delay

A critical delay to a project activity or set of activities affects the overall project completion date, as shown in Figure 17.2.

The analyst should determine whether the delay is excusable or nonexcusable by performing a causation analysis. An excusable delay may entitle the contractor to additional compensation as delay damages. Conversely, a nonexcusable delay

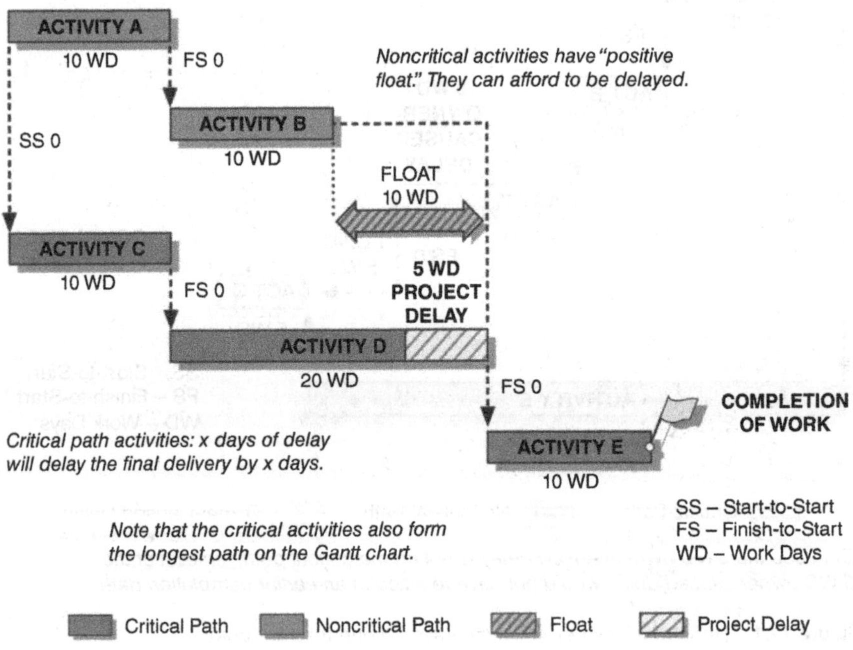

Figure 17.2 Critical path delay.[17]

17 Source: Garrett J. Brown.

may entitle the owner to LDs or actual delay damages, unless the contractor later mitigates and recovers the delay.

Critical, Excusable Delay

Excusable delay may be compensable or noncompensable, depending on causation. Construction contracts often contain *force majeure* clause(s) to remove contractor delay liability for unforeseeable and unavoidable events outside of the contractor's control. Below are examples of *force majeure* events that may excuse the contractor from late completion of the work.

- Acts of God (severe weather conditions)
- Natural disasters (fire, flood, earthquake, etc.)
- War, invasion, terrorist threats, and civil unrest
- Labour disputes
- Government order or law
- Epidemic, pandemic, or similar (*e.g.*, COVID-19)
- Other similar events beyond the contractor's control

If a *force majeure* event causes a delay, the contractor is typically entitled to a time extension but not compensation. If the owner causes the critical delay, the delay is excusable and potentially compensable, absent a "No Damages for Delay" clause. Below are examples of owner-caused delay:

- Suspension of work
- Scope of work changes
- Design changes
- Approval delays
- Site access delays
- Late receipt of owner-acquired permits
- Differing site conditions
- Constructive changes
- Late owner-furnished equipment
- Errors and omissions (owner provided engineering)
- Other owner-caused delays

Critical, Nonexcusable Delay

When a delay is determined to be critical but nonexcusable, the contractor cannot recover delay damages. However, if the contractor and/or the contractor's subcontractors are the sole cause of a delay, the owner may be entitled to LDs or actual damages unless the contractor later mitigates and recovers the delay.

Furthermore, if the contractor materially breaches the contract, the owner may also be entitled to consequential damages, unless the contract excludes such damages. It is important to distinguish between consequential damages and actual (direct) damages. Consequential damages are typically unexpected losses that arise from the breach of a contract and are not actual or direct delay damages.[18]

Concurrent Delay

AACE International addresses concurrent delay in the following ways:

1. Two or more delays that take place or overlap during the same period, either of which occurring alone would have affected the ultimate completion date. In practice, it can be difficult to apportion damages when the concurrent delays are due to the owner and contractor respectively.
2. Concurrent delays occur when there are two or more independent causes of delay during the same time period. The "same" time period from which concurrency is measured, however, is not always literally within the exact period of time. For delays to be considered concurrent, most courts do not require that the period of concurrent delay precisely match. The period of "concurrency" of the delays can be related by circumstances, even though the circumstances may not have occurred during exactly the same time period.
3. True concurrent delay is the occurrence of two or more delay events at the same time, one an employer risk event, the other a contractor risk event and the effects of which are felt at the same time. The term "concurrent delay" is often used to describe the situation where two or more delay events arise at different times, but the effects of them are felt (in whole or in part) at the same time. To avoid confusion, this is more correctly termed the "concurrent effect" of sequential delay events.
4. Concurrent delay occurs when both the owner and contractor delay the project or when either party delays the project during an excusable but noncompensable delay (e.g., abnormal weather). The delays need not occur simultaneously but can be on two parallel critical path chains.
5. The condition where another delay-activity independent of the subject delay is affecting the ultimate completion of the chain of activities. (June 2007)[19]

---

18 "Consequential" damages, which are typically separate from delay damages, occur if one of the contracting parties breaches the contract. Construction contracts typically include waivers for consequential damages, and therefore, this chapter does not discuss in detail delay damages related to consequential damages.

19 See AACE International Recommended Practice No. 10S-90, "Cost Engineering Terminology," Rev. May 15, 2023, p. 29.

Similarly, the Society of Construction Law's Delay and Disruption Protocol provides the meaning of concurrent delay as:

> *True concurrent delay is the occurrence of two or more delay events at the same time, one an Employer Risk Event, the other a Contractor Risk Event, and the effects of which are felt at the same time. True concurrent delay will be a rare occurrence. A time when it can occur is at the commencement date (where for example, the Employer fails to give access to the site, but the Contractor has no resources mobilised to carry out any work), but it can arise at any time.*
>
> *In contrast, a more common usage of the term "concurrent delay" concerns the situation where two or more delay events arise at different times, but the effects of them are felt at the same time.*
>
> *In both cases, concurrent delay does not become an issue unless each of an Employer Risk and a Contractor Risk Event lead or will lead to Delay to Completion. Hence, for concurrent delay to exist, each of the Employer Risk Event and the Contractor Risk Event must be an effective cause of Delay to Completion (not merely incidental to the Delay of Completion).*[20]

Figure 17.3 illustrates an example of concurrent delay.

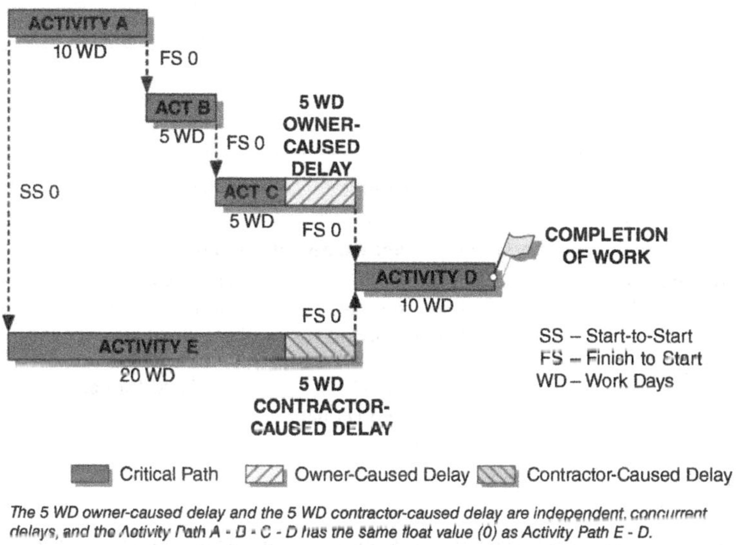

Figure 17.3 Illustration of concurrent delay.[21] Long International.

20 See Society of Construction Law Delay and Disruption Protocol, 2nd Edition, Rev. February 2017, p. 30.

21 See Richard J. Long, "As-Built But-For Schedule Delay Analysis," https://www.long-intl.com/articles/as-built-but-for.

Figure 17.3 shows that the owner delayed Activity C by five working days, and during the same time period, the contractor delayed Activity E by five working days. Because both delays occurred within the same time period,[22] they are concurrent delays. The owner-caused delay may be deemed excusable but not compensable because it did not solely delay the project completion date.

Summary of Types of Construction Delays

Each of the types of delay described above can be summarized as 1) delays within the contractor's control, 2) delays within the owner's control, and 3) delays caused by neither party. Tables 17.1 and 17.2 present the types of delay that may result in delay damages.

Table 17.1 Types of delay for a single delay event.[23]

Single delay event	Excusable		Nonexcusable	Recoverable by contractor	
	Compensable	Noncompensable		Time	Cost
	X			*Yes*	*Yes*
		X		*Yes*	*No*
			X	*No*	*No*

Table 17.2 Types of delay for concurrent delay.[24]

Delay event 1		Delay event 2	Net effect
***Force Majeure* Delay** [Time/No Compensation for Extended Overhead/ No Liquidated/ Stipulated Damage Assessment]	concurrent with	**Contractor-caused Delay** [No Time/No Compensation for Extended Overhead/ Liquidated/ Stipulated Damage Assessment]	**Excusable** [Time/No Compensation for Extended Overhead/ No Liquidated/ Stipulated Damage Assessment]

22 Concurrent delays may be literally concurrent (*i.e.*, they occur at exactly the same time) or functionally concurrent (*i.e.*, they occur during the same time period). Either delay would delay project completion, but because the schedule was not updated on the precise date that the delays occurred, they are called functionally concurrent. See Richard J. Long, "Analysis of Concurrent Delay on Construction Claims," https://www.long-intl.com/articles/concurrent-delay.

23 Source: Garrett J. Brown.

24 See AACE International Recommended Practice No. 29R-03, "Forensic Schedule Analysis," April 25, 2011, Figure No. 15 – Net Effect of Potential Concurrent Delay Combinations, p. 113 of 136. Stipulated damages are synonymous with liquidated damages.

Delay event 1		Delay event 2	Net effect
***Force Majeure* Delay** [Time/No Compensation for Extended Overhead/ No Liquidated/ Stipulated Damage Assessment]	concurrent with	**Owner-caused Delay** [Time/ Compensation for Extended Overhead]	**Excusable** [Time/No Compensation for Extended Overhead/ No Liquidated/ Stipulated Damage Assessment]
Contractor-caused Delay [No Time/No Compensation for Extended Overhead/ Liquidated/Stipulated Damage Assessment]	concurrent with	**Owner-caused Delay** [Time/ Compensation for Extended Overhead]	**Excusable** [Time/No Compensation for Extended Overhead/ No Liquidated/ Stipulated Damage Assessment]

Source: Adapted from AACE International Recommended Practice No. 29R-03. Figure No. 15, p. 113.

Delay Damages

This section discusses types of delay damages and methods for calculating costs resulting from a compensable delay event.

Types of Delay Damages

As previously discussed, delay damages may be categorized as: (1) liquidated damages, (2) actual/direct delay damages, and (3) consequential damages.

Liquidated Damages

Parties most often establish LDs in contracts prior to signing the contract. LDs offset the costs an owner would incur for contractor-caused delay to project completion. A contractor can similarly claim LDs included in a subcontract when the subcontractor fails to complete its work on time. LDs are typically a dollar value per unit of time (*e.g.*, $50,000 per day) and must reasonably represent the actual costs that an owner would incur. They are typically used in lieu of determining actual delay damages and are not intended to "punish" the contractor.

The American Bar Association provides this definition of liquidated damages:

> In general, liquidated damages provisions specify a predetermined amount of money that must be paid as damages if one party fails to meet certain contractual requirements. In construction contracts, this typically manifests as

a fee per unit of time (the "Liquidated Amount") in the event of a missed schedule milestone such as Substantial Completion.[25]

Actual (Direct) Delay Damages

Actual delay damages, also referred to as direct delay damages, may be associated with and result from a project delay event. If the contract does not include language to address delay damages, the contractor may still be liable for actual delay damages that the owner incurs.

As an example, an owner and a contractor enter into a construction contract that does not definitively set out delay damages. The contractor fails to execute and complete the work per the contractual completion date. Because project completion is delayed, the owner may incur delay costs, such as extension of the services of its project management personnel, consultants, engineers, site offices, administrative personnel, etc. Conversely, if the owner causes the delay, costs directly associated with the delay may be compensable to the contractor if it can demonstrate that the delay impacted what was the project critical path at the time that the delay occurred and provide support for its delay costs.

Many delay-related cost components can comprise the contractor's actual delay damages. Below is a list of the more common cost components used to determine these damages.

- Extended field office overhead
  - Field office facilities
  - Field office staff salaries
  - Field office staff vehicles
  - Field office utilities
  - Field office consumables
  - Miscellaneous field office costs
- Unabsorbed home office overhead[26]
  - Home office rental or ownership costs
  - Home office utilities
  - Home office equipment and maintenance costs
  - Corporate taxes
  - Insurance costs not assigned to projects
  - Home office staff salaries
- Idle direct labour costs
- Idle and/or extended construction equipment

25 See Jeremy Wegener and Chad Caplan, "Liquidated Damages – The Basics," Under Construction, ABA Forum on Construction Law, September 20, 2021, https://www.americanbar.org/groups/construction_industry/publications/under_construction/2021/summer2021/liquidating_damages.

26 Unabsorbed home office overhead costs are contractor expenses that are not covered by the project revenue.

- Contractor-supplied temporary utilities
- Equipment preservation
- Additional material storage
- Escalation

Consequential Damages

Consequential damages, also known as special damages or indirect damages, are not always directly caused by a breach of contract. The nonbreaching party of the contract may seek these types of damages for losses that result indirectly from the breach.

Examples of consequential damages might include the following:

- Loss of product
- Loss of profits
- Loss of business opportunities
- Increased financing costs
- Increased operating costs

Because consequential damages may cause significant losses to the breaching party, construction contracts commonly include waivers of consequential damages. The following is an example of a consequential damages waiver:

> 18.2 Consequential Damages Waiver. IN NO EVENT WILL EITHER PARTY BE LIABLE TO THE OTHER PARTY FOR ANY INDIRECT, SPECIAL, CONSEQUENTIAL, PUNITIVE OR EXEMPLARY DAMAGES (INCLUDING, BUT NOT LIMITED TO, DAMAGES BASED UPON LOST PROFITS, BUSINESS INTERRUPTION, LOST BUSINESS, OR LOST SAVINGS (EXCEPT THAT IN CASE OF BREACH BY CUSTOMER, BRAMMER CAN RECOVER ITS LOST PROFITS UNDER THIS AGREEMENT THAT BRAMMER LOST AS A RESULT OF THE CONTRACT NOT BEING FULLY PERFORMED)) FOR ANY ACTS OR FAILURE TO ACT UNDER THIS AGREEMENT, EVEN IF IT HAS BEEN ADVISED OF THEIR POSSIBLE EXISTENCE.[27]

Calculating Delay Damages

When calculating delay damages, it is important to understand the differences among the various cost components discussed in Section "Types of Delay Damages". In the case of a contractor experiencing delay by no fault of its own, certain cost components may comprise an excusable, compensable delay claim. For an owner, delay damages are typically LDs or actual delay damages.

---

27 Development and Manufacturing Services Agreement, Freeline Therapeutics Limited and Brammer Bio MA, LLC, https://www.sec.gov/Archives/edgar/data/1810031/000095012320006919/filename3.htm.

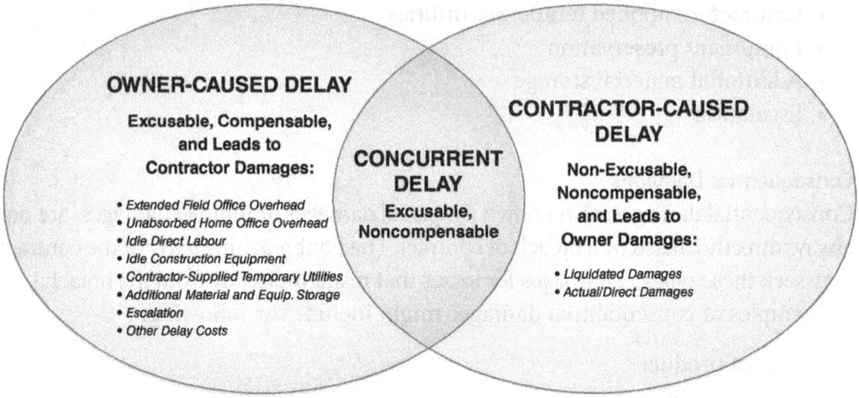

Figure 17.4 Relationship between Owner-caused and contractor-caused delays.[28]

Figure 17.4 shows the relationship between owner-caused delays and contractor-caused delays and the resultant delay damages.

Owner Delay Damages (Contractor-caused Delay)

When a contractor delays a project, the owner may be able to recover delay damages by LDs or actual/direct delay damages, but typically not both.

Owner-assessed Liquidated Damages

When a contractor causes delay and no concurrent delay is established, the owner may be entitled to assess LDs as set forth in the contract. A typical LDs clause may look like the following:

> **Delay Liquidated Damages.** If substantial completion of the liquefaction train occurs after the guaranteed substantial completion date, Bechtel will pay CCL delay liquidated damages at the applicable daily rate as defined in the EPC Contract (T3) until substantial completion of the liquefaction train occurs; provided that, if substantial completion occurs within an initial grace period following the guaranteed substantial completion, no delay liquidated damages would be owed.[29]

28 Source: Garrett J. Brown.

29 See Cheniere Corpus Christi Holdings, LLC, Prospectus Supplement No. 1, Dated December 18, 2017, to the Prospectus Dated November 24, 2017, United States Securities and Exchange Commission Form 8-K, https://www.sec.gov/Archives/edgar/data/1693317/00011931 2517372110/d467315d424b3.htm.

For example, assuming the LDs daily rate in the above excerpt is $50,000 per day, a calculation of delay damages for a delay period of 30 days would be:

$$30 \text{ days} \times \$50,000/\text{day} = \$1,500,000 \text{ as delay damages paid to the owner}$$

Owner Actual (Direct) Delay Damages

If a contract does not set forth liquidated damages, and the project incurs a contractor-caused delay, the owner may be entitled to seek actual damages. Although there is no formula or standard method for calculating actual delay damages that an owner incurs for contractor-caused delay, below are typical examples of owner costs from a delayed project:

- Additional project management personnel, consultants, and engineering costs
- Additional site offices costs
- Additional administrative personnel costs
- Other additional costs

For example, assuming the additional daily costs for the items listed above is $80,000 per day on average, a calculation of delay damages for a delay period of 30 days would be:

$$30 \text{ days} \times \$80{,}000/\text{day} = \$2{,}400{,}000 \text{ as delay damages paid to the owner}$$

Contractor Delay Damages (Owner-caused Delay)

As discussed in Section "Actual (Direct) Delay Damages" above, delay typically impacts several contractor cost types. Below are generally accepted methods for calculating contractor delay damages for each cost type.

Extended Field Office Overhead

The first step in calculating extended field office overhead (FOOH) costs is to determine when a delay occurred in relation to the overall duration of the project. Far too often, contractors establish one daily rate for FOOH and use it to calculate the cost for extended FOOH. However, the daily rate for FOOH costs typically changes over time, following a curve in conjunction with planned progress and direct labour levels. FOOH costs gradually build up during the beginning of a project, level off at peak labour levels when the project is in full swing, and decrease as the project nears completion.

Therefore, to determine the daily rate to apply to a project delay, it may be appropriate to identify phases of field overhead rates. Typically, there are a minimum of three phases: 1) ramp-up phase, 2) construction phase, and 3) ramp-down phase.

Depending on the size and duration of the project, there may be additional phases where FOOH differs from other phases.

If delay occurs during the height of the construction phase (*i.e.*, peak direct labour force), the FOOH daily rate would be greater than the rate during the ramp-up or ramp-down phase. Thus, the resulting extended FOOH costs would be greater during the peak construction phase than they would be if the delay occurred during the beginning or ending phase of the project.

When developing the daily rate for FOOH, a contractor should include applicable cost accounts allocated as FOOH. Section "Actual (Direct) Delay Damages" above includes many of the typical cost accounts. For example, the contractor should divide the monthly cost for field office facilities into a daily rate. Similar calculations should be performed for each of the applicable FOOH cost accounts. However, there may be certain FOOH cost accounts that the contractor should exclude from FOOH daily rate calculation. For example, project mobilization costs are typically coded as FOOH costs, but the contractor should exclude these and other one-time costs from the pool of time-related costs that are subject to prolongation claims.

As an example, assuming the total daily rate for the applicable cost accounts under FOOH is $20,000 per day, a calculation of delay damages for a delay period of 30 days would be:

$$30\,\text{days} \times \$20{,}000/\text{day} = \$600{,}000\ \text{as delay damages calculated by the contractor}$$

Unabsorbed Home Office Overhead

Law Insider provides this definition of a contractor's home office overhead (HOOH) costs:

> Home Office Overhead means that portion of Contractor business costs necessary to operate the business entity that are not specifically related to the Work but are properly allocable to this Contract. Such costs include, but are not limited to, those associated with office and office personnel salaries and related payroll taxes and benefits; costs of office occupancy and maintenance, all supporting services (such as utilities, office equipment, computers, and similar items) related to the home office function; business taxes and licenses; and all such other costs necessary to operate the business entity.[30]

30 See "*Home Office Overhead* definition," Law Insider, https://www.lawinsider.com/dictionary/home-office-overhead.

Unabsorbed home office overhead costs due to project delay could be calculated similarly to extended field office overhead. However, determining all HOOH costs associated with a single project may be difficult, especially if the contractor's organization has multiple projects active simultaneously. While several industry studies exist, the Eichleay formula is a widely used method for calculating HOOH costs. The Eichleay formula first appeared in Eichleay Corp. (1960) when Eichleay, the contractor, sought to recover damages from a US government-caused delay. The original Eichleay formula consisted of three calculations shown in the equations below.

$$\frac{\text{Total Contract Billings}}{\text{Total Company Billings for Contract Period}} \times \text{Total Overhead for Contract Period} = \text{Allocable Overhead}$$

$$\frac{\text{Allocable Overhead}}{\text{Days of Performance}} = \text{Daily Unabsorbed HOOH Rate}$$

$$\text{Daily Unabsorbed HOOH Rate} \times \text{Number of Days of Delay} = \text{Unabsorbed HOOH Amount}$$

Although the above equations provide a reasonable approach to calculating unabsorbed HOOH costs due to project delays, the Eichleay formula is not without shortfalls. For example, it presumes that any delays affect overall project costs uniformly over the duration of a project. As previously discussed, project costs are not linear, and therefore, the application of HOOH costs using the Eichleay formula is not always fair and accurate.

Nonetheless, the Eichleay formula remains frequently used and accepted in court cases to this day. In their scholarly paper "Guidelines for Recovering Home Office Overhead Costs with Emphasis on the Eichleay Formula," Dr. William Ibbs and Benjamin T. Davis provide criteria for applying the formula. The paper states:

> Based on a review of court cases, the following criteria were created for when the Eichleay formula can be applied to recover unabsorbed HOOH. They are ordered by precedence. All seven conditions must be met for Eichleay to apply.
>
> 1. The Eichleay formula is used for primarily prime contractor versus owner disputes. Subcontractor and prime contractor disputes are discussed later in this paper;
> 2. Unabsorbed HOOH caused by force majeure events cannot be recovered;
> 3. The Eichleay formula is only applicable to HOOH when the project's schedule is extended past the contract performance period;

4. Any "wholly unrealistic figure" produced by the Eichleay formula will be disallowed;
5. Oral agreements may be enforceable regardless of a written contractual agreement stating otherwise;
6. Contractual agreements are upheld regardless of actual damages; and
7. Inability to calculate actual damages warrants the use of the Eichleay formula.[31]

Figure 17.5 captures the HOOH recovery formula criteria in a flowchart.

The figure provides a guideline for contractors to determine if the Eichleay formula should be used. When calculating the effects of unabsorbed home office costs, contractors should consider the Eichleay formula unless a fair and equitable daily rate is agreed on or the rate is defined in the contract.

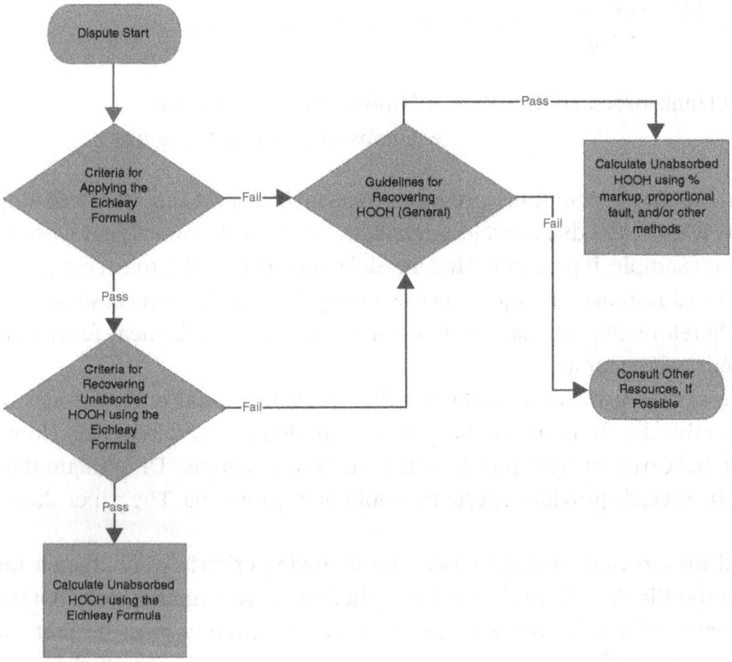

Figure 17.5 Flowchart for navigating unabsorbed HOOH recovery.[32] Long International.

31 See Benjamin T. Davis and William Ibbs, "Guidelines for Recovering Home Office Overhead Costs with Emphasis on the Eichleay Formula," *Journal of Legal Affairs and Dispute Resolution in Engineering and Construction* 9, no. 1, February 2017, p. 2, https://ascelibrary.org/doi/10.1061/%28ASCE%29LA.1943-4170.0000201.

32 See Davis and Ibbs, p. 3.

An example of using the Eichleay formula for a hypothetical project that has experienced owner-caused delay is shown below.

Actual contractor billings for delayed project = \$500,000

Total contractor company billings for the period = \$3,000,000

Total home office overhead costs for contractor company = \$75,000

$$\frac{\$500{,}000}{\$3{,}000{,}000}, \times \$75{,}000 = \$12{,}500 \text{ (HOOH allocable to delayed project)}$$

Days of contractor work through delay period = 90 days

$$\frac{\$12{,}500}{90 \text{ days}} = \$138.89 \text{ (Daily rate contract home office overhead)}$$

Total days of delay on the project = 30 days

\$138.89/day × 30 days = \$4,166.67 (Unabsorbed expenses for additional HOOH costs due to owner-caused delay.)

Idle Direct Labour

Contractors typically include additional costs for direct or craft labour in disruption claims rather than delay claims. However, delays sometimes lead to the idleness of a direct labour force that the contractor cannot reassign to other work or remove from the project. If the contractor can tie these circumstances to a delay, it may properly include the costs in a delay claim.

When calculating the delay damages, it should be determined if the direct labour had to remain on site but could not work due to delay. Contractors typically maintain daily labour force and payroll records for their direct labour personnel and document idle time in daily logs and/or timesheets. The contractor should compile these records for the duration of the delay period, total the actual payroll costs for the idle direct labour force during this time, and include this sum in the overall damages calculation.

Idle and Prolonged Construction Equipment

With a delayed and disrupted project, several factors can complicate the pricing of additional costs for equipment. First, certain equipment, such as welding machines, are more task related than time related, and therefore, the additional costs are more closely tied to craft labour productivity than time. Time and overall project duration typically drive the costs of other equipment, such as tower cranes.

Second, project delay may cause some equipment to sit idle while other equipment is utilized but still on the project longer than planned. Third, hourly, daily, weekly, or monthly equipment rates may be derived from several sources, including rates set out in the contract, estimated ownership costs, rental costs, and other industry sources such as blue book values. Moreover, the contract may include separate rates for idle versus working equipment. Finally, equipment costs can be spread across many different cost accounts in the contractor's records, and the costs may be established using internal methods that may not align with, for example, the rates stipulated in the contract. For example, a contractor may allocate its monthly ownership costs across several projects based on the equipment assigned to those projects, but such allocation may not account for specific contract rates or how the equipment is being utilized on a particular project.

Given these complications, the contractor must clearly establish, often for each type or piece of equipment, the basis for the pricing and the underlying causes for the increased costs. As with idle direct labour costs, the contractor should document and compile costs for idle construction equipment for the duration of the delay period. The contractor should also add ownership and rental costs, and it should assess other project-based equipment for prolongation costs. For example, eligible costs typically include those for site supervision-assigned vehicles, certain large cranes, and other equipment.

Contractor-supplied Temporary Utilities

If the owner causes compensable delay to project completion, and the contractor is required to continue to provide and pay for temporary utilities, then the contractor may be entitled to recover its additional costs for these utilities. Temporary utilities can include but are not limited to electric power, natural gas, drinking water, plant water, nitrogen, hydrogen, other similar gases, etc.

For example, if the daily cost of these temporary utilities is $10,000, and the owner delays the project by 30 days, then the cost of temporary utilities would be:

$$30 \text{ days compensable delay} \times \$10{,}000 \text{ per day} = \$300{,}000$$

Equipment Preservation

If a long, owner-caused delay requires the contractor to protect the plant equipment that it has installed from water damage, corrosion, contamination, cold temperatures, etc., then the costs of this protection may be recoverable. These costs include material costs to protect the equipment, installation of equipment preservation materials, labour costs to maintain the equipment, and labour costs to restore the equipment after construction resumes. If the contractor maintains supportable documentation for all costs associated with equipment preservation and maintenance, it may have a supportable claim for recovery of these costs.

Additional Equipment and Material Storage

Because plant equipment and material storage costs are time dependent, an analyst can determine delay damages using project cost records. The contractor typically pays the costs for facilities required for storing equipment and materials prior to their installation. If a project is delayed and the contractor incurs additional costs for extended equipment and material storage, those costs may be recoverable as delay damages. Alternatively, if the contractor is delayed and elects to postpone the purchase of materials, the differential between predelay and postdelay costs of materials may be recoverable.

Materials used for construction projects are typically stored in onsite or offsite warehousing facilities. Although warehousing may be considered a field office overhead cost, delineating costs associated with material storage increases the likelihood of recovering them.

Below is a simple example for calculating additional material storage delay damages:

Monthly cost for warehousing = \$50,000 per month

$$\text{Daily warehousing cost} = \frac{\$50,000}{30 \text{ days per month}} = \$1,666.67 \text{ per day}$$

$$\begin{aligned}\text{Additional material storage costs} &= 30 \text{ days} \times \$1,666.67 \text{ per day}\\ &= \$50,000\end{aligned}$$

Escalation

Construction contracts often assign the contractor the risk for increases in labor, construction equipment, and material costs during the contracted timeframe of the project. However, if the contractor is delayed through no fault of its own, the costs for escalation resulting from the extended project duration may be recoverable.

Damages resulting from escalation due to delay can be determined by calculating the difference between the cost of labor, construction equipment, or materials that the contractor would have incurred had there been no project delay and the actual cost of the labor, construction equipment, or materials resulting from performing the work later than originally planned. This calculation requires the contractor to provide supporting documentation for its planned and actual labor, construction equipment, and purchased materials costs.

An example of an escalation calculation due to excusable, compensable delay is as follows:

Project costs without any delay:

Labor costs = \$5,000,000
Construction equipment = \$2,000,000
Purchased materials = \$3,000,000

Actual project costs after project delay:
Labor costs = $5,500,000
Construction equipment = $2,500,000
Purchased materials = $3,250,000

The difference between what would have been spent had there been no project delay and the actual costs after project delay equals escalation costs in this example.

Labor escalation: $5,500,000 (actual) – $5,000,000 (planned) = $500,000
Construction equipment escalation: $2,500,000 (actual) – $2,000,000 (planned) = $500,000
Purchased materials escalation: $3,250,000 (actual) – $3,000,000 (planned) = $250,000
Total escalation costs claimed by the contractor due to owner-caused delay = $1,250,000

Other Delay Costs Potentially Recoverable by the Contractor

In the case of a long, owner-caused delay, the contractor may incur additional costs that may be recoverable, including but not limited to:

- Extended equipment warranty costs
- Extended construction insurance costs
- Extended bonding costs

Summary

When a construction project experiences a delay, either the owner or the contractor typically incurs delay damages. An analyst can determine the delay to be:

- Critical or noncritical;
- Concurrent or nonconcurrent;
- Excusable or nonexcusable; and
- Compensable or noncompensable.

If an owner-caused delay falls on the then-current critical path and is nonconcurrent with other contractor-caused delays, the contractor may be entitled to seek delay damages. Conversely, if a contractor-caused delay is not concurrent with owner-caused delays, the owner may be entitled to assess liquidated damages or actual delay damages.

The types of delay damages are: (1) liquidated damages (LDs), (2) actual or direct delay damages, and (3) consequential damages. Parties typically establish LDs in the contract prior to execution of the work, calculating the LDs as a fair representation of the owner's additional costs resulting from contractor-caused delay. Absent an LD provision in the contract, the owner may seek its actual delay damages. If the owner delays the contractor's work, the contractor may calculate its actual delay damages. Consequential damages are often waived in construction contracts or are considered in the owner's LDs.

Contractor delay damages from owner-caused delay can include, but are not limited to, the following cost components:

- Extended field office overhead costs
- Unabsorbed home office overhead costs
- Idle direct labor
- Idle and/or prolonged construction equipment
- Temporary utilities supplied by the contractor
- Equipment preservation
- Additional material storage
- Escalation
- Other delay costs

The contractor typically compiles delay damages from its actual costs during the period of owner-caused delay. It calculates extended field office overhead costs by determining the daily rate of expense that would be incurred during the delay period and multiplying by the number of days of delay. The contractor could determine unabsorbed home office overhead costs similarly, but due to the nature of home office costs, contractors often use the Eichleay formula or other industry formulae. Costs related to idle direct labor, idle construction equipment, and additional material storage are compiled from actual contractor costs from direct labor payroll, paid invoices and/or internal costs for rented or owned construction equipment, and paid invoices or internal cost accruals for additional storage. Escalation costs due to owner-caused delay can be determined by calculating the difference between the cost of labor, equipment, or materials that the contractor would have incurred had there been no project delay and the actual cost of the labor, equipment, or materials resulting from performing the work later than originally planned.

The examples of delay damages discussed above may not be accurately or adequately compiled without robust contemporaneous project documentation. Proof of delay damages is required when submitting a delay claim and timely recordkeeping of project costs associated with any project delay is of utmost importance.

18

Contractor Claims

Laura West

Introduction to Contractor Claims

Contractor claims typically focus on claims for (1) additional time (that is, an extension to the agreed completion date); and/or (2) additional payment.

In respect of the first, the Contractor is seeking to extend the completion date due to events it says are the Owner's risk. Most construction contracts will specify a date by which the works are to be completed. If the Contractor does not complete the works by this date, the contract will usually allow the Owner to recover liquidated damages – a preagreed amount of compensation to be applied often per day or per week to compensate the Owner for its losses for not receiving the works on time. In large complex, projects where timely delivery is critical, these can run for anywhere between £100,000s and million pounds per week. To avoid the Contractor being liable for such liquidated damages construction contracts typically provide a mechanism whereby the Contractor can apply for an extension of time (**"EOT"**) to the completion date for certain delay events for which the Owner takes the risk.[1] Liquidated damages will then apply from the revised completion date.

With such delays, the Contractor may also incur additional costs and so it is common that a Contractor's claim for EOT will also cover a claim for its loss and expense arising from the delay. Although there is not always an automatic right to additional payment where there an EOT is awarded.[2] Likewise, an entitlement

1 See for example, FIDIC Red, Yellow and Silver Books (2017), Sub-Clause 8.5.

2 In the JCT standard form contract used in England, claims for extension of time and loss and expense are dealt with under two separate clauses with different specified trigger events. See for example, the JCT Standard Building Contract with Quantities (2024), clauses 2.26–2.29 and 4.20–4.22.

Major Issues in International Construction Law, First Edition. Mauro Rubino-Sammartano and Troy Harris.
Published 2026 by John Wiley & Sons, Inc.

to additional money does not always lead to the completion date being moved. An example might be where the Owner instructs additional works that have no impact on the completion date because they are not on the critical path for the works. Contractor claims for EOT are described further at Section f (claims for delay) below.

Sources of Liability

In terms of sources of liability to allow the Contractor to recover additional time or money, much will turn on the terms of the contract. It is common for construction contracts to take a standard form contract as a starting point and this chapter will refer to the risk allocation under such standard form contracts focusing, in particular, on the International Federation of Consulting Engineers (FIDIC) forms which are frequently used on international projects.

In addition to the contract, the Contractor may also find rights of recourse under the governing law through relevant legislation or the appropriate civil code. A statute or code may imply certain terms where necessary to do so. For instance, in the UK, the Housing Grants, Construction and Regeneration Act 1996 provides for minimum requirements for payment and for a mandatory form of fast-track dispute resolution - Adjudication - for all "Construction Contracts" (as defined by the Act).[3]

Notification of Claims

This chapter considers some of the typical claims made by a Contractor. Often such claims will be subject to adequate notice being provided by the Contractor to the Owner or its representative. For example, under the FIDIC standard forms, if the Contractor considers that it is entitled to any additional payment and/or EOT the Contractor must "give a Notice to the Engineer, describing the event or circumstance giving rise to the cost, loss, delay...as soon as practicable, and no later than 28 days after [the Contractor] became aware, or should have become aware, of the event or circumstance."[4]

If the Contractor fails to provide this notice, the Owner will often argue that it is not entitled to additional payment or an EOT. Whether such a clause does in fact bar a claim will depend on the circumstances and the governing law of

3 Housing Grants, Construction and Regeneration Act 1996, Section 104.

4 FIDIC Red, Yellow and Silver Books (2017), Sub-Clause 20.2.1.

the contract.[5] The rationale for such notice is to allow for claims to be dealt with during the course of the project giving the Owner certainty of how the project is progressing and to avoid the Contractor saving claims until the end of the project. From a Contractor's perspective, the risk of potentially losing claims makes it important to actively manage the contract throughout the project and is worth bearing in mind when considering making any of the "typical claims" below.

Typical Claims

Nondelivery or Late Delivery of Site

In order to progress the works, the Contractor must have adequate access to site. The construction contract will typically provide that that the Owner must provide access to the site. That will usually require that the Owner also provides access to suitable means for transporting plant and machinery to the site. That means of access does not need to be the least onerous for the Contractor and it is not uncommon for technical schedules to set out key access routes by way of marked up plans or drawings or similar.

The construction contract will usually provide a date by which access will be granted. The FIDIC standard form contracts provide for access at the time stated in the Contract Data.[6] The NEC Engineering and Construction Contract (predominantly used in the UK) provides for an "access date" to be defined and for a date for access to be set out in the Accepted Programme. Access must be provided on or before the later of these two dates.[7]

If no express date is provided, under English law, a term will be implied into the construction contract that access will be provided by the Owner within a reasonable time.[8] FIDIC also provides a fallback if no date is specified that the Owner shall provide access "within such times as may be required to enable the Contractor to proceed in accordance with the Programme."[9]

If the Owner grants access late, it may cause delay to the Contractor. In those circumstances, a Contractor will generally be entitled to an EOT and any losses

5 See for instance, *Obrascon Huarte Lain SA v. AG for Gibraltar* [2014] EWHC 1028 (TCC) at paragraph 312 per Akenhead J – "*I see no reason why this clause should be construed strictly against the Contractor and can see reason why it should be construed reasonably broadly, given its serious effect on what could otherwise be good claims for instance for breach of contract by the Employer.*"

6 FIDIC Red, Yellow and Silver Books (2017), Sub-Clause 2.1.

7 NEC4 ECC, Clause 33.1.

8 *Roberts v. The Bury Improvement Commissioners* (1870) L.R. 5 C.P. 310 at 320 and 325.

9 FIDIC Red, Yellow and Silver Books (2017), Sub-Clause 2.1.

resulting from that. Under FIDIC, the Contractor is entitled to claim Cost plus Profit unless the failure to give access was caused by the Contractor's error or delay.[10] Some standard form contracts domestic to the UK, such as the JCT Standard Building Contract, include provisions allowing the Owner to defer giving the Contractor possession of the site for up to six weeks.[11] The effect of this is to allow the Owner to defer possession without being held to be in breach of contract. However, where the deferment of possession materially affects the progress of the works, the Contractor will be entitled to recover loss and expense from the Owner.[12]

The type of access provided is also important. Generally, the Owner will provide a right of entry rather than ownership or possession though it will be important to be aware of any restrictions or nuances in the local law relevant to the site. The Owner need not provide exclusive access to the Contractor and certainly for larger sites, it is common for the Owner to provide access to only part of the site initially. In FIDIC contracts, phased access will typically be set out in the Contract Data. If the Contractor's works are delayed or disrupted by another Contractor, it may be able to recover these costs from the Owner unless the Owner has passed responsibility for coordinating others to the Contractor.[13]

Lack of Authorization

Lack of authorization broadly falls into two categories, where:

1. the Owner has failed to get the requisite authorization for the works to be carried out e.g., obtaining any necessary planning approval, permit or permission for the works (for which the Owner was responsible for obtaining); or
2. the contract administrator / Engineer appointed by the Owner does not have the required authorization.

The first of these is dealt with further at Section e (Wrong or Late Directions by the Owner) of this chapter. As for the second, the Owner will typically appoint another party to administer the contract. Administration usually includes, issuing payment certificates and determinations on Contractor claims. Under FIDIC contracts, this role is performed by the Engineer.[14] The Engineer's role will be limited

10 FIDIC Red, Yellow and Silver Books (2017), Sub-Clause 2.1.

11 JCT Standard Building Contract with Quantities (2024), Clause 2.5.

12 JCT Standard Building Contract with Quantities (2024), Clause 4.20.

13 For Owner's delay, see for example FIDIC Red and Yellow books (2017), Sub-Clause 8.5(e) and Silver Book (2017), Sub-Clause 8.5(c). For Contractor responsibility for coordinating others, see FIDIC Red Book (2017), Sub-Clause 5.1 and FIDIC Yellow and Silver Books (2017), Sub-Clause 4.4.

14 FIDIC Red and Yellow Books (2017), Sub-Clause 3.

by their contract for services and the Engineer cannot amend the construction contract with the Contractor except as provided for in the contract conditions.[15] Therefore, the Engineer must be careful to act within its authority, particularly when issuing Variations. The Engineer's right to vary is limited to changes to the Works (as defined) and cannot include any omission of any work unless expressly agreed by the Owner and Contractor.[16]

The Engineer can be liable to the Owner if it exercises its duties negligently. Whether the Engineer could also be liable to the Contractor may be more difficult, at least under English law. In *Pacific Associates v. Baxter*,[17] a Contractor sued an Engineer who had rejected the Contractor's claim for additional payment when it had encountered unexpectedly hard materials during dredging works. The Contractor received some payment from the Owner and sought to recover the balance from the Engineer for failure to certify. The Court of Appeal rejected the Contractor's claim finding that the Contractor's recourse was under its contract with the Owner and the Engineer did not owe a duty of care to the Contractor to avoid economic loss.

Lack of Cooperation by Owner/Engineer

There will be certain aspects of the works for which the Contractor may require cooperation by the Owner and/or the Engineer. For example, in accessing site, working alongside other sub-contractors/employees of the Owner or obtaining permissions/permits. It is typical for standard form contracts to provide express duties of cooperation. For instance, under FIDIC Sub-Clause 2.3 of the Red, Yellow, and Silver Books provides:

> The Employer shall be responsible for ensuring that the Employer's Personnel [which includes the Engineer] and the Employer's other contractors (if any) on or near the Site:
>
> **a.** Co-operate with the Contractor's efforts under Sub-Clause 4.6 [Co-operation]; and
> **b.** Comply with the same obligations which the Contractor is required to comply with under sub-paragraphs (a) to (e) of Sub-Clause 4.8 [Health and Safety Obligations] and under Sub-Clause 4.18 [Protection of the Environment].

Where there is no express provision to cooperate, if the Contractor's work requires co-operation from the Owner, such co-operation is likely to be implied,

---

15 FIDIC Red and Yellow Books (2017), Sub-Clause 3.2.

16 FIDIC Red and Yellow Books (2017), Sub-Clause 13.1.

17 *Pacific Associates v. Baxter* [1990] 1 QB 993.

though the precise scope of that duty to co-operate and whether that extends to positive assistance by the Owner will depend on the surrounding circumstances.[18]

If the Contractor can demonstrate that the Owner has failed in a duty to co-operate, it may be able to claim an EOT or additional cost on the basis that the Owner's action (or inaction) has prevented the Contractor's performance.[19]

Repudiation by Owner

Owner repudiation refers to action (or potentially inaction) by an Owner that justifies the Contractor in bringing the contractual relationship to an end. Construction contracts will typically include termination provisions which set out the express grounds on which a Contractor may terminate and may include fault and no-fault grounds. For instance, Sub-Clause 16 of FIDIC allows the Contractor to serve a notice to terminate where the Owner fails to issue a Payment Certificate or to make payment within a specified period; fails to comply with an adjudication decision; the Owner is in material breach of contract or an insolvency event in respect of the Owner.

Under English and Scots law, repudiation refers specifically to a common law right to terminate where a counterparty acts in a way that shows it does not accept the obligations of the contract any further. Depending on the circumstances, this action could amount to what is known as a "repudiatory" breach of contract, entitling the innocent party to accept the repudiation and treat the contract as at an end.

What will amount to a repudiatory breach of contract will be fact dependent. Examples of repudiation by an Owner could include refusal to provide access to site or dismissal of the Contractor from site. Depending on the terms of payment, a failure to make payment could also constitute a repudiatory breach. While there is implied right to suspend works on notice for failure to make payment, many standard forms make express provision for that[20] and it is also enshrined in construction contracts in the UK through Section 112 of the Housing Grants, Construction and Regeneration Act 1996.

However, a note of caution in English and Scots law is needed where a party seeks to terminate either under the contract or at common law without proper basis, this is likely to be a repudiatory breach of contract entitling the party being terminated to treat the contract as at an end and seek damages.[21]

---

18 Under English law, see for example *Mackay v. Dick* (1881) 6 App Cas. 251; *Merton London v. Stanley Leach Ltd* (1985) 32 BLR 51; *Essex County Council v. UBB Waste Ltd* [2020] EWHC 1581.

19 See for example FIDIC Red and Yellow Books (2017), Sub-Clause 8.5(e) and FIDIC Silver Book (2017), Sub-Clause 8.5(c).

20 See for example, FIDIC Red, Yellow and Silver Books (2017), Sub-Clause 16.1 JCT Standard Building Contract with Quantities, 2024 Edition, Sub-Clause 4.13; SBCC Standard Building Contract with Quantities, 2016 Edition, Sub-Clause 4.13.

21 *Architectural Installation Services Ltd v. James Gibbons Windows Ltd* (1989) 46 BLR 91.

Where a repudiatory breach has occurred, the innocent party has a choice, it can affirm the contract or it can accept the repudiation and treat the contract as at an end.[22] Where the innocent party elects to terminate the contract, it will be entitled to damages at common law so long as the contract does not exclude such a remedy.[23]

Wrong or Late Directions from Engineer

The duties of an Owner broadly fall into two categories to:

1. cooperate with the Contractor and do what is necessary to allow it to carry out its works; and
2. make payment for the work carried out by the Contractor and to cooperate.

The Engineer is typically appointed by the Owner to act on its behalf to administer the contract (including certifying payment) and make determinations. If an Engineer issues incorrect or late directions, it will likely prevent the Owner from its duties under (1).

Wrong Instructions

Under the FIDIC Red, Silver, and Yellow books, a Contractor must provide notice within 7 days if it considers that an instruction (1) does not comply with applicable laws; or (2) reduces the safety of the Works or is technically impossible.[24] Under English law, if the Contractor fails to give that notice, it could find itself responsible for any loss or damage that arise out of the incorrect instruction.[25]

If the Engineer does not respond within 7 days after receipt of the notice, then, the Engineer is deemed to have revoked the instruction. If the instruction is withdrawn, then, the Contractor is not obliged to carry out the work. If the Engineer responds within 7 days, the Contractor must comply with the instruction (and the response). If the Contractor disagrees with the Engineer's response, it can refer the matter to dispute resolution.

Late Instructions

Similarly, a late instruction could delay or hinder the works, potentially placing the Owner in breach of a duty to cooperate.

Under Sub-Clause 1.9 of the FIDIC Red Book, the Contractor must give a notice to the Engineer if the Works may be delayed or disrupted because any necessary

---

22 *Woodar Investment Development v. Wimpey Construction UK* [1980] 1 WLR 277.

23 *Architectural Installation Services Ltd v. Jaes Gibbons Windows Ltd* (1989) 46 BLR 91 and *Lockland Builders Ltd v. Rickwood* (1995) 77 BLR 38.

24 FIDIC Red, Yellow and Silver Books (2017), Sub-Clause 3.5.

25 *Plant Construction plc v. Clive Adams Associates* [2000] BLR 205, [17]-[18] (TCC).

drawing or instruction is not issued to the Contractor within a particular, failing which, reasonable time. Any delay or cost suffered by the Contractor due to a delayed instruction may be recovered through a claim under Clause 20. The claim may be reduced proportionally, if the delay in issuing the instruction is partially attributable to the Contractor.[26]

There is often also a general catchall clause in construction contracts entitling a Contractor to an extension of time and cost where it is delayed by an action of the Owner and its agents. Sub-Clause 8.5 of the Red, Yellow, and Silver books provides that the Contractor is entitled to an Extension of Time, if completion is delayed due to "any delay, impediment or prevention caused by or attributable to ... the Employer's Personnel" (which includes the Engineer).[27]

Claims for Delay

Claim for EOT

A construction contract will generally set out a date by which the Contractor must complete the works. FIDIC contracts call this date the "Time for Completion," other contracts refer to the "Completion Date"[28] or similar. If the Contractor fails to complete by this date, it is liable to pay the Owner damages for the Owner's losses for delay. However, more commonly, the construction contract will provide for a preagreed level of liquidated damages that are to apply if the Contractor delivers the work late. This means both parties are clear as to the financial consequences of late completion. If the Contractor is late due to an event that is its risk, the Contractor will pay liquidated damages from the completion date until the works are completed. Liquidated damages are typically agreed at a daily or weekly run rate and can be significant running into £100,000s to million pounds per week.

It is common for a Contractor to encounter issues during the project which delay its works. If these delays impact an activity on the critical path for the works, then they will impact the completion date, exposing the Contractor to liquidated damages. Therefore, standard form construction contracts generally include provisions that entitle the Contractor to an EOT to move the completion date if the delay is caused by certain events that are the Owner's risk. Typically, these "owner risk events" will include variations to the works (see section g (claim for additional works) below), exceptionally adverse weather or force majeure.[29]

If the Contractor is prevented from completing by the completion date by the Owner (e.g., due to its breach of contract) and the construction contract does not

26 FIDIC Red Book 2017, Sub-Clause 1.9.

27 FIDIC Red, Yellow and Silver Books (2017), Sub-Clause 1.1.

28 JCT contracts use "Completion Date."

29 See for example FIDIC Red and Yellow Books (2017), Sub-Clause 8.5.

provide for a mechanism to extend the completion date in those circumstances, time is said to become "at large." In those circumstances, the Owner is no longer entitled to rely on the completion date and the Contractor is instead required to complete the works within a reasonable time.

In order for the Contractor to claim an EOT, such clauses often require it to have given adequate notice. The FIDIC provisions on notice are described in further detail at Section 1 (Sources of Liability). There are also similar notice requirements in respect of "compensation events" under the NEC standard form.[30] The rationale behind these notification provisions is usually to ensure that delay events are dealt with at the time they arise rather than several months or even years later.

At the time of notification or shortly thereafter, the Contractor needs to demonstrate that the delay event relied upon has or will impact the completion date. This will typically require input from specialist project planners to demonstrate with reference to project programs that the delay event has or will occur and has or will impact the project's critical path. To support its claim, it is also important for the Contractor to maintain adequate records documenting the delay, its impact, and any mitigation measures taken to prevent the delay.

The Society of Construction Law publishes a Delay and Disruption Protocol (the "**Protocol**")[31] that provides best practices for dealing with delay and disruption issues. Unless it incorporated into a contract or agreed by the parties, it is not a binding document but it provides a useful reference for some key principles and has received some positive approval from the courts.[32] For example, it provides guidance to parties on appropriate forms of delay analysis[33] as well as a comprehensive list of records to substantiate an EOT claim (including programs, progress reports, resource and cost records).[34]

The method of delay analysis to be used is sometimes expressly set out in the contract. For instance, the NEC standard form envisages that a prospective rather than retrospective form of delay analysis should be used for assessing EOT, requiring that the Project Manager assesses "the length of time that, due to

---

30 NEC4, Clause 61.3.

31 Society of Construction Law's Delay and Disruption Protocol, 2nd Edition, 2917.

32 See for example, the TCC in England *Mirant Asia-Pacific Construction (Hong Kong) Ltd v. Ove Arup and Partners International Ltd* [2007] EWHC 918 (TCC), paragraph 123 and the Australian courts in *Alstom Limited v. Yokogawa Australia Pty Ltd (No 7)* [2012] SASC 49; *Santos Ltd v. Fluor Australia Pty Ltd* [2017] QSC 153; *V601 Developments Pty Ltd v. Probuild Constructions (Aust) Pty Ltd* [2021] VSC 849. However, it is not to be taken as an absolute authority as recognised by the English courts in *Adyard Abu Dhabi v. SD Marine Services* [2011] EWHC 848 (Comm), paragraphs 289–290 and *Thomas Barnes & Sons Plc v. Blackburn with Darwen Borough Council* [2022] EWHC 2598 (TCC), paragraph 110.

33 The Protocol, Section 11.

34 The Protocol, Section 1.

the compensation event, planned completion is later than planned Completion shown on the Accepted Programme."[35]

Another area of debate in delay claims is what happens when two delay events occur at the same time, one of which is a Contractor risk and one of which is an Owner risk. In situations of so called "concurrent delay," should the Contractor be entitled to an extension of time? The Protocol suggests that a Contractor's concurrent delay should not reduce any entitlement to an extension of time due to an Owner risk event. The latest editions of the FIDIC Red and Yellow Books have added provisions to deal with concurrency stating that concurrency should be assessed in accordance with the rules and procedures stated in the Special Provisions or, if nothing is stated, as appropriate, taking due regard of all relevant circumstances.[36]

Claim for Loss and Expense

When a Contractor is delayed, it may also incur additional cost that may be due to:

- **Prolongation costs** – the unavoidable costs of being on site longer such as the cost of plant, labor, and maintaining site facilities for the longer contract period.
- **Disruption costs** – these may relate to having to do the work differently due to a delay event resulting in additional costs.

These costs are separate and distinct from those that may arise from the delay event itself. For example, if the Contractor is delayed due to the Owner instructing a variation, it will typically also be entitled to claim for that additional work. This is discussed in more detail at Section g (claims for additional works) below.

The term "loss and expense" comes from the JCT suite of contracts used in the UK and refers to the additional cost incurred by a Contractor "because regular progress of the Works, or any part of them has been or is likely to be materially affected by any Relevant Matter."[37] Under FIDIC contracts, the Contractor is entitled to additional payment where the direct loss or expense is a "Cost." Cost is defined as "all expenditure reasonably incurred (or to be incurred) by the Contractor in performing the Contract, whether on or off Site, including taxes, overheads and similar charges, but does not include profit."[38] Examples of where the Contractor can recover Cost include where there are archaeological or geological findings,[39] unforeseen physical conditions[40] or the Owner preventing required testing taking place.[41]

---

35 NEC4 ECC, Clause 63.5.

36 FIDIC Red and Yellow Books (2017), Sub-Clause 8.5.

37 JCT Standard Building Contract With Quantities SBC/Q 2024, clause 40.20.1.

38 FIDIC Red and Yellow Books (2017), Sub-Clause 1.1.19.

39 FIDIC Red and Yellow Books (2017), Sub-Clause 4.23.

40 FIDIC Red and Yellow Books (2017), Sub-Clause 4.12.

41 FIDIC Red and Yellow Books (2017), Sub-Clauses 7.4 and 10.3.

For the Contractor to also be able to claim Profit[42] under FIDIC contracts, the ability to claim this must be expressly set out. Examples of where the Contractor can claim Cost Plus Profit include where the Engineer fails to issue a notice or instruction,[43] the Owner fails to obtain the necessary permits[44] or the Owner fails to provide access to the site.[45]

Claim for Additional Works

Construction contracts will set out, often to varying degrees of detail, the type and amount of work required for a project. However, except in the most straightforward of projects, it is almost inevitable that some change will be required as works progress. Examples of where the need for change may arise might include because the Owner's needs change or it becomes apparent that the agreed works cannot be undertaken.

That change or modification to the original scope of work will be dealt with as a "variation" and in standard form contracts there is typically a variation procedure whereby the Owner may instruct variations (usually via its contract administrator). If the variation leads to additional work (as opposed to omitting works), the Contractor is usually entitled to an EOT and additional payment. Some contracts also make express provision for how such variations are to be valued.

However, not all instructions will amount to variations and whether or not an instruction amounts to a change to the original scope of works and, if it is a change, its impact on time and cost are common areas of dispute.

Under the FIDIC Red Book, a Variation is "any change to the Works, which is instructed as a Variation under clause 13."[46] In the Yellow[47] and Silver[48] Books, it also includes a change to the Employer's Requirements. Sub-Clause 13.1 of the Red Book provides a nonexhaustive list of what constitutes a Variation including:

a. Changes to the quantities of any item of work;
b. Changes to the quality of the work;
c. Changes to the levels, positions and/or dimensions of any part of the Works;

42 Profit is a preagreed percentage set out in the Contract Data with a default of 5% (see FIDIC Red and Yellow Books (2017), Sub-Clause 1.1.20).

43 FIDIC Red Book (2017), Sub Clause 1.9.

44 FIDIC Red and Yellow Books (2017), Sub-Clause 1.13.

45 FIDIC Red and Yellow Books (2017), Sub-Clause 2.1.

46 FIDIC Red Book (2017), Sub-Clause 1.1.86.

47 FIDIC Yellow Book (2017), Sub-Clause 1.1.88.

48 FIDIC Silver Book (2017), Sub-Clause 1.1.78.

d. The omission of any work, unless it is to be caried out by others without the agreement of the Parties;

e. Any additional work, Plant, Materials or services necessary for the Permanent Works, including any testing; or

f. Changes to the sequence or timing of the execution of the Works.

There is no equivalent list in the Yellow and Silver Books although both expressly states that omissions are not allowed where the work is to be given to another party unless agreed by the Owner and the Contractor.

Sub-Clause 13.1 of the Red Book also provides that:

> The Contractor shall be bound by each Variation instructed under Sub-Clause 13.3.1 [Variation by Instruction], and shall execute the Variation with due expedition and without delay, unless the Contractor promptly gives a Notice to the Engineer stating (with detailed supporting particulars) that:
>
> **a.** the varied work was Unforeseeable having regard to the scope and nature of the Works described in the Specification;
>
> **b.** the Contractor cannot readily obtain the Goods required for the Variation; or
>
> **c.** it will adversely affect the Contractor's ability to comply with Sub-Clause 4.8 [Health and Safety Obligations] and/or Sub-Clause 4.18 [Protection of the Environment].

The varied work will be "Unforeseeable" if "not reasonably foreseeable by an experienced contractor by the Base Date."[49] The Yellow Book and Silver Book contain the same provisions, with the addition of the following:

> **d.** it will have an adverse impact on the achievement of the Schedule of Performance Guarantees; or
>
> **e.** it may adversely affect the Contractor's obligation to complete the Works so that they shall be fit for the purpose(s) for which they are intended under Sub-Clause 4.1 [Contractor's General Obligations].

After receiving the Notice:

> the Engineer shall respond by giving a Notice to the Contractor cancelling, confirming or varying the instruction. Any instruction so confirmed or varied shall be taken as an instruction under Sub-Clause 13.3.1 [Variation by instruction].

---

49 FIDIC Red Book (2017), Sub-Clause 1.1.87.

Sub-Clause 13.3 provides that an Engineer must initiate a Variation by (1) an instruction; or (2) a request for proposal. This raises the question of whether a Contractor is entitled to an EOT or cost if a variation is not initiated under either of these two procedures. Under English law, the decision Supreme Court decision in *Rock Advertising*[50] suggests that parties are bound by the variation procedure selected. Whereas other common law jurisdictions seem to suggest that party autonomy allows parties to modify their contract irrespective of the contractual procedure.[51] In any event, it is unlikely that an Owner would be able to hide behind the contractual procedures if the Contractor has relied on conduct suggesting that the Owner otherwise accepts the variation.[52]

In terms of valuing the Variation, where it is done by a request for proposal, it is likely that the parties will have agreed a value before proceeding. Where the variation arises from an instruction, there may be more scope for disagreement. The Contractor does not have to make a claim under sub-clause 20 for its additional time or money. Instead, the Engineer is to make a determination and if the Contractor disagrees with that determination, it can refer the matter to dispute resolution.

For additional time, the Engineer is to fix a new completion date "if and to the extent that completion is or will be delayed by the Variation."[53] For additional cost, under the Red Book, the Engineer is to value the Variation as the original works were priced.[54] Under the Yellow and Silver Books, the additional cost is to be determined with reference to the Schedule of Rates and Prices, failing which the rate or price specified for similar work. If there are no comparable rates, the fallback is to determine value with reference to Cost plus Profit.[55]

Claim for Acceleration or Slowing Down

As described at Section f (Claims for Delay) above, the Contractor generally takes the risk on completing the works by the completion date. If the Contractor misses the completion date and is not entitled to an EOT, it is likely to be liable to pay the Owner liquidated damages. Generally, it is for the Contractor to plan its works to meet the completion date. If its progress is slowed due to an Owner risk event, it may claim for an EOT or disruption (see Section j (claim for subsoil conditions) below)

---

50 *Rock Advertising Ltd v. MWB Business Exchange Centres Ltd* [2018] UKSC 24, paragraph 10.

51 J Glover and S Hughes, Understanding the FIDIC Red and Yellow Books, (2018), Sweet & Maxwell, paragraph 13-005.

52 *Rock Advertising Ltd v. MWB Business Exchange Centres Ltd* [2018] UKSC 24, paragraph 13.

53 FIDIC Red, Yellow and Silver Book (2017), Sub-Clause 8.5.

54 FIDIC Red Book (2017), Sub-Clause 12.3.

55 FIDIC Yellow and Silver Books (2017), Sub-Clause 13.3.

or perhaps additional cost for a Variation to the contract. The Contractor's ability to claim will depend on the terms of the contract.

However, there may also be circumstances in which the Owner wishes the Contractor to speed up its progress, giving rise to what is known as a claim for "acceleration." The Protocol describes acceleration as:

> The application of additional resources or alternative construction sequences or methodologies seeking to achieve the planned scope of work in a shorter time than planned or execution of additional scope of work within the original planned duration.[56]

Sub-Clause 8.7 of the FIDIC Red, Yellow, and Silver Books provides that:

> If at any time:
>
> **a.** actual progress is too slow to complete the Works or a Section (if any) with the relevant Time for Completion; and/or
> **b.** progress has fallen (or will fall) behind the Programme (or the initial programme if it has not yet become the Programme) under Sub-clause 8.3 9 Programme.
>
> other than as a result of a cause listed in Sub-Clause 8.5 [Extension of Time for Completion], then the Engineer may instruct the Contractor to submit... a revised programme describing the revised methods which the Contractor proposes to adopt in order to expedite progress and complete the Works or a Section (if any) within the relevant Time for Completion...
>
> the Contractor shall adopt these revised methods, which may require increases in the working hours and/or numbers of Contractor's Personnel and/or the Goods, at the Contractor's risk and cost...
>
> Sub-Clause 13.3.1 [Variation by Instruction] shall apply to revised methods, including acceleration measures instructed by the Engineer to reduce delays resulting from causes listed under Sub-Clause 8.5 [Extension of Time for Completion].

Bearing in mind, the Contractor's obligation to complete by the completion date, it is not uncommon for disputes to arise as to whether accelerative measures are properly a variation or part of the Contractor's obligation to proceed with due expedition[57] and mitigate delays to the Works. For that reason, Contractors often

---

56 The Society of Construction Law, *Delay and Disruption Protocol* (Second Edition), February 2017, Appendix A.

57 FIDIC Red, Yellow and Silver Books (2017), Sub-Clause 8.1.

prefer to have any agreed accelerative measures and corresponding payment documented including provision as to what is to happen to any delays up to the point of the acceleration instruction or future delays to the project.

Where the contract does not provide for acceleration, the Protocol recommends that the basis of payment should be agreed before the acceleration is commenced and that parties should agree the records that are to be kept when acceleration measures are employed.[58]

Claim for Disruption

The Protocol defines "disruption" as:

> disturbance, hindrance or interruption to a Contractor's normal working methods, resulting in lower productivity or efficiency in the execution of particular work activities.[59]

Where disruption occurs, it results in the disrupted work activities costing more to complete reducing the profitability of the works for the Contractor. The Protocol goes on:

> "Disruption" is concerned with an analysis of the productivity of work activities, irrespective of whether those activities are on the critical path to completion of the works.[60]

Disruption is distinct from delay (covered at section f (claims for delay) above) as it does not necessarily lead to a delay to the completion date i.e., disruption can also affect noncritical activities. In that way, disruption does not entitle a Contractor to claim an EOT but it may entitle the Contractor to claim additional costs, depending on the terms of the contract.

Disruption claims can be difficult to prove because they require the Contractor to demonstrate that there was disruption to its activities caused by an event which was the Owner's risk or another breach of contract by the Owner and that the loss claimed flows from that risk/breach. Typically, that requires good records showing the way the works were supposed to be carried out as against how they

---

58 The Society of Construction Law, *Delay and Disruption Protocol* (Second Edition), February 2017, Core Principle 16.

59 The Society of Construction Law, *Delay and Disruption Protocol* (Second Edition), February 2017, Guidance Part A, paragraph 5.

60 The Society of Construction Law, *Delay and Disruption Protocol* (Second Edition), February 2017, Guidance Part A, paragraph 5.

are progressing/have progressed and proving that causal link. The English courts have said there is no set way for Contractors to prove disruption so long as it meets the requisite standard of proof.[61]

Claim for Subsoil Conditions

Subsoil conditions are by their nature hidden and for that reason, it is common for initial investigations to be undertaken into the subsoil conditions on site prior to the Owner issuing a tender for works. A Contractor will then price the works based on those initial investigations. Further investigations may then be undertaken by the Contractor during the tender process, or, more usually, after the Contract has been awarded (to save time and cost during the tender stage).

If adverse subsoil conditions are identified after works commence, further investigations may have to be undertaken, the design for the works may need to be revised and the Contractor may have to alter its methods of working all of which can lead to delays and additional cost. Who bears the risk for those delays and costs will depend on the risk allocation under the contract.

Contracts often allocate this risk depending on what was foreseeable by the parties at the time of entering the contract. Sub-Clause 4.12 of the FIDIC Red and Yellow Books[62] provides that the Contractor shall be entitled to EOT and/or Cost if it "encounters Physical Conditions which the Contractor considers to have been Unforeseeable and that will have an adverse effect on the progress and/or increase the Cost of the execution of the Works."

"Physical Conditions" are defined as "natural physical conditions and physical obstructions (natural or man-made) and pollutants, which the Contractor encounters at the Site during the execution of the Works, including sub-surface and hydrological conditions but excluding climatic conditions at the Site and the effects of those climatic conditions." "Unforeseeable" means "not reasonably foreseeable by an experienced contractor by the Base Date."

Sub-Clause 2.5 of FIDIC[63] provides that "the Employer shall have made available to the Contractor for information, before the Base Date, all relevant data in the Employer's possession on the topography of the Site and on sub-surface, hydrological, climatic and environmental conditions at the Site. The Employer shall promptly make available to the Contractor all such data which comes into the Employer's possession after the Base Date."

The Contractor's claim for additional time and cost is all subject to the notification provisions set out at section 1 (sources of liability) above.

---

61 *Walter Lilly & Co Ltd v. DMW Developments Ltd* [2012] EWHC 1773 (TCC).

62 FIDIC Red and Yellow Books (2017).

63 FIDIC Red and Yellow Books (2017).

The FIDIC Sliver Book,[64] on the other hand, provides that the Contractor takes "total responsibility for having foreseen all difficulties and costs of successfully completing the Works" and is not entitled to any additional cost for any "Unforeseeable or unforeseen difficulties or costs" unless stated in the Particular Conditions.

A similar approach to the FIDIC Red and Yellow Books is taken under NEC4, there the Contractor is entitled to a compensation event if it "encounters physical conditions which are within the site" that "an experienced contractor would have judged at the Contract Date to have such a small chance of occurring that it would have been unreasonable to have allowed from them."[65] Therefore under both forms, disputes often center around what it was reasonable for the Contractor to have assumed from the contract / pre contract information.[66]

If the contract is silent on the risk allocation, the default position under English law is that the Contractor bears the risk.[67]

Claim for Adverse Physical Conditions

Aside from subsoil conditions there may be a variety of other physical conditions that the Contractor encounters that may impact the cost or time it takes to do the works. As with subsoil conditions, the contract should provide whether the Contractor is entitled to an EOT or cost in respect of such conditions.

Another physical condition that it is common for construction contracts to provide for is adverse weather. Under FIDIC Red and Yellow Books, the Contractor is entitled to an EOT (but not additional cost), if completion will be delayed by "exceptionally adverse climatic conditions, which for the purposes of these conditions shall mean adverse climatic conditions at the Site which are Unforeseeable having regard to the climatic data made available by the Employer under Sub-Clause 2.5 and/or climatic data published in the Country for the geographical location of the Site."[68] Under the FIDIC Silver book, the Contractor takes the risk for adverse weather.

As noted in Section k (Claim for Adverse Physical Conditions), under Sub-Clause 2.5, the Owner is required to provide "all relevant data in the [Owner's] possession on the climatic and environmental conditions at the Site."

64 FIDIC Silver Book (2017), Sub-Clause 4.12.

65 NEC4 ECC, Clause 60.1(12).

66 See for example *Obrascon Huarte v. Attorney General for Gibraltar* [2015] EWCA 712 and *Van Oord UK Ltd v. Allseas UK Ltd* [2015] EWHC 3074.

67 Chitty on Contract, 2023, Thomson Reuters, paragraph 40–250.

68 FIDIC Red and Yellow Books (2017), Sub-Clause 8.5.

There is no definition of "exceptionally adverse weather" under FIDIC. In some instances, it may be obvious where the weather conditions are well beyond what could normally be expected such as extreme flash flooding. However, opinions could differ over heavy snow, a prolonged cold snap or strong winds leading to disputes between the Owner and Contractor.

The NEC4 ECC standard form provides a more prescriptive definition of adverse weather, entitling the Contractor to a compensation event where weather occurs which is "shown to occur on average less frequently than once in ten years" with reference to the weather data provided in the contract.[69] In assessing the Contractor's entitlement to an EOT or additional cost, the Project Manager is only to take into account "the difference between the weather measurement and the weather which the weather data show to occur on average less frequently than once in ten years."

The weather measurements typically recorded in the contract data are:

- The cumulative rainfall (mm)
- The number of days with rainfall more than 5mm
- The number of days with minimum air temperature less that 0 degrees Celsius
- The number of days with snow lying at [x] hours

The contract data also encourages parties to specify the source and location of where the measurements were taken. This approach allows parties a greater deal of certainty than the approach taken by FIDIC. However, even if such adverse occurs, the Contractor must still demonstrate that those conditions have affected the works. That may not be the case if the critical path in running through, for example, internal fit out works.

---

69 NEC4 ECC, Clause 60.1(13).

19

Investment Treaty Claims by the Contractor

*Julien Fouret, Camille Dupuy, and Joy Harb**

Introduction

Investment arbitration, often referred to as Investor-State Dispute Settlement (ISDS), is a procedure designed to resolve conflicts between foreign investors and host States.

The right for a natural or legal person to take legal action against a host State is enshrined in a vast network of bilateral and multilateral investment treaties, as well as international agreements and Free Trade Agreement. These instruments codify the current legal framework for the protection and promotion of foreign investment. They require fair and equitable treatment, ensure the protection and security of investments, and prohibit discrimination or expropriation without compensation for foreign investors from signatory States who have made qualifying investments within the territory of another signatory State (the "host State").

In today's rapidly expanding and developing economies, cross-border contracts are multiplying, and complex construction projects are increasingly being implemented, exposing contractors to various risks. Statistics show that out of 1332 investment arbitration cases, 163 involve the construction sector.[1]

In this regard, *Holiday Inns SA & Others v. Morocco*,[2] dispute – popularly dubbed the first World Bank arbitration[3] was the first request for arbitration under the

* Respectively Partner and Global Co-Head of International Arbitration, Senior Associate and Associate, Holman Fenwick Willan (HFW), Paris. The authors want to thank Joëlle Estephan for her assistance in the preparation of this chapter.

1 UNCTAD Investment Policy Hub, Investment Dispute Settlement Navigator, consulted on 8 October 2024, https://investmentpolicy.unctad.org/investment-dispute-settlement.

2 *Holiday Inns S.A. and others v. Morocco*, ICSID Case No. ARB/72/1.

3 Pierre Lalive, "The First 'World Bank' Arbitration (*Holiday Inns v. Morocco*): Some Legal Problems" (1980) 51:1 BYBIL 123–62.

Major Issues in International Construction Law, First Edition. Mauro Rubino-Sammartano and Troy Harris. Published 2026 by John Wiley & Sons, Inc.

ICSID Convention. Registered in 1972, this case revolves around the financing of hotels construction, where the Moroccan government halted the required payments and declined to fulfill the promised incentives. It did not result in a final award on the merits, as the parties eventually settled the dispute after the tribunal issued a decision on jurisdiction and provisional measures. It is however the first decision in the ICSID history confirming that such construction disputes fell within the scope of the ICSID Convention.

This chapter will explore the legal framework and mechanisms through which contractors, or investors in the construction sector, may bring investment treaty claims against States. It will address key jurisdictional considerations, available fora for dispute resolution, and the applicable standards of protection of contractors. Particular focus will be given to the unique features of investment disputes arising in the construction sector, particularly in Libya, a country that faced the largest number of investment arbitration in the construction sector.

Actions and Forum Available to Contractors

In the realm of construction projects, contractors often face challenges that require quick legal recourse. Dispute resolution mechanisms, especially in international projects, provide contractors with multiple avenues for addressing these issues. Below is an overview of some of the key actions and fora available to contractors facing disputes in this sector.

Most Multilateral Investment Treaties (MITs) and Bilateral Investment Treaties (BITs) offer a mechanism for resolving disputes between investors and States, known as investor-state dispute settlement (ISDS). These treaties allow arbitration either under the aegis of ICSID (an entity of the World Bank) or in accordance with the rules of the United Nations Commission for International Trade Law (UNCITRAL).

However, in cases where a BIT does not include an ICSID clause or, if either the host State or the contractor's home State is not a party to the ICSID Convention, contractors can pursue alternative fora, including ad hoc arbitrations, the Permanent Court of Arbitration (PCA), or ICC arbitration. For context, out of the 163 treaty arbitrations in the construction sector, 119 were administered by ICSID, 17 by the PCA, and 12 by the ICC.

It is also worth noting that not all States have embraced ISDS. Several States have avoided concluding international treaties that provide for any form of investor-state dispute resolution. For example, Brazil has opted for a different approach in recent years, preferring to establish cooperation and facilitation investment agreements rather than the more common ISDS model found in other treaties. Moreover, in certain regions, ISDS has lost its appeal. Investment

Protection Agreements (IPAs) and their ISDS provisions are sometimes perceived as offering excessive protections to investors, potentially hindering States, including EU Member States, from regulating in favor of the environment, public health, or public services.

In the European Union (EU), there has been a reassessment of dispute resolution options for investors, steering them away from the traditional ISDS mechanisms found in BITs. The EU has highlighted alternative forums, such as national courts and the EU's own judicial system, which provide a framework for resolving investment disputes while ensuring adherence to EU law. The Court of Justice of the European Union (CJEU) 2018 ruling in the *Achmea v. Slovakia* case represents a landmark decision in the field of Investor-State Dispute Settlement (ISDS) within European Union law. In this case, the CJEU concluded that the ISDS mechanism is incompatible with EU law. The dispute, stemmed from an arbitration clause contained in a BIT between the Netherlands and Slovakia that allowed a Dutch investor to initiate arbitration against Slovakia. The Court found this clause violated Articles 267 and 344 of the Treaty on the Functioning of the European Union (TFEU), asserting that EU law must take precedence over external arbitration mechanisms.[4]

The *Achmea* judgment was further reinforced in the *Moldova v. Komstroy* case, where the CJEU ruled that the ISDS mechanism under Article 26 of the ECT was also incompatible with EU law when applied to intra-EU disputes. This decision further strengthens the CJEU's stance on the autonomy of EU law and restricts arbitration between EU Member States and investors from other Member States under the ECT.[5]

In contrast, several arbitral tribunals, including those in *Masdar Solar v. Spain*[6] and *Vattenfall v. Germany*,[7] have contested the applicability of *Achmea* to the ECT, arguing that multilateral agreements like the ECT should not be affected by EU law in the same way as BITs. These tribunals applied the principle that EU law should not override international treaties like the ECT. For instance, in *Vattenfall v. Germany*, the tribunal concluded that Achmea did not invalidate the ECT's arbitration clause, citing the treaty's specific regime and the principle that a more favorable regime for investors, such as the ECT, should prevail.

---

4 *Achmea B.V. v. The Slovak Republic*, Court of Justice Case C 284/16, Judgment, 6 March 2018, para. 60.

5 Laura Roddy, Spain Asks for ECJ to Rule on ECT, Global Arbitration Review, 31 May 2018.

6 *Masdar Solar & Wind Cooperatief U.A. v. Kingdom of Spain*, ICSID Case No. ARB/14/1, Award, 16 May 2018, para. 679.

7 *Vattenfall AB and Others v. Federal Republic of Germany* (II), ICSID Case No. ARB 12/12, Decision on the Achmea Issue, 31 August 2028.

National courts have also had varied responses. For example, in *Poland v. LC Corp*[8] and *Infrastructure Services Luxembourg v. Spain*,[9] courts in the Netherlands and England have taken positions that downplay the impact of Achmea, prioritizing international obligations under treaties like the ICSID Convention over EU law.

Jurisdictional Considerations

The first step when contemplating to bring an investment claim against a State is to consider any possible jurisdictional hurdle. The most common jurisdictional hurdles to consider are to ensure that the contractor and the project can benefit from investment protection (1) and that the contractor can enforce investment protection against a State (2).

Can the Contractor and the Project Benefit from Investment Protection?

To be entitled to benefit from investment protection, a contractor must qualify as an investor (a) and the project must qualify as an investment (b) both under the terms of a BIT and, where applicable, the ICSID Convention.

The Contractor Must Qualify as an Investor

To benefit from investment protection and for an arbitral tribunal to uphold jurisdiction *ratione personae*, a contractor must qualify as an investor as defined in the BIT on which it intends to base its claim. In addition, when ICSID is the designated institution under the BIT, the contractor must also comply with the criteria set in Article 25 of the ICSID Convention for the arbitral tribunal to uphold jurisdiction on the claim.[10]

Although each BIT can include its own definition of "investor," it is generally accepted that an investor is a natural or a legal person having the nationality of one of the two States party to the BIT and who has made an investment in

8 *LC Corp B.V. v. Republic of Poland*, Judgment of the Amsterdam District Court, 8 March 2023.

9 *Infrastructure Services Luxembourg (Antin) v. Kingdom of Spain*, ICSID Case No. ARB/13/31, Judgment of the UK High Court of Justice, 27 January 2023.

10 For a complete analysis of Article 25, see the chapters by Y. Banifatemi and E. Edson (25(1)), S. Fietta and S. Macrory (25(2)(A)), D. Caron and E. Shirlow (25(2)(B)), A. de Nanteuil (25(3)) and G.M. Alvarez and I. Mira (25(4)) in J. Fouret, R. Gerbay, G.M. Alvarez (eds.), *The ICSID Convention, Regulations and Rules – A practical Commentary, Elgar Commentaries*, Edward Elgar Publishing, 2019, at 102–227.

the other State party (*i.e.*, the host State). Article 25(1) of the ICSID Convention defines an investor as "national of another Contracting State." It results that the key issue to qualify as an investor is nationality.

For a natural person, the main discussion revolves around the status of binational. In this regard, Article 25(2)(a) of the ICSID Convention set the principle that its jurisdiction does not extend to a "person who [...] also had the nationality of the Contracting State party to the dispute." For instance, in the *Mihaljevic v. Croatia* arbitration based on the Germany-Croatia BIT, the ICSID tribunal found that the requirements for jurisdiction *ratione personae* under Article 25(1) of the ICSID Convention are not met considering "the Claimant was a dual citizen of Croatia and Germany on the relevant dates for assessing nationality under Article 25(2)(a) of the ICSID Convention, and the Claimant has not discharged his burden to prove that he had relinquished his Croatian citizenship at any time prior to 18 May 2020." The tribunal ruled that the "ICSID jurisdiction is therefore precluded, and the present case should be dismissed."[11]

In a UNCITRAL case, *Serafín García Armas and García Gruber v. Venezuela*, the arbitral tribunal upheld its jurisdiction over claims by dual nationals against Venezuela, on the basis that the underlying BIT was silent on the issue of dual nationality, and this case was not subject to the ICSID convention.[12] This decision was subsequently confirmed by the Paris Court of Appeal.[13]

For a legal person, BITs contain different criteria or tests to identify the nationality. Three main tests are used: 1. incorporation, 2. main seat, 3. control.[14] The incorporation test is used in the vast majority of investment treaties. It implies that a legal entity meets the nationality definition if it is incorporated or constituted under the laws of one of the contracting parties.

In the *Cairn v. India* arbitration, the tribunal found that it had jurisdiction *ratione personae* as the claimants are companies incorporated in the United Kingdom, and thus qualify as "investors."[15] In other BITs, to be considered as an investor, the legal entity should have its main seat or "*siege social*" in the territory of one of the contracting parties. In a construction case, *Alverley v. Romania*, the ICSID

11 *Marko Mihaljevic v. Republic of Croatia*, ICSID Case No. ARB/19/35, Award, 19 May 2023, para. 139.

12 *Serafín García Armas and Karina García Gruber v. The Bolivarian Republic of Venezuela*, PCA Case No. 2013-03, Decision on Jurisdiction, 15 December 2014.

13 *Serafín García Armas and Karina García Gruber v. The Bolivarian Republic of Venezuela*, Judgment of the Paris Court of Appeal (Department 5 Chamber 16, 22/02752), 27 June 2023.

14 Yannaca-Small K. "Who is Entitled to Claim?," in *Arbitration Under International Investment Agreements*, ed. Yannaca-Small K. (2018), Second Edition, para. 10.38.

15 *Cairn Energy PLC and Cairn UK Holdings Limited v. The Republic of India*, PCA Case No. 2016-07, Final Award, 21 December 2020, para. 668.

tribunal agreed with the analysis made by the tribunal in the *Tenaris v. Venezuela* case,[16] concluding that *siege social* meant "the effective management or some sort of actual or genuine corporate activity."[17]

Finally, the control test is also widely referred to by arbitral tribunals especially when analyzing the incorporation test. As such, it is mostly used as an additional requirement combined with the incorporation test. For instance, in the *Pawlowski v. Czech Republic*, an ISDS construction arbitration, the tribunal explained that for the claimant to qualify as a protected investor under the applicable BIT, it must (i) be a legal entity established under the law of Switzerland; and (ii) be controlled, directly or indirectly, by nationals of Switzerland, or legal entities which are seated and have real economic activities in Switzerland.[18]

The Project Must Qualify as an Investment

To benefit from investment protection and for an arbitral tribunal to uphold jurisdiction *ratione materiae*, a project must qualify as an investment as defined in the BIT on which the contractor intends to base its claim. There is no uniform definition of the term "*investment*." BITs may contain different definitions or list of types of investments they protect. For example, Article 1(2)(b) of the Libya-Turkey BIT reads: "The term investment, in conformity with the hosting Contracting Party's laws and regulations, shall include every kind of asset in particular, but not exclusively: (...) (b) returns reinvested, claims to money or any other rights having financial value related to an investment."[19] On the basis of this BIT, an arbitral tribunal found that it had jurisdiction *ratione materiae* "because the Settlement Agreement is a claim to money related to an investment as stipulated in Article 1(2)(b) of the BIT."[20]

When the term "*investment*" is not defined in the BIT, most arbitral tribunals employ the *Salini* test, or an adapted version of it, to determine whether an alleged investment constitutes a protected investment. This test was said to be established by the tribunal in *Salini v. Morocco*[21] and requires that to be considered

16 *Tenaris S.A. and Talta – Trading e Marketing Sociedade Unipessoal Lda. v. Bolivarian Republic of Venezuela* (I), ICSID Case No. ARB/11/26, Award, 29 January 2016, paras 150–154.

17 *Alverley Investments Limited and Germen Properties Ltd v. Romania*, ICSID Case No. ARB/18/30, Award (Excerpts), 16 March 2022, paras 236–237.

18 *Pawlowski AG and Project Sever s.r.o. v. Czech Republic*, ICSID Case No. ARB/17/11, Award, 1 November 2021, paras 219–222.

19 Libya-Turkey BIT, Article 1(2)(b).

20 *Etrak İnşaat Taahut ve Ticaret Anonim Sirketi v. The State of Libya*, ICC Case No. 22236/ZF/AYZ, Final Award, 22 July 2019, para. 167.

21 It is actually a doctrinal analysis made by Emmanuel Gaillard and re-used by Salini and its followers, see notably J. Fouret, La notion d'investissement, le réalisme objectif du "Test Gaillard," in Liber Amicorum Emmanuel Gaillard, Brill Martinus Nijhoff, Forthcoming 2025.

an investment under Article 25(1) of ICSID Convention, the alleged investment should satisfy four criteria: (i) a contribution, (ii) a certain duration, (iii) a risk, and possibly (iv) a contribution to the economic development of the host State.[22] Non-ICSID tribunals have also applied the *Salini* test to determine whether an alleged investment constitutes an investment under the applicable investment treaty.[23]

Can Contractors Enforce Investment Protections Against a State?

The other side of the coin of jurisdiction *ratione personae* lies on the State's side. For an arbitral tribunal to uphold jurisdiction over an investment treaty claim, the litigious action must have been taken by the State in the exercise of its sovereign authority (a) or if it is taken by a state entity, the litigious action must be attributable to the State (b).

Actions Taken by a State in the Exercise of Its Sovereign Authority

To amount to a treaty violation and therefore trigger the jurisdiction of an investment arbitration tribunal, an action causing damage to an investment must have been taken by the State in the exercise of its sovereign authority. It means that for an arbitral tribunal to have jurisdiction, (i) the claim must first arise under the treaty itself, and (ii) the State must have exercised its sovereign powers, not merely acted in a commercial capacity.

Treaty-based Claims as a Precondition for Arbitral Tribunal Jurisdiction

When an International Investment Agreement (IIA) or an international Treaty is in place, it forces the host State to honor its commitments toward investors and their investments. A State that violates its international obligations in the exercise of its sovereignty will thus be subject to arbitration before an arbitral tribunal designated in the relevant treaty. Consequently, an investment arbitral tribunal has no jurisdiction over a purely contractual claim.[24]

---

22 *Salini Costruttori S.p.A. and Italstrade S.p.A. v. Kingdom of Morocco*, ICSID Case No. ARB/00/4, Decision on Jurisdiction, 23 July 2001, para. 52. See also the analysis of the use and evolution of such test in Gaillard E. and Banifatemi Y. "Chapter 8: The Long March Towards a Jurisprudence Constante on the Notion of Investment," in *Building International Investment Law: The First 50 Years of ICSID, Kluwer Law International: Kluwer Law International*, eds. Meg Kinnear, Geraldine R. Fischer, et al., (2015), 97–126.

23 *Öztaş Construction, Construction Materials Trading Inc. v. State of Libya*, ICC Case No. 21603/ZF/AYZ, Award, 14 June 2018, paras 116–120.

24 *Garanti Koza LLP v. Turkmenistan*, ICSID Case No. ARB/11/20, Award, 19 December 2016.

However, there are also circumstances in which a breach of contract may constitute a violation of an international obligation. Such circumstances arise, for example, if the failure to fulfill a contractual obligation violates the State's duty to observe commitments made (commonly referred to as an "*umbrella clause*").[25] In addition, a dispute resolution provision in a Treaty could grant an arbitral tribunal jurisdiction over disputes related to breaches of contract, regardless of whether those breaches also amount to violations of the BIT itself. This could occur, for example, through a broadly phrased dispute resolution clause that extends to "any disputes related to investments." In such a case, arbitral tribunals have considered in the past that they had the authority to hear both treaty-based claims and contract-based claims, owing to the comprehensive consent clause found in the BIT.[26]

State's Exercise of Sovereign Power as a Requirement for Tribunal Jurisdiction

Furthermore, for an arbitral tribunal to establish jurisdiction under an investment treaty, the actions in question must demonstrate the State's exercise of sovereign power. Investment treaty protections do not apply to ordinary commercial disputes between private parties; they are triggered when the State acts in its public capacity and exercises its public authority. The investor must therefore demonstrate that the State's actions were carried out in its sovereign capacity rather than as a private entity. In the construction context, such claims arise when the owner is a State or a state-controlled entity, or when the State, acting outside the construction contract, takes measures that adversely affect the project, such as through expropriation.

Actions Attributable to the State

When a litigious action was not taken by the State itself, but by a State entity, it is necessary to establish that the action of this State entity is attributable to the State to trigger the jurisdiction of an investment arbitration tribunal.

In assessing the attribution of an action of a State entity to the State, arbitral tribunals rely on the ILC (International Law Commission) Draft Articles on Responsibility of States for Internationally Wrongful Acts, adopted on 9 June 2001, and mainly on the Chapter entitled "Attribution of conduct to a State." Among the

---

25 *Ortiz Construcciones y Proyectos S.A. v. People's Democratic Republic of Algeria*, ICSID Case No ARB/17/1, Award, 29 April 2020; see also Garanti Koza LLP v. Turkmenistan, ICSID Case No. ARB/11/20, Award, 19 December 2016; Strabag SE v. Libya, ICSID Case No. ARB(AF)/15/1, Award, 29 June 2020; Bosh International, Inc. and B&P, LTD Foreign Investments Enterprise v. Ukraine, ICSID Case ARB/08/11, Award, 25 October 2012.

26 *Kontinental Conseil Ingénierie v. Gabonese Republic*, PCA Case No. 2015–25, Award, 23 December 2016; see also Salini Costruttori S.p.A. and Italstrade S.p.A. v. Kingdom of Morocco, ICSID Case ARB/00/4, Decision on Jurisdiction, 16 July 2001.

circumstances contemplated in this Chapter II, the following are the most found by arbitral tribunals to be attributable to states in construction cases:

- The conduct of State organs, such as legislative, executive or judicial organs, but also public sector entities (Article 4 of the ILC Draft Articles)
- The conduct of persons or entities exercising governmental authority (Article 5 of the ILC Draft Articles)
- The conduct directed or controlled by a State (Article 8 of the ILC Draft Articles)
- The conduct acknowledged and adopted by a State as its own (Article 11 of the ILC Draft Articles).

For example, in the ICSID construction case *Pawlowski v. Czech Republic*, the tribunal found that "the Mayor of Benice represents an organ of the Czech Republic at a territorial level, and in accordance with Article 4 of the ILC Articles her conduct must be attributed to the Czech Republic" and concluded that as a result, "the Czech Republic has committed an internationally wrongful act in breach of the BIT."[27] In another construction case, the ICSID tribunal found that "each specific act allegedly in breach of the Treaty was a direct consequence of the decision of the NHA to terminate the Contract, which decision received express clearance from the Pakistani Government." On this basis, the tribunal concluded that "NHA's conduct is attributable to Pakistan under Article 8 of the ILC Articles."[28]

Standards of Protection

Preliminary Observations

Disputes brought by investors operating in the construction sector are becoming increasingly significant in Investor-State Dispute Settlement. Contractors are initiating a growing number of disputes, particularly alleging breaches by the host State of Fair and Equitable Treatment (FET) and indirect expropriation.

Specifically, FET has been alleged in 76 cases, while indirect expropriation has been alleged in 63 cases. Other frequently alleged standards include Full Protection and Security (FPS) (38 cases), the umbrella clause (31 cases),

---

27 *Pawlowski AG and Project Sever s.r.o. v. Czech Republic*, ICSID Case No. ARB/17/11, Award, 1 November 2021, para. 373.

28 *Bayındır İnşaat Turizm Ticaret ve Sanayi A.Ş. v. Islamic Republic of Pakistan* (I), ICSID Case No. ARB/03/29, Award, 27 August 2009, para. 125.

discriminatory measures (27 cases), Most-Favored-Nation (MFN) treatment (20 cases), and National Treatment (17 cases).[29]

Among disputes where investors succeeded, tribunals found breaches of FET in 18 cases and indirect expropriation in 8 cases. Across all ISDS construction cases, 45% involved FET and 37% involved indirect expropriation. The success rate for allegations of FET and indirect expropriation were 23% and 13%, respectively.

Standards of Protection for Investors

The most common standards of investment protection appearing in investment agreements include MFN, National Treatment, FET, FPS, compensation for expropriation, and the umbrella clause.

Fair and Equitable Treatment

FET is one of the most common standards of investment protection, included in most bilateral investment treaties and multilateral agreements. FET is a broad standard protecting the legitimate expectations of investors against unfair actions by the host State. It requires the host State to follow due process or fair procedure requirements, and not to act in a manner that is manifestly inconsistent, non-transparent, unreasonable, or discriminatory.[30]

On the one hand, the FET standard requires the host State to be transparent and to comply with some procedural guarantees. For instance, in *Metalclad v. United Mexican States*, the arbitral tribunal found Mexico liable for administrative misconduct as it did not give the investor an opportunity to defend itself before withdrawing the license.

On the other hand, the FET standard requires the host State to be consistent and predictable *i.e.*, to preserve the investors' legitimate expectations. For instance, in *Tecmed v. United Mexican States*, the arbitral tribunal found that "the foreign investor expects the host State to act consistently, i.e., without arbitrarily revoking any pre-existing decisions or permits issued by the State that were relied upon by the investor to assume its commitments as well as to plan and launch its commercial and business activities."[31]

29 UNCTAD, Investment Dispute Settlement Navigator Advanced Search | Investment Dispute Settlement Navigator | UNCTAD Investment Policy Hub accessed on 3 September 2025.

30 *Saluka Investments B.V. v. The Czech Republic*, UNCITRAL, Partial Award, 17 March 2006.

31 *Técnicas Medioambientales Tecmed S.A. v. The United Mexican States*, ICSID Case No. ARB(AF)/00/2, Award, 29 May 2003, para. 154.

In cases where investors had stabilization clauses protecting them from regulatory changes, tribunals have found FET violations when states changed regulations relied upon by the investors and adversely affecting their investments.[32]

In other cases where investors were not given specific stability guarantees, tribunals also ruled that the State was in breach of the FET obligation when adopting measures that violated the investor's reasonable and legitimate expectations. For instance, in *Sempra v. Argentine Republic*, the tribunal ruled that the host State was required to refrain from affecting the basic expectations that were taken into account by the foreign investor in making the investment, that is, from eviscerating the arrangements upon which the investor relied when deciding to invest.[33]

Nonetheless, arbitral tribunals are also conscious of the need to preserve the host State's right to regulate, as seen in *Saluka v. Czech Republic* where the tribunal stated that "no investor may reasonably expect that the circumstances prevailing at the time the investment is made remain totally unchanged."[34]

Expropriation

Expropriation involves a State's act of taking or destroying rights acquired by a foreign investor. Under international law, expropriation is not by itself unlawful. In accordance with the territorial sovereignty, the State has the right to expropriate property for several reasons, such as political, economic, social etc. However, for it to be lawful, expropriation must be carried out on a lawful basis.

Expropriation is intrinsically lawful as long as certain conditions outlined by international investment treaties are met: (i) that the expropriation be made in the public interest, (ii) that it be nondiscriminatory, (iii) that due process is followed, and (iv) that the expropriation be accompanied by prompt, adequate, and effective compensation.[35]

---

32 *Sergei Paushok, CJSC Golden East Company & CJSC Vostokneftegaz Company v. The Government of Mongolia*, UNCITRAL, Award on Jurisdiction and Liability, 28 April 2011, para. 302; *AES Summit Generation Ltd. & AES-Tisza Erömü Kft v. The Republic of Hungary*, ICSID Case No. ARB/07/22, Award, 23 September 2010; para. 9.3.25; Total S.A. v. The Argentine Republic, ICSID Case No. ARB/04/01, Award, 27 November 2013, para. 309.

33 *Sempra Energy Int'l v. The Argentine Republic*, ICSID Case No. ARB/02/16, Award, 28 September 2007; *CMS Gas Transmission Co. v. The Argentine Republic*, ICSID Case No. ARB/01/8, Award, 12 May 2005, para. 277; *Enron Corp. v. The Argentine Republic*, ICSID Case No. ARB/01/3, Award, 22 May 2007, para. 262.

34 *Saluka Investments v. The Czech Republic*, UNCITRAL, Award, 17 March 2006, para. 305; see also *Lahoud v. Democratic Republic of the Congo*, ICSID Case No. ARB/10/4, Award, 7 February 2014, para. 441; *EDF (Services) Ltd. v. Romania*, ICSID Case No. ARB/05/13, Award, 8 October 2009, para. 217; *Total S.A. v. The Argentine Republic*, ICSID Case No. ARB/04/01, Decision on Liability, 27 December 2010, para. 164.

35 For example, Energy Charter Treaty (ECT), Article 13; Organisation of Islamic Cooperation Investment Agreement, Article 10; Ireland-Czeck Republic BIT, Article 5; Cyprus-Hungary BIT, Article 4; Egypt-Finland BIT, Article 3.

Expropriation can be direct or indirect. Direct expropriation is easy to recognize; it refers to the transfer of title and physical taking or seizure of property by a State, thus depriving the investor of its property. Direct expropriation has become rare and the typical form in which expropriations take place nowadays is indirect expropriations.[36]

Indirect expropriation is more subtle and involves measures that deprive the investor of their property rights without transferring legal title *per se*. Indirect expropriation results in the total or near-total deprivation of an investment that renders the investor's property rights useless. This can arise, for example, through increases in regulation that substantially deprive the investor of the ability to use its property.[37]

In construction cases, expropriation can arise through the direct taking of land, premises, or machinery; through revocation of necessary permits, licenses, or agreements;[38] or through the removal or effective termination of contractual rights.[39] For instance, in *Metalclad Corp. v. Mexico*,[40] the tribunal held that Mexico has indirectly expropriated the claimant's investment in a hazardous waste landfill and that the Municipalities' denial of a construction permit without any legal basis and subsequent administrative actions unlawfully prevented the investor's operation of the landfill. In another construction case, *ADC Affiliate Ltd and others v. Hungary*,[41] the tribunal found that Hungary's legislative measures prevented the investor from operating its investment and collecting the associated revenues and thus constituted expropriation.

Most-favored-nation Treatment

MFN clauses are included in most investment treaties. The purpose of MFN clauses is to ensure equality and nondiscrimination between foreign investors. They require States to treat investors no less favorably than investors from third

---

36 Schreuer, "Chapter 3, The Concept of Expropriation under the ECT and other Investment Protection Treaties," in *Investment Arbitration and the Energy Charter Treaty*, JurisNet, ed. Clarisse Ribeiro (2006), 108.

37 For example, *Compañía* del Desarrollo de Santa Elena, S.A. v. Republic of Costa Rica, ICSID Case No. ARB/96/1, Award, 17 February 2000, 5 ICSID Rep. 153.

38 For example, *Alpha Projektholding GmbH v. Ukraine*, ICSID Case No. ARB/07/16, Award, 8 November 2010.

39 *Siemens AG v. The Argentine Republic*, ICSID Case No. ARB/02/8, Award, 6 February 2007.

40 *Metalclad Corp. v. United Mexican States*, Case No. ARB(AF)/97/1, Award, dispatched 30 August 2000.

41 *ADC Affiliate Ltd and ADC & ADMC Management Ltd v. The Republic of Hungary*, ICSID Case No. ARB/03/16, Award, 2 October 2006.

States. This obligation is a treaty-based obligation; MFN treatment is not required under customary law.

In practice, this means that, through the MFN clause, an investor can import standards of investment protection and other rights from treaties that the host States signed with third States in order to benefit from them.[42] For instance, in *ATA Construction v. Jordan*, a case concerning the construction of a waterway dike, the investor successfully relied on the FET standard and the "obligation to afford the investor treatment no less favorable than that required by international law" in other treaties signed by Jordan, through the MFN provision in the applicable treaty.[43]

National Treatment

As seen with the MFN clause, BITs prohibit discrimination by the host State between foreign investors of different States, but also between foreign investors and domestic investors. Under the national treatment standard, the host State is required to treat the foreign investors no less favorably than domestic investors. This standard of investment protection intends to ensure competitive equality between foreign investors and their local competitors.

Similarly to the MFN clause, the national treatment obligation is viewed as a treaty-based standard, not required under customary law. The tribunal in the case *Methanex v. United States* applied a three-step analysis for national treatment claims: (i) determining whether the domestic investor is in "like circumstances" to the foreign investor; (2) if so, whether the domestic investor received better treatment than the foreign investor; (3) if the treatment is less favorable, whether the discrimination was justified.[44]

In *ADF v. USA*, a case regarding the construction of a highway, the investor alleged that the State introduced subsidies that favored domestic steel manufacturers. However, the tribunal held that the allegation was not substantiated since the measure did not appear to prejudice foreign steel manufacturers,[45] hence falling short of the second step in the *Methanex* test.

42 E.g., *National Grid plc v. The Argentine Republic*, UNCITRAL, Decision on Jurisdiction, 20 June 2006.

43 *ATA Construction, Industrial and Trading Company v. The Hashemite Kingdom of Jordan*, ICSID Case No. ARB/08/2, Award, 18 May 2010, para. 125; See also *KCI v. Gabon*, PCA Case No. 2015-25, Final Award, 23 December 2016, paras. 169-170.

44 *Methanex Corporation v. United States of America*, UNCITRAL, Final Award, 3 August 2005, Part IV, Chapter B [13].

45 *ADF Group Inc. v. United States of America*, ICSID Case No. ARB(AF)/00/1, 9 January 2003 paras 61 and 158-9.

Umbrella Clause

An umbrella clause is a provision in an investment treaty that guarantees the host State will observe obligations or undertakings it has entered into with regard to the investment in question, including obligations assumed under a contract. The purpose of umbrella clauses is to elevate a breach of contract, and breaches of other commitments that the State made to the investor, to a breach of the relevant investment treaty. In a construction treaty-claim context, umbrella clauses will be particularly relevant to parties contracting directly with the host State.

For instance, in the construction case *Strabag v. Libya*, the investor was contracted to carry out several highway and infrastructure construction projects. Following the revolution in Libya, it unsuccessfully sought to recover the amounts it was owed under these contracts. The tribunal held that Libya's failure to fulfill contractual obligations constituted a breach of the investment treaty.[46]

Some tribunals consider that a mere contractual breach cannot be elevated to a treaty breach unless there is very clear language in the relevant investment treaty to that effect.[47] Other tribunals sought to give the umbrella clause an *effet utile* and found that treaty clauses should be interpreted so as to render them effective rather than ineffective.[48]

In *Sempra v. Argentina*, the tribunal noted that the dispute arose from "how the violation of contractual commitments with the licensees [...] impacts the rights the investor claims to have in the light of the provisions of the Treaty and the guarantees on the basis of which it made the protected investment." The tribunal thus recognized that these contractual claims were also treaty claims.

In any event, the umbrella clause does not extend to any contract claims when such claims do not rely on a violation of the standards of protection of the BIT, such as national treatment, MFN clause, fair and equitable treatment, and FPS.

In cases involving a host State directly as a party to any contracts on construction projects, they are likely to remain an important protection that investors would be well-advised to seek out.

---

46 *Strabag SE v. Libya*, ICSID Case No. ARB(AF)/15/1, Award, 29 June 2020, paras 173–188.

47 *SGS Société Générale de Surveillance S.A. v. Islamic Republic of Pakistan*, ICSID Case No. ARB/01/13, Decision of the Tribunal on Objections to Jurisdiction, 6 August 2003, paragraph 167, and SGS Société Générale de Surveillance S.A. v. Republic of the Philippines, Decision of the Tribunal on Objections to Jurisdiction, 29 January 2004, para. 125.

48 *Sempra Energy International v. Republic of Argentina*, ICSID Case No. ARB/02/16, Decision on Jurisdiction, 11 May 2005; *Eureko B.V. v. Republic of Poland; Bureau Veritas, Inspection, Valuation, Assesment and Control, BIVAC B.V. v. Republic of Paraguay*, ICSID Case No. ARB/07/9; *SGS Société Générale de Surveillance S.A. v. Republic of Paraguay*, ICSID Case No. ARB/07/29.

Full Protection and Security

A FPS obligation relates to the physical and legal security of the investment. The host State is required to protect the investor's investment from property damage, physical violence or harassment, regardless of whether these actions are carried out by state entities or private parties.[49] Indeed, in some cases, harmful actions may originate from nongovernmental sources, such as protesters, employees, or business partners.[50]

This standard is particularly relevant in construction projects located in regions where security risks are prevalent. A violation of the FPS standard can occur not only through direct State action but also through the State's inaction, making it applicable to a wide range of scenarios.[51]

It is generally accepted that the FPS standard does not subject the host State to strict liability but rather imposes an obligation of due diligence, also referred to as an "obligation of vigilance." This distinction between strict liability and due diligence is crucial, in particular in cases where the harm in question is caused by actions not directly attributable to the State.

Despite being frequently alleged in construction disputes, the FPS standard has only succeeded in one notable case – *Cengiz v. Libya* – where the tribunal held Libya accountable for failing to protect the claimant's construction equipment during civil unrest.[52]

Other Standards of Investment Protection

Some obligations of States stem not from investment treaties, but from international law directly. Many of the aforementioned standards of investment protection, such as FET, are codified principles of customary international law. However, a treaty may for some reason not contain a FET provision. If so, a State could be deemed bound by it regardless as it stems not from the treaty but from custom.[53]

---

49 HE Zeitler, "The Guarantee of 'Full Protection and Security' in Investment Treaties Regarding Harm Caused by Private Actors" (2005) 3 Stockholm Intl Arbitration Rev 1.

50 *Parkerings-Compagniet AS v. Republic of Lithuania*, ICSID Case No. ARB/05/8, Award, 11 September 2007, para 355; *Pantechniki SA Contractors & Engineers (Greece) v. the Republic of Albania*, ICSID Case No ARB/07/21, Award, 30 July 2009; Omar Moussly, Same concept, Different Interpretation: The Full Protection and Security Standard in Practice, Kluwer Arbitration Blog, 27 October 2019.

51 *Cengiz İnşaat Sanayi ve Ticaret A.S v. Libya*, ICC Case No. 21537/ZF/AYZ, Award, 7 November 2018, paras 402–403.

52 *Cengiz İnşaat Sanayi ve Ticaret A.S v. Libya*, ICC Case No. 21537/ZF/AYZ, Award, 7 November 2018, paras 435 and 451–2.

53 *Eg Koch Minerals Sàrl and Koch Nitrogen International Sàrl v. Bolivarian Republic of Venezuela*, ICSID Case No. ARB/11/19, Award, 30 October 2017.

Additionally, some investment agreements may include clauses on State performance requirements, setting regulatory conditions under which they admit investments. These obligations are imposed on the investors through domestic law and may require them to achieve certain economic and social goals relating to the establishment or operation of their investments.[54] For instance, the investor might be obliged to hire local employees or only obtain supplies from domestic producers.

However, an increasing number of international investment agreements expressly prohibits Performance Requirements and thus limit the ability of States to insist on performance requirements in order to avoid States overly burdening the investors with such rules.

Counterclaims by Host States

Investment treaties primarily aim to encourage foreign investment and protect investors. As such they often remain silent on the rights of States vis-a-vis investors and on the obligations of investors vis-a-vis States.[55]

Nonetheless, the right to bring counterclaims is recognized in all arbitration rules.[56] For instance, Article 46 of the ICSID Convention provides in this respect that:

> Except as the parties otherwise agree, the Tribunal shall, if requested by a party, determine any incidental or additional claims or counterclaims arising directly out of the subject-matter of the dispute provided that they are within the scope of the consent of the parties and are otherwise within the jurisdiction of the Centre.

Similarly, Article 21(3) of 2013 UNCITRAL Arbitration Rules also acknowledges the right to bring counterclaims:

> In its statement of defence, or at a later stage in the arbitral proceedings if the arbitral tribunal decides that the delay was justified under the

54 *Cargill, Incorporated v. Republic of Poland*, ICSID Case No. ARB(AF)/04/2, Final Award, 29 February 2008, para. 541; see also Ahnlid, A., "Performance Requirements and Investment Incentives," *The Multilateral Agreement on Investment: State of Play*, (OECD, 2017), 28.

55 Treaty Concerning the Reciprocal Encouragement and Protection of Investment. U.S.-Arg., art. VII, Nov. 19, 1991, 31 I.L.M 124 (1992).

56 AAA International, art. 3(2); ICC, art. 5(5); ICSID Arbitration Rules, Rule. 40; ICSID Convention, art. 46; SCC Rules, art. 10(3); 1976 UNCITRAL Rules, art. 19; 2013 UNCITRAL Rules, art. 21(3); SIAC Rules 2013, Rule 20.4; HKIAC Rules 2013, art. 17(3).

> circumstances, the respondent may make a counterclaim or rely on a claim for the purpose of a set-off provided that the arbitral tribunal has jurisdiction over it.

While counterclaims from host States remain relatively uncommon,[57] an increasing number of host States are pursuing counterclaims, fostering a more balanced dynamic in treaty-claim arbitration. Host States have initiated counterclaims against investors based on violations of domestic law and public international law. Within the latter category, States have leveraged various sources such as multilateral treaties on human rights, environmental protection, corporate social responsibility standards, and principles of good faith, as well as prohibitions against corruption and fraud.

In the construction sector, host States have based counterclaims against investors on multiple grounds. Common issues include time for delivery of the project, the budget, the quality of the work delivered, termination and payment in accordance with the contractual specifications.[58]

States have also ascertained counterclaims for damages in the construction sector. For example, in *Desert Line v. Yemen*, the host State sought damages for (a) breaches of undertakings in the Settlement Agreement; and (b) unfulfilled obligations decided in the Arbitral Award, including damages for the investor's unfulfilled construction obligations and its obligation to maintain the bank guarantees.[59]

Paradigmatic Example of ISDS Claims by the Contractor: The Libyan Cases

Investor protection measures that enable claims against states have been well-established throughout the MENA region. Most MENA countries have signed and ratified the ICSID Convention, with notable exceptions such as Iran and Libya.

Although Libya is not a party to the ICSID Convention, it has nonetheless entered into bilateral investment treaties (BITs) with 32 countries, half of which

57 Preeti Bagnani, Revisiting the Countermeasures Defense in Investor-State Disputes: Approach and Analogies, in Andrea K. Bjorklund, *Yearbook on International Investment Law & Policy 2013–2014, 424* (Oxford University Press, 2015).

58 Scott Stiegler and Stephanie Archer, Contractual Dispute Resolution in Construction Contracts, Global Arbitration Review, 12 October 2023.

59 *Desert Line Projects LLC v. Republic of Yemen*, ICSID Case No. ARB/05/17, Award, 6 February 2008.

are currently in force. These include treaties with Egypt, France, Germany, Italy, and Switzerland. Additionally, since 2021, Libya has been a signatory to the Agreement on Promotion, Protection, and Guarantee of Investments among Member States of the Organization of the Islamic Conference (OIC Investment Agreement).[60]

Since 2012, Libya has seen a notable increase in investment treaty arbitrations, largely triggered by the aftermath of the Arab Spring. The construction industry has been particularly affected by political instability and civil conflict, resulting in numerous disputes between contractors and the Libyan government or state-owned entities. Arbitration has become the preferred method for resolving these disputes due to its efficiency and the enforceability of awards across borders.

However, Libya is not a member of the New York Convention on the Recognition and Enforcement of Foreign Arbitral Awards of 1958. As a result, arbitral awards issued outside of Libya are treated as foreign judgments, as per Article 761 of the Libyan Civil Procedure Code. Once recognized by Libyan courts, enforcement of awards is governed by Articles 405–411 of the Civil Procedure Code.

Construction Arbitration Cases in Libya

Libya has been involved in several high-profile construction arbitration cases brought by foreign investors in recent years. Out of the 1401 treaty-based investor-state dispute settlement (ISDS) cases worldwide, 23 involve Libya, a notable figure for a single country. Remarkably, 16 of these cases pertain to the construction sector.

Routes and Fora for Investors in Construction Disputes Involving Libya

International arbitration is frequently used in Libya for dispute resolution. Article 83 of Libya's Administrative Contracts Regulation No. 563/2007 permits parties to agree to international arbitration, subject to approval from the Council of Ministers. Consequently, many disputes have been resolved through international arbitration, often under the auspices of institutions like the International Chamber of Commerce (ICC) and the London Court of

60 Agreement on Promotion, Protection and Guarantee of Investments amongst the Member States of the Organization of the Islamic Conference, 1981, entered force in 1988 OIC Investment Agreement (1981) | International Investment Agreements Navigator | UNCTAD Investment Policy Hub.

International Arbitration (LCIA). London, Paris, Geneva, and Cairo are popular seats for these arbitrations.[61]

Article 24 of Law No. 9/2010 (regarding the encouragement of foreign capital investment) allows for dispute settlement under multilateral or bilateral treaties between Libya and the investor's home country, or through arbitration as provided in the investment contract.

There are no specific courts in Libya dedicated to hearing disputes related to international arbitration. Disputes are handled by courts of first instance, which have jurisdiction over requests for the recognition ("*exequatur*") of foreign awards and any objections or annulment requests.

Standards of Protection Claimed and Granted (Example of Notable Construction Cases Involving the State of Libya)

The most common breaches alleged and found in construction cases involving Libya include violations of the FET standard, the umbrella clause, and indirect expropriation.

Özer and Dirlik v. Libya[62]

In this case, the claimants, Mustafa Orhan, Özer and Nurettin Mendost Dirlik, alleged that Libya unlawfully expropriated their construction projects, violating the 2009 Turkey-Libya BIT and the 1981 OIC Treaty. They seek compensation for their losses while Libya, on the other hand, disputes the applicability of the OIC Treaty. The arbitration process, initiated on 2 November 2020, is still ongoing, with both parties awaiting a resolution.

Güriş v. Libya[63]

In an ICC arbitration, Ankara-based Güriş İnşaat ve Mühendislik A.Ş. filed a claim against Libya under the Libya-Turkey BIT. The dispute stemmed from a suspended project, including the construction of a public park in Tripoli, halted due to the civil war.

In a partial award on jurisdiction and liability, rendered in February 2020, the tribunal found that Libya had breached the BIT by interfering with the investor's contractual rights and failing to provide them FPS.

---

61 International Arbitration 2021 – Libya | Global Practice Guides | Chambers and Partners.

62 *Mustafa Orhan Özer and Nurettin Mendost Dirlik v. The State of Libya.*

63 *Güriş İnşaat ve Mühendislik A.Ş. v. Libya*, ICC Case No. 22137/ZF/AYZ, Partial Award on Jurisdiction and Liability, 4 February 2020.

Strabag v. Libya[64]

The dispute was over disruptions to construction contracts, signed by a joint venture between Strabag and an entity owned by the Government of Libya, that concerned a suburban development in Libya. Strabag initiated arbitration under the ICSID Additional Facility Rules, pursuant to the Austria-Libya BIT. In June 2020, the tribunal upheld jurisdiction and ruled that Libya breached the war and umbrella clauses.

Libya's involvement in international arbitration, particularly in construction-related disputes, highlights the complex challenges faced by States dealing with political instability. Despite not being a party to key international conventions such as ICSID or the New York Convention, Libya remains deeply involved in international arbitration, underscoring the importance of arbitration as a tool for resolving disputes in volatile environments.

Twelve percent of all Investment Treaty Arbitrations have been in the Construction industry.[65] Thus the protections offered by such treaties are a fundamental part of the international legal framework that contractors consider when contracting in a given country:[66] arbitration proceedings are an important potential tool for any investor, with such advantages as the absence of time limitation provisions hence possibly offering a viable claim in circumstances where a classical contractual claim under local law would be impossible.

Indeed, and notwithstanding the hurdles with respect to the jurisdiction of arbitral tribunal in ISDS procedures, contractors have now understood that construction projects are considered protected investment in most, if not all, BITs and MITs. These conventional instruments thus provide reassurance for contractors engaged in cross-border projects.

With the extremely significant investments of time and money in modern construction projects and the fact that contractors are generally foreigners from all around the world, their constant exposure to domestic risks from the actions of the host State or from political instability and civil conflict is a continuous source of potential problems. The proper structuration, at the outset, of the Contractor's investment in such host state is highly advisable to ensure that the protection offered by investment treaty arbitration is efficient and offers the proper insurance against governmental interference with the private construction sector.

64 *Strabag SE v. Libya*, ICSID Case No. ARB(AF)/15/1, Award, 20 June 2020.

65 UNCTAD, Investment Dispute Settlement Navigator Advanced Search | Investment Dispute Settlement Navigator | UNCTAD Investment Policy Hub accessed on 8 October 2024.

66 Walker R. and Randhawa J. "The Resolution of Construction Claims Through Investor-State Dispute Settlement: Alternative Opportunities for Relief for International Contractors," *International Construction Law Review* (2019): 255–83.

20

Investment Treaty Claims by the Owner

Nicholas Lingard, Rohit Bhat, and Yoonji Lee

Introduction

The construction sector encompasses a vast array of activities such as building of residential, commercial, and industrial structures, and large infrastructure projects such as roads, bridges, and public utilities. This sector is a fundamental pillar of any nation's economy, playing a critical role in driving economic growth and creating employment opportunities.

Large construction projects are complex and involve multiple parties (with each playing a distinct role). For construction projects to succeed, it is important to secure sufficient investment. It is equally important for investors to have mechanisms in place to protect their investment from potential wrongful conduct by the State. This chapter explores one such mechanism for protection, i.e., the network of investment treaties that provide investors with the ability to bring claims against States – and, more rarely, States against investors – for wrongful conduct before independent international tribunals.

Investment treaties – which could be bilateral ("**BITs**") or multilateral ("**MITs**")[1] – are formal agreements between countries that protect investments by investors from one State (the "**Home State**") in the territory of the other State party to the treaty (the "**Host State**"). There are currently 2221 BITs and 394 MITs with investment provisions that are in force.[2]

1 One notable example of a multilateral investment treaty is the Energy Charter Treaty, which aims to promote energy security through the establishment of open and competitive energy markets.

2 *See* the International Investment Agreements Navigator published by the United Nations Conference on Trade and Development ("**UNCTAD**") at https://investmentpolicy.unctad.org/international-investment-agreements (last accessed on 17 December 2024).

The real muscle behind investment treaties is that they typically give an investor the ability to submit a dispute with the Host State to binding international arbitration if the investor believes that the State through its acts or omissions has failed to comply with the investment protections in the relevant treaty. These protections may include guarantees of fair and equitable treatment, full protection and security, protection against expropriation except in accordance with due process of law and against prompt payment of full compensation, and most-favored-nation treatment. Claims are usually brought by investors under the ICSID Convention[3] or under the ad hoc UNCITRAL Arbitration Rules.[4]

Investment treaty disputes often arise in the construction sector. According to the International Centre for Settlement of Investment Disputes (**"ICSID"**), there have been 100 construction cases (68 of which have concluded) amounting to nearly 10% of the total caseload of the ICSID.[5] These cases could potentially be brought by any of the parties involved in a construction project, including owners, project managers, contractors, engineers, or designers. As the title suggests, the focus of this chapter is on "owners" of construction projects. Owners can be categorized into several types depending on the nature of the project and the source of funding. For example, in a real-estate project, an owner is usually the entity that owns the land and is in charge of developing the land. In an Engineering, Construction and Procurement (**"EPC"**) contract, the owner is the project company that engages the contractor to oversee the construction of the project. Often in large government-run infrastructure projects for the construction of bridges, highways and ports, for instance, the State-owner grants concessions to the project company which oversees financing, building and operating of the project.

Therefore, owners could be both private entities who, as investors, may be entitled to claim treaty protection, or States and State-owned entities who may be subject to claims against them for their acts as owners of construction projects.

This chapter focusses on the particular scenario where a State-owner of a construction project raises counterclaims against an investor-contractor in an investment treaty arbitration.

The construction sector, characterized by large-scale projects often located in environmentally and socially sensitive areas, is inherently vulnerable to counterclaims in investment treaty arbitrations. States frequently utilize counterclaims

---

3 Convention on the Settlement of Investment Disputes between States and Nationals of Other States (**"ICSID Convention"**), adopted on 18 March 1965; in force on 14 October 1966.

4 UNCITRAL (United Nations Commission on International Trade Law) Arbitration Rules (2021) (**"UNCITRAL Arbitration Rules"**), adopted on 9 December 2021. Some institutions such as the Singapore International Arbitration Centre are now developing bespoke investment arbitration rules.

5 *See* the ICSID Caseload – Statistics, Issue 2024, p 11.

to protect public interests and enforce compliance with domestic laws and contractual obligations. These counterclaims commonly involve allegations of environmental damage, labor and human rights abuses, missed deadlines impacting public welfare, and faulty construction posing safety risks. For instance, in a highway project traversing a mountainous region, a State might allege that the investor disregarded environmental impact assessments, leading to deforestation and ecological damage. Similarly, significant delays in a major airport construction project could prompt a State to counterclaim for economic losses arising from disruptions to trade and tourism. In another scenario, if sub-standard materials are used in a hospital construction project, jeopardizing the safety of patients and medical staff, the State could demand rectification and compensation for any harm caused. These examples illustrate the diverse range of counterclaims States may pursue to hold investors accountable in the construction sector.

However, and as we discuss below, bringing counterclaims in an investment-treaty arbitration is not straightforward. There are various substantive hurdles: namely, that the counterclaim must be within the scope of the parties' consent and the tribunal's jurisdiction; and must arise directly out of the subject-matter of the dispute. As we discuss below, much will depend on the language of the treaty or the arbitration agreement at hand.

The rest of this chapter is structured as follows:

1. The first section titled "Threshold Requirements for Claims Under Investment Treaties" explores the threshold requirements for investors to bring a claim under investment treaties.
2. The second section titled "Can Counterclaims Be Brought by State-owners of Construction Projects in Investment Treaty Arbitration?" addresses the substantive requirements for counterclaims to be brought by State-owners of construction projects in investment treaty arbitration.
3. The third section titled "Case Study: State-owners of Construction Projects as Counterclaimants in Investor-treaty Arbitrations" looks into *Burlington v. Ecuador*,[6] *Urbaser v. Argentina*,[7] *Desert Line v. Yemen*,[8] and *Salini v. Morocco (II)*[9] where the arbitral tribunals upheld their jurisdictions to hear counterclaims brought by States.

6 *Burlington Resources, Inc. v. Republic of Ecuador*, ICSID Case No. ARB/08/5 ("***Burlington v. Ecuador***").

7 *Urbaser S.A. and Consorcio de Aguas Bilbao Biskaia, Bilbao Biskaia Ur Partzuergoa v. Argentine Republic*, ICSID Case No. ARB/07/26 ("***Urbaser v. Argentina***").

8 *Desert Line Projects LLC v. Republic of Yemen*, ICSID Case No. ARB/05/17 ("***Desert Line v. Yemen***").

9 *Salini Costruttori S.p.A. v. Kingdom of Morocco*, ICC Case No. 16550/ND ("***Salini v. Morocco (II)***"), Award, 5 December 2011.

Threshold Requirements for Claims Under Investment Treaties

To benefit from treaty protection, the investor must, at a minimum, satisfy the definitions of "investor" and "investment" in the relevant treaty. Investment treaties typically define who qualifies as an investor eligible to bring a claim under the treaty's provisions. The specific criteria for qualifying as an investor vary depending on the treaty, but they commonly include a nationality requirement, *i.e.*, that the investor must be a "national" of the Home State.[10]

Whether an investor has a qualifying investment is a fact-specific question that depends on the provisions of a particular treaty. A qualifying investment under an investment treaty is typically defined by a broad and inclusive set of criteria that encompass diverse forms of assets. These often include an enterprise, equity interest in an enterprise, debt security, loans, or real estate.[11] Of relevance in the construction context is the specific reference in many investment treaties, such as in the North American Free Trade Agreement ("**NAFTA**"), to "contracts involving the presence of an investor's property in the territory of the Party, including turnkey or construction contracts, or concessions";[12] or in the Regional Comprehensive Economic Partnership Agreement ("**RCEP**"), to "rights under contracts, including turnkey, construction, management, production, or revenue-sharing contracts."[13] Investors from a Home State can avail of substantive protections granted by investment treaties against the *Host State*.[14]

10 To avoid shell companies being established in protection-friendly jurisdictions and qualifying as "investors" merely by virtue of their incorporation, some modern BITs also require that the investor undertake a genuine commercial activity in the Home State, before it can obtain protection by the Host State. In cases where the relevant treaty provides for substantial business activities, Tribunals have held that "substantial" means "of substance, and not merely of form" and that it "does not mean 'large,' and the materiality not the magnitude of the business activity is the decisive question." *See Limited Liability Company Amto v. Ukraine*, SCC Case No 080/2005, Final Award, 31 October 2005, para 69.

11 *See*, for example, the definition of investment in the NAFTA, at Art. 1139.

12 NAFTA, adopted on 17 December 1992, in force on 1 January 1994, Art. 1139, definition of "investment," para (h)(i).

13 RCEP, adopted on 15 November 2020; in force on 1 January 2022, Art. 10.1(c)(iii).

14 Some investment treaties provide clarity on which entities form part of the state. Art. 11.1.3 of the Free Trade Agreement between the United States of America and the Republic of Korea, signed on 30 June 2007, in force on 15 March 2012, provides that the investment chapter of the treaty applies to "Measures adopted or maintained by: (a) central, regional, or local governments and authorities; and (b) non-governmental bodies in the exercise of powers delegated by central, regional, or local governments or authorities."

The International Law Commission's Draft Articles on Responsibility of States for Internationally Wrongful Acts (the "**ILC Articles**")[15] and the commentary thereto[16] provide the appropriate framework for determining whether a conduct is attributable to a State. Under ILC Article 4, the conduct of any State organ will be attributed to the State.[17] Article 5 of the ILC Articles provides that the State is responsible for conduct by a non-State entity "which is empowered by the law of that State to exercise elements of the governmental authority [...], provided the person or entity is acting in that capacity in the particular instance."[18] Finally, under Article 8 of the ILC Articles, the conduct of a person or group of persons "acting on the instructions of, or under the direction or control of," a State shall be considered an act of that State under international law."[19]

Can Counterclaims Be Brought by State-owners of Construction Projects in Investment Treaty Arbitration?

Until recently, investment treaties rarely imposed substantive obligations on investors vis-à-vis Host States. Consequently, counterclaims brought by Host States against investors were few and far between, and only a handful of them have succeeded. As we explain in detail below, Host States' counterclaims based instead on the investors' alleged breach of local laws or obligations under investment contracts have more often than not been dismissed by investment treaty arbitration tribunals for lack of jurisdiction.

In the new generation of investment treaties, States are increasingly imposing express obligations on investors to balance against the protections offered to investors. These obligations include protection of environment, corporate social responsibility, and protection of human rights. For instance, the

---

15 International Law Commission Draft Articles on the Responsibility of States for Internationally Wrongful Acts, text adopted at 53rd Session, *Yearbook of the International Law Commission, Volume II (Part Two)*, 9 June 2001.

16 International Law Commission Draft Articles on the Responsibility of States for Internationally Wrongful Acts, with commentaries, text adopted at 53rd Session, *Yearbook of the International Law Commission, Volume II (Part Two)*, (9 June 2001).

17 ILC Articles, Art. 4 ("1. The conduct of any State organ shall be considered an act of that State under international law, whether the organ exercises legislative, executive, judicial or any other functions, whatever position it holds in the organization of the State, and whatever its character as an organ of the central Government or of a territorial unit of the State. 2. An organ includes any person or entity which has that status in accordance with the internal law of the State").

18 ILC Articles, Art. 5.

19 ILC Articles, Art. 8.

United Arab Emirates-India BIT (2024)[20] and Brazil-India BIT (2020)[21] dedicate an entire section to extensively set out investor obligations, including the obligations to "comply with the provisions of Law of the Parties concerning taxation"[22] and to "endeavo[r] to voluntarily incorporate internationally recognized standards of corporate social responsibility in their practice and internal policies."[23] Similarly, Nigeria-Morocco BIT (2016) expressly imposes obligations on investors to uphold human rights and respect the environment, labor rights and corporate social responsibility standards.[24] The socioeconomic and environmental impact on the local community as well as labor issues arising from large construction projects may potentially be covered by these new-age treaty provisions in the near future.

In this section, we analyze the jurisprudence on the existing procedural and substantive requirements for States to bring counterclaims in investment-treaty arbitrations.

Procedural Rules

The ICSID Convention and the UNCITRAL Arbitration Rules do envision the possibility of counterclaims being brought by Host States. Article 46 of the ICSID Convention stipulates that "[E]xcept as the parties otherwise agree, the Tribunal shall, if requested by a party, determine any incidental or additional claims or *counterclaims arising directly out of the subject-matter of the dispute* provided that they are within the scope of the consent of the parties and are otherwise within the jurisdiction of the Centre."[25]

Similarly, Article 21(3) of the UNCITRAL Arbitration Rules expressly sets out the Host State's right to bring a counterclaim. This Article provides: "In its statement of defence, or at a later stage in the arbitral proceedings if the arbitral tribunal decides that the delay was justified under the circumstances, the respondent may make a counterclaim or rely on a claim for the purpose of a set-off provided that the arbitral tribunal has jurisdiction over it."[26]

---

20 Bilateral Investment Treaty Between the Government of the Republic of India and the Government of the United Arab Emirates, signed on 13 February 2024 (in force on 31 August 2024) (**"United Arab Emirates-India BIT (2024)"**), Chapter III.

21 Investment Cooperation and Facilitation Treaty between the Federative Republic of Brazil and the Republic of India, signed on 25 January 2020 (**"Brazil-India BIT (2020)"**), Part III.

22 United Arab Emirates-India BIT (2024), Art. 12(iii). *See also* Brazil-India BIT (2020), Art. 11(c).

23 United Arab Emirates-India BIT (2024), Art. 13. *See also* Brazil-India BIT (2020), Art. 12.1.

24 Reciprocal Investment Promotion and Protection Agreement Between the Government of the Kingdom of Morocco and the Government of the Federal Republic of Nigeria, signed on 3 December 2016 (**"Nigeria-Morocco BIT (2016)"**), Arts. 18, 19.

25 ICSID Convention, Art. 46 (emphasis added).

26 UNCITRAL Arbitration Rules, Art. 21(3) (emphasis added).

However, tribunals have held that the existence of these procedural rules does not itself mean that counterclaims can be raised in arbitrations governed by these rules. In *Oxus Gold v. Uzbekistan*, Uzbekistan raised a counterclaim against Oxus Gold, a company incorporated in the United Kingdom, in an investment-treaty arbitration based on the UK-Uzbekistan BIT. The parties agreed that the question of jurisdiction and admissibility of the counterclaims were to be determined based on Article 8(1) of the UK-Uzbekistan BIT and Article 21(3) of the UNCITRAL Arbitration Rules.[27] The tribunal held that Article 21(3) of UNCITRAL Arbitration Rules does not create jurisdiction where there is none, and that counterclaims are admissible only to the extent that they already fall under the scope of jurisdiction of the arbitral tribunal as conferred by the BIT.[28] Article 8(1) of the UK-Uzbekistan BIT referred only to disputes "concerning an obligation of the latter [*i.e.*, a Contracting Party] under [the BIT]," and not the obligation of the investors.[29] From this, the Tribunal decided that the parties' consent to arbitration under the BIT only covered claims from investors and not from the Host State.[30] The Tribunal therefore held that it lacked jurisdiction over Uzbekistan's counterclaims.

Substantive Requirements

ICSID tribunals dealing with Host State counterclaims have focused on the substantive requirements derived from Article 46 of the ICSID Convention quoted above, namely:

- the counterclaim must fall "within the scope of the consent of the parties";
- it must arise "directly out of the subject-matter of the dispute";
- it must be "within the jurisdiction of [ICSID]"; and
- in addition, parties are allowed to "otherwise agree" – *i.e.*, to contractually exclude the arbitrability of counterclaims.

The PCA tribunal in *Saluka v. Czech Republic* held that the UNCITRAL Rules "reflect essentially the same requirement" as Article 46 of the ICSID

27 *Oxus Gold plc v. Republic of Uzbekistan, the State Committee of Uzbekistan for Geology & Mineral Resources, and Navoi Mining & Metallurgical Kombinat* (**"Oxus Gold v. Uzbekistan"**), Final Award, 17 December 2015, para 943.

28 *Oxus Gold v. Uzbekistan*, Final Award, 17 December 2015, paras 944–945.

29 Agreement between the Government of the United Kingdom of Great Britain and Northern Ireland and the Government of the Republic of Uzbekistan for the Promotion and Protection of Investments, signed on 24 November 1993; in force on 24 November 1993 (**"UK-Uzbekistan BIT"**), Art. 8(1).

30 *Oxus Gold v. Uzbekistan*, Final Award, 17 December 2015, para 948.

Convention.[31] The tribunal held that the requirement of "the close conne[ct]ion which a counterclaim must have with the primary claim if a tribunal with jurisdiction over the primary claim is to have jurisdiction also over the counterclaim" is a "general legal principle" applicable even to UNCITRAL-administered cases.[32] Subsequent tribunals in cases administered by the UNCITRAL Rules have largely followed the same approach.[33] This section will therefore focus on the substantive requirements of Article 46 of the ICSID Convention. We explore each of these requirements below.

Parties' Consent to Arbitrate Counterclaims

The threshold question for the arbitrability of counterclaims is no different from that of any arbitration claim: the parties must have consented to the arbitration.

The scope of the State parties' consent is set out in the applicable investment treaty, or the investment agreement concluded between the State and the investor that contains the relevant arbitration agreement. As such, tribunals have paid particular attention to the language of the applicable investment treaties and the arbitration agreements. In the construction context, the applicable arbitration agreements may be found in turnkey or concession contracts between a State-owner and a contractor that govern the construction project. For example, in *Rasia v. Armenia*, the concession agreements between Armenia and Rasia for the construction and maintenance of a high-speed road and a railway link contained arbitration clauses whereby the parties consented to submit to ICSID any dispute arising out of or relating to those contracts.[34]

On the one hand, where the applicable treaty refers only to the State parties' obligations to investors and not vice versa, tribunals have shown reluctance to uphold their jurisdiction to hear counterclaims. For example, in *Roussalis v. Romania*,[35] a Greek individual claimant purchased controlling interest in a frozen food warehousing facility from Romania's privatization authority. Spyridon Roussalis brought a claim against Romania under the Greece-Romania BIT for, among other things, Romania's alleged failure to comply with its post-investment

---

31 *Saluka Investments BV v. The Czech Republic*, PCA Case No. 2001–04 ("***Saluka v. Czech Republic***"), Decision on Jurisdiction over the Czech Republic's Counterclaim, 7 May 2004, para 76.

32 *Saluka v. Czech Republic*, Decision on Jurisdiction over the Czech Republic's Counterclaim, 7 May 2004, para 76.

33 *See, e.g., Sergei Paushok, CJSC Golden East Company and CJSCVostokneftegaz Company v. The Government of Mongolia ("**Paushok v. Mongolia**")*, Award on Jurisdiction and Liability, 28 April 2011, para 693.

34 *Rasia FZE and Joseph K. Borkowski v. Republic of Armenia*, ICSID Case No. ARB/18/28 ("***Rasia v Armenia***"), Award, 20 January 2023.

35 *Spyridon Roussalis v. Romania*, ICSID Case No. ARB/06/1 ("***Roussalis v Romania***").

obligation related to the sale, and for imposing tax liabilities and penalties to the investor's company. Romania brought counterclaims against Roussalis and his companies for damages for breach of investment obligation; damages for misappropriation of funds; damages for breach of contractual pledge to transfer shares; and declaratory relief and damages in relation to invalidity of resolution to increase share capital. The majority of the tribunal dismissed Romania's counterclaims for lack of jurisdiction, on the basis that there was no consent to arbitrate the State's counterclaims given that the BIT referred only to the obligations of the State, not the investor.[36] In his minority declaration, Professor Michael Reisman dissented on the basis that "when the States *[sic]* Parties to a BIT contingently consent, *inter alia*, to ICSID jurisdiction, the consent [...] is *ipso facto* imported into any ICSID arbitration which an investor then elects to pursue."[37]

The majority of the tribunal in *Gavazzi v. Romania* followed a path similar to that of the majority in *Roussalis v. Romania*. In that case, two Italian nationals, Marco Gavazzi and Stefano Gavazzi, acquired a majority shareholding in a Romanian steel manufacturing entity under a privatization agreement they entered into with Romania. The Claimants brought a claim against Romania under Italy-Romania BIT, alleging that Romania has breached its obligations under the privatization agreement which resulted in the steel plant's liquidation. Romania brought a counterclaim for damages for the Claimants' alleged failure to comply with their investment obligations imposed by the applicable Romanian law. The majority rejected jurisdiction for counterclaims, on the basis that the Italy-Romania BIT only grants investors the right to claim against the Host State.[38] The majority held that its jurisdiction had to be provided by the wording of the BIT, not merely from the "spirit" of the BIT.[39] Mauro Rubino-Sammartano issued a dissenting opinion, noting that "the omission of any express mention of the Host State's right to file a counterclaim may be due to the fact that the drafters of the BIT focused on the protection of the investor," and that "due process includes the right to defend a claim and [...] natural justice requires that such defence may include making a counterclaim related to such issues."[40]

On the other hand, where the applicable treaty is broadly drafted allowing either the State or the investor to commence arbitration, tribunals have often held

---

36 *Roussalis v. Romania*, Award, 7 December 2011, para 869.

37 *Roussalis v. Romania*, Declaration of W. Michael Reisman (Award).

38 *Marco Gavazzi and Stefano Gavazzi v. Romania*, ICSID Case No. ARB/12/25 ("***Gavazzi v. Romania***"), Decision on Jurisdiction, Admissibility and Liability, 21 April 2015, para 151.

39 *Gavazzi v. Romania*, Decision on Jurisdiction, Admissibility and Liability, 21 April 2015, para 154.

40 *Gavazzi v. Romania*, Dissenting opinion by arbitrator Mauro Rubino-Sammartano, para 42.

that the States' counterclaims are admissible provided other conditions are also met. In *Urbaser v. Argentina*, for instance, Argentina's counterclaim was allowed on the basis that Spain-Argentina BIT contains a broadly formulated dispute resolution clause applicable to "disputes arising between a Party and an investor of the other Party" allowed either party to commence arbitration.[41] Detailed examination of this case is set out in Section "*Urbaser v. Argentina*."

In *Saluka v. Czech Republic*, the tribunal followed a similar approach. In that case, the Dutch claimant acquired ownership of controlling shares from a state-owned bank. Subsequently, the Czech National Bank imposed a forced administration upon the investor's banking enterprise. The Claimant brought a claim under the Slovak Republic-Netherlands BIT for, among other things, failure to grant fair and equitable treatment and denial of justice. The Czech Republic brought counterclaims for, among other things, the Claimant's alleged breach of a share purchase agreement and violation of the Czech Commercial Code. The tribunal made clear that the Czech and Slovak Republic-Netherlands BIT's requirement for a dispute to be "between one Contracting Party and an investor of the other Contracting Party" carries with it no implication that only an investor can initiate claims under the treaty.[42] Nonetheless, the tribunal unanimously dismissed its jurisdiction over the counterclaims on the basis that the share purchase agreement was between different parties and contained a mandatory arbitration provision,[43] and that the claims based on violation of Czech law fell outside of the scope of the BIT.[44]

A tribunal may find it easier to uphold its jurisdiction where the State and the investor parties agree in an investment agreement to arbitrate counterclaims raised by the State. This was the case in *Burlington v. Ecuador*. In that case, the tribunal derived its jurisdiction to hear counterclaims from the parties' express agreement that the ICSID arbitration is the "appropriate forum for the final resolution of the Counterclaims arising out of the investments made by Burlington Resources and its affiliates."[45] Detailed examination of this case can be found in Section "*Burlington v. Ecuador*".

41 *Urbaser v. Argentina*, Award, 8 December 2016, para 1143. *See also* Section "*Urbaser v. Argentina*."

42 *Saluka v. Czech Republic*, Decision on Jurisdiction over the Czech Republic's Counterclaim, 7 May 2004, para 39.

43 *Saluka v. Czech Republic*, Decision on Jurisdiction over the Czech Republic's Counterclaim, 7 May 2004, paras 56, 57.

44 *Saluka v. Czech Republic*, Decision on Jurisdiction over the Czech Republic's Counterclaim, 7 May 2004, para 79.

45 *Burlington v. Ecuador*, Decision on Counterclaims, 7 February 2017, para 60. *See also* Section "*Burlington v. Ecuador*."

In the same vein, where the parties expressly agreed on the exclusive jurisdiction of another tribunal or the local courts, the tribunal refused to hear States' counterclaims. In *Muhammet Çap & Sehil v. Turkmenistan*, a Turkish company was awarded a large number of construction projects by Turkmenistan State organs and State-owned companies in relation to construction of buildings in Turkmenistan. The claimants initiated an ICSID arbitration against the Turkmenistan government under the Turkey-Turkmenistan BIT for, among other things, alleged unlawful expropriation of the construction projects, including defaulted payments and the termination of some of the construction contracts at issue. The State raised counterclaims against the Claimants, for alleged breach of those construction contracts, including substantial delays in performance, failure to meet fundamental obligations under a contract, and undue payments for unperformed works. The tribunal dismissed both the primary claims and the counterclaims that arose out of the disputed contracts for lack of jurisdiction, as those contracts provided for the exclusive jurisdiction of the Arbitration Court of Turkmenistan.[46]

Counterclaims Must Arise Directly Out of the Subject-matter of the Dispute

Another hurdle that State-owners often face in bringing counterclaims is the subject-matter requirement: the counterclaim must have a direct connection with the principal claim raised by the investor.

The tribunal in *Klöckner v. Cameroon* established the relevant principles. In that case, Cameroon and Klöckner established a joint venture company to build, supply and manage a fertilizer plant for a minimum of five years. Cameroon prematurely closed down the plant, for lack of profitability. Klöckner brought an investment treaty claim against Cameroon on the basis of a framework agreement that it entered into with Cameroon, which provided ICSID jurisdiction. Cameroon brought a counterclaim against Klöckner, on the basis that Klöckner misrepresented its management capabilities which allegedly rendered the joint venture company unprofitable. The management duties were not governed by the framework agreement which was the basis of Klöckner's claim. Instead, they were governed by another contract between Cameroon and Klöckner's locally incorporated subsidiary which provided ICC jurisdiction. Nevertheless, the tribunal found that it had jurisdiction over Cameroon's counterclaims, on the basis that the parties' claims were directly connected such that they comprised an

46 *Muhammet Çap & Sehil Inşaat Endustri ve Ticaret Ltd. Sti. v. Turkmenistan*, ICSID Case No. ARB/12/6, Award, 4 May 2021, paras 711, 712.

"indivisible whole,"[47] with "a common origin, identical sources, and an operational unity" assumed for the "accomplishment of a single goal."[48]

Subsequent tribunals have relied on this principle as a reason to dismiss State parties' counterclaims. In *Saluka v. Czech Republic*, the tribunal adopted the principle established in *Klöckner v. Cameroon*. It dismissed Czech Republic's counterclaim on the basis that it lacked a "close connexion [*sic*] with the primary claim to which it is a response."[49] The "indivisible whole" standard was interpreted narrowly by the tribunal. As *Klöckner v. Cameroon* emphasized, the main question was whether the tribunal had jurisdiction *rationae materiae* over the application and interpretation of the agreement that was the subject matter of Cameroon's counterclaims.[50] On the basis that that agreement and the framework agreement that was the subject matter of Klöckner's claim were closely connected, the tribunal upheld its jurisdiction.

Counterclaims Must Be Within the Jurisdiction of the Tribunal

Early attempts by States to counterclaim damages from investors who allegedly breached their obligations found in local laws were mostly dismissed, as tribunals held that those claims fall outside of their jurisdiction. In *Saluka v. Czech Republic*, the tribunal made clear that the appropriate course to deal with claims arising from the investor's noncompliance with the general law of the Czech Republic was through the relevant procedures under the Czech law, not investment treaty dispute settlement procedure.[51] Similarly, in *Paushok v. Mongolia*, the tribunal dismissed Mongolia's counterclaims based on Mongolian domestic laws on the basis that it could not "acquiesc[e] to a possible exorbitant extension of Mongolia's legislative jurisdiction without any legal basis under international law to do so."[52]

More recently, however, in *Burlington v. Ecuador* and *Perenco v. Ecuador*, the tribunals took note of Article 42(1) of ICSID Convention, which stipulates that "the Tribunal shall apply the law of the Contracting State party to the dispute

47 *Klöckner Industrie-Anlagen GmbH and others v. United Republic of Cameroon and Société Camerounaise des Engrais*, ICSID Case No. ARB/81/2 ("***Klöckner v. Cameroon***"), Award, 21 October 1983, para 21.

48 *Klöckner v. Cameroon*, Award, 21 October 1983, para 136.

49 *Saluka v. Czech Republic*, Decision on Jurisdiction over the Czech Republic's Counterclaim, 7 May 2004, para 61.

50 *Klöckner v. Cameroon*, Award, 21 October 1983, para 19.

51 *Saluka v. Czech Republic*, Decision on Jurisdiction over the Czech Republic's Counterclaim, 7 May 2004, para 79.

52 *Paushok v. Mongolia*, Award on Jurisdiction and Liability, 28 April 2011, para 695.

(including its rules on the conflict of laws) and such rules of international law as may be applicable." Relying on this provision, the tribunals in the two cases applied Ecuadorian tort law and domestic regulatory criteria to assess the levels of environmental harm and the consequent liabilities.[53] The counterclaims were eventually upheld and both Burlington and Perenco were found liable. We assess the *Burlington v. Ecuador* tribunal's findings in detail in Section "*Burlington v. Ecuador.*"

Consequences of Bringing Unsuccessful Counterclaims

The rarity of State counterclaims in investment treaty arbitrations can be traced not only to their limited success rate to date, but also the ancillary consequences of bringing unsuccessful counterclaims.

A counterclaiming State may be met with adverse cost consequences even if the investor's claim against it is unsuccessful. In *RSM Production Corporation v. Grenada*, even when the investor was unsuccessful in its claims against the State, the tribunal decided to split the costs between the parties on the basis that both the investor and the State had been unsuccessful in their claims/counterclaims.[54] The converse may also be true. In *Atlantic Triton v. People's Revolutionary Republic of Guinea*, even though the State's counterclaim was rejected, the tribunal decided to reduce the amount of award to the claimant in consideration of the claimant's failure to carry out its undertaking as alleged by the State in its counterclaim.[55]

Further, a State may find itself contradicting its own case on attribution by bringing a counterclaim against the investor. A State-counterclaimant is unable to recover loss suffered by the parastate entity unless it concedes that that entity is an extension of the State itself. If the State concedes this point, it must also concede that the damage and/or loss caused to the investor by that parastate entity is attributable to itself.

53 *Burlington v. Ecuador*, Decision on Liability, 14 December 2012, paras 177 to 179; *Perenco Ecuador Limited v. Republic of Ecuador*, ICSID Case No. ARB/08/6 (**"Perenco v. Ecuador"**), Decision on Remaining Issues of Jurisdiction and on Liability, 12 September 2014, paras 532, 533.

54 *RSM Production Corporation v. Grenada*, ICSID Case No. ARB/05/14, Award, 13 March 2009, paras 495, 498.

55 *Atlantic Triton Revolutionary Company Limited v. People's Republic of Guinea*, ICSID Case No. ARB/84/1, Award, 21 April 1986.

Case Study: State-owners of Construction Projects as Counterclaimants in Investor-treaty Arbitrations

Have State-owners succeeded in their counterclaims brought against the investor-contractors in construction cases? Again, the success rate has so far been limited. We look into the details of four cases below.

Burlington v. Ecuador

The Claimant in this case, Burlington, was a Delaware-incorporated entity. Burlington's wholly-owned subsidiary and Ecuador entered into production sharing contracts for the exploration and exploitation of certain oil reserves in Ecuador. Under these contracts, Ecuador was the owner of the exploration and exploitation project, and Burlington was one of the contractors. In response to the spike in oil price, Ecuador enacted a law imposing a 99% levy on foreign oil revenues. Burlington brought a claim under Ecuador-United States of America BIT before an ICSID tribunal against Ecuador for, among other things, direct expropriation of its investment.

Ecuador filed a counterclaim alleging that Burlington caused significant environmental damage due to its oil operations. Ecuador claimed that Burlington failed to comply with environmental regulations, and that Burlington's operations caused extensive contamination of soil and water resources through oil spills, improper waste disposal, and inadequate infrastructure maintenance. Ecuador sought compensation for the costs of environmental remediation and restoration of the affected areas.

Burlington did not raise any jurisdictional objections to Ecuador's counterclaims. The tribunal upheld its jurisdiction over them on the basis of an agreement between the parties referring resolution of counterclaims to the ICSID tribunal.[56] The tribunal further noted that the conditions under Article 46 of the ICSID convention were satisfied, given that (*i*) the counterclaims arise directly out of the subject-matter of the dispute, namely Burlington's investment in the right to explore and exploit the aforementioned oil reserves; (*ii*) they are within the scope of the parties' consent to ICSID arbitration; and (*iii*) they also fall within the jurisdiction of the Centre as circumscribed by Article 25 of the ICSID Convention.[57] Eventually, the tribunal upheld Ecuador's counterclaims and ordered Burlington

56 *Burlington v. Ecuador*, Decision on Counterclaims, 7 February 2017, para 60.

57 *Burlington v. Ecuador*, Decision on Counterclaims, 7 February 2017, para 62.

to pay USD 41 million plus interest in compensation to Ecuador for the environmental and infrastructural damage that it had suffered.[58]

Urbaser v. Argentina

The Claimant was a Spanish company with a direct and indirect minority shareholding in an Argentinean company. The province of Greater Buenos Aires entered into a concession contract with that company, whereby the company as the concessionaire would provide drinking water supply and sewerage services in Buenos Aires, and the province of Greater Buenos Aires would be the grantor of that concession. During the 2001–2002 economic crises, Argentina enacted certain emergency measures, which allegedly breached the concession agreement and interfered with the tariff regime applicable to Urbaser's investment.

Argentina raised a counterclaim that Urbaser failed to provide the level of investment into the concession that they had undertaken to make and thereby violated their commitments under the concession. Argentina further alleged that Urbaser, in failing to make the necessary investments into the concession, violated the human right of access to water under international law. Argentina claimed damages arising from these alleged breaches by Urbaser. Urbaser challenged the tribunal's jurisdiction to hear the counterclaims.

The tribunal rejected Urbaser's categorical characterization of BITs as asymmetric, and instead pointed to the language of Spain-Argentina BIT which "is completely neutral as to the identity of the claimant or respondent in an investment dispute" and therefore does not prevent a State party from suing an investor in relation to a dispute concerning an investment.[59] The tribunal held that the scope of the parties' consent is not restricted without express exclusion of counterclaims agreed by both parties.[60] The tribunal further observed that the factual link between the two claims was "manifest," given that they were "based on the same investment, or the alleged lack of sufficient investment, in relation to the same Concession."[61] The tribunal also took note that Argentina's counterclaim was based on "the very purpose of the investment agreed upon" in the contracts

58 *Burlington v. Ecuador*, Decision on Counterclaims, 7 February 2017, para 1075.

59 *Urbaser v. Argentina*, Award, 8 December 2016, para 1143. *See also* Article X(1) of Spain-Argentina BIT ("1. Disputes arising between a Party and an investor of the other Party in connection with investments within the meaning of this Agreement shall, as far as possible, be settled amicably between the parties to the dispute.").

60 *Urbaser v. Argentina*, Award, 8 December 2016, para 1187.

61 *Urbaser v. Argentina*, Award, 8 December 2016, para 1151.

which formed the basis of Urbaser's claims against Argentina.[62] Ultimately, however, the counterclaim was dismissed for lack of merits.[63]

Desert Line v. Yemen

Desert Line Projects, an Omani construction company, entered several contracts with the Yemeni government for road construction in Yemen.[64] After completing most of the construction works, Desert Line Projects demanded payment, which Yemen disputed.[65] Yemen then offered a settlement (the "Settlement Agreement"), which Desert Line Projects agreed to under alleged duress, as their personnel were arrested.[66] Subsequently, Desert Line Projects initiated an ICSID arbitration against Yemen under the Oman-Yemen BIT, seeking payment for the roadworks performed and damages for losses suffered due to Yemen's breaches of the construction contracts and its obligation under the BIT to provide fair and equitable treatment.[67]

Yemen counterclaimed, seeking set-off for (*i*) the cash amounts already received by Desert Line Projects under the Settlement Agreement, (*ii*) the value of the unperformed remedial and other works, and (*iii*) the value of the relinquished bank guarantees.[68]

The tribunal found the Settlement Agreement to be "internationally ineffective" but took into account the amount paid under it when quantifying the amount due to Desert Line Projects.[69] The other counterclaims, including the claimant's alleged failure to execute maintenance and repair obligations and to maintain the two bank guarantees, were dismissed. The tribunal concluded that the Settlement Agreement provided for the release of these obligations, and Yemen was estopped from relying on the nullity of the Settlement Agreement, which it had imposed on Desert Line Projects.[70]

---

62 *Urbaser v. Argentina*, Award, 8 December 2016, para 1151.
63 *Urbaser v. Argentina*, Award, 8 December 2016, para 1221.
64 *Desert Line v. Yemen*, Award, 6 February 2009, paras 1–49.
65 *Desert Line v. Yemen*, Award, 6 February 2009, paras 18–20.
66 *Desert Line v. Yemen*, Award, 6 February 2009, paras 39–49.
67 *Desert Line v. Yemen*, Award, 6 February 2009, para 50.
68 *Desert Line v. Yemen*, Award, 6 February 2009, para 64.
69 *Desert Line v. Yemen*, Award, 6 February 2009, para 223.
70 *Desert Line v. Yemen*, Award, 6 February 2009, paras 224–225.

Salini v. Morocco (II)

Two Italian claimants, Salini Costruttori S.p.A and Italstrade S.p.A, entered a public procurement agreement with a State-administered entity in Morocco for highway construction. Under the contract, Morocco as the owner had the obligation to pay the contractor for the performance of the works. There were significant delays to the project. When the claimants sent the final account to the Moroccan government and requested for payment, the Moroccan government did not respond. The claimants commenced an arbitration under Italy-Morocco BIT.[71] Morocco brought counterclaims, demanding payment of an indemnity for delay to the works, and for loss allegedly suffered due to the fact that it had to conclude a contract with a third party in order to complete the construction.[72] While the tribunal upheld these heads of counterclaim, it found that Morocco's calculation was exaggerated and awarded a small part of the amounts claimed.[73]

Conclusion

Construction projects are complex and involve several parties working together. State-owners may play a pivotal role in driving construction projects and requires substantive protections of its interests and mitigation of risks. As this chapter has demonstrated, States have little room to make a counterclaim against investors in investment treaty arbitrations and much of the chance of success will depend on the text of the treaty concerned. As mentioned above, the new generation of investment treaties includes express obligations on investors to balance against the protections offered to investors, but it remains to be seen whether these express obligations would pave the way for State-owners to successfully bring counterclaims – or standalone claims – against investors in construction cases.

---

71 *Salini v. Morocco (II)*, Award, 5 December 2011.

72 *Salini v. Morocco (II)*, Award, 5 December 2011, paras 177–194.

73 *Salini v. Morocco (II)*, Award, 5 December 2011, paras 181, 188, 193.

21

The Duty of Solidarity and Its Limits

Alexandra Michalopoulou and Eleni Tsakri

Introduction

In international construction law, the notion of solidarity most notably arises in the context of contractual arrangements and pertinent relationships. Nowadays, international construction contracts involve mega-size, complex and long-term projects (e.g., for the construction of motorways and transportation (airports, ports, etc.) infrastructure, stadiums, energy plants, etc.), and a large number of participants in their planning and execution, in addition to the project owner: designers, consultants, financiers, architects and civil engineers, contractors and sub-contractors, technical experts and supervisors, suppliers of materials and equipment, etc. In that sense, such contracts may be viewed as constituting "common projects", with all participants sharing responsibility for the orderly and timely execution of the overall works and the project's delivery, and creating unique, interdependent relationships.

Construction contracts are filled with uncertainty; in the course of their execution, several and varied contingencies may arise, requiring all above participants to collaborate, as the case may be, for addressing any issue arising therefrom. Even in the absence of such unforeseen circumstances, successful implementation of this type of contracts depends on proper communication and collaboration among the participants in discharging their (predicted) obligations under the contract and attaining the desired final result in a regular and orderly manner.

In this context, acknowledging the existence of solidarity as a value imbuing the contract and as a set of specific practical undertakings and substantive obligations (in other words, as a contractual substratum that surfaces on occasions, as required, and manifests in the form of given actions or omissions) is key for the proper and unfussed evolvement of the construction process. The substance of the duty of solidarity, its limits, and the importance of implementing it in particular ways in the course of executing a construction contract are explored hereunder.

Major Issues in International Construction Law, First Edition. Mauro Rubino-Sammartano and Troy Harris.
Published 2026 by John Wiley & Sons, Inc.

The Concept of Solidarity in International Construction Law

The notion of solidarity in contractual relationships first appeared at the end of the nineteenth century, pursuant to the realization that a number of contracts was not negotiated and were instead concluded on terms provided unilaterally by the strongest contracting party; most often these were housing, employment, or transportation contracts. At the beginning of the twentieth century, a number of legislative interventions attempted to restore the balance of powers between contracting parties in such socially sensitive contracts. Thereafter, case law took over to emphasize that contracting parties ought to act in a spirit of trust and good faith in all their contractual relationships; for instance, the parties to a distribution contract ought to collaborate in order to assist the distributor to face competition.[1]

Through these processes, the "contractual solidarity" theory emerged. This theory adopted new ideas toward enhancing the social aspect of contracting, with cooperation and ethics prevailing in the relations between parties to the agreement. The French jurist, René Demogue, was one of the first to propose this modern concept and another French jurist, Denis Mazeaud, has since been one of its defenders. This theory makes loyalty, respect, and cooperation a new basis for the contract.[2] Demogue saw the contract as some kind of microcosm – a small company organization, which by its nature obligates the parties to cooperate in order to achieve common goals. This requires: (i) each party to take into account the legitimate interests of the other(s) and (ii) to act in a manner that protects such interests and brings them all together on a common ground.[3]

Thus, the first of the solidarity components is a fraternizing bond between the contracting parties, in the sense of each party showing consideration for the interests of the other. The second component is the conciliation of those interests, in the sense of making a fair allocation of contractual obligations and benefits among the parties and safeguarding the ultimate superior interest of fulfilling the

1 Colloque "Le solidarisms contractuel: mythe ou réalité?," Centre d' Études Juridiques de la Rochelle (CEJLR), 9/7/2024 (https://cejlr.univ-larochelle.fr/Le-solidarisme-contractuel-Mythe,143.html).

2 Abdulraoof S.E., Al-Bsherawy A.K., Laffa N.S., *Contractual Solidarity and Future Contractual Relations: a Comparative Study* (Central Asia and the Caucasus, 2022) (vol. 1, Issue 1) (English ed.), p. 686.

3 Gh. Tabi Tabi, "Ajustement necessaire du volontarisme contractuel: du voluntarisme au solidarisme?," (2014) 44 R.D.U.S., p. 80. See also colloque "Le solidarisme contractuel: mythe ou réalité?," Centre d' Études Juridiques de la Rochelle (CEJLR), 9/7/2024 (https://cejlr.univ-larochelle.fr/Le-solidarisme-contractuel-Mythe,143.html). See further, A. Cayol, "La réforme du droit des contrats, une consecration du solidarisme contractuel?," HAL (https://hal.science/hal-04022874), 20 Feb 2024, p. 3.

purposes of the project. Under the light of solidarity, parties see each other not only as independent, individual entities but also mainly as partners, motivated by the desire to bring to fruition a joint overarching outcome.

A Legal Norm, a Legal Principle, or a Practice?

The legal nature of contractual solidarity has been the subject of debate and varied opinions: depending on the degree of bindingness attributed to it by the parties, it has been widely discussed whether it has the force of a standard legal rule or is a general principle or even a mere behavioral practice.

A Legal Norm

Although unusual, in some cases, the parties may agree to express terms of dealing fairly, in good faith and in cooperation with one another in the context of the construction contract. Such express terms should be treated as rules, setting an objective standard of action and generating legally binding obligations, recognized and enforceable by courts, such as:

- negative obligations, not to engage in certain activities hindering or preventing the other parties from performing and
- positive obligations, to carry out certain activities/take all steps reasonably necessary for enabling/facilitating performance by the other parties and the project's progress.

A Legal Principle

More often, solidarity is viewed and treated as a value permeating the construction contract and its functioning. Although such treatment affords it a lower normative status, it does not render it devoid of legal weight. Even as a general proposition, solidarity does not have a merely "inspirational" character but is instrumental in the effective implementation of the construction contract.

At the very least, it is a tool for interpreting other contractual provisions (a "hermeneutic tool"): it can guide how such provisions should be construed (offering the justificatory reasons and *ratio* for their introduction in the contract) and applied.

In the context of unforeseen circumstances arising during the contract's implementation, affecting the contractual equilibrium and the schedule of works and payments, it can act as a "mechanism" to outline the manner in which parties should perform, protecting the parties' legitimate expectations and facilitating an efficient performance of the contractual goals.

In its most dynamic form, it may also be the *"ius variandi,"* providing the ground for modifying existing contractual provisions and establishing particular exceptions to agreed terms, when situations that had not been contemplated by the parties arise.

An Autonomous Principle or an Ancillary One Linked to That of Good Faith?

The principle of good faith in the performance of contracts (especially of contracts of duration that require a certain level of cooperation between the parties, such as construction contracts) encapsulates the essence of the duty of solidarity.

"Good faith," which has moral underpinnings, is an elusive concept that is approached differently by the various legal systems of the Continent and common law. In general terms, in the context of contracts, it involves compliance with honest standards of conduct (acting without malice or fraud) and cooperation by the parties to achieve the contractual objectives.[4]

In the context of construction contracts, in particular, good faith lies at the core of the contractual performance; it imposes on each party the overarching obligation to act in such a way as to preserve the interest of the other parties, to the extent that such behavior does not imply a major sacrifice on its part.[5] This general obligation encompasses a cluster of individual imperatives, which are all reflections of the parties' solidarity in servicing the contract. More specifically, according to these:

- Any discretionary powers conferred on a party in the contract to make decisions that affect all parties' commercial interests (e.g., to extend time for completion or grant a variation) should be exercised reasonably to promote the overriding purpose of the construction contract (to build the works with certain standards, within a specific time and budget). In other words, such discretionary powers should be exercised "honestly and on proper grounds and not in a manner that is arbitrary, irrational or capricious," and the relevant party should not engage in conduct that is misleading or deceptive.[6]
- All parties are prevented from acting (or willfully defaulting in acting) in a way that would render performance by another party/the other parties impossible or more difficult (e.g., the Owner from making excessive changes to the design or works); on the contrary, parties should do everything in their power to facilitate the other parties' performance of their obligations.
- Parties must adopt a coherent attitude and not behave in a contradictory manner to the detriment of the other contracting parties. For example,

4 Rosher P., "Good faith in Construction contracts under French law and some comparative observations with English law," *The International Construction Law Review (ICLR)* 32 (2015): 303.

5 Construction Law International, *The IBA Construction Projects Committee of the Energy, Environment* (Natural Resources and Infrastructure Law Section, 2020), (vol. 15, issue 2, June); [see the contribution by Arianna Perotti, FIDIC (International Federation of Consulting Engineers) around the world: Italy].

6 Tamburro P., "Controlling Contractual Discretion: the Limits placed on discretionary powers in Construction contracts," *The International Construction Law Review (ICLR)* 37, (October 2020): 394, quoting Edwards-Stuart J in Portsmouth City Council v. Ensign Highways Ltd (QBD (TCC) [2015], par. 112.

a party that by its actions and omissions has given another party to understand that it will not pursue a particular course of conduct (e.g., impose a penalty) should not subsequently act contrary to that legitimate expectation it created in the other party's mind.

According to some French jurisprudence, the concept of good faith during the execution phase of the contract means that the parties look after their interests, while exhibiting a minimum of transparency and a commitment to transacting with honesty and integrity.[7] Particularly with reference to (international) construction contracts, this last set of responsibilities essentially means that parties protect and defend their interests within the framework of the contract, while maintaining a standard of morality in their behavior that is necessary to guarantee respect for the legitimate interests of the other parties (e.g., in the form of disclosing an action that might hurt another party).[8]

In the context of another civil law system, that of Greece, the rule of *bona fide* performance by a party in the context of a contract (as enunciated in Article 288 of the Greek Civil Code) essentially obligates such party to act with forthrightness and fairness; to curtail the relentless pursuit of its individual interests and not be indifferent to the interests of the other parties that may be affected as a result of its actions/failures. Basically, each contracting party is expected to behave honorably and not engage in dishonest/deceitful conduct that undermines the interests of the other parties. Moreover, same contracting party is expected to not simply confine itself to mere fulfillment of its own obligation, but to show its inclination to collaborate with the other parties in them fulfilling their respective obligations.[9]

From the English, common law perspective, good faith in the context of a contract, in general, means that it is expected that "a contracting party will act honestly towards the other party and will not conduct itself in a way which is calculated to frustrate the purpose of the contract or which will be regarded as commercially unacceptable by reasonable and honest people."[10]

7 Rosher P., "Good faith in Construction contracts under French law and some comparative observations with English law," *The International Construction Law Review (ICLR)* 32, (July 2015): 306.

8 Abdulraoof S.E., Al-Bsherawy A.K., Laffa N.S., *Contractual Solidarity and Future Contractual Relations: a Comparative Study* (Central Asia and the Caucasus, 2022) (English ed.) (vol. 1, Issue 1), p. 688. See also Gh. Tabi Tabi, "Ajustement necessaire du volontarisme contractuel: du voluntarisme au solidarisme?," (2014) 44 R.D.U.S., p. 91–92.

9 See Agallopoulou P., Basic Concepts of Greek Civil Law, (2005): 256–257. See also Stathopoulos M., General Law of Obligations, (2004): 202.

10 Bailey J., "Faith, Hope, and Charity": the Role of Good Faith in Construction – a Common Law Perspective, 55 Revista Derecho & Sociedad, p. 20 quoting Leggatt J. in *Astor Management AG v. Atalaya Mining Plc* [2017] EWHC (Comm) at [98].

In *Sheik Al Nehayan v. Kent*, photographing (international) construction contracts and the demonstration of solidarity in their context in the form of the parties' behaving in good faith, Leggatt L. J. drew attention "to a category of contract in which the parties are committed to collaborating with each other, typically on a long term basis, in ways which respect the spirit and objectives of their venture but which they have not tried to specify, and which it may be impossible to specify, exhaustively in a written contract. Such 'relational' contracts involve trust and confidence but of a different kind from that involved in fiduciary relationships. The trust is not in the loyal subordination by one party of its own interests to those of another. It is trust that the other party will act with integrity and in a spirit of cooperation. The legitimate expectations which the law should protect in relationships of this kind are embodied in the normative standard of good faith."[11]

It is reported that the trend lately in English contracts is to expressly include good faith (*i.e.*, solidarity) duties;[12] a trend most useful for all participants in an international construction contract.[13]

A Contractual Practice/A Standard of Contractual Behavior

Another way of viewing and treating solidarity in the context of (international construction) contracts is as a behavioral command, prescribing a certain line of conduct [*i.e.*, solidarity both as an attitude and as practice(s) that may derive from such attitude]: fidelity to the agreed common purpose (concluding the project), fair responsibility/burden sharing (having regard to the nature of the project); fair and equitable distribution of costs (and benefits), among the contracting parties.

Again, although this approach to contractual solidarity attributes to it a lower degree of normativity and bindingness, it does not negate its strategic usefulness as a benchmark fixing the system of the contract's operation and guiding the parties' actions in the context thereof.

Solidarity may be present in the (international) construction contract under any of the above guises, even if no express reference is made to it as a notion, and govern the contract's entire lifecycle.

---

11 Darling QC P. & Bradford J., Contracts: Formation and Interpretation, in *Wilmot-Smith on Construction Contracts*, 4th ed. ed. R. Wilmot-Smith QC & P. Darling QC (Oxford University Press, 2021), par. 1.191, p. 72.

12 See, for instance, the reference in NEC forms to "the duty to act in a spirit of mutual trust and cooperation."

13 Klee L., *International Construction Contract law*, 2nd ed. 2018, 83. See also Prof. David Mosey, "Good faith in English Construction Law – What does it mean, and does it matter" in The International Construction Law Review (ICLR), vol. 32 (October 2015), p. 393, quoting from a range of standard form contracts, including the NEC3: "The Employer, the Contractor, the Project Manager and the Supervisor shall act as stated in the contract and in a spirit of mutual trust and cooperation."

Solidarity Rights or Obligations?

Solidarity is most commonly perceived and associated with concrete (direct and indirect) duties; nonetheless, on the reverse side of duties, lies each party's right to claim the benefits resulting from the affirmation and practice of solidarity in the context of an (international construction) contract. The relevant duties and rights can show up either at the stage of the contract's formation (negotiations and drafting), or at the stage of its execution.[14]

More specifically:

Solidarity Rights

In the form of rights, solidarity may be found, among else, in:

- a party's right to be provided with all necessary information at the contract's formative stage in order to form a free and clear-minded judgment of the contract's scope and terms, based on serving its own interests. Each party is bound to inform its prospective partner of all matters that would enlighten the latter in reaching its decision and which could be crucial in giving its consent.
- a party's right to be recognized as an equal partner in the project, despite any economic or other asymmetry of power with the other contracting parties. Construction contracts (especially of an international nature) mix players of different size and sophistication, yet all critical in attaining the best project outcome.

Solidarity Obligations

As a commitment or undertaking, solidarity can be traced, *inter alia*, in:

- the prohibition of unjustified rupture of the negotiations. If negotiations have been lengthy and conducted with a real likelihood that they would lead to a binding agreement, solidarity dictates that parties do not cease them abruptly or "abusively," without there being legitimate reasons.[15]
- the prohibition of including in the contract terms that are excessively onerous for a party to bear;
- the parties' obligation not to exploit an advantage deriving from a literal but mutually unintended reading of a contractual term;

---

14 Cayol A., "La réforme du droit des contrats, une consécration du solidarisme contractuel?," HAL (https://hal.science/hal-04022874), 20 Feb 2024, pp. 4–5.

15 Rosher P., "Good faith in Construction contracts under French law and some comparative observations with English law," *The International Construction Law Review (ICLR)* 32, (July 2015): 310–312.

- the parties' obligation to avoid unilateral, coercive and aggressive behavior in the performance of works or tasks;
- the parties' obligation to exercise any discretion they are afforded in a reasonable manner, in consideration of what they contemplated at the time of the contract's formation;
- the parties' obligation to consult and co-operate with one another and do all such things as are reasonably necessary for the successful execution of the contract in case of unforeseeable change of the original economic circumstances, having a material adverse effect on a party/parties [including renegotiating and modifying the contract so that the party(ies) is/are not materially financially disadvantaged because of the change].

Cognate Notions – Aspects of Solidarity

Being a multi-faceted concept, solidarity knows several forms of manifestation in the context of an international contractual relationship shaped for executing a construction project and various aspects thereof make their appearance during such contract's lifetime. The following are some aspects of the concept of contractual solidarity and expressions of the relevant duty.

Cooperation

Given the multiple bases, levels, determinants and situations (international) construction contracts have, which involve unknown contingencies, uncertain outcomes, personal conflicts and problem solving, parties are required, on the one hand, to co-operate actively, to the extent necessary, to perform each own's obligations and, on the other hand, to avoid conducting themselves in a way that would interfere with the other parties performance.

Active cooperation essentially consists in practicing optimal communication to define each party's needs, searching for adequate solutions and acting on the basis of mutual trust. The extent of cooperation required will depend on the contractual obligations imposed on the parties and also the surrounding circumstances.[16] The Owner's giving of complete, timely, and accurate instructions to the Contractor so that the latter can execute the work on time and uninterruptedly is a manifestation of such active cooperation.

On the other hand, not acting in a way that would prevent performance of another party's contractual obligations, consists in each party doing nothing to cancel the set of circumstances that are necessary for the contractual agreement to be implemented; e.g., the Owner preventing the Contractor from taking proper

16 Darling QC P. & Bradford J., Contracts: Formation and Interpretation, in *Wilmot-Smith on Construction Contracts*, 4th ed. ed. R. Wilmot-Smith QC & P. Darling QC (Oxford University Press, 2021), par. 1.145–1.146, p. 57.

possession of the site or providing erroneous or misleading information required for the necessary permits for construction works to be issued.[17]

Especially in the event of obstacles hindering implementation of the contract, cooperation involves each of the party being obliged to take all available reasonable measures to eliminate such obstacles. It also involves the less or non-affected party(ies) to take into consideration the (corporate, financial, etc.) situation of the other parties to the contract that are encumbered by such obstacle in performing and to take action to alleviate the unexpected imbroglio encountered therefrom and re-balance the contractual interests.

Mutual Assistance

In international construction contracts, parties are required to interact over time, in a variety of settings. In this context, they ought to provide mutual assistance as a matter of moral duty (as opposed to as a matter of law).

Nonetheless, such a purely moral obligation can have legal implications and give binding force to a non-legally binding act; for instance, such would be the obligation to provide and exchange critical information required for a party to comply with its contractual undertakings (e.g., to apply for and obtain necessary licenses and permits).

Equity

This is essentially the aspect of solidarity consisting in dealing fairly with those contractually bound: "a thread runs through our contract law that effect must be given to the reasonable expectations of honest men" [Lord Steyn, LQR 1997, 133 (Jul)].

Fair dealing in the context of a (international) construction contract can be translated into: (i) the parties facilitating and supporting each other's performance tasks and (ii) a party not insisting on the performance of an obligation it has prevented the other party from performing.

The Purpose and Role of Solidarity in International Construction Contracts

Given the special nature and requirements of international construction contracts, the acceptance of solidarity as a value permeating the entire contract and its application both as standard behavior and as course-correcting mechanism when need arises play a key role to the contract's successful carrying out. The purpose

17 P. Darling QC & J. Bradford, Contracts: Formation and Interpretation in: Wilmot-Smith on Construction Contracts, 4th ed., R. Wilmot-Smith QC & P. Darling QC (editors), Oxford University Press 2021, par. 1.147–1.148, p. 57.

of solidarity is (i) to enable and facilitate the smooth and efficient performance of the contractual scope at normal times, and (ii) to safeguard the contract's survival under adverse circumstances, which put contractual relationships and the contract's scope (execution of the project) to the test.

Acknowledgment of the importance of solidarity and systematic practice thereof in the context of international construction contracts has been seen to have positive effects on the operation of the contractual mechanism both at normal times and under extraordinary circumstances.

Solidarity in Ordinary Circumstances

In the course of the contract's ordinary evolution, solidarity:

- contributes to the effective performance of the contracting parties' obligations (through reinforcing commitment to cooperation and good interface). Parties become result-oriented and willing to be more flexible in their approaches to practical issues, while maintaining schedule and cost.
- improves relations among participants to the construction project. Solidarity comes from the Latin term *"solidum"* that connotes shared responsibility for a collective objective.[18] Practiced in the context of a contract, solidarity nurtures the sense of all parties sharing responsibility for making the contract work and for the project's successful completion. Authors have spoken of fostering a contractual "we."[19] Solidarity is the reminder and guardian of the existence of the "common contractual interest" which binds the parties and which the parties have consciously decided to serve by joining their wills and resources, putting aside the costs and sacrifices/compromises to their own private interests that may be required in this respect.[20] Solidarity highlights that all participants are key players in the construction process. It encourages the different parties that are summoned to interface with each other in this process (the owner, the designer, the contractor) to understand each other's individual problems, expectations and values and work as a team for

---

18 According to Laitinen and Pessi [A. Laitinen & AB Pessi, Introduction in Solidarity: Theory and Practice, Laitinen & Pessi eds, 2018], in Roman Law *"obligatio in solidum"* involved the group liability of joint debtors. According to them, the term gradually came to be used in a broader manner as "motivated readiness for mutual support."

19 Gh. Tabi Tabi, "Ajustement necessaire du volontarisme contractuel: du voluntarisme au solidarisme?," (2014) 44 R.D.U.S., pp. 100–101. See also Laitinen and Pessi [in A. Laitinen & AB Pessi, Introduction in Solidarity: Theory and Practice, Laitinen & Pessi eds, 2018] who argue that "solidarity is often based on 'we' thinking."

20 Gh. Tabi Tabi, "Ajustement necessaire du volontarisme contractuel: du voluntarisme au solidarisme?," (2014) 44 R.D.U.S., p. 120.

the benefit of the project, focusing on solutions to problems. Endorsement of that attitude is a sure way to a lasting, balanced and satisfactory contractual relationship. It creates a more harmonious working environment, involving less paperwork and confrontations and resulting in greater efficiency, cost effectiveness and strengthened quality of the project.

- leads to effective risk-sharing and resulting benefits. Construction projects are particularly prone to risks that can affect price and time; some relate to the parties (e.g., the Owner's risk to secure funding, the Designer's risk to deliver the design, the Contractor's performance risk) and some to external circumstances (e.g., civil unrest, weather, availability of site, materials, and labor). Solidarity at the level of risks means that risks are embraced by all parties, rather than cast on a single one. Appropriate and adequate risk sharing can result in price reduction; if contractors can eliminate certain risks from their pricing analysis (such as force majeure, environmental, schedule risks etc.), pricing reductions can be effected. For instance, if at the pre-tender stage, the Owner was charged with obtaining information on site conditions and passing them on to the Contractor, the risk of local conditions, access to the site, availability of resources, etc. would be substantially reduced for the Contractor; in this case, the latter would not only be able to make a lower bid at the tender, but also to maintain a positive cash flow and achieve reasonable profitability during performance of the contract works.[21]
- resolves routine problems and disagreements (while they are live) and/or prevents them from escalating into disputes that could jeopardize the contract's streamlined execution. Proper implementation of the duty of solidarity may prevent initial disagreements from developing into major conflicts. In this respect, solidarity expressed through (i) the contracting parties communicating their requirements and expectations openly, clearly and effectively, in a timely and straightforward way and (ii) all parties' staff exercising good conflict management skills can help make the construction process more fluent and coordinated.

Solidarity in Light of Extraordinary Circumstances

Under such circumstances solidarity is paramount for:

- effectively addressing unforeseen events or occurrences entirely beyond the parties' control, utterly or majorly preventing them from performing (*force majeure* events). The longer the contract period, the higher are the chances

21 Yeang F., "Fixed Price Design and Build Contracts: Risk Allocation and Analysis," *The International Construction Law Review (ICLR)* 13, (July 1996): 303–304.

for such events to occur and put a strain on the contractual bond. In those cases where one of the parties is placed in a situation of vulnerability and the need for restoring contractual equilibrium arises, solidarity becomes most relevant.

- In such cases, especially if it has been agreed that such events or occurrences amount to grounds for exempting the inflicted party from liability and the obligation to indemnify the other party, the duty of solidarity dictates that the party facing such major difficulties should inform the other party promptly. Following such notification, all parties should meet and consider the optimal reaction and measures for containing the adverse effects of the unforeseen circumstance, while assisting resumption of the works. In the same context, the parties should communicate, meet and engage into discussions for fairly allocating among them the burden of the unforeseen forced changes to the contractual parameters and balance, in order to protect the integrity of the contract's interest and the viability of the contractual relationship.
- effectively responding to unforeseen circumstances that do not amount to force majeure events yet impact the contract's implementation (agreed time schedule and contract price) and warrant a departure from the strict application of a contractual provision. Particularly when operating in an international context, construction contracts are prone to contingencies that can change the premises on which risk was allocated by the parties at the contract formation stage (e.g., unforeseen ground conditions, change in the availability of materials and their cost, etc.). In such cases, compliance with the duty of solidarity is vital for the project's potential success. Where a change in the circumstances results in the contractual equilibrium getting substantially shaken and in the legitimate expectations of the party finding itself in a vulnerable position being overthrown or severely restricted, solidarity steps in to restore the upset balance and preserve the contract intact and immutable on the outside, while apace with the revised contractual reality in terms of its content. In these situations, the duty of solidarity translates into the parties taking action to ensure that the conditions that originally favored the contract's conclusion remain essentially the same; just as the will of the parties to stay in a contractual relationship for the achievement of the contract's scope.
- First and foremost, this is achieved through the parties exchanging information in respect of such circumstances and their impact and communicating their requirements and expectations clearly and effectively. On a second level, solidarity dictates that the party in a position of authority and less affected by the unexpected event avoids to further destabilize the contractual equilibrium by, for example, demanding the performance by

the other party of obligations that would be overwhelming in the circumstances. Instead, there is an obligation to consult/confer with the afflicted party and to seek to agree on a set of specific measures that would, on one hand, give the hampered party some leeway (e.g., by extending a deadline or suspending performance and/or related penalties or instalments) and, on the other hand, assist in advancing the project (e.g., by sub-contracting part of the works to a third party) and preserving the contractual relationship (e.g., by executing the requisite contractual amendments to adapt to any change in external circumstances which renders continuation of the contractual relationship unsatisfactory for either party).

The Limits of the Duty of Solidarity

The Express Contractual Provisions

The duty of solidarity is meant to reinforce the parties' obligation to comply with contractual provisions and respect agreements they have lawfully entered into - not to substitute or override them. Solidarity does not require self-sacrificing behavior; parties are not required to refrain from exercising their respective contractual rights, nor are they precluded from promoting their own legitimate interests by reason of this duty.

Contractual certainty in the performance and enforcement of the terms of the contract and the parties' freedom to determine the risk allocation among themselves and exercise their discretions and rights pursuant to the contract are important to be preserved and are not supposed to be sacrificed on the altar of contractual morality.

The Measure of Solidarity

Solidarity is closely connected to the ideas of necessity and commercial sensibility: exercise of the duty of solidarity is required to the extent that this is necessary to make the contract workable.

As regards the manner in which this should take place, reasonableness is a key parameter. Reasonableness helps identifying the degree of solidarity to be demonstrated in a given factual situation by way of specific actions being taken by the parties. The idea is for parties to be prepared and take fair measures, which adequately respond to the situation at hand and protect the other parties' legitimate interests, while also taking into consideration their own interests and the long-term purpose of the contract, so as to leave the latter's internal coherence undisturbed.

It is one thing to act in a manner consistent with honesty, avoid opportunistic behavior and respect the other parties' reasonable expectations and quite another to completely subordinate one's own interests to the other party's (parties') position and/or the common objective of the contract. As Vinelott J. put it in *Merton London Borough Council v. Stanley Hugh Leach Ltd*, a party is not required "to go further than what might be called 'enlightened self-interest,'" such being the implied obligation, on one hand, not to do anything to prevent or delay the other party from performing under the contract and, on the other hand, to do whatever necessary to enable a contract to be carried out.[22]

Applying the principle of proportionality is helpful in determining such appropriate measures for discharging the duty of solidarity. The above notions (reasonableness, commercial sensibility, necessity, proportionality, and coherence) form the internal boundaries of solidarity.

Implementation of the Duty of Solidarity in International Construction Contracts

Broad Categories of Solidarity Manifestations

In the context of executing international construction contracts, conscious implementation of the duty of solidarity is required both:

- in case of indivisible shared obligations between the contracting parties (i.e., where there is an obligation of result), whereby the parties' fates are inherently tied together and the obligation can only be fulfilled by all of them simultaneously; and
- in case the shared obligations are divided and each contracting party has its own "share" of the performance.

Depending on the particular contractual arrangements, the duty of solidarity can manifest in a contracting party:

- remaining passive (abstaining from specific conduct or actions, including in the form of exercising rights) in order not to hinder the other party's/parties' performance (e.g., the Owner refraining from performing an activity at the worksite that would obstruct the proper execution of the Contractor's work, from taking physical control of the execution of the works);

22 Prof. Mosey D., "Good Faith in English Construction Law – what does it mean, and does it matter," *The International Construction Law Review (ICLR)* 32, (October 2015): 397.

- engaging in specific actions that may be required to enable or facilitate another party's performance (e.g., the Owner providing necessary information or clarifications or correcting a false assumption made by the Contractor). On certain occasions, it could be performing an obligation the party had not expressly committed to;[23] and
- undertaking certain actions in order to induce the other(s) contracting party/ parties to comply with the obligations resting upon it/them.[24]

Specific Forms of Solidarity

At a micro, practical level, in the context of executing a construction contract, solidarity is articulated in specific forms of conduct (specific "ways of doing business") responsive to specific factual instances, such as in:

- notices from one party to let the others know of the existence of special circumstances requiring all parties' consideration and determination. Such would be, for instance, the Contractor notifying the Owner promptly and providing full details upon encountering adverse conditions, in order for the Owner to make itself available for discussions to address any relevant matters ensuing from those conditions.
  - meetings – mutual consultations, to discuss solutions for resolving issues in the most efficient way before generating disputes. Such would be, for instance, meetings agreed to be held among the Owner, the Contractor's supervisory staff, and the project-wide consultants on a regular basis (e.g., monthly) or at moments of critical construction events, in order to address any issues arising in the execution of the Project (e.g., schedule, payments, performance conditions, potential disputes, claims or other controversies) and proposed solutions.
  - assistance/support to each party to properly discharge its obligations. This could be, for instance, in:
    - providing information/directions. The Owner (project owner/ developer) is required to disclose all relevant data that are necessary for the proper completion of the project by the Contractor/Engineer [provide details on the project at stake, its specifications, the precise goals that have to be achieved and the contract's general purpose, as well as any

23 Candida Leone, What Solidarity? A Look behind the veil of Solidarity in "Corona Times" Contractual Relations, Netherlands, *Journal of Legal Philosophy* no. 50, 2021: 241.

24 Natasha Nedeski, Concepts of Solidarity in the International Law of Obligations, Amsterdam Law School Legal Studies Research Paper No. 2020-11, Amsterdam Center for International Law No. 2020-09.

report/findings on any special conditions existing at the beginning of the works or discovered/arisen in the course of their performance (e.g., adverse ground conditions)]. Similarly, the Contractor should inform the Owner when it considers that executing the contract in accordance with the latter's specifications might, in fact, be harmful or advise the Owner on the choice of materials[25] to prevent the latter from incurring damage (duty of examination and warning).
   - applying for/obtaining licenses/permits/authorizations to the extent necessary for the Contractor/Sub-contractors to perform their work. The Owner must effect all administrative procedures and formalities that are necessary for the commencement and carrying out of the works.
   - providing access the Worksite (as well as a full report of the site's investigation);
   - delivering any necessary equipment and materials, etc.
- coordination and good interface of all project participants. In this context, solidarity could appear in the Owner facilitating the coordination of the various sub-contractors, employees and representatives working together on the project, their continuous participation in the testing of the works and, ultimately, participation in the final take over. This could be the result of a more collective action effected through a Management Team, made up by the Project Manager, the Owner's and the Contractor's Construction Managers and Managers of other disciplines (e.g., Design Manager, Environmental Issues Manager, etc.) for coordinating interaction of the crews, directing and inspecting the various teams, supervising their work, taking initiatives to facilitate the best solutions for the project, monitoring its progress and making recommendations to the appropriate parties, supporting in-between consultations, etc.
- cessation [of improper action(s)] and reparation. Solidarity can also take the form of the parties putting in place protocols and taking initiatives to stop early enough conduct that falls short of the project's quality standards or threatening to derail the parties' time schedule, budget, and legitimate expectations. Such initiative could be, for instance, the Contractor communicating, timely and adequately, to the Owner any departures from the original scope of works or the materialization of any risks. In said set of cases, solidarity could also take the form of rectification costs being paid to the Owner for not having received prompt and adequate notice of the occurrence of any risk events that affect the works and their proper completion.

25 Rosher P., "Good faith in Construction contracts under French law and some comparative observations with English law" in The International Construction Law Review (ICLR), vol. 32 (July 2015), p. 308.

- employment of preliminary ADR procedures for attempting an initial resolution of disputes though amicable discussion/structured negotiation – mediation.[26] These are personalized, consensual conflict-resolution procedures, appropriate for opening channels of communication among the parties to the contract and preserving their on-going relationship. Their advantage lies in being informal, confidential, expedient, and low-cost techniques by which the parties can attempt to resolve their differences at a primary phase before facing binding arbitration or litigation as the final, but perhaps necessary, choice for conclusive resolution. Under this approach, parties are called to commit to providing open and timely disclosure of the relevant facts, information and documents to facilitate the resolution of each dispute. They are also called to fully investigate the issues in dispute and explore possible settlement scenarios. The most productive use of that process can lead the parties to adapt the contract to take account of realities.
- renegotiation of the critical contractual terms that are impacted by external circumstances, which have altered the underlying economic relationship among the parties and have rendered performance completely unprofitable for a party. Essentially, what solidarity dictates in this context is that the parties engage in active attempts to find constructive ways to re-balance the shaken economic relationship. Such attempts could be in the direction of revising the agreed risk allocation criteria, amending/suspending a contractual clause or reducing any sanctions originally agreed for failure to perform a certain obligation in a given way or by a given time. Of course, this does not mean that each party is under an obligation to accept the other party's proposals or to sacrifice its private or financial interest in this respect.

Conclusion

Construction projects are generally affairs of considerable magnitude and technical complexity, requiring extensive (time, human, material and financial) resources over a lengthy period of time and involving a multitude of specialists with divergent needs and interests that need to be bridged for the sake of successfully completing the project.[27]

---

26 Taylor R. G. and Hinkle, B. Jr., "How to use ADR clauses with standard form Construction industry contracts." *ICLR* 13, (January 1996): 61 noting that mediation "is especially suited for construction disputes, which frequently involve multiple parties and complex technical issues that are difficult to explain at trial."

27 Tamburro P., "Controlling Contractual Discretion: the Limits placed on discretionary powers in Construction contracts," *The International Construction Law Review (ICLR)* 37, (October 2020): 403 and footnote 92.

The relevant contact is in itself "a carrier of solidarity,"[28] in the sense that it sets out the common goal toward which the various parties involved are to work and defines the practical commitments each party undertakes in contributing to the achievement of that common goal.

Contractual solidarity makes that common objective the main objective of the parties' endeavors under the contract; the epicenter of their efforts, over and above their self-interests. Furthermore, it brings to focus the parties' legitimate expectations, dictating that each party is obligated to consider, in addition to its own, the interests of the other parties and their materialization[29] and to act in a manner consistent with honesty, respecting fair dealing in sharing the risks and rewards in the process of effectuating the contract.

The above dictates could take the form of an express provision in the contract or be read as underwriting its spirit, permeating all its facets and entering the picture to ensure harmonious cooperation and to moderate the scope of obligations, which could become overwhelming for a party found in a vulnerable position in given circumstances.

Under normal circumstances, solidarity operates in a formative way, forging the parties' collaborative consciousness in the direction of prioritizing the interest of the project itself and allowing for more flexible behaviors in the context of the complex construction project contract.

Contractual solidarity acquires special value when the contract's implementation during the project's execution faces challenges on account of external factors or unforeseen circumstances in respect of which there may either be no contractual provisions, or their strict application may have such an impact on either party's position or interests as to make the contract and the project (highly) onerous and unappealable. In such instances, the duty of solidarity steps in to correct the balance of contractual powers and afford adequate protection to the legitimate expectations of any contracting party that finds itself in a vulnerable position.

In that sense, contractual solidarity interpolates a humane dimension in the construction contract and qualifies the application of the principle of *pacta sund servanda* in light of fluid/mutable factual situations.

Solidarity can show up in many forms: as a right, just as well as an obligation, as specific action(s) or intentional abstention from engaging in an activity, as a modality creating, modifying, or altogether extinguishing an undertaking

---

28 Abdulraoof S.E., Al-Bsherawy A.K., Laffa N.S., *Contractual Solidarity and Future Contractual Relations: a Comparative Study* (Central Asia and the Caucasus, 2022), (English ed.) (vol. 1, Issue 1 2022). 689.

29 Abdulraoof E., Al-Bsherawy A.K., Laffa N.S., *Contractual Solidarity and Future Contractual Relations: a Comparative Study* (Central Asia and the Caucasus, 2022), (English ed.) (vol. 1, Issue 1 2022), p. 695.

the performance of which may be overwhelming for a party under the circumstances, etc.

Regardless of the various forms it can take, these should all meet the exigency of reconciling the parties' interests toward serving the common contractual objective, without extinguishing them. Practicing solidarity cannot and should not be in a manner as broad as to destroy the certainty of agreed terms under the contract. Solidarity is not subversive of the contractual agreements but reinforcing good partnership for the benefit of the project.

22

Contract Termination

Marlena Harutyunyan and Réka Àgnes Papp

Introduction

A construction contract ends naturally when parties have performed all their obligations.

It is also possible for parties to make use of their contractual freedom and to terminate their contract by mutual agreement.[1] However, termination by mutual agreement happens rarely in practice. A mutual agreement to terminate is more likely to occur in cases of "descoping," when the scope of the original contract is reduced, and parties choose to enter into a new one, rather than to amend the initial, structurally off balance, contract.

The most common occurrence remains when a party decides to unilaterally terminate the contract before it is fully performed. Indeed, in practice, most construction disputes concern unilateral termination and its consequences.

Given the important legal, practical, and financial consequences of termination, parties to construction contracts will invariably include termination clauses to manage the risks related to its early ending. These clauses, also referred to as "forfeiture clauses," are contractual provisions that allow one or both parties to bring their contractual obligations to an end before they have been fully performed.

Termination clauses are less standardized than many other clauses in international contracts and are generally subject to negotiations between parties.[2] This is because termination rights are closely related to the core of each contract and to the specificities of parties' contractual relationship.[3]

1 Article 115 of the Swiss Code of Obligations ('SoC'); Müller, 2023, para. 65; Chaix, 2021, para. 20, Fontaine and de Ly, 2006, p. 577; Article 1193 of French Civil Code.

2 Fontaine and de Ly, p. 565.

3 Fontaine and de Ly, p. 565.

Major Issues in International Construction Law, First Edition. Mauro Rubino-Sammartano and Troy Harris.
Published 2026 by John Wiley & Sons, Inc.

No matter how sophisticated parties and their lawyers, the construction of any given contractual clause will most probably be debated among parties once a dispute arises. As will be shown below, not only the governing law plays an important role in the interpretation of termination clauses or in defining general rules on the statute of limitations but national laws also specifically regulate parties' unilateral right to put an end to their contractual relationship. These national laws, for the most part, do not impose mandatory rules regarding contract termination, and parties may in general define the conditions and consequences of termination.[4]

This chapter will first examine the formal preliminary steps that must be completed before a contract can be terminated (Section "Formal Requirements for Termination"). It will then analyze the grounds for termination (Sections "Grounds for Termination"), as well as the interplay between them (Section "Interplay Between Different Grounds for Termination"). Finally, the consequences of termination will be set out (Section "Consequences of Termination").

The authors have analyzed the topic from a comparative law perspective. Most of the cases described herein were decided in application of Swiss, English or French law.[5] On the one hand, this can be explained by the authors' legal and professional backgrounds. On the other hand, these laws are often chosen by parties as the applicable law to their construction contract. For instance, Swiss law frequently applies to international construction contracts because of its perceived neutrality, easy accessibility, and conciseness.[6]

This contribution will not touch upon specific issues that may be related to termination, but that are general causes for release from performance, such as hardship or *force majeure*. Termination as a result of nullity or voidability will also not be considered. Termination clauses regarding nullity or voidability remain rare, mostly because rules on contract formation are traditionally considered mandatory.[7]

Formal Requirements for Termination

Termination of construction contracts gives rise to significant practical issues, such as demobilization of resources, dismantling of equipment, return of guarantees, notice to sub-contractors, insurance issues, etc. Considering the practical

4 Fontaine and de Ly, p. 565.

5 The analysis under French law concerns only private construction contracts and does not include public contracts.

6 Scherer and Schneider, 2007, p. 559.

7 Fontaine and de Ly, 2006, p. 574.

ramifications of termination, construction contracts typically provide formal requirements for the notice of unilateral termination, i.e., a form for the notice to comply with and a time-period upon the expiration of which termination becomes effective. In some cases, a prior step to issuing a notice of termination may also be required – a "show cause" formally requesting the other party to rectify the breaches or a notice of default requesting the other party to cure the breach within a given period and provide evidence of the correction.[8]

The form requirements often include a notice in writing announcing grounds for termination, namely whether it is for cause or for convenience. They may also include provisions on who should send the notice or specific persons to whom the notice must be sent and/or copied in addition to the contractor or the owner. The notice period may vary depending on whether termination is for cause or for convenience. In exceptional circumstances, the contract may also allow immediate termination without prior notice.

For example, in the latest 2017 Red Book edition of International Federation of Consulting Engineers (known as FIDIC), which is the most commonly used form for international construction contracts,[9] termination for convenience by the owner must be in writing and state that it is given under Subclause 15.5, i.e., for convenience. The notice takes effect 28 days after the later of the dates on which the contractor receives the notice or the owner returns the performance security.[10] When it comes to termination by the owner for default, the formalities depend on the specific default on which the owner intends to rely. In some cases, such as contractor's failure to comply, abandonment of works, failure to proceed, etc., the owner must first issue a notice of intention to terminate and give the contractor a period of 14 days to cure the default. It is only if the default persists upon the expiration of this period that the owner may issue a notice for immediate termination of the contract.[11] In other cases, namely unauthorized subcontracting or assignment, bankruptcy or insolvency, and corrupt practices of the contractor, the owner may terminate the contract immediately.[12] The same applies for the contractor's termination for default, which is the only ground available to the contractor. The contractor must give a written notice of its intention to terminate, specifically mentioning the ground for termination, and giving the owner an opportunity to remedy the matter within 14 days of receiving the notice, and only in case of the owner's failure to do so, terminate the contract. Again, in some

8 Bailey, 2016, para. 9.83.

9 Seppälä, 2023, p. 1.

10 FIDIC Red Book, Subclause 15.5.

11 FIDIC Red Book, Subclauses 15.2.1 and 15.2.2.

12 FIDIC Red Book, Subclause 15.2.

exceptional circumstances, such as the owner's insolvency, prolonged suspension of works, assignment of the contract without the required agreement and the owner's corrupt, fraudulent, collusive or coercive practice, the contractor may terminate the contract immediately.[13]

While the formal requirements for termination should in principle be complied with, the specific consequences arising from a party's failure to do so would need to be assessed based on the contract itself and the applicable governing law. English law is particularly stringent when it comes to the application of contractual provisions relating to the timetable of giving notice. Any notice that is given too early or too late will be ineffective in law.[14] Another peculiar feature of English law is that if a party fails to comply with the conditions of giving termination notice, this might be construed, depending on the circumstances, as a repudiation of the contract by the otherwise innocent party.[15]

In addition, formal requirements for termination may also be imposed on parties by the governing law and imply significant practical consequences. For instance, under French law, the specific insurance for the defects affecting the soundness of the structure or rendering it unfit for its intended use[16] is actionable only after a demand letter to cure the defects has remained unsatisfied and the construction contract has been terminated for contractor's default.[17] The *Cour de cassation*[18] specified that to trigger the implementation of the insurance, such a prior notice, must be issued by the owner itself or its agent. Notably, the *Cour de cassation* refused the possibility of the notice being issued by the contract administrator, even if the latter was entitled to give directions to the contractor for the performance of the works.[19]

In practice, this means that the party giving the notice must be either expressly vested with the authority of notifying termination of the contract by the owner and act as the owner's agent or its authority to give notice must come from the contract between the owner and the contractor. Furthermore, the *Cour de cassation* considers this formality as an essential one: parties may abstain from issuing prior notice to termination for default only in exceptional cases when it is impossible or useless, for example in case of the contractor's insolvency.[20]

---

13 FIDIC Red Book, Subclause 16.2.

14 Hopkins, 2023, para. 3.15; see also in this sense *Eminence Property Developments Ltd. v. Heaney* [2010] EWCA Civ 1168, para. 65.

15 Hopkins, 2023, para. 315, for a contrary example see *Eminence Property Developments Ltd. v. Heaney* [2010] EWCA Civ 1168, para. 65.

16 Article 1782 of French Civil Code.

17 Article L.242-1 of French Insurance Code.

18 The Supreme Court in the French judicial order.

19 See Cass. Civ. 3ème, 7 September 2022, No. 21-382.

20 Cass. Civ. 3ème, 7 September 2022, No. 21-382.

Finally, it should be noted that in many civil law countries influenced by the old French law (the original Civil Code of 1804), unilateral termination of a bilateral contract, such as a construction contract, may be unavailable in law.[21] The formal requirement for the termination in such a case involves obtaining a court order from the competent court to be able to terminate the contract, unless parties have expressly agreed otherwise.[22] An illustration of a "way out" from the contract without complying with such a formal requirement may be found in the ICC case No. 22899/DDA.

A contract for the construction of a sports stadium, governed by Gabonese law and French law on a subsidiary basis, was unilaterally terminated by the contractor for the owner's failure to fulfil its obligations to finance the project and make the payments under the contract. The contractor then left the site and initiated arbitration proceedings against the owner to obtain payments for the works performed and compensation for the loss suffered. The owner objected to the validity of such termination. Considering that the contract expressly stipulated that termination must be ordered by the competent court, the arbitral tribunal found that the contractor's notice for termination was in breach of the contract and not valid. However, the tribunal pronounced rescission of the contract based on the owner's serious contractual breaches – a ground available in law but also stipulated in the contract by a reference to the public procurement code containing the provision. Noting that parties agreed that the contract *de facto* ceased to exist when the contractor gave notice of termination, albeit invalidly, the tribunal considered that rescission should take effect at the same date.

Thus, failure to comply with formal requirements was excused by relying on a different ground for termination.

Grounds for Termination

Termination for Cause

Termination for cause usually involves the default of one of the parties. Put differently, it is preceded by a breach of the contract giving the right to the innocent party to terminate it. Parties may also define a cause for termination, which is not strictly speaking a breach. For instance, construction contracts frequently provide that insolvency of the other party may justify immediate termination by the other party.[23]

21 Seppälä, 2023, pp. 77, 1001.

22 Seppälä, 2023, p. 1001.

23 See the 2017 FIDIC Red Book, Subclauses 15.2.1 (g) and 16.2.1 (i).

Parties have the freedom of defining the breaches that may justify the contract's termination. In practice, however, because of the substantial consequences of termination, only the most serious breaches in parties' view result in termination. The most common grounds for termination by the owner are contractor's unsatisfactory performance or inability to perform,[24] such as nonconformities with the industry best practices or standards defined in the contract, and undue delays or insufficient rate of progress. For example, in the 2020 ICC Model contracts, material breach by the contractor, that may result in termination, includes but is not limited to:

Persistently failing to produce Design documents in accordance with the Schedule of Contractual Dates or any agreed revised Schedule;

A substantial and continuing failure to proceed with the Works in accordance with the Schedule of Contractual Dates and Good Practice despite the warning given 30 days in advance that it is failing to do so in respect of specified areas of the Works;

Failure to achieve Taking-Over/Provisional Acceptance of the Works after the period when the maximum amount of Delay Damages applies and there has been a continuing failure despite the warning given 30 days in advance;

Failure to correct any Defects notified by the Employer within 30 days from the Employer's Notice thereto;

Failure to comply with labor, health, safety, and environmental protection laws and regulations;

Failure to achieve Minimum Performance of the Works and make good such nonfulfillment within 30 days from the Employer's Notice thereto.[25]

Concerning termination by the contractor, the event that gives rise to termination usually involves a grave breach of the contract by the owner, such as a failure to pay for the works performed or unlawful interference with the performance, which seriously affects the contractor's ability to perform as per the original agreement.[26] By way of illustration, in the 2020 ICC Model, material breach of the owner includes but is not limited to:

Persistent failure or refusal to complete the Employer's Requirements;

Frequently instructing changes to the Employer's Requirements or the Works without acknowledging that such change constitutes a Variation under Article 35;

---

24 Bailey, 2016, para. 9.70.

25 2020 ICC Model contracts, Art. 55.10.

26 Bailey, 2016, para. 9.79.

Failing for whatever reason to give possession of the Site or Sections thereof at the agreed time or times of handover and/or failing thereafter to agree to extensions of time;[27]
Failing to have access to the funds necessary to pay sums due under the Contract as they are or become due; or
Persistently or seriously hindering or delaying the Contractor; or.
Exceeding the period to evidence financial arrangements under Article 14.2.[28]

Grounds for default termination may also be found in all the FIDIC Books. It is worth noting that Subclauses 15.2 and 16.2 of the 2017 Red Book expressly provide that termination of the contract for the causes provided therein "*shall not prejudice any other rights*" of the terminating party under the contract or otherwise. Consequently, neither party should be precluded from relying on grounds for termination available in law. This is an important clarification since in certain jurisdictions, the specific grounds for termination in the contract may result in the exclusion of the right to terminate at law based on these specific grounds or the circumstances defined in the contract.[29] The only caveat is that, as already mentioned, before terminating the contract for default, the party in breach should be given an opportunity to correct the default. The ability to rely on another available ground may therefore depend on whether this formality has been complied with.[30]

In any case, the validity and enforceability of these clauses depend on the law governing the contract.

For instance, under French law, the clause whereby the insolvency of the other party will result in immediate termination may not be enforceable. The court appointed authority to represent the insolvent company may indeed require that the contract be continued despite the express contractual clause.[31] Furthermore, national laws themselves determine the circumstances under which contracts can be unilaterally terminated be it in general or specifically in the construction context. Parties must therefore take care to consider these grounds under the applicable law and to exclude them in their contract, if possible, whenever they do not wish them to apply.

27 Here performance is hindered by the owner insofar as the owner refuses to grant an extension of time to the contractor, which would be necessary as a result of the owner's breach to grant timely access to the Site or Sections.

28 2020 ICC Model, Art. 55.9.

29 Baker, Mellors, Chalmers, Lavers, 2009, para. 8.157.

30 Seppälä, 2023, pp. 999, 1030.

31 Art. L.622-13, L.631-14 and L.641-10 of French Commercial Code.

Under French law, safe certain exceptions,[32] termination clauses are not regulated by any specific provision and are in principle valid.[33] Regardless of termination clauses, French law also allows unilateral rescission (résolution), provided that the breach is "sufficiently serious,"[34] which is subject to the court's appraisal if objected to. Article 1224 is not considered a public policy provision and may be validly excluded,[35] although some authors disagree.

Swiss law enumerates in detail both the grounds for termination as well as their financial consequences. For instance, Article 368(1) SoC gives the right to the owner to rescind the contract when the work suffers from a substantial lack of conformity and the owner cannot reasonably be required to accept the nonconforming work.

Under English law, if one party's breach of the contract is sufficiently important and goes to the "root of the contract," it is to be considered a repudiatory breach, which entitles the innocent party to accept the repudiation of the contract and thereby bring the contract to an end. In the construction law context, poor workmanship is the most frequently invoked repudiatory breach.[36] Other examples include situations when the owner prevents access to the site, or the contractor abandons works. An accumulation of relatively small breaches can also constitute a repudiatory breach when taken together.[37]

Termination for Convenience

Construction contracts frequently give the owner the right to determine without the fault of the contractor (termination for convenience clauses).[38]

All the FIDIC Books reserve this right for the owner.[39] Subclause 15.5 of the 2017 Red Book reads as follows:

> The Employer shall be entitled to terminate the Contract at any time for the Employer's convenience, by giving a Notice of such termination to the Contractor (which Notice shall state that it is given under this Subclause 15.5).
>
> After giving a Notice to terminate under this Subclause, the Employer shall immediately:

---

32 E.g. termination of a lump-sum contract for convenience discussed below.

33 Hage-Chahine, 2021, para. 381.

34 Art. 1224 of French Civil Code.

35 Chantepie, Latina, 2024, para. 656.

36 Hopkins, 2021, para. 13.25.

37 Hopkins, 2021, para. 13.25.

38 Hopkins, 2021, para. 13.12.

39 Baker, Mellors, Chalmers, Lavers, 2009, para. 8.161.

a. have no right to further use any of the Contractor's Documents, which shall be returned to the Contractor, except those for which the Contractor has received payment or for which payment is due under a Payment Certificate;
b. if Subclause 4.6 [Co-operation] applies, have no right to allow the continued use (if any) of any Contractor's Equipment, Temporary Works, access arrangements and/or other of the Contractor's facilities or services; and
c. make arrangements to return the Performance Security to the Contractor.

The owner has thus the broad discretion to terminate the contract at any time without giving any reasons.[40] In addition, Subclause 15.5 in the Red Book explicitly allows the owner to perform the works or arrange their performance by third parties provided the contractor has been compensated. Such compensation includes, under Subclauses 15.5 and 15.7, not only the value of the work done but also the contractor's lost profit or other losses caused by termination.

Given the substantial compensation due to the contractor, some of the most common reasons for termination for convenience by the owner are when the project is abandoned,[41] has become economically unviable, obsolete, or is no longer possible.[42]

This being said, in practice, parties may also contractually exclude the contractor's compensation. The validity of the clause allowing termination for convenience and its consequences need to be analyzed under the law governing the contract.

English law does not recognize the owner's right to terminate at will, unless specifically provided for in the contract. In general, even if such a clause is included in the contract, it must expressly allow the owner to give the work to another contractor. In absence of clear wording to this effect, the contractor could claim that the owner had repudiated the contract and could recover damages from the owner.[43]

Nevertheless, in *TSG Building Services Plc v. South Anglia Housing Ltd*, the High Court of Justice has accepted the early termination of a maintenance contract by the owner despite (i) the fact that the future maintenance work would necessarily

40 Seppala, 2023, p. 1010.

41 Baker, Mellors, Chalmers, Lavers, 2009, paras. 8.163-8.164.

42 Seppälä, 2023, p. 1010 citing Bailey.

43 Bell and Choat, 2024, question 17; Hopkins, 2021, para. 13.13.

have had to be done by a replacement contractor and (ii) the absence of specific contractual language allowing the completion of works by another contractor.[44]

The owner relied on a "termination for convenience" clause to terminate, after thirteen months, the maintenance contract that was entered into for a period of four years. The contractor, in turn, invoked Clause 1.1 of the contract, which provided that parties "shall work in the spirit of trust, fairness and mutual co-operation for the benefit of the Term Programme, within the scope of their agreed roles, expertise and responsibilities."[45]

Justice Akenhead considered that the "termination for convenience" clause gave an unconditional right to all parties to terminate the contract for convenience (or without any already given reason).[46] This right is not subject to any conditions or constraints imposed by Clause 1.1 or an implied duty of good faith.[47] Consequently, any clearly worded clause allowing the termination of a contract for convenience will be upheld by the English courts, even if the right to terminate is apparently unfettered.

Under Swiss law, the owner's, and only the owner's, right to terminate for convenience is enshrined in Article 377 SoC: "The employer may withdraw from the contract at any time before the work is completed provided he pays for work already done and indemnifies the contractor in full."

The owner is free to terminate the contract for whatever reason or even without giving any reason.[48] In fact, if the owner entrusts the remaining work to a third-party in an apparent manner or when the owner carries the work out without being authorized to do so by substitution, such conduct might constitute termination in accordance with Article 377 SoC by conclusive actions.[49] The owner's right under Article 377 expires the moment the works are completed, even if they are defective.[50]

The majority of Swiss doctrine considers that Article 377 SoC is nonmandatory in nature, although this question has not yet been settled by case law.[51] Therefore, parties can make the exercise of this right easier, more difficult or exclude it

---

44 *TSG Building Services Plc v. South Anglia Housing Ltd* [2013] EWHC 1151, Hopkins, 2021, para. 13.14.

45 *TSG Building Services Plc v. South Anglia Housing Ltd* [2013] EWHC 1151 para. 4.

46 *TSG Building Services Plc v. South Anglia Housing Ltd* [2013] EWHC 1151 paras. 31 and 42.

47 *TSG Building Services Plc v. South Anglia Housing Ltd* [2013] EWHC 1151 paras. 42 and 51.

48 Müller, 2023, para. 1799.

49 Müller, 2023, para. 1801.

50 Decision of the Swiss Supreme Court, No. 4A_96/2014, 2 September 2014, para. 4.1.

51 Müller, 2023, para. 1836.

altogether in their contract.[52] Parties may also settle the consequences of the termination pursuant to Article 377 SoC differently in their contract.

However, parties' freedom has two important limits. First, termination rights stipulated in contracts can never be exercised in an abusive manner.[53] For instance, under certain circumstances if the owner withdraws from the contract shortly before the completion of works, it could be considered contrary to the rules of good faith, and thus an abuse of right.[54]

Second, the owner always retains the possibility to terminate the contract for valid reasons as per Article 19(2) SoC[55] and despite contrary contractual provisions.[56] Under Swiss law, termination for valid reasons is open to all parties to a construction contract, if one cannot reasonably expect from the party to remain bound by the contract due to changed circumstances, either for economical or for personal reasons.[57] What constitutes a valid reason will depend on the circumstances of the case, but a simple loss of trust in the contractor,[58] poor workmanship, or delays attributable to the contractor, which fall under Article 366 SoC, will in general not suffice.[59] A serious contractual breach or less serious breaches that repeatedly occur despite previous warnings are generally considered valid reasons to terminate a construction contract.[60] This is a mandatory provision, which applies to all construction contracts governed by Swiss law.

Nevertheless, certain Swiss authors argue that when Swiss law applies to a construction contract that has no link with Switzerland, the law should not be applied as strictly as in a domestic context, and especially not when it would lead to a result that none of the parties anticipated or that would be contrary to their common intention.[61]

This is an argument that could be validly made before an arbitral tribunal. A similar argument has indeed succeeded in the context of a purchase agreement before an arbitral tribunal seated in France. The parties chose French law to their contract, but the arbitral tribunal decided not to give effect to a mandatory rule of French law, as it considered that the parties could not have wanted to defeat

52 Müller, 2023, paras. 1836–1837; Chaix, 2021, para. 20.

53 Chaix, 2021, para. 8.

54 Müller, 2023, para. 1810.

55 Müller, 2023, para. 1838.

56 Tercier and Pichonnaz, 2019, para. 1224; Müller 2023, para. 65.

57 Müller 2023, para. 65; Decision of the Swiss Supreme Court, No. 4A_59/2017, 28 June 2017, para. 4.1.1.

58 Müller, 2023, para. 1835.

59 Decision of the Swiss Supreme Court, No. 4C. 393/2006, 27 April 2007, para. 3.3.3.

60 Decision of the Swiss Supreme Court, No. 4A_59/2017, 28 June 2017, para. 4.1.1.

61 Scherer and Schneider, 2007, p. 562.

several clauses in their contract by choosing French law, which contained this specific mandatory rule.

Under French law, Article 1794 of Civil Code expressly allows the owner to terminate a lump-sum contract, even after commencement of the works. This possibility is however reserved to the owner (the contractor is not allowed to terminate for convenience, even vis-a-vis subcontractor when there is a sub-contract) and is only available for lump-sum contracts.[62] It is indeed a derogation to the general principle that premature termination of a specific-term contract, such as construction contracts, is allowed only for default.[63] The owner may terminate the contract at any time, even before performance of the works, the only time limit being when the works subject to the contract are practically finished.[64] As described in more detail below, Article 1794 is not a public policy provision and may be excluded and/or amended by parties.

The authors have encountered several "termination for convenience" clauses in their practice that were very favorable to the owner as they did not provide for any formal requirements as to the exercise of the withdrawal right other than informing the contractor of the fact of the withdrawal and its date. Some expressly allowed the owner to have the works completed by another contractor or to withdraw even when the works were already completed, thereby avoiding potential problems with contrary provisions of national laws.

Interplay Between Different Grounds for Termination

Another issue with important practical and legal consequences concerns the choice parties face between the termination rights available to them. Can they rely on both contractual and statutory termination rights, and should owners invoke termination for cause and/or termination for convenience provisions when facing a default in the contractor's performance?

In general, unless there is express contractual provision to the contrary, the existence of a forfeiture clause does not exclude the statutory or common law rights to terminate, which will subsist in parallel.[65]

For instance, when faced with an overly technical "termination for cause" clause, parties might consider invoking statutory provisions or common law in parallel to establishing a contractual breach or they might prefer relying on statutory or contractual provisions establishing a right to terminate for convenience. This is a

62 Boubli, 2016, para. 137.

63 Articles 1212 and 1226 of French Civil Code; Latina, 2023, paras. 130, 132.

64 Cass. Civ. 3^{e}, 18 February 1976, No. 74-13.480.

65 Hopkins, 2021, para. 13.01.

valid approach; however, parties must tread carefully and be aware of the limits and consequences their choice has under the applicable national law.

Under Swiss law, if the owner (i) decides to terminate for convenience under Article 377 SoC, when it could have invoked another ground for termination, or (ii) if the owner invokes another ground for termination, but the conditions are not complied with or fulfilled, Article 377 SoC will automatically apply and the owner will have to fully indemnify the contractor in accordance with Article 377 SoC, notwithstanding the damage the owner might have suffered as a result of the contractual breach.[66] In other words, if the owner fails to establish a breach of contract justifying termination, Article 377 SoC will automatically apply, unless otherwise provided by the contract.

The solution is similar under French law. If the owner terminates the contract for contractor's default, a ground which ultimately is dismissed in court, the court may recharacterize it as termination for convenience under Article 1794 of Civil Code provided the requirements set forth therein and discussed above are met.[67] Conversely, even if termination for convenience may be exercised in the absence of any default, the owner who terminates the contract for convenience may still rely before the court on the contractor's breaches to recharacterize termination as termination for default and seek compensation for its loss.[68]

Concerning the distinction between a termination clause for default and unilateral rescission, the *Cour de cassation* ruled that in the latter case, the party at breach was not entitled to invoke the contractual stipulations governing the conditions and consequences of unilateral termination by the other party, such as payment of an indemnity.[69] However, it has decided otherwise in a subsequent case, where a liability limitation clause was applied despite unilateral rescission.[70]

As a matter of English law, termination can occur (i) if a party exercises a contractual termination right or (ii) if one party accepts the other party's repudiatory breach.

There is a Court of Appeal authority suggesting that the right to terminate in accordance with the terms of the contract and the right to treat the contract as discharged by reason of repudiation at common law are in effect one and the same,[71] unless of course clear contractual language provides otherwise. This means that in cases where the contract provides a right to terminate which corresponds to

66 Tercier, Bieri and Carron, 2016, para. 4153.

67 Boubli, 2024, para. 285; see Cass. Civ. 3e, 15 June 2000, No. 98-21.011.

68 Cass. Civ. 3e, 9 March 1988, No. 86-18.464.

69 Cass. Com. 3 May 2012, No. 11-17.779.

70 Cass. Com., 7 February 2018, No. 16-20352; Malaurie, Aynès, Stoffel-Munck, 2020, para. 559.

71 *Stocznia Gdynia SA v. Gearbulk Holdings Ltd* [2009] EWCA Civ 75, para. 20; see also Hopkins 2021, para. 13.19.

a right under general law, no election is necessary. Even when the injured party gives a bad reason for terminating the contract, its action remains effective if the circumstances support it.[72]

It is also possible to argue that the same should apply if the contract contains a generous "termination for convenience" clause and one party nevertheless relies on its common law right to repudiate the contract. In this case too, the termination should be valid, as the party would have had the right to terminate for convenience in any event.

Consequences of Termination

Construction contracts usually set out in detail both the financial and other consequences of termination. Such consequences include:

- payment obligations;
- determination of the ownership of the material already paid and/or delivered on site;
- removal of equipment and withdrawal of staff;
- clean-up of the site;
- handing over of drawings and specifications as well as other work results by the contractor, etc.

Termination clauses also frequently contain provisions requiring the contractor to either terminate its contracts with the subcontractors or assign those contracts to the owner with the same terms and conditions.

As far as the financial consequences of both termination for cause and convenience are concerned, national law rules are divergent. The only common denominator remains the large freedom granted to parties to regulate themselves the consequences of termination in their contract.

For instance, under Swiss law, when a party invokes substantial lack of conformity and decides to rescind the contract in accordance with Article 368(1) SoC, as a result of the rescission, each party is released from its obligations and must return to the other what it has already received. In other words, it becomes a winding-up relationship.[73] In addition, if the contractor is at fault, the owner can seek damages to compensate the loss resulting from the lack of conformity.[74]

72 *Stocznia Gdynia SA v. Gearbulk Holdings Ltd*, para. 44.

73 Müller, 2023, para. 1552.

74 Article 368(1) SoC; Müller, 2023, paras. 1561–1565.

Swiss law provides that in case of termination for convenience, under Article 377 SoC, the owner must not only indemnify the contractor for the works already done but must compensate the contractor in full. Full compensation means "positive interest damages," covering the interest that the contractor had in performing the contract in full, including lost profit.[75] The compensation may be reduced or eliminated if the contractor, through faulty conduct, has contributed to a significant extent to the event which led to the owner's withdrawal from the contract.[76]

Similarly, if the owner terminates the contract for valid reasons, and not in application of any contractual or other statutory rights of termination for cause or convenience, the contractor's otherwise existing right to compensation in full might be reduced or even eliminated for the same reasons as when the termination occurs for convenience.[77] Whether a valid reason for termination exists and what should be its impact on the obligation to compensate in full is a matter to be decided according to the rules of law and fairness (*"selon les règles du droit et de l'équité"*).[78] As a reminder, parties' right to terminate for valid reasons, as well as the consequences of such termination, are mandatory provisions of Swiss law, which cannot be excluded by parties.

As set out above, under Swiss law, parties can depart from Article 377 SoC and foresee in their contract different financial consequences in case of termination for convenience. It has thus been accepted that in case of termination for convenience, parties can agree on liquidated damages instead of the statutory regime of full compensation.[79] The same is true when termination occurs for cause, although it is more likely that parties will foresee a penalty clause in case of termination following a breach of contract. Parties can also decide to cap the damages due, be that for cause or convenience.

However, it must be borne in mind, that Swiss law reserves the right for judges and arbitrators to reduce liquidated damages and penalty clauses if the liquidated damages or the penalty are excessive.[80] Moreover, in accordance with Article 100 SoC, any agreement excluding or limiting a party's liability is null and void in case of willful misconduct or gross negligence.

Under French law, as a result of termination for convenience, in accordance with Article 1794 of Civil code, the owner must compensate the contractor for all

75 Müller, 2023, para. 1823.

76 Müller, 2023, para. 1833; Decision of the Swiss Supreme Court, No. 4C. 393/2006, 27 April 2007 paras. 3.3.1 and 3.3.3.

77 Decision of the Swiss Supreme Court, No. 4C.393/2006, 27 April 2007, paras. 3.3.1 and 3.3.3.

78 Müller, 2023, para. 1835; Chaix 2021, para. 17.

79 Pichonnaz, 2023, p. 277.

80 Pichonnaz, 2023, p. 278.

its costs, works and lost profit. Costs are understood to include the price of the equipment but also of installation works unless they have been performed against the owner's will or the contract's specifications.[81] In addition, if the circumstances of termination cause moral damage to the contractor, *i.e.*, a nonmaterial loss or injury, for example, a reputational damage for a company, they may give rise to additional compensation.[82] However, the right to compensation is not absolute.

First, as described above, even if the contract was terminated for convenience, the owner may rely on the contractor's breaches to obtain compensation for its own loss.

Second, Article 1794 of Civil Code protects private interests and is not considered as a public policy provision.[83] Parties may therefore exclude its application or modify the scope by excluding any compensation of the contractor.[84] This should not concern the payment for the works performed by the contractor until termination, which is not a compensation but the consideration for the works performed.

In addition, parties may also keep the compensation but derogate from the provisions of Article 1794 Civil Code by agreeing on an amount defined in advance, for example, a percentage of the portion of the price not yet used on the day of termination.[85] Put differently, it's a right to exit the contract by paying a certain amount agreed in advance. French courts consider that in this case, compensation is not triggered by a default and, therefore, may not be analyzed as contractual penalty. Rather, it's an indemnity to exit the contract ("*clause de dédit*"). Consequently, its amount may not be reduced by a court even if it is excessive.[86]

However, the outcome may be different if compensation is stipulated by reference to the contractor's loss. For example, a clause whereby the indemnity was stipulated "in compensation for the contractor's costs incurred and the profit the contractor has been derived of" was interpreted as a contractual penalty.[87] The wording of the clause is therefore critical for the analysis of whether the indemnity may be reduced.

Concerning compensation arising from termination for default, the situation is the contrary. The clauses whereby parties fix damages for nonperformance in a manner to compel the debtor to perform, such as indemnities stipulated in case

---

81 Cass. Com., 22 February 1961, Bull. civ. III, n.° 101.

82 Thioye, 2021, para. 89; see CA Paris, 30 September 2009, SARL Edifice c/ Ste Enox: compensating for a moral damage caused by an abrupt termination.

83 Auby, Périnet-Marquet, Noguellou, 2020, para. 836; Gibirila, 2024, para. 104; CA Paris, 29 October 1992, D. 1993.

84 Boubli, para. 139.

85 E.g., Cass. Civ. 3, 30 January 2019, No. 17-25.952.

86 Cass. Civ. 3, 9 January 1991, No. 89-15.780; CA Paris, 29 October 1992, D. 1993.

87 See CA Nîmes, 12 January 2023, No. 21/02295.

of a failure to complete the works within the contractual milestones, are analyzed as a contractual penalty clause. These clauses are valid, but their amount may be revised by the court when it is manifestly excessive in comparison to the damage suffered.[88]

Limitation of liability clauses or caps for damages, which are also very frequent in practice, are valid under two limits. First, they are not enforced when the default constitutes (i) a willful misconduct or (ii) gross negligence.[89] Second, they are not valid if they deprive the debtor's essential obligation of its substance,[90] *i.e.*, when the exclusion or the cap is so important that it contradicts the very scope of the essential obligation of the party that relies on it.[91] For example, the contractor may not fully or substantially exclude its liability for nonperformance of the works subject to the contract as it contradicts its essential obligation to carry out the works.

Finally, if parties have not anticipated the financial consequences arising from termination for breach, French law allows compensation for all foreseeable loss arising therefrom.[92] The loss should be foreseeable in its nature at the date of the contract's conclusion rather than in its quantum.[93] Unforeseeable loss may, however, be compensated in cases of willful misconduct and gross negligence.

Under English law, as termination for convenience is an unknown concept under common law, parties must set out clearly the consequences of such a termination in their contract. In addition, clear contractual language is also required to exclude those remedies that would ordinarily be available in common law in case of repudiation.[94]

It is important to note that the accrual of liquidated damages comes to an end with the termination of the contract, after which event, parties must seek damages for breach of contract under the general law.[95]

For instance, when the contract provides for liquidated damages in case of delay and a party decides to terminate the contract, liquidated damages for delay will be due up until the termination, unless specifically provided otherwise in the contract. It was decided by the UK Supreme Court that this rule shall apply if

88 Article 1231-5 of Civil Code; Cass. Com. 3 December 2002, No. 99-18.349.

89 Cass. Civ. 1re, 24 February 1993, No. 91-13.940; Malaurie, Aynès, Stoffel-Munck, 2020, para. 590.

90 Article 1170 French Civil Code.

91 Le Tourneau, 2023, para. 3225.12.

92 Articles 1231-2, 1231-3 Civil Code.

93 Cass. Com. 4 March 1965, No. 60-12 767.

94 *Stocznia Gdynia SA v. Gearbulk Holdings Ltd* [2009] EWCA Civ 75, paras. 19–25; see also Hopkins 2021, para. 13.33.

95 *Triple Point Technology Inc v. PTT Public Company Ltd* [2021] UKSC 29, para. 35.

the liquidated damages clause in a contract foresees that liquidated damages for delay must be paid up to the date the undelivered work is accepted by the owner and the work is not accepted before termination.[96] This was found to be the only approach consistent with commercial common sense and the accepted function of liquidated damages. Any other approach would mean that the owner's accrued rights in relation to delay would simply disappear upon termination.[97]

Naturally, the case turned on the specific contractual provisions, but it is clear from the findings of the UK Supreme Court that if parties want to defeat the principle according to which liquidated damages accrue until termination, specific wording to this effect will be required. It is submitted that the same solution should apply under Swiss law.[98]

Similarly, and largely in contrast with Swiss and French laws, English law accepts the exclusion or limitation of liability even in case of deliberate wrongdoing, if the construction of the corresponding clause leads to this conclusion.[99]

Care should be taken that unlike under French or Swiss law, penalty clauses are prohibited under English law, only liquidated damages clauses being considered as enforceable. There is no possibility for judges or arbitrators to reduce the amount of liquidated damages.

Bibliography

Auby, J-B. Périnet-Marquet, H. Noguellou, R. *Droit de l'urbanisme et de la construction*, 12th ed. (LGDJ, Lextenso, 2020).

Bailey, J. *Construction Law*, 2nd ed. (London: Informa law from Routledge, 2016).

Baker, E. Mellors, B. Chalmers, C. and Lavers, A. *FIDIC Contracts: Law and Practice*, (London: Informa law from Routledge, 2009).

Bell, A. and Choat, R. "Construction Arbitration: United Kingdom – England & Wales," *Global Arbitration Review* [online]. available at:, https://globalarbitrationreview.com/insight/know-how/construction-arbitration/report/united-kingdom, (2024).

Boubli, B. "*Contrat d'entreprise*," in Répertoire de droit civil, (Dalloz, 2016).

Chantepie G, Latina M, *Le nouveau droit des obligations. Commentaire théorique et pratique dans l'ordre du Code civil*, 3rd ed. (Dalloz, 2024).

---

96 *Triple Point Technology Inc v. PTT Public Company Ltd* [2021] UKSC 29.

97 *Triple Point Technology Inc v. PTT Public Company Ltd* [2021] UKSC 29, paras. 34–37.

98 See in this sense Tercier, Bieri and Carron, 2016, para. 4165.

99 *Frans Maas (UK) Ltd v. Samsung Electronics (UK) Ltd* [2004] EWHC 1502 (Comm), paras. 145 and 152.

Chaix, F. "Art. 377 CO" in *Commentaire romand Code des obligations I*. 3rd ed. Thévenoz, L. and Werro, F. (Basel: Helbing Lichtenhahn Verlag, 2021).
Decision of the Swiss Supreme Court, No. 4C.393/2006, 27 April 2007.
Decision of the Swiss Supreme Court, No. 4A_96/2014, 2 September 2014.
Decision of the Swiss Supreme Court, No. 4A_59/2017, 28 June 2017.
Decision of Cour de cassation, Cass. Civ. 3ème, 7 September 2022, No. 21-382.
Decision of Cour de cassation, Cass. Civ. 3e, 18 February 1976, No. 74-13.480.
Decision of Cour de cassation, Cass. Civ. 3e, 15 June 2000, No. 98-21.011.
Decision of Cour de cassation, Cass. Civ. 3e, 9 March 1988, No. 86-18.464.
Decision of Cour de cassation, Cass. Com. 3 May 2012, No. 11-17.779.
Decision of Cour de cassation, Cass. Com., 7 February 2018, No. 16-20352.
Decision of Cour de cassation, Cass. com., 22 February 1961, Bull. civ. III, n° 101.
Decision of Cour de cassation, Cass. Civ. 3, 30 January 2019, No. 17-25.952.
Decision of Cour de cassation, Cass. Civ. 3, 9 January 1991, No. 89-15.780.
Decision of Cour de cassation, Cass. Com. 3 December 2002, No. 99-18.349.
Decision of Cour de cassation, Cass. Civ. 1re, 24 February 1993, No. 91-13.940.
Decision of Cour de cassation, Cass. Com. 4 March 1965, No. 60-12 767.
Decision of Paris Court of Appeal, 30 September 2009, SARL Edifice c/ Ste Enox.
Decision of Paris Court of Appeal, 29 October 1992, D. 1993.
Decision of Nîmes Court of Appeal, 12 January 2023, No. 21/02295.
Eminence Property Developments Ltd. v. Heaney [2010] EWCA Civ 1168.
Fontaine, M., and de Ly F. *Drafting International Contracts*, (Leiden: Brill, 2006).
Frans Maas (UK) Ltd v. Samsung Electronics (UK) Ltd [2004] EWHC 1502 (Comm).
Gibirila, D. (2024) *Louage d'ouvrage et d'industrie.- Contrat d'entreprise*. J-cl, Fasc. 40.
Hage-Chahine N. (2021) *Résolution-Résiliation*. Répertoire de droit civil, Dalloz.
Hopkins, D. (2021) "Determination or Forfeiture Clauses," in *Wilmot-Smith on Construction Contracts* 4th ed. Wilmot-Smith, R. and Darling, P. (Oxford: Oxford University Press, 2021) 355–369.
ICC case No. 22899/DDA.
Jenkins J., *International Construction Arbitration Law*, 3rd ed. (Wolters Kluwer, 2021).
Latina M., "*Contrat : généralités.*" *Répertoire de droit civil*, (Dalloz, 2023).
Le Tourneau P. (2023) *Dalloz action Droit de la responsabilité et des contrats*. Dalloz.
Müller, C. *Contract Law in International Commercial Arbitration: A Commentary*, (Cambridge: Cambridge University Press, 2023).
Malaurie, P., Aynès, L., Stoffel-Munck, P. *Droit des obligations*, 11th ed. (LGDJ Lextenso, 2020).
Pichonnaz, P. "La jurisprudence récente en droit privé: Le contrat de vente et le contrat d'entreprise" in *Journées suisses du droit de la construction 2023*, ed. Zufferey, J.-B. (Fribourg: Institut pour le droit suisse et international de la construction, 2023), 269–291.

Scherer, M., and Schneider, M. (2007). "*International construction contracts under Swiss law: An introduction*," *Construction Law Journal*, 23, no. 8(2007): 559–569.

Seppälä, C. R. *The FIDIC Red Book Contract: An International Clause-by-Clause Commentary*, (Kluwer Law International, 2023).

Stocznia Gdynia SA v. Gearbulk Holdings Ltd [2009] EWCA Civ 75.

Tercier, P., Bieri, L., and Carron, B. *Les contrats spéciaux*. 5th ed. (Geneva-Zurich-Basel: Schulthess Éditions Romandes, 2016).

Tercier P., and Pichonnaz P. *Le droit des obligations*. 6th ed. (Geneva-Zurich-Basel: Schultess Éditions Romandes, 2019).

Triple Point Technology Inc v. PTT Public Company Ltd [2021] UKSC 29.

TSG Building Services Plc v. South Anglia Housing Ltd [2013] EWHC 1151.

Thioye, M. *Louage d'ouvrage et d'industrie.- Contrat d'entreprise.- Marché à forfait*, (Jurisclasseur, Lexis, 2021).

23

Dispute Boards

Roberto Hernández-Garcia

Introduction

The construction industry is full of participants and risks. Participants can voluntarily take risks that they can foresee and are capable of facing from a technical or contractual point of view. Participants will also face unforeseen risks because of the incapacity of identifying them or because such risks are unforeseeable. In any event, when risks become a reality, facing them means that a party or participant in the contract /project has to pay for such an event. As expected, very few people are interested in recognizing their responsibility and paying for their faults.

When there are differences in opinion of who is responsible for something that happens in the project, and who has to pay for it, a dispute arises. Disputes are costly in construction. All projects are limited in terms of available time and money. If disputes extend to unlimited horizons, the involved parties will start creating "spider webs" to protect themselves. These spider webs can be contract and legal actions or other kind of aggressive measures that cost more money and put the project at risk. Sometimes, disputes can extend for years and years with unlimited costs for their actors. Sometimes, a dispute can lead to the bankruptcy of participants. But most important of all, disputes put the projects at risk, which is the main essence of a construction contract.

Ironically, the reality is that (most of the time) owners, contractors, subcontractors, suppliers, and other stakeholders are professionals who only want to complete the project, get what they asked for (the project or money), and not get into trouble, and leave for the next business.

The issue is that when participants of a construction project enter a dispute during the course of a project, they only have in their contract dispute limited dispute resolution provisions: Negotiation (which will often mean escalating the noise of the aggressions), mediation, or litigation and arbitration.

While all of these methods have demonstrated their benefits, none of them have two essential characteristics that differentiate them from the topic of this chapter: (i) none of them are intended to prevent a dispute, (ii) none of them are intended to work contemporaneously to the project and the contract and with closeness to the site.

Dispute Boards ("DB") have these two elements: they are intended to prevent disputes, and resolve them if necessary, (ii) contemporaneously to the project performance with absolute closeness to the site. Under these circumstances, DBs are extraordinary tools for avoiding disputes in construction projects, helping the parties to be more focused and efficient on the performance of the project, maintaining relationships in a small industry.

This chapter provides a practical general overview of DBs to help potential users understand their benefits. This is not an academic chapter and therefore, it does not deliberately contain reference to books and footnotes different than what can complement ideas. This chapter is the result of almost 20 years of experience of its author as a DB. At the end of the chapter, we are adding a list of invaluable bibliographies for those who want to get deeper into the world of DBs.

As we will see, DBs are an excellent tool in construction projects. Of course, as with everything in life, DBs are not perfect, but they are far better than an eternal fight among parties without a completed project. In this chapter, we will explain what DBs are, and some tips on how to use them, hoping that readers not only understand the concept but also give the opportunity to DBs for their projects, for the sake of their good completion.

Definition of Dispute Boards

Dispute Boards (DBs), also known as Dispute Resolution Boards (DRBs) or Dispute Adjudication Boards (DABs), depending on their nature and functions (which will be explained later), are independent bodies typically formed by one or three persons as a consequence of two parties deciding in a clause of a Contract, to include such prevention and resolution of disputes mechanism.

Dispute Boards are established on construction projects to help parties prevent and resolve disputes that may arise during the project. DBs are comprised of impartial and experienced professionals appointed at the beginning of a project (or during the project as we will learn) by the two parties (Owner and Contractor) to proactively manage, prevent and resolve disputes in a timely and cost-effective manner.

Purpose of Dispute Boards

The primary function of DBs is to prevent disputes from escalating into costly and time-consuming legal battles by providing a forum for the parties to present their arguments, receive an impartial evaluation of the issues, and ultimately reach a fair and mutually acceptable resolution during the performance of a project.

DBs are also considered as a risk management tool, since parties can benefit from the experience and impartiality of the DB Members to help the parties focusing on their projects.

Finally, DBs have the task of solving disputes which are raised during the course of the project, in order to solve them contemporaneously allowing the Parties to keep on going on the construction contract, rather than being stuck by unresolved issues.

DBs help to maintain the project momentum, foster open communication between parties, and preserve working relationships, ultimately contributing to the successful completion of construction projects.

Brief Notes on the History of DBS[1]

DBs have a history that dates back several decades. DBs originated in the construction industry to resolve conflicts and disputes that commonly arise during complex construction projects.

Here is a brief history of DBs:

- Origins in the United States (1970s): The modern concept of DBs can be traced back to the United States in the 1970s.
- Evolution and Adoption in the Construction Industry: DBs gained popularity in the construction industry as a proactive approach to dispute resolution. By having a neutral panel of experts involved from the early stages of a project, potential conflicts could be addressed and resolved in a timely manner, thus helping to prevent disputes from escalating into costly and time consuming legal battles.
- International Expansion: The success of DBs in the United States led to their adoption in other countries around the world. Organizations such as FIDIC (International Federation of Consulting Engineers), the International Chamber of Commerce (ICC), the Dispute Resolution Board Foundation (DRBF), and the American Arbitration Association International Centre of Dispute Resolution (ICDR) promote the use of DBs as a best practice in construction projects.

1 For a more detailed and complete information: Chapman, Peter H.J. "DBs." https://www.fidic.org/sites/default/files/25%20Dispute%20Boards.pdf.

- Standardization and Guidelines: Over time, standard forms of contract incorporating DB provisions were developed to provide a framework for their use. Organizations like FIDIC (International Federation of Consulting Engineers) include DB clauses in their standard construction contracts, further promoting their use globally. In addition, since Multilateral Development Banks such as the World Bank and the Interamerican Development Bank include FIDIC Contracts, the use of DBs in financed projects has been fostered exponentially.

Importance of DBS in Construction Projects

There are several studies that show how costly disputes are for the construction sector worldwide. While some of such disputes are raised after a project is completed, on many occasions, projects get stuck and are not completed, failing to provide the final user its benefits, creating an endless number of side effects. That is why DBs are widely regarded as an effective means of preventing and resolving disputes in a timely and cost-effective manner, helping to keep projects on track and minimize the risk of costly delays and legal battles. The history of DBs demonstrates their evolution from a niche concept in the construction industry to a widely accepted best practice in project management and dispute resolution.

Types of DBS

The types of DBs can be divided into two categories:

- First category – Related to the time of appointment and intervention of DBs.
- Second category – Related to the substantial function of DBs.

First Category – Related to the Time of Appointment and Intervention of DBs

DBs, which are established to manage and resolve disputes in construction and engineering projects, can be classified into two main types: Ad-hoc DBs and Standing DBs.

Ad-hoc DBs

These DBs are created for a specific purpose or a particular dispute and are typically formed only when a disagreement arises. They may be appointed for a single issue, and their existence is temporary, often concluding once the dispute is resolved. These boards tend to take a reactive approach to disputes, stepping in only when specific conflicts have been identified that require resolution. Their involvement is typically limited to the time frame of addressing individual disputes. The parties may not have a continuous relationship with the members of an Ad-hoc Board, as the board is only engaged for particular disputes. This can lead to less familiarity with the project context and stakeholder relationships and the impossibility of working on dispute avoidance. Members are typically appointed as disputes arise and may not have a pre-established relationship with the parties involved. The selection process may be focused specifically on the issue at hand rather than on the overall project context. Costs may be associated with forming a new board for each dispute, which could include the appointment of new members and the fees incurred during the resolution process.

Standing DBs

Standing DBs are established at the outset of a project and remain in place for its entire duration. They provide ongoing oversight and support, allowing them to address issues as they arise continuously. Standing Boards play a proactive role, regularly engaging with the project and monitoring its progress. They seek to identify potential disputes early on and provide recommendations, aiming to prevent issues from escalating into formal conflicts. Members of a Standing Board are engaged throughout the life of the project, which allows them to build a deeper understanding of the project dynamics, stakeholder interactions, and the evolving circumstances. This familiarity can lead to more informed decision-making and recommendations. Members are usually appointed at the project's beginning and expected to work together throughout its duration. This creates a cohesive team with shared knowledge and understanding of the project's goals and challenges. While there may be ongoing costs for maintaining a Standing Board (which are far from the costs of a legal battle), this can often lead to overall cost savings for the project. Their proactive approach to preventing disputes can reduce the need for costly arbitration or litigation.

Second Category – Related to the Substantial Function of DBs

There are generally three types of DBs that are commonly used in construction projects and other industries.

- DABs consist of one or three independent and impartial members who may or not be appointed at the beginning of a project (Standing DBs or Ad Hoc DBs). The primary role of a DAB is to make decisions on disputes that arise during the project. Parties involved in a dispute present their arguments to the DAB, which then issues a contractually binding decision. DAB decisions are interim and binding, meaning that they must be followed until the dispute is finally resolved through arbitration or litigation in the case that such a decision is challenged on the time established in the contract (Notice of Dissatisfaction). These kinds of DBs are included in the 1999 FIDIC Forms of Contract and in the ICC Dispute Board Rules).

- Dispute Avoidance and Adjudication Boards (DAABs) consist of one or three independent and impartial members who are appointed at the beginning of a project (Standing DBs). The primary role of a DAAB is to prevent disputes, promote collaboration and if needed, make decisions on disputes that arise during the course of the project. Parties share with the DAAB concerns and issues that may generate future disputes on regular meetings and site visits. The Parties can request to the DAAB informal opinions that can guide the parties to an agreed resolution. In addition, parties can submit disputes to the DAAB, which then issues a binding decision. DAAB decisions are interim and binding, meaning that they must be followed until the dispute is finally resolved through arbitration or litigation if necessary. This is the type of DBs included in the 2017 Version of the FIDIC Suite of Contracts. These kinds of DBs are included in the 2017 FIDIC Forms of Contract and the ICC Dispute Board Rules does not consider it as a kind of DB, but implicitly motivates dispute avoidance in DBs).
- Dispute Review Boards (DRBs) consist of three members who are jointly selected by the parties at the beginning of a project. The members are neutral and impartial experts in the relevant fields. The primary role of a DRB is to provide nonbinding recommendations or advice on disputes that arise during the project. The parties involved in a dispute present their arguments to the DRB, which then issues a recommendation on how the dispute should be resolved. While DRB recommendations are nonbinding, they are often highly respected by the parties involved and can help facilitate the resolution of disputes without the need for formal arbitration or litigation. This type is often used in the United States. These kinds of DBs are not considered in any of the FIDIC Forms of Contract. The ICC Dispute Board Rules expressly recognizes them).
- Combined Dispute Boards (CDBs) incorporate elements of both DRBs and DABs. Parties can choose from the same DB, which kind of benefits they want (either DAB or DRB). The choice among a DAB, a DAAB, and a DRB often depends on the specific contractual requirements and the preferences of the parties involved in the project. These kinds of DBs are not considered in any of the FIDIC Forms of Contract. The ICC Dispute Board Rules expressly recognizes them).

Composition of DBS

The selection of DB members is a crucial aspect of setting up an effective DB. The members of a DB shall be independent and impartial professionals who have the necessary qualifications and experience to understand and thus, help to prevent and resolve effectively, disputes that may arise during the construction project. Here are some key considerations for the selection of DB members:

- **Independence and Impartiality:** It is essential that DB members are independent and impartial, meaning that they have no conflicts of interest and can make decisions objectively without favoring any party involved in the project. Members should not have any financial or personal interest in the project or any of the parties, as this could compromise their neutrality and undermine the credibility of their decisions.
- **Expertise and Qualifications:** DB members should have the relevant expertise and qualifications to effectively resolve the types of disputes that are likely to arise in the project. Members are typically experts in fields such as law, engineering, construction, project management, or other relevant disciplines, depending on the nature of the project.
- **Experience:** It is beneficial for DB members to have prior experience serving on DBs and resolving disputes in similar projects. Experienced members are better equipped to handle complex disputes, understand industry practices, and effectively facilitate the resolution process.
- **Neutrality and Fairness:** The selection process should ensure that all parties involved in the project have confidence in the neutrality and fairness of the DB members. Parties may have the opportunity to jointly select members or agree on a list of qualified candidates from which members will be chosen to ensure that the process is transparent and fair.
- **Communication and Interpersonal Skills:** Effective DB members should have strong communication and interpersonal skills to facilitate constructive discussions, manage conflicts, and build trust with the parties involved in the project. Members should be able to encourage open dialogue, listen to different perspectives, and guide the parties toward mutually acceptable solutions.
- **Training and certifications:** Parties can benefit from those members that belong to specialized organizations such as the DRBF, FIDIC, ICC, and other associations, which provide training and credentials to members who have received specialized training and certifications.

Qualifications of DB Members

Qualifications of DB members are essential for ensuring that they have the necessary expertise and experience to effectively prevent and resolve disputes in a fair and impartial manner. DB members are typically selected based on their professional background, qualifications, and experience in relevant fields. Here are some key qualifications that are typically sought in DB members:

- Legal and Technical Expertise: DB members should have a strong background in law, engineering, construction, project management, or other relevant technical fields related to the industry in which the DB will operate. Legal/Contractual expertise is important for interpreting contracts, laws, and regulations governing the project, while technical expertise is crucial for understanding the technical aspects of the project and the nature of the disputes that may arise.
- Industry Experience: Members should have substantial experience working in the industry in which the project is taking place, whether it be construction, energy, infrastructure, or another sector. Industry experience helps members understand industry best practices, standards, regulations, and common challenges faced in similar projects, enabling them to provide informed and practical advice on dispute resolution.
- Dispute Prevention and Resolution Skills: DB members should have a strong background in the understanding of dispute prevention and resolution techniques, such as mediation, arbitration, negotiation, and conflict management. Members should be skilled in facilitating discussions, managing conflicts, and guiding parties toward mutually acceptable solutions in a collaborative and constructive manner. It is, however, important to make clear that good arbitrators, litigators, and former judges must understand the very different focus and conduct that they should perform when acting as a DB.
- Communication and Interpersonal Skills: Effective communication and interpersonal skills are crucial for DB members to build trust with the parties involved, encourage open dialogue, and maintain a positive working relationship throughout the dispute resolution process. Members should be able to listen actively, communicate clearly, and foster a respectful and collaborative environment for resolving disputes.
- Neutrality and Impartiality: DB members must demonstrate a high level of neutrality and impartiality in their decision-making process. Members should be free from conflicts of interest, biases, and preconceptions that could influence their judgment, and they should be committed to making decisions based on the facts and merits of each dispute.

- Continuing Education and Training: DB members should stay current with industry trends, legal developments, and best practices in dispute resolution by engaging in continuing education and training programs. Ongoing professional development helps members enhance their skills, knowledge, and expertise, ensuring that they are well-equipped to effectively resolve disputes in a rapidly changing environment.

Overall, the qualifications of DB members play a critical role in the success of the DB. By selecting members with the right mix of legal, technical, industry, dispute resolution, communication, and interpersonal skills, parties can ensure that their disputes can be prevented and/ or resolved in a fair, impartial, and efficient manner, ultimately contributing to the successful completion of the project.

Appointment of DB Members

The Selection and appointment of the DB is crucial. The process of appointment usually follows any of the following guidelines. The first one is the common selection of members, by both parties, Owner and Contractor. The second is, in the case of a three members DB is that each party selects one member and the third member, that will act as a Chair (President), may be selected by the other two members usually with the consent of the Parties. In some cases, Parties select the Chair by themselves and just confirm that there is no conflict with the other two. Finally, in the case of any conflict in the selection of any member (including the Chair), the Parties shall use the provisions of the contract that contain the organization in which the Parties should refer to present the appointment. In any case, the drafting of the contract provisions should be followed to ensure that the appointment strictly adheres to the agreement of the Parties.

Engagement of DB Members

Once selected, every member of a DB is required to sign a formal agreement with both the Owner and the Contractor, known as a Tri-party Agreement. This document serves a crucial role in establishing a clear understanding among all parties involved. The Tri-Party Agreement outlines all pertinent provisions related to the selection of the DB members, detailing their specific duties and rights as well as the responsibilities of the Owner and the Contractor. It ensures that each party is aware of their obligations and expectations, fostering an environment of transparency and cooperation. Additionally, this agreement addresses the financial arrangements between the parties and the DB members. It specifies how the costs related to the DB's functioning – such as fees for the services rendered and any other related expenses – will be managed and allocated.

Roles and Responsibilities of DB Members

DB members play a crucial role in the effective resolution of disputes that may arise during a construction project or in any other industry where DBs are utilized. Their responsibilities include various tasks to ensure that disputes are addressed in a fair, impartial, and timely manner. Here are the key roles and responsibilities of DB members:

- Impartiality and Neutrality: DB members must maintain a high level of impartiality and neutrality throughout all their performance of the Project and especially during all the dispute prevention and resolution processes. They should approach each possible or actual dispute with an open mind, free from biases or conflicts of interest that could impact their judgment.
- Preventing Disputes: Members are responsible for helping the parties to prevent disputes before the disputes are submitted to the DB. It is important to say that this duty is only possible in a Standing DBs.
- Regular Meetings and Site Visits: DB Members shall keep themselves informed directly of the progress of the works through regular meetings (monthly) and regular visits to the project site, in order to get a firsthand view of what happens on the project. It is important to say that this duty is only possible in a Standing DBs.
- Follow the dispute procedures including the Conduction of Hearings: DB members shall be very careful on understanding and managing effective processes during disputes submissions, including the conduction of hearings or meetings with the parties to discuss the issues in dispute, gather additional information, and allow each party to present their case with the same opportunities. While the process is far formalistic and more flexible than arbitration, DB members shall ensure that the proceedings are conducted in a fair, respectful, and confidential manner.
- Issuing Decisions, Recommendations or Informal Assistance: Depending on the type of DB (DAB or DRB), members are responsible for either issuing binding decisions (DAB), nonbinding recommendations (DRB) or Informal Assistance, on how the dispute should be resolved. Members must provide clear, reasoned, and well-supported decisions or recommendations based on the facts and applicable contract provisions.

- Facilitating Settlement: DB members may actively work with the parties to facilitate settlement negotiations and help them reach a mutually acceptable resolution. They should encourage open dialogue, explore potential solutions, and guide the parties toward a settlement that is fair and equitable.
- Continuing Education and Training: Members should engage in ongoing professional development activities to enhance their knowledge, skills, and expertise in dispute resolution. Staying current with industry trends, legal developments, and best practices is essential for effective DB performance.

Ethical Obligations of the DB Members

The DRBF provides guidelines and ethical standards for members of DBs to ensure integrity, fairness, and transparency in the dispute resolution process. While the specific provisions can vary over time or with updates to their guidelines, the ethical rules generally address the following key principles:[2]

> **Canon 1 – Conflict of Interest and Disclosure**
> Board members must avoid the appearance of, or any actual, conflict of interest during the term of the DB. Board members must disclose, before their appointment, any interest, past or present relationship, or association that could reasonably be considered by a contracting party as likely to affect that member's independence or impartiality. If, during the term of a DB, a Board member becomes aware of any fact or circumstance that might reasonably be considered by a contracting party as likely to affect that Board member's independence or impartiality, the Board member must inform the other Board members and disclose the matter to the contracting parties.
> **Canon 2 – Confidentiality**
> Board members must ensure that information acquired during the term of the DB remains confidential and must not be disclosed, unless such information is already in the public domain. Any such confidential information may only be disclosed if approved by the contracting parties or if compelled by law. Board members must not use such confidential information for any purpose beyond the activities of the DB.

2 Code of Ethical Conduct. Revised in March of 2018, the DRBF Code of Ethical Conduct consists of the four canons below that all DRBF members agree to adhere to in practice. The DRBF recommends that these canons be adopted by all users of the process as a matter of best practice. Available at: https://www.drb.org/code-of-ethical-conduct. The exact reproduction of these canons was kept in order to avoid any misinterpretation of them due to their high importance and it was authorized by writing by the Dispute Resolution Board Foundation management.

Canon 3 – Board Conduct and Communications
Board members must conduct all Board activities in an expeditious, diligent, orderly, and impartial manner. Board members must act honestly, with integrity and without bias. There must be no unilateral communications as between a Board member and a contracting party, except as permitted under the DB operating procedures.
Canon 4 – Board Procedures
All Board meetings and hearings must be conducted in accordance with the applicable contract provisions and operating procedures, in a manner that provides procedural fairness to the contracting parties. DB recommendations and decisions must be made expeditiously on the basis of the provisions of the contract, the applicable law and the information, facts and circumstances submitted by the contracting parties.

The ethical rules established by the DRBF are designed to foster an environment of trust, fairness, and professionalism within DBs. By upholding these principles, members can effectively manage conflicts while maintaining the integrity of the dispute resolution process, ultimately supporting successful project delivery and positive working relationships among all stakeholders involved.

DBS Dispute Avoidance/Prevention Process

Dispute Boards (DBs) use a variety of tools and techniques to help prevent disputes in the construction industry. These tools focus on proactive communication, continuous engagement, and early identification of potential issues. Here are some specific tools that DBs employ:

- Regular Meetings: DBs typically hold regular meetings with the project stakeholders to discuss progress, potential issues, and any concerns that may arise. This allows for open communication and helps address problems before they escalate. It is important to say that this duty is only possible in a Standing DBs. Meetings usually happen virtually in a video platform such as Teams® or Zoom®.
- Site Visits: DB members often conduct site visits to observe the project's progress firsthand. By being physically present, they can better understand the situation, assess challenges, and provide recommendations based on real-time conditions. It is important to say that this duty is only possible in a Standing DBs.
- Progress Reports: DBs may require periodic progress reports from the project team. These reports can highlight any issues that might lead to disputes, allowing the DB to intervene early. It is important to say that this duty is only possible in a Standing DBs.

- Facilitative Role: DBs can act as facilitators in negotiations between parties, helping them to communicate effectively and work through disagreements in a constructive manner. It is important to say that this duty is only possible in a Standing DBs.
- Dispute Resolution Protocols: Establishing clear protocols for how disputes will be handled can prevent issues from escalating. This includes defining the process for raising and addressing concerns, timelines, and responsibilities.
- Issue Identification and Monitoring: DBs help in identifying and monitoring potential disputes by being involved in the project timeline. By tracking key milestones and any deviations, they can address concerns before they develop into formal disputes.
- Recommendations and Guidance: DBs can provide nonbinding recommendations on contentious issues that may arise. Their expert advice can guide parties toward mutually agreeable solutions.
- Training and Awareness: Providing training sessions for project stakeholders about potential dispute triggers and effective communication strategies can foster a better understanding of the collaborative process.
- Documentation Practices: Encouraging proper documentation of all project-related communications and agreements helps to maintain clarity and accountability, reducing the likelihood of misunderstandings that may lead to disputes.
- Conflict Management Training: Offering workshops on conflict resolution for project teams can equip them with the skills to address issues constructively, minimizing disputes.

By integrating these tools into the project management process, DBs aim to create a collaborative environment that reduces the likelihood of disputes escalating into formal arbitration or litigation.

DBS' Dispute Avoidance and Resolution Process

DBs play a crucial role in effectively managing and resolving disputes that may arise during construction projects. The dispute resolution process typically consists of three main phases: the pre-dispute phase, the dispute resolution phase, and the post-dispute phase. Here is an explanation of each phase:

- Pre-dispute Phase: In the pre-dispute phase, the DB is established at the beginning of the project, typically through the project contract. During this phase, the DB members are selected, and the parties involved in the project agree on the procedures and rules that will govern the dispute resolution process. The DB may familiarize themselves with the project, review the contract documents, and establish lines of communication with the parties to promote early collaboration and prevent disputes from arising. The pre-dispute phase is proactive in nature, focusing on establishing a framework for effective dispute prevention and resolution and fostering a cooperative relationship among the parties. In this phase Parties can benefit from getting advise and informal assistance from the DB members on issues that are of their concern during the performance of the project. The DB has to be not only attentive to the requests of the Parties, but also of any actions, conducts, and behaviors that the DB detects, in order to proactively provoke a common understanding and the prevention of any issue or dispute before it explodes within the parties. It is important to say that this duty is only possible in a Standing DBs.

- Dispute Resolution Phase: The dispute resolution phase begins when a dispute arises between the parties involved in the project. The parties may submit the dispute to the DB, which will review the issues, gather relevant information, and may hold hearings to allow each party to present their case. Depending on the type of DB (DAB or DRB), the members will either issue binding decisions (DAB) or nonbinding recommendations (DRB) on how the dispute should be resolved. The DB plays a crucial role in facilitating discussions, guiding the parties toward mutually acceptable solutions, and ultimately resolving the dispute in a fair and efficient manner.
- Post-dispute Phase: The post-dispute phase begins once a decision or recommendation has been issued by the DB. If the DB's decision is binding (in the case of a DAB), the parties are typically required to comply with the decision and take necessary actions to resolve the dispute. If the DB's recommendation is nonbinding (in the case of a DRB), the parties may use the recommendation as a basis for negotiating a settlement or may choose to pursue formal arbitration or litigation. The DB may monitor the implementation of its decision or recommendation, follow up with the parties to ensure compliance, and assist in addressing any issues that arise during the implementation phase.

By following a structured process that encompasses the pre-dispute, dispute resolution, and post dispute phases, DBs help parties address conflicts and disputes in a proactive, collaborative, and efficient manner. Effective dispute resolution not only contributes to the successful completion of projects but also helps maintain positive relationships among project stakeholders.

DBS Decision: Binding But Not Final?

A DB decision is considered binding but not final due to the nature of the dispute resolution process and the specific context in which DBs operate.

Below are the key reasons that clarify this distinction:

- Binding Nature: The decisions made by a DB, particularly in the case of DABs, are binding on the parties involved in the dispute. This means that the parties must comply with the decision of the board as if it were a contractually agreed determination. This binding characteristic encourages parties to adhere to the decision and helps maintain project continuity.
- Temporary Resolution: While a DB decision is binding, it is often seen as a preliminary resolution intended to provide immediate guidance and keep the project on track. The binding decision is not the final conclusion of the dispute but serves to temporarily resolve issues while allowing the possibility for further review. Parties can challenge the binding decision in subsequent arbitration or litigation if they believe there is a valid reason to do so. This provision allows for further examination of the issue by a different legal avenue, which can lead to a more comprehensive consideration of the facts and legal arguments.
- Facilitating Ongoing Relationships: DBs operate with the understanding that fostering ongoing business relationships is important, especially in industries like construction, where projects can span several years. The binding yet non-final aspect allows for disputes to be managed in a way that does not entirely shut down discussions or cooperation between the parties. This encourages a collaborative approach to conflict resolution.
- Encouragement of Compliance: By making the DB's decisions binding, parties have an incentive to comply with the rulings, promoting smoother project operations. However, allowing for subsequent reviews ensures that parties still feel they have recourse if they strongly disagree with the board's findings.

A DB decision is binding in that it must be complied with by the parties since it is a contractual obligation, yet it is not final because it can be challenged in subsequent arbitration or litigation. This duality allows for efficient dispute resolution while still providing a method for deeper judicial review, thus balancing immediate project needs with the equitable consideration of disputes. Some people question whether this lack of finality is positive or negative. However, by understanding the nature and reason of DBs, the questions and doubts can be addressed.

Benefits of DBS

- Time and Cost Savings: DBs can lead to significant time and cost savings by providing a mechanism for preventing and resolving disputes quickly and efficiently. By having a structured process in place for addressing issues as they arise, parties can avoid lengthy and costly litigation or arbitration proceedings. This can help keep the project on track and on budget.

- Early Detection and Resolution of Issues: One of the primary benefits of DBs is their ability to facilitate early detection and resolution of issues before they escalate into larger disputes. By having a neutral third party involved in the process, parties can address concerns as they arise and work toward finding mutually acceptable solutions. This proactive approach can help prevent conflicts from derailing the project and causing delays.
- Maintenance of Project Relationships: DBs can also help maintain positive project relationships by providing a forum for constructive dialogue and problem-solving. By fostering open communication and collaboration between the parties, DBs can help preserve working relationships and prevent disputes from causing long-term damage to professional connections. This can be particularly important in the construction industry, where successful projects often rely on strong partnerships between various stakeholders.

DBS in FIDIC Contracts

The International Federation of Consulting Engineers (FIDIC) is a widely recognized organization that develops standard forms of contracts for the construction and engineering industry. FIDIC contracts are commonly used in international construction projects and often include provisions for dispute resolution mechanisms, including DBs.

In FIDIC contracts, the use of DBs is typically recognized after 2017 in the figure of the DAAB provisions. The DAAB is a type of dispute resolution body appointed at the beginning of a project to help parties prevent and resolve disputes that may arise during the course of the project. Some key points about DBs in FIDIC contracts are:

- Appointment and Composition: In FIDIC contracts, the DAAB is usually composed of one or three members, who are appointed by the parties or as specified in the contract. The members of the DAAB are typically experienced professionals with relevant expertise in the construction industry. FIDIC has a very prestigious list of DB members called the President's list that publicly available.
- Roles and Responsibilities: The primary role of the DAAB in FIDIC contracts is to provide a forum for the early identification and resolution of disputes. The DAAB members shall have regular meetings, visit the project site, review documents, hear arguments from both parties, and help on prevention, issue informal consultations, issue decisions or recommendations to resolve disputes in a timely manner.

- Decision-Making Process: The decisions or recommendations of the DAAB are binding on the parties unless and until the dispute is referred to arbitration. Parties are expected to comply with the DAAB's decisions in good faith to maintain the progress of the project.
- Timeframes and Procedures: FIDIC contracts typically include specific provisions outlining the procedures, timeframes, and requirements for the operation of the DAAB. These provisions help ensure that disputes are addressed promptly and fairly according to the contract terms.
- Enforcement and Finality: The decisions of the DAAB in FIDIC contracts are intended to be enforceable, unless and until they are challenged in arbitration or litigations. This provides a level of certainty and efficiency in resolving disputes without the need for lengthy court proceedings.

Recently, FIDIC has done extraordinary work on the implementation, promotion and good use of DAABS in FIDIC Contracts. Several important documents on the matter can be found on the organization's website including the Dispute Avoidance Documents that were recently issued.

DBS Ruled by the ICC Rules of DBS

The International Chamber of Commerce (ICC) has established rules and procedures to govern the operation of DBs as a means of resolving disputes in international construction and engineering projects. The ICC's rules on DBs provide a comprehensive framework for resolving disputes in construction projects through proactive engagement and decision-making. By incorporating these rules into contracts, parties can establish a structured process that promotes early intervention, collaboration, and efficient resolution of conflicts, ultimately contributing to the successful completion of projects. An important note is that, different than Arbitration, DBs are not administered by the ICC, but the later only intervenes in specific moments such as the appointment of a DB, Challenge of a DB Member or the Scrutiny of a Decision. Below are some key elements of the ICC Rules on DBs:

- Appointment: Parties to the contract agree on the composition of the DB, which can consist of one or three members. If the parties cannot agree, the ICC can assist in appointing the members. Members are typically required to have relevant expertise in the construction industry and dispute resolution. They should also be impartial and independent.
- Functions and Powers: Proactive Role: DBs have a proactive role in monitoring the project and advising parties on potential disputes. By being involved throughout the project, they can help identify and address issues

early on. Decision-Making. In the case of a DAB, the board has the authority to issue binding decisions on disputes that arise. This ensures that disputes are resolved efficiently, minimizing disruptions to the project.

- Procedural Rules and Meetings: The rules outline the procedures for holding meetings, including regular site visits and hearings. The boards can issue decisions based on the information presented during these meetings. Parties are typically required to provide relevant documentation to the board when submitting a dispute for resolution. This helps facilitate informed decision-making. The proceedings before the DB are generally confidential, ensuring that parties can communicate openly without concern for public exposure.
- Recommendations and Decisions: For DRBs, the boards provide nonbinding recommendations that parties can choose to follow. This encourages collaboration and prevents disputes from escalating. For DABs, decisions are binding on the parties unless and until they are challenged in arbitration or litigation.

The International Center for Dispute Resolution Rules on DBS

The International Center for Dispute Resolution (ICDR), which is the international division of the American Arbitration Association (AAA), provides a set of rules specifically governing the establishment and functioning of DBs. These rules are designed to facilitate effective and efficient dispute resolution in international construction and engineering projects. Below are the key aspects of the ICDR rules on DBs:

- Types of DBs: The ICDR rules allow for different types of DBs, including: DRBs and DABs.
- Establishment and Composition: The parties to the contract typically agree to the formation of the DB at the outset of the project. If the parties cannot agree on the composition or appointment, the ICDR can assist in appointing the board members. DB members are usually appointed based on their expertise and experience in relevant areas, such as construction, engineering, or dispute resolution. A DB may consist of one or three members.
- Powers and Functions: DBs are designed to take a proactive approach in identifying potential disputes early. They engage with the project and monitor progress to help mitigate conflicts. For DABs, the decisions made by the board on disputes are binding on the parties, unless and until they are either modified by an arbitration ruling or a judicial decision.

- Procedure and Processes: The rules outline procedures for the DB to meet with the parties and conduct site visits. Regular interactions and hearings allow the board to stay informed about the project and any potential issues. The rules detail how parties can submit disputes to the DB and the documentation required for the board's consideration.
- Confidentiality: The ICDR emphasizes the importance of confidentiality in DB proceedings. All discussions, documents, and communications with the board are treated as confidential to encourage open dialogue among the parties.
- Fees and Costs: The ICDR rules specify how the fees for the DB members and associated costs will be managed. This transparency helps parties understand the financial implications of engaging a DB.
- Termination of the Board: The rules also set forth conditions under which a DB may be terminated, typically upon the completion of the project or as agreed by the parties.
- Ethical Standards: DB members are expected to adhere to high ethical standards, including independence and impartiality. Members must disclose any potential conflicts of interest and avoid situations that could compromise their neutrality.

Challenges and Limitations of DBS

Challenges and limitations of DBs can arise in various areas, including enforcement of decisions, lack of awareness and understanding, and cultural and legal differences:

- Enforcement of DB Decisions: One of the key challenges with DBs is the enforcement of their decisions. While DB "recommendations" are typically nonbinding, parties are expected to act in good faith and comply with the board's findings. On the contrary, if a DB issues a "decision," these are absolutely binding since parties have contractually agreed so. However, there may be instances where one party refuses to accept the board's decision, leading to difficulties in enforcing the outcome. In such cases, parties may need to resort to formal dispute resolution mechanisms, such as emergency arbitration, regular arbitration or litigation, to enforce the board's decision, depending on the dispute resolution mechanism included in the contract.
- Lack of Awareness and Understanding: Another challenge with DBs is the lack of awareness and understanding among stakeholders in the construction industry. Many parties involved in construction projects may not be familiar with the concept of DBs or how they operate. This lack of awareness can lead to resistance to using DBs or misunderstandings about their role and

effectiveness. Educating stakeholders about the benefits of DBs and how they can help resolve disputes early on is essential to increasing their acceptance and usage.
- Cultural and Legal Differences: Cultural and legal differences can also pose challenges for DBs, especially in international projects involving parties from different jurisdictions. Different cultural norms and legal systems may impact how parties perceive and engage with dispute resolution processes. For example, some parties may prefer more adversarial approaches to dispute resolution, while others may prioritize consensus-building and relationship preservation. Bridging these cultural and legal differences to ensure effective use of DBs can be a complex task requiring sensitivity and flexibility.

Therefore, while DBs offer many benefits for resolving disputes in construction projects, they also face challenges and limitations related to enforcement of decisions, lack of awareness and understanding, and cultural and legal differences. Overcoming these challenges requires proactive measures, such as clear communication, education, and cultural sensitivity, to enhance the effectiveness of DBs and promote their successful implementation in construction projects.

DBS are Still Questioned by the Industry. Why?

While DBs offer many benefits in resolving conflicts, there are several reasons why some people may not prefer or like DBs:

- Perceived Lack of Independence: Some parties may question the independence and impartiality of the DB members. Concerns about potential bias or conflicts of interest could undermine trust in the dispute resolution process and the decisions made by the board. This can be easily worked out by selecting the right members and requesting them to inform any matter than can affect such independence for any reason.
- Cost and Resources: Engaging in DB proceedings may require additional time, resources, and costs for the parties involved. Some may view this as an unnecessary burden or expense, especially if they believe the dispute can be resolved through other means more efficiently or cost-effectively. In any case, it is important to understand the preventive nature of DBs and the way to solve their matters.
- Perceived Ineffectiveness: There may be a perception that DBs are not always effective in resolving disputes or that their decisions may not be respected or enforced. If parties do not have confidence in the process or the outcomes of the DB, they may be reluctant to engage with it.

- Preference for Adversarial Processes: Some individuals or organizations may prefer adversarial processes such as litigation or arbitration, where parties can present their case in a more formal and confrontational manner. DBs, with their emphasis on collaboration and consensus-building, may not align with this preference for adversarial approaches to dispute resolution.
- Lack of Awareness or Understanding: A lack of awareness or understanding about the benefits and procedures of DBs could lead to resistance or skepticism toward using this dispute resolution mechanism. Parties may not fully appreciate the value that DBs can bring in resolving disputes efficiently and maintaining project relationships.
- Cultural or Legal Factors: Differences in cultural norms, legal traditions, or business practices could influence attitudes toward DBs. Some parties may be more accustomed to traditional dispute resolution methods or may have specific legal requirements that make them less inclined to utilize DBs.
- Concerns about Confidentiality: In some cases, parties may have concerns about the confidentiality of the DB proceedings and decisions. They may worry about sensitive information being disclosed or shared during the dispute resolution process, leading to hesitancy in engaging with the DB.

Overall, while DBs can offer numerous advantages in resolving conflicts, individual preferences, perceptions, and circumstances can influence whether parties choose to utilize this particular dispute resolution mechanism. Addressing concerns, providing education, and promoting transparency in the dispute resolution process may help improve acceptance and understanding of DBs among stakeholders.

Comparison with Other Dispute Resolution Mechanisms

Dispute resolution mechanisms play a crucial role in resolving conflicts in various contexts, including construction projects. Here is a comparison of DBs with arbitration, litigation, mediation, and expert determination:

- DBs: DBs are typically used in the construction industry to provide ongoing guidance and assistance in resolving disputes that may arise during a project. They are proactive in nature, aiming to prevent disputes from escalating by addressing issues as they arise. DB decisions are contractually binding as previously discussed, and recommendations are nonbinding, but parties are encouraged to comply with the board's recommendations in good faith. It is important for parties to understand that these outcomes are necessary for the development of the project and that keeping an eternal conflict may not be a real option for them.

- Arbitration: Arbitration is a formal dispute resolution process where parties present their case to an arbitrator or panel of arbitrators, who then make a binding decision called award, which is absolutely enforceable per law and international treaties. Arbitration proceedings are typically private, less formal, and more flexible compared to litigation. Arbitration awards are enforceable in courts, providing parties with a legally binding resolution to their dispute.
- Litigation: Litigation involves resolving disputes through the court system, where parties present their case before a judge or jury. Litigation proceedings are formal, public, and can be time-consuming and costly. Court decisions in litigation are binding and enforceable by law.
- Mediation: Mediation is a voluntary process where a neutral third party (the mediator) helps parties in conflict reach a mutually acceptable agreement. Mediation focuses on facilitating communication, negotiation, and compromise between parties. Mediated agreements are nonbinding unless parties choose to formalize them into a legally enforceable contract.
- Expert Determination: Expert determination involves appointing a neutral expert to make a decision on a specific issue in dispute based on their expertise. Expert determination is often used for technical or complex matters where specialized knowledge is required. The decision of the expert is usually binding, unless parties have agreed otherwise.

In summary, each of these dispute resolution mechanisms has its own advantages and limitations, and the choice of which mechanism to use will depend on various factors such as the nature of the dispute, the preferences of the parties, the desired level of formality, and the enforceability of the decision. DBs, arbitration, litigation, mediation, and expert determination each offer unique approaches to resolving conflicts and can be tailored to meet the specific needs of parties involved in a dispute.

Legal Recognition and Support

Many countries have recognized the value of DBs in preventing and resolving construction disputes. Some jurisdictions have even enacted legislation to support the use of DBs in construction contracts. Especially in Latin America, countries such as Peru, Ecuador, and Costa Rica have now specific law that recognize DBs as an official mechanism for dispute avoidance and resolution in public projects. This has helped DBs to gain traction but also to face new challenges. Specialized organizations have been crucial for the development, recognition and understanding of DBs. The DRBF is a very active international organization that promotes the good use of DBs. The societies of Construction Law around the world are also excellent promoters of good practices.

Future Trends in DBs

The use of DBs is gaining traction in the construction industry and other sectors as a proactive and efficient means of resolving conflicts. Some issues to consider are:

- Increased Adoption Across Industries: DBs have traditionally been popular in the construction sector, but there is potential for their adoption to expand into other industries such as energy, infrastructure, and technology. As organizations seek more efficient and cost-effective ways to resolve disputes, they may turn to DBs as a proactive and collaborative approach.
- Digital Transformation: The digital transformation of dispute resolution processes is likely to impact DBs as well. Online platforms and tools could be developed to streamline communication, document sharing, and decision-making within DB proceedings, making the process more efficient and accessible.
- Specialization and Diversity: Future trends may see more specialized DBs being established to address specific types of disputes or industries. Additionally, there may be a push for greater diversity and inclusion within DBs to ensure a broader range of perspectives and expertise in resolving conflicts.
- Integration with Technology: The integration of technologies such as artificial intelligence (AI), blockchain, and data analytics could enhance the effectiveness of DBs by providing data-driven insights, predictive analytics, and automated decision-making support.
- Globalization and Cross-Border Disputes: With the increasing globalization of businesses and projects, there may be a rise in the use of DBs for resolving cross-border disputes. This trend could lead to the development of standardized practices and procedures for international DBs.
- Emphasis on Sustainability and Environmental Disputes: As sustainability and environmental concerns become more prominent in project development, there may be a growing need for DBs to address disputes related to sustainability practices, environmental impact assessments, and compliance with regulations.
- Enhanced Training and Accreditation: Future trends may involve the establishment of standardized training programs and accreditation processes for DB members to ensure consistency, quality, and professionalism in their roles.

Overall, the future of DBs is likely to involve continued evolution and adaptation to meet the changing needs of businesses, projects, and stakeholders. By embracing innovation, technology, specialization, and diversity, DBs can continue to serve as effective mechanisms for resolving conflicts and promoting successful project outcomes.

Conclusion: Why DBS are an Excellent Option for Dispute Prevention and Resolution

Including a DB in construction contracts offers several significant benefits that can enhance the project's overall success. Here are five reasons why construction contracts should incorporate a DB:

- Proactive Dispute Prevention and Resolution: A DAAB helps to identify and address potential disputes before they escalate into significant conflicts. By being actively involved throughout the project, the board can facilitate discussions, providing guidance and recommendations that foster collaboration, which helps maintain a positive working relationship among stakeholders.
- Early Intervention: The presence of a DAAB allows for early intervention when issues arise, helping the parties to resolve problems promptly. By addressing disputes at an early stage, the DAAB can help minimize delays, reduce escalating costs, and prevent potential disruptions to the project schedule.
- Expertise and Experience: DB members are typically chosen for their expertise in construction, contract management, and dispute resolution. Their knowledge can bring valuable insights into the project's challenges and ensure that decisions are based on industry best practices. This expert advice can lead to more informed and fair outcomes for all parties involved.
- Binding and Nonbinding Options: DBs can provide nonbinding recommendations (in the case of DRBs) or binding decisions (in the case of DABs). This flexibility allows parties to benefit from the board's recommendations while also having recourse to binding decisions when needed. Such flexibility helps to strike a balance between collaboration and enforcing accountability.
- Reduced Time and Costs: Incorporating a DB can lead to significant time and cost savings for the project. By resolving disputes more quickly and efficiently through a structured process, parties can avoid lengthy arbitration or litigation procedures, which can be costly and time-consuming. This ultimately protects the project's budget and timeline, allowing stakeholders to focus on project delivery rather than protracted disputes.

In summary, including a DB in construction contracts enhances proactive risk management, facilitates timely resolutions, leverages expert knowledge, and contributes to overall project efficiency. This collaborative approach promotes successful project outcomes, minimizes disruptions, and fosters positive relationships among all parties involved.

Resources

International Organizations:

- Dispute Resolution Board Foundation. https://www.drb.org/
- International Chamber of Commerce. https://iccwbo.org/
- FIDIC. https://iccwbo.org/
- International Centre for Dispute Resolution. https://www.icdr.org/

International Books:

- *Chern on DBs.* Cyril Chern.
- *DBs: Procedure and Practice.* Gwyn Owen
- *Construction Arbitration and Alternative Dispute Resolution: Theory and Practice Around the World*, Renato Nazzini
- *DBs: Teoría y Práctica*, Roberto Hernández-Garcia

Specialized Documents:

- Practice Note I Dispute Avoidance – focusing on DBs. https://fidic.org/publications/practice-notes
- Code of Ethical Conduct. Dispute Resolution Board Foundation. https://www.drb.org/code-of-ethical-conduct
- Guide on Effective Conflict Management. International Chamber of Commerce. https://iccwbo.org/wp-content/uploads/sites/3/2023/09/2023_Effective-Conflict-Management-901-1.pdf
- DB Rules. International Chamber of Commerce. https://iccwbo.org/dispute-resolution/dispute-resolution-services/adr/dispute-boards/dispute-board-rules/
- AAA-ICDR® Dispute Avoidance and Resolution Board. https://www.adr.org/sites/default/files/Construction_DARB_Specifications_Rules_Procedures.pdf

24

Statutory Adjudication*

R. Bruce Reynolds, Nicholas Reynolds, and Kathryn Irwin

Introduction

Background

Construction disputes that arise while a project is ongoing can have a significant impact on the ability of the project to run efficiently, and can in fact have wider-reaching ramifications for a construction company's entire business. Without a fast and efficient mechanism of dispute resolution, disputes can have the impact of removing "working capital and human resources from the economy unnecessarily and on a significant scale,"[1] and can drive under-capitalized companies to the brink of, or into, insolvency in especially severe circumstances.

As a policy response to this issue, statutory adjudication has been rapidly growing in popularity.

Statutory adjudication was first introduced in England in 1998.[2] Since then, several other common law countries have incorporated statutorily mandated adjudication into their construction law legislation as a quick, flexible, and interim binding dispute resolution mechanism that permits funds to flow and cannot be waived contractually.

Adjudication is a "proven, pragmatic solution for projects gridlocked by disputes, a solution that frees cash flow and resources, while at the same time striking an appropriate balance among competing interests."[3]

* The information contained in this chapter is current to September 1, 2024. Given the possibility of legislative changes since that date, and for the most up-to-date information, readers are encouraged to review the legislation in question directly.

1 David I. Bristow, Duncan W. Glaholt, R. Bruce Reynolds, Howard M. Wise, "Construction, Builders' and Mechanics' Liens in Canada," 8th ed. § 13:2.

2 *Housing Grants, Construction and Regeneration Act 1996*, 1996 c. 53 (the "***English Act***").

3 R. Bruce Reynolds & Sharon C. Vogel, "Striking the Balance: Expert Review of Ontario's Construction Lien Act" (Ontario Ministry of the Attorney General, 2016), Ch. 9, s. 1.

Key Features of Statutory Adjudication

In an adjudication, one of the parties to a construction contract submits a dispute to an adjudicator, who is often a qualified person with experience in the construction industry (not necessarily a legal professional – it can be an experienced engineer, quantity surveyor, etc.), to render a quick determination of the dispute (typically within a statutorily-mandated timeline of a certain number of days or weeks). The adjudicator then makes a determination that is both interim binding and enforceable.

As statutory adjudication is limited to disputes arising out of construction contracts, it is nevertheless rooted in privity of contract. Participation in an adjudication process is limited to the parties to the construction contract.

Another key feature of adjudications is their fast timelines, which are generally prescribed by the enacting legislation. While the timelines are quick, adjudication is also a flexible mechanism, whereby much of the process can be determined by the parties at either the outset of a construction contract or when a dispute arises, as long as the minimum requirements prescribed by the legislation are met.

Statutory adjudication in some jurisdictions only entitles parties to refer certain categories of disputes to adjudication. Often, adjudications are focused on disputes related to nonpayment or valuation of services or materials, although the enabling legislation (including the original English legislation) may allow the parties to submit any matter to adjudication by agreement.

As there is a quick turnaround and relatively limited evidentiary and procedural requirements (as compared to litigation or arbitration), adjudication is a simpler and more cost-effective mechanism for dispute resolution while a project is ongoing.

Jurisdictions That Have Adopted Adjudication

Several common law countries have adopted or introduced legislation to implement adjudication. These countries include: Canada (including several of its provinces)[4]; England[5]; various states and territories of Australia including

4 *Federal Prompt Payment for Construction Work Act*, S.C. 2019, c. 29, s. 387 (the "***Canadian Federal Act***"); *Construction Act*, R.S.O. 1990, c. C.30 (the "***Ontario Act***"); *Prompt Payment and Construction Lien Act*, RSA 2000, c P-26.4 (the "***Alberta Act***"); *The Builders' Lien Act*, SS 1984–85–86, c B-7.1 (the "***Saskatchewan Act***").

5 *U.K. Act*.

New South Wales,[6] Victoria,[7] the Australian Capital Territory,[8] Queensland,[9] Tasmania,[10] South Australia,[11] and Western Australia[12] (collectively, "**Australia (South)**"); the Northern Territory[13] of Australia (collectively, "**Australia (North)**"); Singapore[14]; Malaysia[15]; Ireland[16]; and New Zealand.[17]

In Canada, there is both federal legislation, which regulates construction contracts that fall under the federal government's jurisdiction, and provincial legislation for three different common law provinces: Ontario, Alberta, and Saskatchewan.[18] Legislation for statutory adjudication has also been introduced or considered in other common law provinces (Nova Scotia, British Columbia, Manitoba, and New Brunswick), but has either not yet passed or not yet come into force. As the federal and provincial legislation prescribe different procedural requirements, they will be addressed separately in the comparison below.

In addition to these countries, several other countries have taken steps toward enacting statutory adjudication, and commonly use contractual adjudication as an alternative method of dispute resolution.[19] Although this chapter will focus on the countries with statutory adjudication currently enacted, readers should be aware that this area of law is still developing internationally.

Comparisons Between Jurisdictions with Adjudication

For each jurisdiction above, this section will address the following questions related to adjudication:

---

6 *Building and Construction Industry Security of Payment Act 1999* (the "***New South Wales Act***").

7 *Building and Construction Industry Security of Payment Act 2002* (the "***Victoria Act***").

8 *Building and Construction Industry (Security of Payment) Act 2009* (the "***ACT Act***").

9 *Building Industry Fairness (Security of Payment) Act 2017* (the "***Queensland Act***").

10 *Building and Construction Industry Security of Payment Act 2009* (the "***Tasmania Act***").

11 *Building and Construction Industry Security of Payment Act 2009* (the "***South Australia Act***").

12 *Building and Construction Industry (Security of Payment) Act 2021* (the "***Western Australia Act***").

13 *Construction Contracts (Security of Payments) Act 2004* (the "***Northern Territory Act***").

14 *Building and Construction Industry Security of Payment Act 2004* (the "***Singapore Act***").

15 *Construction Industry Payment and Adjudication Act 2012* (the "***Malaysia Act***").

16 *Construction Contracts Act 2013* (the "***Ireland Act***").

17 *Construction Contracts Act 2002* (the "***New Zealand Act***").

18 In this comparison, we will be focusing on the common law provinces. In addition to the three common law provinces, Quebec has also recently adopted adjudication in its legislation, however, as Quebec is a civil law country, we will not be discussing its legislative scheme.

19 For example, Hong Kong and South Africa.

Who is entitled to invoke adjudication?

1. What training is required for an adjudicator and how is the adjudicator nominated?
2. What types of disputes are subject to adjudication?
3. What are the procedural steps and timelines in an adjudication?
4. How are adjudication decisions enforced?

The Party Entitled to Invoke Adjudication

In order to commence the adjudication process, a party must refer a dispute to an adjudicator. This is typically done by the party issuing an application to adjudicate, or delivering a notice of intent to commence adjudication (or simply a "notice of adjudication," depending on the jurisdiction).

With respect to which party is entitled to invoke adjudication, there is a split among jurisdictions that have implemented adjudication – specifically, those jurisdictions employ one of two different approaches: (1) any party can refer a dispute to an adjudicator; or (2) only the unpaid party can refer a dispute to an adjudicator.

The English model of statutory adjudication provides that a "party to a construction contract has the right to refer a dispute arising under the contract."[20] Similarly, New Zealand, Ireland, Malaysia, and Australia (North) also allow for any party to refer the dispute to adjudication.[21]

In contrast, Singapore and Australia (South) both limit the entitlement of the parties to refer a dispute to adjudication to only the payee.[22]

In Canada, the federal legislation and the common law provincial legislation differs. The provincial legislation is less restrictive and allows "a party to a contract or subcontract" to refer a dispute to adjudication.[23] In contrast, the federal legislation restricts the entitlement of referring a dispute to adjudication to the contractor or subcontractor who has not been fully paid for its construction work within the payment time limits.[24]

The various approaches as to which party is entitled to commence an adjudication are summarized below:

---

20 *U.K. Act*, s. 108(1).

21 *New Zealand Act*, s. 25(1); *Malaysia Act*, s. 7(1); *Ireland Act*, s. 6(1); *Northern Territory Act*, s. 27.

22 *Singapore Act*, s. 13(1); *New South Wales Act*, s. 17(1); *Victoria Act*, s. 18(1); *ACT Act*, s. 19(1); *Queensland Act*, s. 79(1); *Tasmania Act*, s. 21(1); *South Australia Act*, s. 17(1); *Western Australia Act*, s. 28(1).

23 *Ontario Act*, s. 13.5(1); *Alberta Act*, s. 33.4(1); *Saskatchewan Act*, s. 21.21(1).

24 *Canadian Federal Act*, s. 16(1).

England[25]	Canada (Federal)[26]	Canada (Provincial)[27]	Australia (South)[28]	Australia (North)[29]	Singapore[30]	Malaysia[31]	Ireland[32]	New Zealand[33]
Any party	Payee only	Any party	Payee only	Any party	Payee only	Any party	Any party	Any party

Training and Nomination of Adjudicators

Eligibility Requirements for Adjudicators

With respect to adjudicators' eligibility, the different common law jurisdictions fall on a spectrum regarding their respective eligibility requirements of adjudicators.

Nearly all of the jurisdictions require that the adjudicator successfully complete training provided by a specified jurisdictional institute. These institutes are known as either Adjudicator Nominating Bodies ("**ANBs**") or Authorized Nominating Authorities ("**ANAs**"), depending on the jurisdiction.

The majority of jurisdictions require that the adjudicator have certain credentials, in addition to not having a conflict of interest. These credentials often include: a degree or diploma in one of a number of specified fields (e.g., law, architecture, engineering or other construction-related field); a minimum of 5 to 10 years of experience working in the construction industry, depending on the jurisdiction; and the completion of training from the applicable ANB or ANA. These requirements are often prescribed in the associated regulations to each jurisdiction's applicable legislation.

For example, the regulation associated with the *Singapore Act* requires that an adjudicator:

- possess a degree or diploma in architecture, building studies, engineering, environmental studies, law, planning, real estate or urban design, or "such other qualification, as may be recognized by the authorized nominating body";

25 *U.K. Act*, s. 108(1).

26 *Canadian Federal Act*, s. 16(1).

27 *Ontario Act*, s. 13.5(1); *Alberta Act*, s. 33.4(1); *Saskatchewan Act*, s. 21.21(1).

28 *New South Wales Act*, s. 17(1), *Victoria Act*, s. 18(1); *ACT Act*, s. 19(1); *Queensland Act*, s 79(1); *Tasmania Act*, s. 21(1); *South Australia Act*, s. 17(1); *Western Australia Act*, s. 28(1).

29 *Northern Territory Act*, s. 27.

30 *Singapore Act*, s. 13(1).

31 *Malaysia Act*, s. 7(1).

32 *Ireland Act*, s. 6(1).

33 *New Zealand Act*, s. 25(1).

- have at least 10 years of work experience related to the "building and construction industry in Singapore"; and
- complete the assessment and training course administered by the authorized nominating body.[34]

Falling on the other end of the spectrum for adjudicator's eligibility requirements is New Zealand. The *New Zealand Act* only requires that an adjudicator is not a party to the construction contract at issue, and that they do not have an undisclosed and un-waived conflict of interest.[35] Beyond this requirement for not having a conflict of interest, the *New Zealand Act* does not require that the adjudicator have any additional credentials.

In contrast to the above, neither the English *Act* nor its associated regulations specifically set out requirements for adjudicators. However, while there are no specified requirements, the English ANBs have implemented and required training for any new adjudicators to allow them to be available for nomination.

In practice, however, it is far more common (and unsurprisingly so) for parties to select experienced construction professionals to act as their adjudicator.

Nomination Process for Adjudicators

The nomination process for an adjudicator is a highly prescribed process that is expressly set out in each of the relevant acts and has strict and fast timelines. This process varies from jurisdiction to jurisdiction, although there are common trends throughout.

In general, jurisdictions will either (1) allow the parties the option to select the adjudicator, failing which, the nomination process will be passed to the applicable ANB or ANA, or (2) the adjudicator is solely appointed by the ANB or ANA and the parties have no input into the appointment.

England provides the broadest options for the nomination process for adjudicators. In England, the appointment of the adjudicator can be made by:

- Prior agreement between the parties;
- Agreement between the parties when the dispute arises; or
- Unilaterally referring the dispute to an ANB to nominate the adjudicator.[36]

The majority of the other common law jurisdictions allow the parties the option of selecting the adjudicator by agreement or by nomination by the selected

---

34 *Building and Construction Industry Security of Payment Regulations*, s. 11(1).

35 *New Zealand Act*, ss. 34(2) and (3).

36 John L. Riches and Christopher Dancaster, *Construction Adjudication*, 2nd ed. (Blackwell Publishing, 2004) at 146.

ANB or ANA. These jurisdictions include: New Zealand,[37] Ireland,[38] Malaysia,[39] Western Australia,[40] Australia (North),[41] Canada (Federal),[42] and Canada (Provincial).[43]

For example, the *Canadian Federal Act* provides that the "parties to a dispute must jointly appoint an adjudicator to determine the dispute," and "if the parties are unable to jointly appoint an adjudicator, either party may request that the Adjudicator Authority appoint an adjudicator for them."[44]

In contrast, there are a few jurisdictions, such as Singapore and the remaining states and territories in Australia (South), that limit the appointment of the adjudicator to only those nominated by the ANB or ANA authorized to appoint the adjudicators.[45] In these jurisdictions, the parties are not provided with the option of selecting the adjudicator themselves.

Disputes Subject to Adjudication

The *English Act* establishes the broadest scope of disputes that can be adjudicated. Subject to a few exclusions imposed by case law and orders made by the Secretary of State, the *English Act* provides that a party to the construction contract is entitled to "refer **a dispute arising under the contract** for adjudication" [emphasis added].[46] The *English Act* specifies that for this purpose, the term "dispute" includes any difference.[47]

The *New Zealand Act* similarly creates a broad entitlement to refer a dispute under a construction contract to adjudication.[48] The *New Zealand Act* expressly provides that an example of a dispute is a disagreement between the parties to a construction contract about "(a) whether an amount is payable under the contract (for example, a progress payment) or the reasons given for non-payment of that amount; or (b) whether there has been a breach of a term of the contract."[49]

37 *New Zealand Act*, s. 33(1).

38 *Ireland Act*, ss. 6(3) and (4).

39 *Malaysia Act*, s. 21.

40 *Western Australia Act*, s. 29(1).

41 *Northern Territory Act*, s. 28(1).

42 *Canadian Federal Act*, s. 17.

43 *Ontario Act*, s. 13.9(2); *Saskatchewan Act*, s. 21.32; *Prompt Payment and Adjudication Regulation*, Alta Reg 23/3033, s. 22.

44 *Canadian Federal Act*, s. 17.

45 *Singapore Act*, s. 14(1); *New South Wales Act*, s. 17(6); *Victoria Act*, s. 18(7); *ACT Act*, s. 19(4); *Queensland Act*, s. 79(5); *Tasmania Act*, s. 22(1); *South Australia Act*, s. 17(6);

46 *U.K. Act*, s. 108(1).

47 *U.K. Act*, s. 108(1).

48 *New Zealand Act*, s. 25(1).

49 *New Zealand Act*, s. 25(2).

In contrast to the above, many other jurisdictions limit the scope of the disputes that can be referred to adjudications to those that are payment related.

For example, in Malaysia, a party may only refer “a dispute arising from a payment claim” to adjudication.[50] In Singapore, Ireland, Canada (Federal), and Australia (North and South), parties are also only allowed to refer payment matters or disputes relating to payment to adjudication.[51]

While the Canada (Provincial) legislation does not provide as broad a scope for adjudication as the *English Act* or the *New Zealand Act*, it does not prevent the parties from referring nonpayment related disputes to adjudication. Each of the Canadian Provincial Acts are slightly different, but all expressly provide a list of the types of disputes that may be referred to adjudication.[52] For example, the *Ontario Act* provides:

> A party to a contract may refer to adjudication a dispute with the other party to the contract respecting any of the following matters:
>
> **1.** The valuation of services or materials provided under the contract.
> **2.** Payment under the contract, including in respect of a change order, whether approved or not, or a proposed change order.
> **3.** Disputes that are the subject of a notice of non-payment under Part I.1.
> **4.** Amounts retained under section 12 (set-off by trustee) or under subsection 17 (3) (lien set-off).
> **5.** Payment of a holdback under section 26.1 or 26.2.
> **6.** Non-payment of holdback under section 27.1.
> **7.** Any other matter that the parties to the adjudication agree to, or that may be prescribed.[53]

While the majority of the disputes that are referrable to adjudication under the *Ontario Act* are payment-related, the *Ontario Act* broadens the party’s entitlement to adjudication by allowing parties to agree to adjudicate any other matter. All

---

50 *Malaysia Act*, s. 7(1).

51 *Singapore Act*, s. 12; *Ireland Act*, s. 6(1); *Canadian Federal Act*, s. 16(1); *New South Wales Act*, s. 17(1); *Victoria Act*, s. 18(1); *ACT Act*, s. 19(1); *Queensland Act*, s. 79(1); *Tasmania Act*, s. 21(1); *South Australia Act*, s. 17(1); *Western Australia Act*, s. 28(1); *Northern Territory Act*, s. 27.

52 *Ontario Act*, s. 13.5(1); *Saskatchewan Act*, s. 21.21(1); *Prompt Payment and Adjudication Regulation*, Alta Reg 23/2022, s. 19.

53 *Ontario Act*, s. 13.5(1).

three of the Canada (Provincial) Acts include similar deferential language, which allows the parties to determine which matters may be subject to adjudication.[54]

The Adjudication Process

The procedural rules applicable to adjudication for each jurisdiction tends to fall into one of two categories at either end of a procedural spectrum: adjudication will lean either toward (1) a procedurally flexible approach, or (2) a prescriptive one.

England uses a procedurally flexible approach, such that the *English Act* does not prescribe a detailed and rigid process that the adjudicator must follow. Instead, while there are minimum requirements with which the adjudicator must comply, the adjudicator is given the flexibility to determine the process that they will follow in order to reach a determination.

Other jurisdictions, such as New Zealand, Australia (North and South), Malaysia, and Singapore, take a more prescriptive approach such that the conduct of the adjudication proceeding is more narrowly prescribed by the legislation. Perhaps most importantly, the legislation will generally prescribe the timelines for each step of the adjudication, administrative requirements for each step of the adjudication process (e.g., requirements for the application of adjudication), and the procedural powers of the adjudicator.

To oversimplify somewhat, the adjudication process will generally proceed as follows:

1. The claimant will issue an application or a notice of its intention to adjudicate.
2. Either the parties will mutually agree upon an adjudicator, or the adjudicating body will nominate an adjudicator.
3. Depending on the jurisdiction, the claimant will then issue its claim and supporting documents to the adjudicator and the other party.
4. Depending on the jurisdiction, the respondent will usually have an opportunity to respond to the claimant's application, and deliver additional documents.
5. Depending on the jurisdiction, the claimant may have an opportunity to provide a reply.
6. The adjudicator will review the materials and issue their determination.

54 *Ontario Act*, s. 13.5(1); *Saskatchewan Act*, s. 21.21(1); *Prompt Payment and Adjudication Regulation*, Alta Reg 23/2022, s. 19.

A summary of the adjudication process timelines for each jurisdiction is provided below:

Jurisdiction	Timeframe to Apply for adjudication	Timeframe for a response	Timeframe for adjudication decision
England	At any time.[55] Within 7 calendar days of the Notice of Adjudication, the claimant must send associated documents.[56]	Adjudicator will determine whether a response is necessary and the timetable for the response.[57]	Within either 28 calendar days after the referral or 42 days after the referral if the parties consent (or any other period of time if both parties agree).[58]
Singapore	Within 7 days after entitlement arises.[59]	Within 7 days of receipt of the application.[60]	Within 7 days of the commencement of the adjudication or 14 days in any other case.[61]
Malaysia	At any time following the payment claim response period, subject to limitations legislation.[62]	Within 10 working days from receipt of claim.[63]	Within 45 working days of expiry of the period for response.[64] This may be extended by agreement of the parties.[65]
Ireland	At any time.[66] Within 7 days of appointment, the claimant must provide the dispute and the documents to the adjudicator.[67]	Adjudicator will determine whether a response is necessary and the timetable for the response.[68]	Within either 28 calendars days of the referral or 42 days by extension of the adjudicator and claimant consent (or any other period of time if both parties agree).[69]

55 *U.K. Act*, s. 108(2).
56 *The Scheme for Construction Contracts (England and Wales) Regulations 1998*, s. 7.
57 *The Scheme for Construction Contracts (England and Wales) Regulations 1998*, s. 13.
58 *U.K. Act*, s. 108(2).
59 *Singapore Act*, s. 13(3).
60 *Singapore Act*, s. 15(1).
61 *Singapore Act*, s. 17(1).
62 *Malaysia Act*, ss. 7(2) and (3).
63 *Malaysia Act*, s. 10(1).
64 *Malaysia Act*, s. 12(2).
65 *Malaysia Act*, s. 12(2).
66 *Ireland Act*, s. 6(2).
67 *Ireland Act*, s. 6(5).
68 *Construction Contracts Act 2013 (Code of Practice Governing the Conduct of Adjudications)*, s. 24.
69 *Ireland Act*, ss. 6(6) and (7).

Jurisdiction	Timeframe to Apply for adjudication	Timeframe for a response	Timeframe for adjudication decision
New Zealand	At any time, subject to limitations legislation.[70] Within 5 days of receipt of adjudicator's appointment, the claimant must serve its claim.[71]	Within 5 days of the claim.[72]	Within 20 working days of the response or 30 working days if the adjudicator extends the timeline (or any further time the parties all agree to).[73]
Canada (Federal)	Within 21 days of either the certificate of completion or the expiration of the time limit for the payment of the final proper invoice.[74] Within 5 days of the adjudicator's appointment, the claimant must provide additional documents to the adjudicator and other party.[75]	Within 20 days after receiving documents from the claimant.[76]	Within 20 days after receiving the response, or within 20 days after the date the response was to be provided.[77] This timeline may be extended by 5 days by the adjudicator or for a longer period if all parties agree.[78]

(Continued)

70 *New Zealand Act*, ss. 25(1) and 71(1).

71 *New Zealand Act*, s. 36(1).

72 *New Zealand Act*, s. 37(1).

73 *New Zealand Act*, s. 46(2).

74 *Canadian Federal Act*, s. 16(2).

75 *Federal Prompt Payment for Construction Work Regulations (Dispute Resolution)*, SOR/2023-271, s. 16.

76 *Federal Prompt Payment for Construction Work Regulations (Dispute Resolution)*, SOR/2023-271, s. 18.

77 *Federal Prompt Payment for Construction Work Regulations (Dispute Resolution)*, SOR/2023-271, s. 22(1).

78 *Federal Prompt Payment for Construction Work Regulations (Dispute Resolution)*, SOR/2023-271, s. 22(2).

(Continued)

Jurisdiction	Timeframe to Apply for adjudication	Timeframe for a response	Timeframe for adjudication decision
Canada (Provincial)			
Ontario	Before the date the contract or subcontract is complete, unless the parties agree otherwise.[79] Within 5 days of appointment, the claimant must provide additional documents to adjudicator and other party.[80]	The timeline will be decided by the adjudicator.[81]	Within 30 days after receiving the claimants documents.[82] This timeline may be extended by 14 days by the adjudicator with consent of the parties, or for the period of time agreed upon by the parties.[83]
Alberta	Before the date the contract or subcontract is complete, unless the parties agree otherwise.[84] Within 5 days of appointment, the claimant must provide additional documents to adjudicator and other party.[85]	Within 12 days of receiving claimants documents.[86]	Within 30 days after receiving the claimants documents or within 40 days of receiving the claimants documents if the adjudicator extends the timeline.[87]

79 *Ontario Act*, s. 13.5(3).

80 *Ontario Act*, s. 13.11.

81 *O. Reg. 306/18: ADJUDICATIONS UNDER PART II.1 OF THE ACT*, s. 17(3).

82 *Ontario Act*, s. 13.13(1).

83 *Ontario Act*, s. 13.13(2).

84 *Alberta Act*, s. 33.4(2)

85 *Prompt Payment and Adjudication Regulation, Alta Reg 23/2022*, s. 23.

86 *Prompt Payment and Adjudication Regulation, Alta Reg 23/2022*, s. 24(2).

87 *Prompt Payment and Adjudication Regulation, Alta Reg 23/2022*, s. 26(1).

Jurisdiction	Timeframe to Apply for adjudication	Timeframe for a response	Timeframe for adjudication decision
Saskatchewan	Before the date the contract or subcontract is complete, unless the parties agree otherwise.[88] Within 5 days of appointment, the claimant must provide additional documents to adjudicator and other party.[89]	Within 5 days of receiving the claimants documents.[90]	Within 30 days after receiving the claimants documents.[91] This timeline may be extended by 14 days by the adjudicator with consent of the parties, or for the period of time agreed upon by the parties.[92]
Australia			
New South Wales	Within 10 or 20 business days depending on the applicable circumstances.[93]	Within 5 business days of receiving copy of application or 2 business days after receiving notice of adjudicator's acceptance.[94]	Within 10 business days of notifying the parties of the adjudicator's acceptance.[95]
Victoria	Within 7 to 10 days depending on whether a payment schedule was received.[96]	Within 5 business days of receiving copy of application or 2 business days after receiving notice of adjudicator's acceptance.[97]	Within 10 business days of notifying the parties of the adjudicator's acceptance.[98] With claimants agreement, this timeline could be extended to 15 business days.[99]

(Continued)

88 *Saskatchewan Act*, s. 21.21(3).
89 *Saskatchewan Act*, s. 21.42(1).
90 *The Builders' Lien Regulations*, RRS c B-7.1 Reg 1, s. 5.61(1).
91 *Saskatchewan Act*, s. 21.5(1).
92 *Saskatchewan Act*, s. 21.5(2).
93 *New South Wales Act*, s. 17(3).
94 *New South Wales Act*, s. 20(1).
95 *New South Wales Act*, s. 21(3).
96 *Victoria Act*, s. 18(3).
97 *Victoria Act*, s. 21(1).
98 *Victoria Act*, s. 22(4).
99 *Victoria Act*, s. 22(4).

(Continued)

Jurisdiction	Timeframe to Apply for adjudication	Timeframe for a response	Timeframe for adjudication decision
Queensland	Within 20 or 30 business days depending on whether a payment schedule was received.[100]	For a standard claim, within either 10 business days of receiving application or 7 business days of receiving notice of adjudicator's acceptance.[101] For a complex claim, the timeline above are extended to 15 and 12 business days respectively, with an option of extending by 15 business days.[102]	For a standard claim, within 10 business days after receiving response.[103] For a complex claim, 15 business days.[104]
South Australia	Within 15 or 20 business days depending on whether a payment schedule was received.[105]	Within 5 business days of receiving copy of application or 2 business days after receiving notice of adjudicator's acceptance.[106]	Within 10 days after receiving the response, or within 10 days after the date the response was to be provided, or within a timeline parties agreed to.[107]
Tasmania	Within 10 or 20 business days depending on whether a payment schedule was received.[108]	Within 10 business days of receiving copy of application or 5 business days after receiving notice of adjudicator's acceptance.[109]	Within 10 business days of receiving the respondent's response.[110]

100 *Queensland Act*, s. 79(2).
101 *Queensland Act*, s. 83(1).
102 *Queensland Act*, ss. 83(2) and (3).
103 *Queensland Act*, s. 85(1).
104 *Queensland Act*, s. 85(1).
105 *South Australia Act*, s. 17(3).
106 *South Australia Act*, s. 20(1).
107 *South Australia Act*, s. 21(3).
108 *Tasmania Act*, s. 21(3).
109 *Tasmania Act*, s. 23(2).
110 *Tasmania Act*, s. 24(1).

Jurisdiction	Timeframe to Apply for adjudication	Timeframe for a response	Timeframe for adjudication decision
Australian Capital Territory	Within 10 or 20 business days depending on whether a payment schedule was received.[111]	Within 7 business days of receiving copy of application or 5 business days after receiving notice of adjudicator's acceptance.[112]	Within 10 business days of receiving the respondent's response.[113]
Western Australia	Within 20 business days after the dispute arises.[114]	Within 10 business days of being served with an application for adjudication.[115]	Within 10 business days of receiving the respondent's response.[116]
Northern Territory	Within 65 working days after the dispute arises.[117]	Within 15 working days after being served.[118]	Within 10 working days of receiving the respondent's response.[119]

Enforcement of Adjudicators' Determinations

If an adjudicator determines that an amount must be paid to the claimant, the respondent will be required to pay that amount to the claimant promptly.

In all of the common law jurisdictions reviewed in this chapter, a determination provided by an adjudicator is binding on the parties on an *interim* basis, and is enforceable through court processes (indeed, determinations are to some extent functionally equivalent to court orders, and to the extent a party refuses to comply with a determination, the other party may apply to court for the determination to be recognized as a formal court order). On the other hand, the adjudicator's decision is considered interim binding insofar as it can be subsequently displaced by either a court order or an arbitral award, if agreed to by the parties or prescribed by the corresponding legislation.

111 *ACT Act*, s. 19(3).
112 *ACT Act*, s. 22(1).
113 *ACT Act*, s. 23(3).
114 *Western Australia Act*, s. 28(4).
115 *Western Australia Act*, s. 34(1).
116 *Western Australia Act*, s. 37(2).
117 *Northern Territory Act*, s. 28(1).
118 *Northern Territory Act*, s. 29(1).
119 *Northern Territory Act*, s. 33(3).

In general, each jurisdiction is deferential to an adjudicator's determination, and often will enforce the decision unless there has a breach of natural justice or a want of jurisdiction (in which case the dissatisfied party may be able to seek judicial review of the determination).

The Relationship Between Adjudication and Other Dispute Resolution Processes

In addition to statutory adjudication, the construction industry typically relies upon the following six mechanisms of dispute resolution, among others: (1) litigation; (2) arbitration; (3) construction liens; (4) dispute resolution boards; (5) mediation; and (6) negotiation. While there are other dispute resolution mechanisms that are used with varying degrees of regularity, such as expert determinations or independent certifiers, the following section will consider the relationship between adjudication and these six types of dispute resolution processes. The below review is only summary in nature, given that several of these methods of dispute resolution are addressed in detail in other chapters of this publication.

Litigation

Litigation is the process of advancing a civil claim in a court of law. In the construction industry, litigation is used less frequently for a variety of reasons. Most notably, this includes the fact that in many jurisdictions, there is no specialized construction court; because the parties cannot select which judge will hear the dispute, and because the pool of judges often has very few construction law experts, there is a strong likelihood that the judge assigned to the dispute will not have expertise in the issues at hand.

In comparison to adjudication, litigation has strict procedural and evidentiary rules that the parties must follow. As discussed above, while international jurisdictions will generally prescribe certain procedural and administrative requirements for adjudication, these rules are often limited and flexible. For example, several jurisdictions provide the parties with the ability to agree to extended timelines for adjudication. In contrast, parties are not able to contract out of the rules prescribed for civil litigation, and failure to strictly comply with the rules could have significantly adverse impacts on a party's ability to enforce its rights.

Another key difference between the two methods of dispute resolution is the time frame within which parties can expect to reach a decision. While adjudication is a fast method of resolving disputes,[120] a claim in litigation will remain in the court process for months or (more likely) years before a decision on the merits is rendered by the court. Litigation is widely known as the most time intensive form of dispute resolution.

Confidentiality is another significant difference between the two methods of dispute resolution. While an adjudicator's decision is initially private between the parties and will rarely make it into the public domain, a court's decision (as well as the trial record and all the documents contained therein) is public, unless a litigant is able to obtain a sealing order (which is difficult to obtain). While the parties to an adjudication can agree to enter into a confidentiality agreement, the parties are not able to contract out of the public element of a court proceeding.

Arbitration

Arbitration is, by now, arguably the dominant contractual mechanism of dispute resolution in the global construction industry.

Construction contracts will often contain an arbitration provision requiring the parties to proceed by way of arbitration prior to commencing litigation, or will indeed foreclose the possibility of litigation (except, for example, for any appeals of arbitral awards, or set-aside applications, or enforcement proceedings, permitted by the applicable curial legislation).

In an arbitration, the parties will mutually agree on the arbitrator(s) and will set out specific procedural rules for the conduct of the arbitration. If the parties cannot agree on the arbitrator(s), then similar to an adjudication, the parties can refer the dispute to an arbitral institute and request that the institute nominate the arbitrator(s). A key difference from adjudication, though, is that many of the jurisdictions discussed in this chapter – particularly those whose arbitration legislation is based on the UNCITRAL *Model Law* – provide for court appointment (or appointment by a nominating authority[121]) of the arbitrator(s) where the parties have not agreed on the procedure to appoint the arbitral tribunal.[122] By contrast, none of the jurisdictions discussed herein provide for the court appointment of adjudicators, although all of them contemplate nominating authorities (as discussed above).

120 As discussed above, the timeline for an adjudication may span from a very intensive 14-day turnaround for certain jurisdictions, up to a maximum total of 133 working days in one jurisdiction (Malaysia). However, it should be noted that the maximum 133 working days for adjudication is an outlier internationally, as most jurisdictions have adjudication processes that span between 14 days to 50 days.

121 See, for example, Article 6 of the UNCITRAL *Model Law*.

122 For example, Ontario's *Arbitration Act, 1991*, S.O. 1991, c. 17, s. 10(1).

While there are limited statutory requirements for an arbitration – and indeed, many *Model Law* jurisdictions have intentionally adopted arbitration legislation that prescribes little if any procedural rules – most arbitral institutes publish their own sets of rules by which the parties can agree to be bound (or which they can disregard by agreement). While certain minimum requirements are prescribed, an arbitration is much more flexible than litigation and allows parties to select the majority of the process for arbitration.

As the parties will be able to select which arbitrator(s) will hear their dispute, the parties can (and most often do) select arbitrators that have construction industry experience. This means that, similar to an adjudication, the decision-maker will be well-versed in construction issues and will more easily be able to understand and correctly decide complex construction disputes. On the other hand, arbitration and adjudication therefore share the same concern in this regard, in that the appointed decision-maker might enter the process with their own preconceived ideas and biases on particular issues.

Finally, another similarity that arbitrations and adjudications share is the privacy of the dispute resolution process. Both arbitrations and adjudication are typically kept confidential, unless a party submits the decision by the arbitrator or the adjudicator to the court system for enforcement or appeal.

Construction Liens

A construction lien is another form of security which is registered against the property on which the construction work was provided, and which can be enforced through litigation in order to enforce payments of amounts owed in respect of the property in question – in the event of nonpayment of a lien judgment, by the judicial sale of the property in question.

Adjudication is a process for enforcing payment under construction legislation. In comparison, a construction lien provides security for the enforcement of payment. While the lien acts as a security to ensure that the contractor gets paid, it is not itself a method of enforcement.

The majority of the jurisdictions with statutory adjudication do not have, and have not had, lien legislation. These jurisdictions include: the UK, Ireland, Singapore, and Malaysia.

Other jurisdictions at one point had lien legislation, which provided a measure of security for payees, but repealed it when adjudication was introduced. Both New Zealand and certain Australian states and territories had lien legislation that was later repealed. For example, the Northern Territory of Australia elected to repeal its lien legislation when adjudication was introduced. When it was repealed, commentators noted that the lien legislation was "outdated and largely ineffective, in light of modern security of payment legislation."[123]

In other countries, including certain other Australian states and territories, there is a form of lien that exists that provides protection only for specific circumstances (e.g., unpaid progress payments or amounts due pursuant to an adjudication award). These liens are not directly comparable to a traditional construction lien.

For example, the legislation in Queensland creates a subcontractor charge over the money payable when a subcontractor has worked or supplied goods and services but has not been paid within the contractual timeframe.[124] Instead of having a lien on the property, the *Queensland Act* works to create a charge on a monetary amount such that the payee becomes a type of secured creditor.

In Western Australia and Tasmania, a contractor has a right to a lien but only in relation to "the unpaid amount over any unfixed plant or materials supplied by the claimant" for use in connection with the carrying out of building work or construction work for the respondent.[125]

Finally, there are countries that have maintained their construction lien legislation, even after adopting a statutory adjudication scheme. Canada, at the provincial level,[126] and South Australia[127] continue to have both construction liens and adjudication available to payees in the construction industry.

In South Australia, the construction lien legislation and the *South Australia Act* work side-by-side.[128] Payees have to choose which procedure in the two Acts that they would like to follow in order to enforce their rights to progress payments.[129]

In Canada, construction lien proceedings and adjudication can be used in parallel. Most often, a party will need to register a lien on the property while other dispute resolution mechanisms are used so as to preserve their lien rights under the relevant legislation.

Ultimately, though, construction liens have one distinct advantage over other forms of dispute resolution – including adjudication. Depending on the jurisdiction, the relevant legislation and case law may confirm that (1) lien rights

---

123 Julian Bailey, *Construction Law* (Routledge: London, 2011) at 944, footnote 308.

124 *Queensland Act*, s. 109.

125 *Western Australia Act*, s. 64(1); *Tasmania Act*, s. 14(1).

126 *Ontario Act*, Part III; *The Builders' Lien Act*, SS 1984–85–86, c B-7.1, Part III; *Alberta Act*, Part 2.

127 *Worker's Lien Act 1893*.

128 MinterEllison, "Construction Law: Chapter 15 Security of Payment Legislation South Australia" (online).

129 MinterEllison, "Construction Law: Chapter 15 Security of Payment Legislation South Australia" (online).

are not waivable by contract (the United States are a notable exception to this, insofar as multiple states permit waiver of lien rights by contract or otherwise), and/or (2) lien rights survive the bankruptcy or insolvency of the payor, such that payees may have recourse to relief in circumstances where their payor (or the payor's payor) has initiated bankruptcy or insolvency proceedings. By comparison, adjudication in circumstances of bankruptcy or insolvency could be impossible (depending on the jurisdiction), insofar as the bankruptcy court may impose a stay of all of the bankrupt/insolvent's proceedings.

Dispute Resolution Boards

Dispute resolution boards are creatures of contract that are largely ungoverned by international conventions or legislation (although they are commonly used as part of the FIDIC suite of contracts). Because of this lack of a legislative or soft law framework, the parties must provide all of the rules and procedures for the dispute resolution board in their construction contract.

There are a number of different formats that a dispute resolution board can take: (1) a dispute review board; (2) a dispute adjudication board; or (3) a combined dispute board. Depending on the type of board the parties choose to employ, the process may be more or less formal and the determinations made by the board may either be binding (typically on an interim basis) or nonbinding.

Dispute resolution boards are typically constituted at the outset of a construction project, prior to any disputes arising, and serve as standing entities. The board will generally consist of three impartial professionals. Similar to adjudication, the members of the board will generally have construction experience or technical expertise which will help them understand and decide complex construction issues.

A key difference between adjudication and dispute resolution boards is that the board will follow the construction progress throughout the project and assist in the quick resolution of any disputes that arise. Since a dispute resolution board follows the progress of the project, the board will be familiar with the project and the matter at issue *prior* to it being formally raised as a dispute. This is a significant difference, insofar as a decision maker will typically have no prior knowledge of the project or the dispute.

The determinations made by dispute resolution boards will vary depending on the type of board being used on a project. A dispute review board will typically only issue nonbinding recommendations or determinations, whereas a dispute adjudication board will make binding interim decisions (a combined dispute board can make either binding or nonbinding determinations). An appeal of a decision made by a dispute adjudication board would require either arbitration or litigation, similar to a decision made by an adjudicator, unless the contract provides for an internal appeal mechanism (e.g., to an independent certifier).

As dispute resolution boards are employed continuously throughout a project and can address issues informally with the parties immediately as they arise, dispute resolution boards are often seen as a more collaborative method of dispute resolution. In contrast, adjudication is adversarial in nature, as the parties must apply to have a dispute resolved by an adjudicator and must typically follow the prescribed process, which allows for limited collaboration.

Mediation

Mediation is the process in which the parties submit a dispute to a neutral and independent third party to facilitate negotiations between the parties. It is often described as a collaborative and confidential process that is used to settle disputes between the parties, and it may be required by the contract or be agreed to on ad hoc basis.

Generally speaking, parties can agree to submit a dispute to mediation at any time during a project, similar to adjudication in the majority of jurisdictions. As mediation is a collaborative mechanism of dispute resolution in comparison to submitting a dispute for a binding decision by a third party, parties will typically use mediation while a construction project is ongoing as, in theory, it better preserves the relationship between the parties. In practice, however, experienced practitioners will know that this is not necessarily the case, and that mediation can sometimes be almost as acrimonious as contentious litigation.

Similar to the adjudication processes prescribed in the majority of jurisdictions discussed herein, the mediator will be jointly selected by the parties. If the parties cannot agree on a mediator, they can agree (either at the time of the dispute or upon entering into the construction contract) that a certain institution will appoint the mediator. Both adjudication and mediation have jurisdictional institutes which regulate and nominate appointees.

A key difference between mediation and adjudication is that there are typically no statutory requirements for the mediation of construction disputes. As such, the parties and the mediator have full discretion to determine how the mediation will proceed, including the documentation requirements, timelines, and general approach of the mediation (e.g., whether the mediator should be purely facilitative or they should take a more deliberative approach and provide their opinion on the dispute).

Another key difference between these two mechanisms is that when the parties participate in mediation, there is no guarantee that a resolution will be reached. Further, unless the parties enter into binding minutes of settlement, the entirety of the mediation is nonbinding on the parties and will have no effect on the dispute or any parallel resolution proceedings. Unlike most other dispute resolution mechanisms (apart from without-prejudice negotiations), mediations are without prejudice and cannot be referred to in later forms of dispute resolution.

In comparison, when parties submit a dispute to an adjudicator, the parties are guaranteed that an interim binding and enforceable decision will be reached. As adjudications are not conducted on a without prejudice basis, participating in this form of dispute resolution will create a record that may affect the substance of the dispute (especially to the extent that the record from the adjudication may then be used in litigation or arbitration).

Negotiation

Negotiation is the least formal mechanism of dispute resolution and typically, occurs solely between the two parties under the construction contract at issue.

Unless a binding agreement is reached between the parties, the content of a negotiation is typically privileged and cannot be used to impact the substance of the dispute. As adjudications do result in a binding and enforceable decision that is not issued on a without prejudice basis, this a key difference between a negotiation and an adjudication.

In general, and especially on larger projects, the construction contract will often include a provision requiring that the parties engage in negotiations prior to a dispute being escalated to one of the other mechanisms of dispute resolution. Negotiation provisions in construction contracts are sometimes stepped, in that they may involve different levels of management at different stages of a dispute (e.g., the negotiations may first occur between project-level personnel before being escalated to executives).

A significant difference between adjudication and negotiation is that the issues subject to a negotiation may evolve over time and the negotiation process may be ongoing throughout a project. In contrast, adjudication will typically deal with only one specific issue, or grouping of issues, at a set point in time on the project. As adjudications have a set time frame for the determination to be reached, the adjudication of an issue will not be ongoing throughout the project.

Another distinguishing feature of negotiation is that unlike other dispute resolution mechanisms, negotiation can be, and often will be, ongoing while the parties engage in one of the other dispute resolution mechanisms. Parties may take advantage of this feature to exert commercial pressure within a negotiation. For example, if one party starts the adjudication process, the other will be forced to make any settlement decisions quickly prior to the adjudicator releasing their determination.

Points to Remember

Statutory adjudication is a quick and flexible mechanism of dispute resolution that is growing increasingly popular with parties in the construction industry in many jurisdictions across the world.

In general, all the countries discussed above use adjudication in a similar way to encourage swift resolution of payment-related disputes in the construction industry. Although the process, and even the types of disputes to be addressed, differs across jurisdictions, the general parameters of adjudication remain largely similar.

In that regard, while there can be significant differences in each jurisdiction's legislation, there are a number of common key features or adjudication that are seen throughout:

1. The time frame is typically prescribed by the legislation and requires the adjudicator to make a determination of the issues quickly.
2. The adjudicator's determination will be binding on an interim basis and enforceable.
3. Adjudication is used by construction parties to address issues related to nonpayment.

In comparison to other dispute resolution mechanisms, statutory adjudication is a fast and efficient process that construction parties can use when they want to ensure that an interim binding and enforceable resolution will be achieved.

25

Mediation

R. Zachary Torres-Fowler, Albert Bates, and Jamey B. Collidge

Introduction

Construction disputes that are not resolved by the project participants or senior management often follow a typical path. One party will file for litigation in court or, if allowed, for arbitration. The parties will then engage in time-consuming and costly proceedings that will end with a decision by a neutral judge, arbitrator, panel of arbitrators, or even in some cases, a jury. Once the participants close their respective cases, they have no control over the outcome – a judge, jury, or arbitrator(s) will render a decision that will often declare one party a winner and the other a loser. A challenge to the decision (if permitted at all) will risk prolonging the dispute and further increasing costs. As most outside and in-house counsel will readily acknowledge, the time and expense associated with pursuing litigation or arbitration as a means of formally and finally resolving a party's claims cannot be understated.

Construction disputes are particularly prone to lengthy and costly disputes because, among other things, the projects at issue often take years to complete, are technically complex and fact-intensive, and can involve hundreds of discrete issues. Construction projects are also expensive, and disputes often involve amounts in controversy upwards of hundreds of millions of dollars and beyond. Indeed, in today's world, the cost of construction frequently reaches hundreds of millions, with billion-plus dollar projects becoming far more commonplace. In the United States, between 2012 and 2019, at least 320 megaprojects – projects with a total project value of $1 billion or more – were awarded, and another 670 were then in the planning stages.[1] Research estimates that by 2029, megaprojects will grow from $50 billion to $350 billion in terms of "construction put-in-place."[2]

In addition, setting aside the length and expense of resolving a construction dispute, parties often underestimate the amount of time that they will need to invest in a proceeding and how this process can disrupt the day-to-day operations of their businesses. As construction projects increasingly involve global participants, disputes become more costly due to the number of parties, the necessity for additional and diverse counsel to litigate in unfamiliar jurisdictions, and the logistical hurdles and travel required for key personnel to assist with lengthy and protracted international disputes.

Given the substantial time and effort required to pursue litigation or arbitration, many construction-related entities prefer to consider alternative methods for resolving their disputes before proceeding to litigation or arbitration. Historically, construction industry participants have been willing to explore creative new means to resolve disputes quickly and cost-effectively. For example, the involvement of third-party decision makers during the project via dispute boards has been one of the most novel dispute resolution methods to have emerged from the construction industry.[3]

While relatively common to the United States and other common law jurisdictions, another form of dispute resolution known as mediation has gathered attention from international construction industry participants in recent years. Indeed, the international community has recognized the proliferation of mediation as a viable dispute resolution process. Various recent efforts have been made to create mediation centers and to standardize mediation rules and procedures, such as the Dubai International Arbitration Centre DIAC Mediation Rules. Further, efforts have been made to develop a framework for the uniform enforcement of negotiated settlement agreements through the Singapore Convention.

---

1 Hoover S., "Megaprojects: Changing the Conversation," *FMI Construction Industry Round Table*, 4, (2019). Available at https://fmicorp.com/insights/industry-insights/megaprojects-changing-the-conversation (Accessed: 26 September 2024).

2 U.S. Census Bureau, "Construction Spending." (2022). Available at https://www.census.gov/construction/c30/c30index.html (Accessed: 26 September 2024). ("Construction Put in Place" cost data includes 1) labor and materials, 2) architectural and engineering services, 3) overhead, 4) interest and taxes paid during construction, and 5) contractor's profits). Hoover, S. (2019) "Megaprojects: Changing the Conversation," *FMI Construction Industry Round Table*, 4. Available at https://fmicorp.com/insights/industry-insights/megaprojects-changing-the-conversation (Accessed: 26 September 2024).

3 Bates A. & Torres-Fowler Z., "Dispute Boards: A Different Approach to Dispute Resolution," in *Comparative Law Yearbook of International Business* (2020), p. 237–64 (outlining history of Dispute Boards in construction industry and providing primer on various structures of practices and procedures applicable to Dispute Boards). Available at: https://www.constructlaw.com/2020/04/02/dispute-boards-a-different-approach-to-dispute-resolution/.

Mediation is a nonbinding process where parties work with a third-party neutral to achieve a mutual settlement. Despite its nonbinding nature, mediation is a powerful dispute resolution tool and can effectively steer even reluctant parties toward middle ground. But mediation is voluntary and requires willingness and engagement to be successful. It is ultimately up to the parties to drive the parameters outlining a mutually acceptable resolution to their dispute.

With the risk posed by costly, drawn-out litigation or arbitration proceedings, mediation has proven to be a viable alternative method to assist parties in bringing their disputes to a close. Mediation can be quick, efficient, confidential, and, at times, even amicable. Mediation can allow those who partnered on a construction project to resolve disputes on their own terms and on a timeline that fits their needs. Most importantly, the nature of mediation can help to avoid a combative litigation or arbitration and preserve relationships so that parties to the dispute can move on and continue their partnerships on future projects.

The following chapter outlines the mediation process, offers an international comparative view of mediation, and describes how construction industry participants use mediation to resolve their disputes.

Overview of Mediation

How Does Mediation Work?

Mediation is a dispute resolution process where parties select a single mediator (typically) to assist the parties with reaching a negotiated and mutually acceptable resolution to a dispute. Experienced mediators come prepared with various tools and strategies to work to build consensus and explore opportunities for common ground among the parties.

While a mediator helps to facilitate negotiations and offers neutral perspectives to drive the parties together, mediation is still a voluntary process. Mediation requires all parties to agree to mediate – parties may be skeptical, but mediation requires mutuality and engagement. Indeed, even if the parties are compelled to mediate their dispute pursuant to some form of dispute resolution provision, if one party is reluctant or unwilling to take mediation seriously, there is a risk that the mediation will fail.

Because mediation is a voluntary process, the decision to mediate can be made at any time. Parties commonly mediate even after the initiation of litigation or arbitration. The choice to mediate can be required by contract or formally documented through standalone agreements. Indeed, it is common for parties to include mediation clauses in their construction contracts. Mediation can merely be an option available to the parties should they so choose, or a compulsory step in the dispute resolution process that serves as a condition precedent to the right to pursue formal dispute resolution (i.e., litigation or arbitration), or anything in between.

Before the mediation date, parties often submit mediation statements. These can be exchanged between the parties, submitted confidentially to the mediator, or sometimes, both. These statements typically provide brief summaries of the factual and legal issues in dispute and offer an opportunity for both the mediator and opposing party to hear the submitting party's side of the story. Confidential statements provide the parties an opportunity to open up to the mediator and explain factors at play that they may not wish to share with the opposing party.

Once at the mediation, the mediator may wish to hold a joint session to introduce themselves and set ground rules for the mediation. If agreed to, the parties may make opening statements or presentations. In all instances, however, the parties will eventually part ways into separate breakout rooms. The mediator will meet with each party one-on-one (commonly referred to as a caucus session) during which the mediator will learn more about the party's story and claims, possibly ask questions or offer observations, and relay information from the other parties. The mediator often will shuffle back and forth between the parties throughout the day (or more) until a deal is reached (or not).

Mediation can be an effective process that allows willing participants to put their stories forward and utilize a neutral third-party to facilitate a potential deal. But mediation also can serve as an opportunity for parties to pressure-test their claims and legal theories with a neutral mediator (often a judge or experienced industry professional) before committing to a lengthy and costly litigation or arbitration.

In the sections below, the authors provide a high-level overview of the mediation process and various considerations for those considering mediation.

Structuring the Mediation

After deciding to mediate, the parties typically develop a framework or guidelines for the mediation. Mediations between private parties are commonly organized by the parties themselves on an ad hoc basis. Very often, the parties or their counsel have worked with mediators in the past with whom they are comfortable and recommend. From there, the parties, sometimes with the assistance of the mediator, can develop a general structure and procedure for the mediation (often based on the mediator's preferred approaches).

Parties may also engage a dispute resolution institution to help oversee the mediation and whose rules the parties will use for the mediation. There are many dispute resolution organizations that offer mediation services. In the United States, two of the largest organizations are the American Arbitration Association

("AAA")[4] and JAMS.[5] There are also a host of international organizations that offer mediation services, such as the International Chamber of Commerce ("ICC")[6] and the International Centre for Dispute Resolution ("ICDR").[7] These institutions offer ancillary services and institutional rules to assist parties with their mediations.

While ad hoc mediations tend to be slightly more common in practice, both ad hoc and institutional mediations offer advantages and disadvantages.

Management and Cost. An institution can offer an established structure and logistical help, which may be particularly important for parties who lack familiarity with the mediation process. Institutions may have physical centers where parties can hold the mediation and other support that removes the need for such coordination and a potential point of disagreement before the mediation.

However, parties well versed in mediation with experienced counsel may have the tools in place to negotiate the ground rules and self-select the mediator. In those instances, parties may wish to retain the flexibility of an ad hoc mediation with a mutually selected mediator and avoid oversight by an institution. By doing so, the parties can avoid the administrative costs that come with an institutional mediation and instead only compensate the mediator for their time.

Vetted Mediators. Construction projects often involve highly-technical issues with which practitioners or former judges – who often serve as mediators in the United States – may not be familiar. Institutions typically have deep rosters of prevetted mediators with details of their backgrounds and relevant experiences to allow parties to make an informed decision and select a mediator that best fits their needs and dispute. Selecting a mediator with the right mix of technical expertise, familiarity with the industry, and temperament can be the difference between an early, cost-effective resolution and a lengthy, expensive litigation or arbitration. These vetted rosters can be particularly beneficial for participants who are in unfamiliar territory or unfamiliar with the mediation process. Conversely, experienced parties and counsel may have their own institutional

---

4 American Arbitration Association *What We Do*, (2024). Available at https://www.adr.org/Mediation (Accessed: 26 September 2024).

5 Judicial Arbitration and Mediation Services Inc. ("JAMS"), *Mediators & Mediation Services*. (2024). Available at https://www.jamsadr.com/mediation?tab=overview (Accessed: 26 September 2024).

6 International Chamber of Commerce, *Mediation*. (2024). Available at https://iccwbo.org/dispute-resolution/dispute-resolution-services/adr/mediation/ (Accessed: 26 September 2024).

7 International Centre for Dispute Resolution, *Services*. (2024). Available at https://www.icdr.org/services/international-mediation (Accessed: 26 September 2024).

knowledge and a deep bench of potential mediators that obviate the need for an institution's roster.

Standardized Rules. Organizations provide standard rules that the parties may choose to govern their mediations, which can help provide the parties with structure to their mediation process and avoid unnecessary fights over logistics.[8] For example, selecting a mediator can be contentious. Organizational rules typically set forth procedures for appointing a mediator where the parties cannot agree. For example, the ICC Rules allow the ICC to select a mediator after consultation with the parties.[9] The AAA Rules allow the AAA to provide a list of potential mediators from which the parties may rank their top choices and the AAA will select a mediator based on the parties' rankings.[10]

Institutional rules, however, are not strictly necessary. Parties routinely develop mediation agreements either on their own or with the help of a mutually selected mediator. Governing the mediation need not be complicated or be overly formalistic. Parties can simply agree to a basic set of ground rules or grant the mediator discretion to oversee the process efficiently and cost-effectively. And while organizational rules are helpful, they need not be treated as set instructions. Indeed, such rules are flexible and typically defer to the parties' agreements. Participants should feel free to mold the rules and process to best meet the needs of the parties and the dispute.

Sample Clauses. Institutional organizations provide sample clauses that parties may include in their contracts if they wish to utilize a particular organization to oversee their mediation. These standard clauses help protect against ambiguity and avoid disputes over uncertain language that may hinder or entirely undo an agreement to mediate before the parties even start. Again, however, parties familiar with the mediation process with experienced counsel often draft effective mediation clauses.

The AAA Rules provide a standard clause the parties may insert into a contract to compel the parties to engage in mediation under the AAA Rules before resorting to formal dispute resolution:

---

8 *See, e.g., AAA Rules*; JAMS, *JAMS Mediation International Mediation Rules.* (2024). Available at https://www.jamsadr.com/international-mediation-rules/ (Accessed: 26 September 2024) (hereinafter "*JAMS Rules*"); International Chamber of Commerce, *ICC Mediation Rules.* (2014). Available at https://iccwbo.org/dispute-resolution/dispute-resolution-services/adr/mediation/mediation-rules/ (Accessed: 26 September 2024) (hereinafter "*ICC Rules*"); International Centre for Dispute Resolution, *ICRD International Dispute Resolution Procedures (Including Mediation and Arbitration Rules).* (2016). Available at https://www.adr.org/sites/default/files/ICDR%20Rules_0.pdf (Accessed: 26 September 2024) (hereinafter "*ICDR Rules*").

9 *ICC Rules*, Article 5.

10 *AAA Rules*, Rule M-5.

> If a dispute arises out of or relates to this contract, or the breach thereof, and if the dispute cannot be settled through negotiation, the parties agree first to try in good faith to settle the dispute by mediation administered by the American Arbitration Association under its Construction Industry Mediation Procedures before resorting to arbitration, litigation, or some other dispute resolution procedure.[11]

The ICDR Rules provide a similar clause:

> In the event of any controversy or claim arising out of or relating to this contract, or a breach thereof, the parties hereto agree first to try and settle the dispute by mediation, administered by the International Centre for Dispute Resolution under its International Mediation Rules, before resorting to arbitration, litigation, or some other dispute resolution procedure.

When and how mediation might be utilized to resolve disputes is a sliding scale. The ICC provides different standard clauses that highlight the various ways in which mediation can fit into a dispute resolution process:[12]

- The parties may elect to simply have available to them mediation at any time in accordance with the ICC Rules: "The parties may at any time, without prejudice to any other proceedings, seek to settle any dispute arising out of or in connection with the present contract in accordance with the ICC Mediation Rules."[13] This option merely provides that mediation is always an available option. It does not commit the parties to engage in a mediation or any other settlement procedure. Such a clause, however, may be helpful to rely on should one party wish to propose mediation to another, and can provide a basis for a party seeking mediation to contact the ICC's ADR services to assist with that process.
- The parties may take it a step further and commit to working together to discuss mediation and consider referring a dispute to the ICC as a first step when a dispute arises: "In the event of any dispute arising out of or in connection

11 American Arbitration Association *Construction Industry Arbitration Rules and Mediation Procedures*. (2022). p. 12. Available at https://www.adr.org/sites/default/files/ConstructionRules_web.pdf (Accessed: 26 September 2024) (hereinafter "*AAA Rules*")

12 International Chamber of Commerce, *Mediation Clauses*. (2024). Available at https://iccwbo.org/dispute-resolution/dispute-resolution-services/adr/mediation/mediations-clauses/ (Accessed: 26 September 2024).

13 International Chamber of Commerce, *Mediation Clauses*. (2024). Available at https://iccwbo.org/dispute-resolution/dispute-resolution-services/adr/mediation/mediations-clauses/ (Accessed: 26 September 2024).

with the present contract, the parties agree in the first instance to discuss and consider referring the dispute to the ICC Mediation Rules."[14] This option also does not commit the parties to engage in a mediation, but it evidences an intent that the parties agree to seriously consider mediation as a first step in the dispute resolution process.

The parties may also elect to compel mediation as part of their dispute resolution process. Mandatory mediation can take two forms – it can be compulsory, but allowed to work in parallel with formal dispute resolution proceedings like litigation or arbitration, or it can be a condition precedent to the initiation of litigation or arbitration.

- Mandatory mediation in parallel with formal dispute resolution proceedings:
  (x) In the event of any dispute arising out of or in connection with the present contract, the parties shall first refer the dispute to proceedings under the ICC Mediation Rules. The commencement of proceedings under the ICC Mediation Rules shall not prevent any party from commencing arbitration in accordance with Subclause y below.
  (y) All disputes arising out of or in connection with the present contract shall be finally settled under the Rules of Arbitration of the International Chamber of Commerce by one or more arbitrators appointed in accordance with the said Rules.[15]

Allowing mediation to work in parallel with litigation or arbitration can prevent potential delays that may result if mediation is a condition precedent. Parties may also benefit from exchanges of information and other discovery that typically occurs in more formal proceedings. But there is also a risk that the mediation becomes secondary to a more contentious litigation or arbitration. Parties may become more hostile as proceedings move forward and forget the reasons they chose to mediate in the first place.

- Mandatory mediation as a condition precedent:
  In the event of any dispute arising out of or in connection with the present contract, the parties shall first refer the dispute to proceedings under the ICC Mediation Rules. If the dispute has not been settled pursuant to the said Rules within [45] days following the filing of a Request for Mediation

14 International Chamber of Commerce, *Mediation Clauses.* (2024). Available at https://iccwbo.org/dispute-resolution/dispute-resolution-services/adr/mediation/mediations-clauses/ (Accessed: 26 September 2024).

15 International Chamber of Commerce, *Mediation Clauses.* (2024). Available at https://iccwbo.org/dispute-resolution/dispute-resolution-services/adr/mediation/mediations-clauses/ (Accessed: 26 September 2024).

> or within such other period as the parties may agree in writing, such dispute shall thereafter be finally settled under the Rules of Arbitration of the International Chamber of Commerce by one or more arbitrators appointed in accordance with the said Rules of Arbitration.[16]

Where mediation is a condition precedent, parties will typically have less information about the other's claims and positions and may be reluctant to settle with so many unknowns. But successful mediation at early stages promotes one of mediation's primary benefits: an efficient and cost-effective resolution that allows the parties to move on quickly.

Selecting a Mediator

In complex construction disputes, practitioners generally prefer to select someone with relevant experience in the construction industry. A mediator who is familiar with the industry, the relationships and roles on a project, and even the technical concepts at issue can be particularly beneficial. A mediator with relevant experience may need less time getting up to speed on industry terms and the logistics of the project, saving the participants money. An experienced mediator also may have a greater ability to quickly identify and deeply understand the intricacies of the relationships and disputed issues. They also may more keenly isolate the true roadblocks to resolution and develop ideal strategies to bridge the disagreement and achieve a resolution.

Industry experience or not, mediators vary widely. They may be aggressive or more passive, forceful or subdued, or opinionated or compromising. The mediator's style and ability to build rapport with the participants can go a long way to helping the parties reach a resolution. Practitioners selecting a mediator should evaluate a potential mediator's style for the fit with both the needs of the client and the dispute.

There are generally three styles of mediator: (i) facilitative, (ii) evaluative, and (iii) transformative.[17] There is, however, no strict delineation between these styles. Experienced mediators will often adapt their style to the needs or circumstances and toggle between or use a combination of different approaches.

Facilitative. A facilitative mediator is often viewed as a referee, there to promote fair play while otherwise staying largely on the sidelines. They will generally attempt to foster an environment of open and direct discussion between the parties. A facilitative approach may include joint sessions with the participants to allow the

---

16 International Chamber of Commerce, *Mediation Clauses*. (2024). Available at https://iccwbo.org/dispute-resolution/dispute-resolution-services/adr/mediation/mediations-clauses/ (Accessed: 26 September 2024).

17 Zumeta Z, "Styles of Mediation," *International Mediation Institute*. (2000). Available at https://imimediation.org/resources/styles-of-mediation/. (Accessed: 26 September 2024).

open exchange of viewpoints and to allow the parties to learn information and an understanding of the other's position. The mediator may ask questions to promote discussion, but the facilitative mediator typically will refrain from offering their own views or opinions. Instead, the goal is to position the participants to drive the discussions and parameters of a mutual resolution.[18] Indeed, facilitative mediators generally defer to the parties to outline and structure the mediation process.

Evaluative. An evaluative mediator is more akin to a judge or arbitrator in that their goal is to provide a third-party evaluation of the dispute to help the parties reach resolution. Whereas a facilitative mediator typically will keep views and opinions to themselves unless prompted, an evaluative mediator's role is to evaluate the parties' respective positions. These mediators may discuss the strengths and weaknesses of the parties' respective positions and the costs and benefits of reaching resolution at mediation versus litigation or arbitration. Evaluative mediators are generally more hands-on, preferring caucuses with individual parties than long and open joint-sessions. Evaluative mediators typically will define the mediation structure.[19]

Transformative. A transformative mediator will, in some ways, mirror a facilitative approach. But instead of focusing on facilitating discussions between the parties, a transformative mediator will attempt to transform the relationship itself. A transformative approach can be helpful to level the balance of power (whether actual or perceived) and assist communications by attempting to place the parties on a level playing field. Through transforming the relationship, the parties' views, interests, and values may change and more closely align, allowing a resolution to break through. Like the facilitative approach, a transformative mediator will generally allow the parties to structure the mediation.[20]

Independently from the preferred mediator approach(es), it is important to evaluate potential mediators for the traits that are important to all disputes. The ability to listen and impartiality are paramount. A mediation will end quickly if one or more of the parties feels it is not being heard or the mediator favors the other's position. Flexibility and imagination are also helpful traits. Resolution need not always be limited to the flow of money from one party to another. A creative mediator can help the parties chart many paths to a potential resolution. Ultimately, the mediator must be one who the parties can trust and speak with openly and comfortably.

Mediation Submissions

Mediators often ask the parties to submit a memorandum or statement in advance of the mediation. A mediation statement is generally a brief summary of the

18 Zumeta 2000.

19 Zumeta 2000.

20 Zumeta 2000.

issues in dispute, the parties' legal theories, and offers the mediators and opposing parties an opportunity to hear your side of the story. While the parties will likely have some familiarity with the opposition's position, the mediation statement is usually the participants' first opportunity to provide the mediator with their view of the dispute. It will frame the issues, educate the mediator on the important facts and nuances of the dispute, and sometimes draw attention to particular legal authorities. Beyond the written words, however, the mediation statement will set the tone for the ensuing discussions and negotiations.

Mediation statements can be confidential or shared. Confidential mediation statements are submitted to the mediator only and are not exchanged between the parties. In contrast, shared mediation statements are submitted to the mediator and exchanged with all parties in advance of the mediation.

Shared Mediation Statements. If mediation is designed to foster an open dialogue between the participants, then the exchange of mediation statements is a first step toward achieving that goal. Shared mediation statements can help to promote discussion and eliminate surprises, allowing the parties to start the mediation on a shared understanding of the issues in dispute. An open exchange may also help to ease tensions among suspicious parties who may have concern about what is being said about the other to the mediator.

Confidential Mediation Statements. On the other hand, confidential statements allow for openness without fear of consequence. Participants may be more candid with the mediator about strengths and weaknesses without fear that opposing participants will seize on a purported weakness. Confidential statements may also lead to more openness about settlement options and underlying personal or business interests that a party does not wish to disclose to the other side. Any information about the dispute and the parties can help a sophisticated mediator craft a strategy to reach resolution. Confidential statements may also avoid escalating tensions in disputes with particularly adversarial party relationships.

Mediation Logistics

The mediation itself is entirely flexible and can be molded to what best fits the needs of the dispute and a participant's own overall strategy. Below are some items to consider when developing the structure for a mediation:

- *Whether the mediation will be in-person or virtual.* Virtual mediations are more convenient and typically cheaper as no travel is necessary. However, virtual mediations may be taken less seriously or result in less engagement than in-person mediations that require travel, coordination, and face-to-face discussions.
- *Who will attend the mediation.* Parties should carefully consider who to bring to the mediation. Mediations sometimes serve as forced opportunities

for business-to-business discussions that can be more productive than the mediation itself. Indeed, mediations have settled simply because both parties brought representatives who worked well together during the project and were presented with an opportunity to meet face-to-face. Parties that are hopeful for settlement may wish to leave behind antagonistic representatives;

- *Whether the parties will hold a joint opening session or proceed directly to individual caucuses with the mediator.* Joint sessions with the mediator can set the stage for an amicable day of negotiations. But in heavily contentious disputes or palpable animosity between the parties, joint sessions may simply antagonize the parties and reduce participation and the willingness to find common ground. Parties should consider whether a joint session will enhance opportunity for an ultimate resolution or create more hurdles;
- *Whether the parties will make opening statements or presentations.* Opening statements or presentations can be an effective way for parties to make their case in a neutral setting and prompt questions from the mediator that are helpful for all parties to consider. Presentations are useful where the parties have brought experts to help explain highly technical issues. But in instances where the parties and mediator are zeroed in on the drivers of the dispute, lengthy presentations may wear out the parties' patience before they have had the chance for caucus sessions and to begin negotiations in earnest.
- *Will experts be involved and, if so, to what extent.* Parties should weigh whether it makes sense to bring experts. Experts can help explain technical issues to the mediator or other parties, but doing so may tip off the other side that experts have been retained well before arbitration or litigation. Further, experts may overly complicate the issues at an early stage and cause an impediment to resolution.

The Mediator's Introduction

The mediation may begin with an opening session with all participants present together in one room. The mediator may wish to provide some background about themselves and their approach to mediation, as well as the benefits of a negotiated resolution. Openings from the mediator are often strategic and designed to promote civility and defuse animosity at the outset. As described by the American Arbitration Association, the mediator's opening statement is often designed to address three essential questions in the participants' minds:[21]

---

21 Coleman H., "Opening 'Mindfully': Mediator Methods," *AAA-ICDR Blog*. (2021). Available at https://www.adr.org/blog/Mediation-Mind-Shifts-Series-Part-3a/Opening-Mindfully-Mediator-Methods (Accessed 26 September 2024).

- *Can the mediator help me?* Parties will evaluate whether the mediator has the temperament, requisite skill and experience, and understanding of the disputed issues to be effective.
- *Can the mediator be trusted?* Mediation often comes down to trust. Parties expect the mediator to maintain the confidentiality of the information shared with them, both from the public and from the other side, unless explicitly told the information may be disclosed. Parties also expect the mediator to maintain impartiality and professionalism.
- *Does the mediator truly care?* The mediator provides an impartial conduit to allow the parties to reach resolution. If the mediator appears uninterested or unmotivated, the parties will likely follow suit.

The Mediation

Whether the parties elect to hold an opening session and make statements or presentations, the parties will, at some point, split ways and prepare for individual caucus sessions with the mediator.

An experienced mediator will drive the format of the caucus sessions to fit their current mediation strategy. A mediator may use the initial caucus session to hear out a party who feels aggrieved, helping provide the mediator with insight into underlying issues beyond the parties' factual and legal claims and defenses. Other times, the mediator may wish to understand the story through the party representative's eyes. A mediator who has had the benefit of detailed mediation statements may instead come with specific questions about discreet factual or legal issues that will drive the negotiations. As the mediation progresses and gaps are bridged, the caucus sessions often become shorter and more targeted, with less time in between, particularly once the parties begin exchanging potential settlement offers.

The mediation itself is a marathon, not a sprint. The participants should be prepared for downtime during this process. Particularly where multiple parties are involved, the mediator will want to hear from everyone. Participants should not be discouraged by initial discussions with the mediator, or opening positions or settlement offers from the other participants. The mediator's job is to help guide the parties to a mutual resolution, and this process can take time. The authors of this chapter have participated in successful mediations that have lasted hours, and others that have lasted days.

If an agreement is reached during the mediation, the parties will then typically document the key terms in writing through a memorandum of understanding or other writing. The parties should outline the key settlement points, which may include, among other things, the settlement amount and who is paying what to whom, the terms of any release, any indemnity obligations, and any other special agreements between the parties. Parties may also wish to include a term that the memorandum of understanding is itself an agreement and can form the basis of a settlement agreement. It is understandable that parties who have just successfully reached agreement may wish to quickly disperse and skip this step, but documenting the key agreements is critical to avoid parties later backing out or attempting to revise central terms. Most importantly, it can prevent the loss of hours or days of hard work and nullifying a successful mediation.

Advantages and Disadvantages of Mediation

Like any process, the use of mediation has its advantages and disadvantages. Perhaps most importantly in favor of mediation is that it is effective. The American Bar Association reports that 70–80% of mediations resulted in a settlement,[22] and other organizations report similar success rates.[23] Consistent with the ABA's findings, the Centre for Effective Dispute Resolution in the United Kingdom found that 72% of its mediations resulted in a settlement on the day of mediation and 20% resulted in a settlement shortly thereafter, bringing the UK's reported total settlement rate to 92%.[24] The overall strong settlement rate reflects the effectiveness of mediation in resolving disputes. The following section outlines the various advantages and disadvantages of mediation.

Mediation Advantages

Confidentiality. Mediation is confidential and private, protecting sensitive information from becoming public unless the parties agree otherwise. Confidentiality agreements are normally signed at the beginning of the mediation process, ensuring that all parties and the mediator are committed to maintaining privacy. Participants can engage in mediation with confidence that sensitive financial information, trade secrets, proprietary information, or even their participation in mediation will not be publicly disclosed. While the mediator generally will require the parties to sign such an agreement, it is not strictly mandatory in all instances. But engaging in a mediation without a confidentiality agreement risks the dissemination of sensitive information disclosed at the mediation and negates a primary advantage of mediation--a confidential and private process. Also, at least in

22 American Bar Association, *How Courts Work, Mediation*. (2019). Available at https://www.americanbar.org/groups/public_education/resources/law_related_education_network/how_courts_work/mediation_advantages/ (Accessed 26 September 2024).

23 American Arbitration Association, *A Guide to AAA Disaster Recovery Claims Mediation Procedures*. (2014). Available at https://www.adr.org/sites/default/files/document_repository/A-Guide-to-AAA-Disaster-Recovery-Claims-Mediation.pdf (Accessed 26 September 2024) (reporting statistics show 85% of commercial matters end in written settlement agreements).

24 CEDR Services Ltd, "A survey of commercial mediator attitudes and experiences in the United Kingdom," *The Tenth Mediation Audit*. (2023). Available at https://www.cedr.com/foundation/mediation-audit/ (Accessed 26 September 2024).

the United States, state and federal rules typically provide further protection by restricting a party from using settlement offers and discussions during mediation in court or arbitration if the dispute is not resolved at mediation.[25]

Control and Flexibility. Parties retain control over the outcome of a mediation as the mediator cannot impose a decision. This allows for solutions to be tailored to the specific needs of the parties, facilitating project completion, and preserving relationships. The flexibility of mediation enables creative problem-solving that might not be possible in a more rigid legal framework. This flexibility can lead to more satisfactory and sustainable resolutions that address the underlying interests of both parties.

Cost-effective and Time-efficient. Mediation is generally faster and more affordable than litigation or arbitration. Sessions typically last a day or two, compared to months or years for litigation or arbitration. A quick resolution facilitates continued cooperation and minimizes disruptions to the project and the participants' businesses. The reduced costs and time commitment make mediation an attractive option for resolving disputes without incurring significant financial or temporal burdens.[26]

Relatedly, while some discovery or document exchange may occur during the course of a mediation to enable the parties to gather more information regarding the evidentiary support (or lack thereof) for the other party's case, it will generally pale in comparison to what one would expect in court litigation and even international arbitration, reducing both time and costs. By focusing on the most relevant information, parties can avoid the exhaustive and often contentious discovery process typical of litigation.[27]

Preservation of Relationships. The construction industry is tightknit. Members of the industry often work together on numerous projects over years. Allowing participants to manage the mediation process and drive the framework of a resolution encourages cooperation and communication, reducing animosity and fostering a collaborative environment. An amicable resolution can allow longtime partners who find themselves involved in a temporary dispute to continue to work together productively into the future.

Mediation Disadvantages

Nonbinding. Mediation does not result in a binding decision and is not a judgment. If one party subsequently refuses to honor the terms of a settlement agreement, the agreement typically must be enforced through the courts.[28] With international

25 Kessler L., "Using Mediation to Resolve Construction Disputes," *The Lexis Practice Advisor Journal*, (2017).

26 Kessler 2017.

27 Kessler 2017.

28 Kessler 2017.

disputes, this could require moving to enforce a settlement in a foreign country with an unfamiliar, and potentially unforgiving, court system. The nonbinding nature of mediation can lead to a lack of commitment to the process, making it less likely a settlement is reached before litigation or arbitration ensues.

Requires Mutual Commitment. Mediation is a voluntary process, and a participant cannot be forced into a settlement. Success in mediation often depends on both parties' willingness to negotiate and settle in good faith. Indeed, genuine commitment from both sides is essential for mediation to be effective. If one party is participating only due to contractual obligations or per court mandate, the process may not be productive.

Potential for Incomplete Resolution. Mediation may not address all issues comprehensively, especially if parties are not fully prepared or willing to compromise. Complex disputes may require multiple sessions or additional legal proceedings to fully resolve all aspects of the conflict. This can sometimes lead to a piecemeal approach that leaves mediation as only a preliminary step rather than a final solution.

Comparative View of Mediation

It is well-recognized that no country has established a unified general law governing all forms of mediation across all areas of law, and the construction industry is no different.[29] As a result, mediation and the enforcement of settlement agreements arising from mediation are managed by a disparate collection of rules, procedures, and laws.[30] The use of mediation and views concerning its efficacy as a dispute resolution tool vary throughout the world. In the following sections, the authors provide a concise overview of how mediation is perceived and practiced in a select group of countries and continents, offering insights into diverse approaches and cultural perspectives on mediation.

The United States

Mediation is an often-utilized tool by US practitioners, whether through private or court-mandated mediations. Disputants typically view mediation as a cost-effective and efficient alternative or supplement to what is often a time-consuming and

29 Hopt K. and Steffek F, *Mediation, Principles and Regulation in Comparative Perspective*, (2013), p. 70.

30 *But see* Section "The Singapore Convention," infra, discussing the United Nations Convention on International Settlement Agreements Resulting from Mediation.

expensive litigation or arbitration process. Court-imposed or court-sponsored mediation is typically viewed as less costly than the more in-depth, lengthy, and expensive private mediations.[31] But private mediations generally afford the participants more control over the structure and logistics of the mediation.[32] Although mediation is a voluntary process, it can be compelled by United States courts. Courts in the United States. often have heavy caseloads and view mediation as an effective early resolution tool. Some jurisdictions within the United States even threaten to impose penalties on parties who refuse to participate in mediation in good faith.[33]

South America

Mediation remains in its relative infancy as a dispute resolution means in South America, but it is gaining recognition as a viable alternative to traditional litigation and arbitration. This trend is driven by the need to address the backlog of cases in various countries and the high costs associated with binding dispute settlement methods.[34] For instance, Brazil has enacted laws mandating mediation for disputes with public, State-owned entities, and its Code of Civil Procedure requires the parties to pass through a mediation or conciliation phase before a defendant files their answer in certain circumstances.[35] Bolivia, Colombia, Ecuador, and Peru have also adopted legislation that attempts to give greater enforcement options for negotiated settlement agreements.[36] These countries also have legislation that provides for the treatment of mediation settlement agreements with the same enforceability as arbitral awards.[37] The International Development Bank has also played a significant role in promoting mediation throughout Latin America by supporting about 230 mediation centers and training over 2,000 mediators since 2004.[38]

---

31 Kessler 2017.

32 Kessler 2017.

33 *See, e.g., U.S. Bank Nat. Ass'n v. Vasquez,* 10 N.Y.S.3d 386, 388–89 (Sup. Ct. 2015); Del. Super. Ct. Civ. R. 16.

34 Burnett H., "Recent Developments in Key Latin American Jurisdictions to Attract International Commercial Arbitration," *Am. U. Bus. L. Rev.* 5, (2015): 387, 398–99. Available at https://advance.lexis.com/api/permalink/4d121ec8-0188-46a3-9c51-353802352bf5/?context=1000516.

35 Siqueira Castro C.R., "The mediation agreement in Brazil and its essential requirements, recognition and enforcement," *Int'l Bar Assoc.* (2021). Available at https://www.ibanet.org/article/90750472-5198-4900-9E2F-063150D5768A (Accessed: 26 September 2024).

36 Burnett 2015, p. 399.

37 Burnett 2015, p. 398–99.

38 Burnett 2015, p. 398.

China

Mediation has been used in the Chinese legal system for many years.[39] But one criticism of the use of mediation in mainland China is the development of various frameworks and inconsistent requirements resulting in a lack of uniformity and clarity. For instance, the China Council for the Promotion of International Trade Mediation Centre uniquely requires that all mediations be completed within 30 days, a requirement not standard across the country. Some critics have suggested that the backlog of disputes in China's judiciary system places pressure on parties to a mediation to settle, leading to parties not voluntarily honoring the terms of purportedly voluntary settlement agreements or submitting applications for coercive enforcement from the courts.[40]

Middle East – UAE, Saudia Arabia, and Qatar

In recent years, countries like the United Arab Emirates (UAE), Saudi Arabia, and Qatar have taken steps to empower mediators and mediation and continue to formalize the process.[41] For example, the Dubai International Arbitration Centre recently issued the DIAC Mediation Rules 2023, effective October 1, 2023. As made clear, DIAC's core objective is to promote the use of mediation as a fair, impartial, efficient, and proportionate dispute resolution tool.[42] The UAE also recently passed Federal Decree Law No. 40 of 2023, promoting the use of court-imposed and private mediation in civil and commercial disputes.[43] Mediation is viewed similarly in other major economic hubs throughout the Middle East, including Saudi Arabia and Qatar. This is evidenced by both countries' signature and ratification of the Singapore Convention in 2020.

---

39 Du G., "A Decade of Rise: How Commercial Mediation in China Evolved in the 2010s," *China Justice Observer*. (2023). Available at https://www.chinajusticeobserver.com/a/a-decade-of-rise-how-commercial-mediation-in-china-evolved-in-the-2010s (Accessed 31 July 2024). Chen, M. (2023) "Commercial Mediation in Mainland China: Pitfalls & Opportunities," *23 Pepperdine Dispute Resolution L.J. I.3 A.3*. Available at https://digitalcommons.pepperdine.edu/cgi/viewcontent.cgi?article=1532&context=drlj.

40 Chen 2023, p. 198–200.

41 Al Nahyn K. B. Z., "Fed. Law No. 6 of 2021 on the Establishment of Mediation and Conciliation Centres in Civil and Commercial Disputes," *United Arab Emirates*. (2021). Available at https://uaelegislation.gov.ae/en/legislations/1501/download.

42 Dubai International Arbitration Centre, *DIAC Mediation Rules 2023*, (2023). Art. 2, § 2.5. Available at https://www.diac.com/en/adr-services/mediation/diac-mediation-rules-2023/.

43 Saji 2024.

Europe

Mediation is generally treated favorably throughout Europe, but the rules and views of mediation vary widely by country.

In Italy, mediation is well-regulated and defined, and it is compulsory for many commercial disputes.[44] A unique aspect of Italy's framework is that lawyers are required to advise their client that the possibility of using mediation exists for their particular matter.[45] Mediators are encouraged to provide a final settlement proposal to the extent the parties cannot amend their differences, and parties may elect to accept or reject the mediator's final proposal.[46] Mediations in Italy result in an agreement 50% of the time.[47]

In Poland, mediation was more formalized in 2005 through the introduction of its Code of Civil Procedure and other acts into Poland's legal system.[48] Mediation in Poland is voluntary, but like Italy, Polish lawyers are obliged to advise clients of the possibility of mediation.[49] Given its recent development in Polish law, Poland has a low usage rate, and like many countries within the EU, Poland has not joined the Singapore Convention.[50]

In Spain, mediation has progressively integrated into its legal system, particularly with the enactment of Law 5/2021, which regulates mediation in civil and criminal matters. The law is designed to promote faster and cost-effective alternatives to litigation through mediation, and in some cases, mediation is a prerequisite to litigation. About 70% of mediations in Spain result in an agreement.[51]

Turkey has a variety of laws defining, mandating, or otherwise regulating mediation. By law, mediation is a preferred method for settling commercial and civil disputes quickly and confidentially.[52] While mediation is voluntary for

---

44 European Union, "Italian Legislation on Mediation," *EU Directorate-General, Policy Department*, (2011): 8. Available at https://www.europarl.europa.eu/RegData/etudes/note/join/2011/453175/IPOL-JURI_NT(2011)453175_EN.pdf.

45 European Union, 2011.

46 Gaultier T., "Cross-Border Mediation: A New Solution for International Commercial Dispute Settlement?", *N.Y.S.B.A. International Law Practicum* 26, (2013). Available at https://nysba.org/NYSBA/Sections/Dispute%20Resolution/Materials/DRS%20Fall%20Meeting%20Materials/Cross-Border%20Mediation%20-%20A%20New%20Solution%20for%20International%20Commercial%20Dispute%20Settlement.pdf.

47 Hopt and Steffek, 2013, p. 103.

48 European Union, "The Development of mediation in Poland," *EU Directorate-General, Policy Department*. (2011). Available at https://www.europarl.europa.eu/document/activities/cont/201105/20110518ATT19605/20110518ATT19605EN.pdf.

49 Hopt and Steffek, 2013, p. 24.

50 Hopt and Steffek, 2013, p. 95.

51 Hopt and Steffek, 2013, p. 103.

52 *See* Turkish Law No. 6325 (2012).

general civil commercial disputes, it is mandatory for labor disputes and disputes under the Turkish Commercial Code (TCC).[53] If an agreement is reached in a Turkish mediation, a document of understanding is signed, and the parties are precluded from filing a lawsuit on the issue.[54] The Mediation Department of the Ministry of Justice of Turkey reports that about 70% of mediations conclude in a settlement.[55]

In the UK, mediation is voluntary except for certain small claims (i.e., less than £10,000), but its use is strongly encouraged in practice.[56] When used, mediation typically involves a one-day time-limited session wherein both parties present their case.[57] The limited time frame is intended to promote settlement and efficiency, contrasting to drawn out litigation proceedings. With strong encouragement and time constraints, two UK mediators suggest that mediation boasts about a 72% success rate just on the day of, with additional success in the days following the mediation.[58] The UK has demonstrated further commitment to the use and benefits of mediation both in the UK and abroad. In May 2023, the UK became a signatory to the Singapore Convention.

India

Mediations in India generally proceed in different stages with the parties working separately over a longer period of time than what is more commonplace in the United States, where a mediation generally consists of one or a few all-day

---

53 Turkish Law. No. 7036 (2018); Turkish Law. No. 7155 (2019).

54 Ikhlayel N., "The Mediation: Legal Comparison Between Mediation in the United States of America and the Turkish Republic," *Indiana University, Maurer Global Forum*. (2024). Available at https://blogs.iu.edu/maurerglobalforum/2024/04/12/the-mediation-legal-comparison-between-mediation-in-the-united-states-of-america-and-the-turkish-republic/ (Accessed: 31 July 2024); Cetinkaya O. and Baydar B., "Mandatory Mediation for Commercial Disputes," *IBA Litigation Newsletter*, (2019): 65. Available at https://www.aija.org/images/uploads/IBA newsletter_Litigation (September 2019).pdf.

55 Ikhlayel, 2024; Soylu M., Ozilhan C., İyem D., and Yılmaz C., "Mandatory Mediation for Commercial Disputes," *Mondaq*. (2019). Available at https://www.mondaq.com/turkey/arbitration-dispute-resolution/770986/mandatory-mediation-for-commercial-disputes (Accessed: 26 September 2024).

56 Smith D. and Sejpal P., "Mediation: A Perspective from India and the UK," *Law.com Commentary*. (2024). Available at https://www.law.com/international-edition/2024/06/26/mediation-a-perspective-from-india-and-the-uk/; Lambert D, "Mediation in the United Kingdom," *Penningtons Manches Cooper LLP*. (2019). Available at https://www.lexology.com/library/detail.aspx?g=02ee5416-79ba-484b-bd62-eb26318d330b.

57 Smith and Sejpal, 2024.

58 Lux J, 2023 GTDT: Mediation United Kingdom & Wales, Lexis (2023).

sessions.[59] Mediations in India are premised upon being private and confidential.[60] With the world's largest population and a back-logged court system, India continues to turn to mediation as a way to help businesses, alleviate the courts, and provide legal access to those in need.[61] While quite new, India has shown a recent dedication to the promotion of mediation as a dispute resolution means with the adoption of the Mediation Act 2023, which directly encourages the use of mediation to reduce the courts' caseloads and promotes business efficiency. It provides comprehensive rules and regulations on governing voluntary mediation for individual and business entities.[62]

Africa

South Africa frequently uses mediation, but one study revealed the perception from parties to mediations in the country was that formal early conflict management processes like mediation or conciliation were seldomly successfully employed in the country.[63] South Africa draws its mediators from the very trade that the dispute is centered upon with designers, contractors, and engineers all acting as mediator when necessitated.[64] In Northern and Central Africa, mediation is used more sparingly due to the lack of uniformity in its practice and finality in its outcomes.[65] Despite its limited use, countries such as Ghana, Gabon, Chad, and Nigeria have joined the Singapore Convention (discussed below), which offers at least some indication that mediation will continue to be promoted in the region.[66]

---

59 Smith and Sejpal, 2024.

60 Vikram Singh, *2023 GTDT: Mediation India*, LEXIS (2023) (lexis.com); Smith and Sejpal, 2024.

61 Vikram Singh, *2023 GTDT: Mediation India*, LEXIS (2023) (lexis.com).

62 Bhola M, "India: Transforming Dispute Resolution In India: An Overview Of The Mediation Act, 2023," *Mediate.com*. (2024). Available at https://mediate.com/news/india-transforming-dispute-resolution-in-india-an-overview-of-the-mediation-act-2023/.

63 Lu M., "Mapped: Breaking Down the $3 Trillion African Economy by Country," *Visual Capitalist*. (2024). Available at https://www.visualcapitalist.com/breaking-down-african-economy-by-country/; Povey A.L, "Mediation practice in the South African construction industry," *J. of S. Africa Institution of Civil Engineering*, (2005), p. 60; Rwelamila, P.D. and Brown, S.T., *Implementation of Early Conflict Management Measures in Construction Projects – the Case of South Africa*. Available at https://www.irbnet.de/daten/iconda/CIB_DC28870.pdf.

64 Povey 2005, pp. 51–52.

65 *See generally*, Povey, 2005.

66 Chief Rotimi Williams Chambers, *2023 GDT: Mediation Nigeria*, LEXIS (2023); Victoria Barth, et al., *2023 GDT: Mediation Ghana*, LEXIS (2023).

The Singapore Convention

In December 2018, the international community took a significant step toward the promotion and use of mediation to resolve international commercial disputes through the passing of the United Nations Convention on International Settlement Agreements Resulting from Mediation (the "Singapore Convention") by the United Nations General Assembly.[67] The Singapore Convention entered into force in September 2020.

The Singapore Convention is touted as a "uniform and efficient framework for international settlement agreements resulting from mediation."[68] Its primary objective is to enable the enforcement of mediated settlement agreements among its signatories. The Singapore Convention's significance lies in its harmonized enforcement mechanism. Prior to the Singapore Convention, parties who successfully resolved their disputes through negotiated settlements had little recourse against a party who later reneged on the agreement or refused to pay in accordance with a settlement. Parties looking to enforce their agreements had to resort to breach of contract actions, often in multiple, and perhaps unfamiliar and unfavorable, jurisdictions. The lack of an enforcement mechanism served as a deterrent to negotiated settlements, leaving litigation or arbitration as a more favorable dispute resolution mechanism.

To date, the Singapore Convention applies only to "international commercial settlement agreements."[69] This means the mediation settlement agreement must involve an international dispute as defined by the Singapore Convention, and it must be commercial in character. The Singapore Convention explicitly excludes transactions for personal, family, or household purposes, or relating to family, inheritance, or employment law.[70] Additionally, the party seeking enforcement must show the existence of a settlement agreement and evidence the settlement agreement resulted from a mediation. There are several options to make such a showing, including obtaining the mediator's signature on the settlement agreement, a

67 Singapore Convention on Mediation, "Background to the Convention," *Singapore International Dispute Resolution Academy*. (2024). Available at https://www.singaporeconvention.org/convention/about.

68 Singapore Convention on Mediation, "Convention Text," *Singapore International Dispute Resolution Academy*. (2024). Available at https://www.singaporeconvention.org/convention/text.

69 Singapore Convention on Mediation, "Convention Text," *Singapore International Dispute Resolution Academy*. (2024). Available at https://www.singaporeconvention.org/convention/text.

70 Singapore Convention on Mediation, "Convention Text," *Singapore International Dispute Resolution Academy*. (2024). Available at https://www.singaporeconvention.org/convention/text.

separate document signed by the mediator or the institution overseeing the mediation, or other evidence acceptable to the competent authority.[71]

The Singapore Convention serves two purposes: (i) to allow for the direct enforcement of a settlement agreement by a competent authority of a Party state; and (ii) to allow a party to introduce a settlement agreement in any subsequent proceeding as evidence the matters captured in the settlement have been resolved.[72]

The Singapore Convention is in its relative infancy, and its effectiveness is still being understood. Its reach and power currently falls well short of its parallel convention governing international arbitration – the Convention on the Recognition and Enforcement of Foreign Arbitral Awards (the "New York Convention"). The New York Convention regulates both agreements to arbitrate and arbitration awards, meaning parties can rely on uniform enforcement by Party states of not only arbitration awards and judgments, but also of agreements to arbitrate. It also has a wide-global reach, with 24 signatories and 172 parties to date.[73] In contrast, while still young, the Singapore Convention has 58 signatories and 18 parties to date, illustrating its more limited reach.[74] Further, the Singapore Convention does not currently provide for the enforcement of agreements to mediate, only settlement agreements resulting from a mediation. Parties seeking to enforce an agreement to mediate thus still must look to the domestic laws of the enforcing state. In theory, however, the Singapore Convention offers international legitimacy to the use of mediation as a dispute resolution tool, and offers international investors some confidence their negotiated settlements will be enforced in Party states.

Conclusion

Much more can be said about the viability and use of mediation as a dispute resolution mechanism. The preference for and effectiveness of mediation varies widely by culture, country, and continent. But where mediation is used, it is generally

71 Singapore Convention on Mediation, "Convention Text," *Singapore International Dispute Resolution Academy*. Article 4. (2024). Available at https://www.singaporeconvention.org/convention/text.

72 Singapore Convention on Mediation, "Convention Text," *Singapore International Dispute Resolution Academy* Article 3, (2024). Available at https://www.singaporeconvention.org/convention/text.

73 New York Convention, "Contracting States," *The NewYorkConvention.org*. (2024). Available at https://www.newyorkconvention.org/contracting-states.

74 Singapore Convention on Mediation, "Jurisdictions," *Singapore International Dispute Resolution Academy*, (2024). Available at https://www.singaporeconvention.org/jurisdictions.

effective at achieving earlier resolutions and helping parties to avoid costly and time consuming litigation and arbitration. With relatively recent moves like the Singapore Convention and the DIAC Mediation Rules 2023, it is also apparent that there will continue to be a global push to promote and improve the use of mediation as a viable alternative, particularly for international projects. As construction projects continue to grow in size and global reach, and disputes become increasingly costly and complex, expect mediation to continue to take center stage as an efficient and effective dispute resolution mechanism.

26

Construction Commercial Arbitration Peculiarities

*Mauro Rubino-Sammartano**

Introduction

This chapter deals with commercial arbitration of construction disputes. If a Contractor may establish that the construction contract has given rise to an investment for him, he may consider also to institute an investment treaty arbitration if there are the other premises and requirements for it.

Arbitration: Not Always the Same Dispute Resolution Mechanism

Referring a dispute to commercial arbitration does not mean that one always submits to the same dispute resolution mechanism. In fact, the term "commercial arbitration" covers various arbitral proceedings, which, while they are based on a large common background, may differ even considerably due to the applicable law, mandatory provisions of the seat of the proceedings and the arbitration regulations, which have been selected or in any event apply to that specific arbitration.

Domestic and Foreign Arbitration: the Issue of Its Nationality

A distinction is made between domestic and foreign arbitration, which is based on their nationality. The existence of foreign arbitration is confirmed by the Geneva

* Member of the Milan Bar, Ch.ArbFCIArb (Lifetime), former member of the Paris Bar.
The sub chapter on Anti-Suit Injunctions was co-written by Virginie Colaiuta and Basil Thévignot.

Major Issues in International Construction Law, First Edition. Mauro Rubino-Sammartano and Troy Harris.
Published 2026 by John Wiley & Sons, Inc.

Convention.[1] Likewise the New York Convention[2] refers to the exequatur to foreign awards.

Various criteria have been advocated to establish the nationality of an arbitration. For example, the English Arbitration Act[3] defines as domestic arbitrations those that take place in the UK. In *Nuovo Pignone*,[4] the Supreme Court of Brazil has confirmed that arbitration, which takes place in Brazil, is Brazilian. This is the general criterion. Chinese law[5] further distinguishes within its domestic arbitrations those which are strictly domestic from those which are still domestic but involve a foreign element.

Geographic Criterion

As earlier mentioned, nationality is generally decided based on the territory of the State where the award has been made. In turn the award is or should be made at the seat of the arbitral proceedings. However, this criterion may give rise to difficulties if the award is made in a State which is different from the one of the seat of the arbitral proceedings or from the different place in which that arbitration has taken place. In *Hiscox*,[6] the arbitral proceedings have taken place in England, where their seat was, but the award has been signed in Paris using the following words "*dated_______, Paris France.*" The House of Lords has held, with a decision which has induced its Parliament to amend the law, that the award was French.

Even as to the place where the arbitral proceedings take place, difficulties may arise if some hearings take place in a State, but other ones are held in a different State.

In similar case, like it was held by the Court of Appeal of Sweden in *Titan*,[7] there may be no automatic link with a given State.

The Economic Criterion

Based on an economic criterion,[8] an arbitration is not domestic if it concerns a transaction, which involves more than one State, independently from the

1 Convention pour l'éxécution des sentences arbitrales étrangères, Geneva, September 26, 1927.

2 Convention on the recognition and enforcement of Foreign Arbitral Awards (New York June 10, 1958).

3 Arbitration Act April 4 1979, which has later been replaced by Arbitration Act 1996.

4 *Nuovo Pignone S.p.A. v. Petromec Inc. et al.*, Supreme Court of Justice of Brazil, May 24, 2011, Yearbook international arbitration, 2011, 260.

5 The Baker and McKenzie International Arbitration Yearbook 2011–2012, New York, JurisNet 2012.

6 *Hiscox v. Outhwaite*, House of Lords, July 24, 1991 3WLR [1991] 297–307.

7 The *Titan Corp. v. Alcatel CIT S.A.*, Court of Appeal, Sweden, Stockholm, February 28 2005, Yearbook Commercial Arbitration, 2005, 142.

8 Sec. 1504, Civil Procedure Code (France).

nationality of the parties, the substantive law and the seat of the arbitration. In France an arbitration is then international if made under such criterion *"s'il met en cause des intérêts du commerce international."*[9]

Procedural Criterion

The Court of Appeal, Paris, has dealt with this issue in *Götaverken*[10] affirming that an award made under a procedural law that is not French, and that has no connection with French public policy, since the parties are not French and the contract has been entered and has to be performed abroad, may not be held as being French. According to this criterion, if the procedural law is Turkish, the award is Turkish, even if it has taken place in Canada. Along the same line is *Ghezzi*[11] in which the German Supreme Court has held that, in line with its previous decisions, an award is foreign when the arbitral tribunal has based its award on a foreign procedural law.

The procedural criterion is confirmed by the New York Convention 1958, which defines as foreign not only the awards made in another State but also a second category of awards, which, even if made in a given country, has been governed by the procedural law of another State.

It must be pointed out that even if according to this criterion, the nationality of an award may be provided by the applicable procedural law, this nationality may be different from the one that will be taken into account for the recognition and enforcement of that award in another State.

International Arbitration

The distinction between domestic and foreign awards might be seen as covering all possible awards. It is then proper to wonder whether there is room for an international award. In order to respond to this query, one has to search whether there are elements which would allow the existence of this *"tertium genus."*

In some jurisdictions, such as France, the domestic law governs both domestic awards and awards made in such a country having a foreign element. In these situations, the reference to an "international" and not to a foreign arbitration on the one hand shows that there is a French award's subdivision international, and on the other hand shows that this does not help to identify the notion of internationality. In fact, even this second class of awards is treated as domestic in France from that point of view.

9 *"if it involves international trade."*

10 *Götaverken Arendal A.B. v. Lybian General National Maritime Transport Co.*, Court of Appeal, Paris, February 21, 1980, Clunet 1980 at 660.

11 *Ghezzi*, Bundesgerichtshof, June 30, 1961 (BG HZ 21, 365) Yearbook Commercial Arbitration, 1990, XV, at 450.

The Subjective Criterion

An arbitration is frequently referred to as international, when the parties have a different nationality. It is submitted that, based on the comparison with Court proceedings, this is not appropriate. Court proceedings in fact are not influenced by the nationality of the parties. Even if the parties are respectively from Singapore and from Greece, the judgment made by a court remains a judgment of that jurisdiction. The same should apply to awards. The nationality of the parties does not then allow to hold that the award is international. The Hong Kong Supreme Court has held in *Fung Sang*[12] that an arbitration is international even if the parties have their center of businesses in the same State.

The Economic Criterion

The earlier discussed theory of linking arbitration to internationality based on such criterion has given rise even in this respect to the economic criterion.

It is submitted that even this criterion is not appropriate to establish the internationality of an award. The parallel between judgments in court proceedings and awards demonstrates that court proceedings, for example, in Germany remain German even if the object of the dispute is a matter of international trade.

The Procedural Criterion

Even if the earliest discussed procedural criterion is followed, one has to establish if and when the procedural law is international. Even if it is not frequent, the parties to an arbitration may, in choosing the applicable procedural law select a given procedural law, for example Venezuela law but selecting also, as to the taking of evidence, a different law, for example English law. This possible plurality of procedural laws may grant to such proceedings a supranational or international character.

Supranational Arbitration Rules

If the parties make no reference to a national procedural law and refer only to a supranational set of arbitration rules, even such proceedings may be treated as procedurally international. This selection in fact is possible since it is a well-established principle that arbitration has no *lex fori* and the freedom of the parties to select the law is recognized.

Arbitrations Based on International Conventions

Depending on the wording of the arbitration agreement and of international conventions, an arbitration may be treated as supranational. For example,

---

12 *Fung Sang Trading Ltd. v. Kai Sun Sea Products & Food Co. Ltd.*, Supreme Court, Hong Kong, October 29, 1992, Yearbook Commercial Arbitration 1992, at 280.

the Washington Convention 1968[13] by providing that an award made under its Rules must be treated in each Member States as a judgment of that State, affirms its supranationally and internationality.

This does not mean that in this way the award would place itself in the stratosphere of subjectivism.[14]

Transnational Arbitration

Arbitral proceedings have a tendency to be delocalized and this may produce inter alia the result that the applicable procedural law is different from the procedural law of its seat.

Such delocalization satisfies the wide tendency to detach arbitration from national legal systems. It must be recognized that, in spite of this tendency, there is also the opposite risk that an international arbitration be nationalized.[15] It is suggested that the reference to international arbitration should frequently amended to "transnational arbitration" when it refers to arbitrations, which – while they are not international from a systematic point of view – contain a foreign element such as to the parties or the object of the dispute.[16]

International Arbitration Law

According to an old approach,[17] no international arbitration would exist. Along the same line is *Banque Ottomane*[18] according to which no principles of an alleged international public policy exist, not being recognized by any legal system. This approach has changed in the subsequent years, such as in *Uni-Kod*[19] and in *Pom'alliance*[20] which refer to a material rule of international arbitration law. A doctrine that has been opposed on the grounds that arbitration between private

13 Convention on the Settlement of Investment Disputes between States and Nationals of other States, Washington March 28, 1968.

14 F.A. Mann, *England Rejects Delocalized Contracts and Arbitration*, International and Comparative Law Quarterly 1984 at 193.

15 Otto Sandrock, *To Continue Nationalizing or to de-nationalize? That Is Now the Question in International Arbitration*, Am. Rev. Intl. Arb. 2011 at 12.

16 P.C. Jessup, *Transnational law, Storrs Lectures on Jurisprudence*, New Haven, Yale University Press 1956.

17 Résolution de l'Institut du droit international de 1957, in Gary B. Born, *International Commercial Arbitration*, 3rd ed. (Kluwer, 2020) at 473.

18 *Bakaljan et Hadjthmar v. Banque Ottomane*, Court of Appeal Paris, March 19 1966, Clunet, 1968 at 137.

19 *Sté Uni-Kod v. Sté Ouralkali*, Court of Cassation, France March 30, 20'04, Rev. Arb. 2017, at 851.

20 *M. P. Facchi v. SAB Pom'Alliance*, Court of Appeal, Paris, January 10, 2017, Rev. Arb. 2017, at 332.

parties has its source only on national legal systems, while international law has its source on conventions among States. However, some international conventions deal also with arbitration between private parties. The conventions being binding on their Member States, it is suggested that they amount to international arbitration law. One might have expected a set of provisions applying to all the legal systems. Nevertheless, a large number of differences is to be found in the various legal systems. Sometimes such differences are due to local approaches.

The Arbitration Agreement

The first source of international arbitration is the agreement of the parties. The intention of the parties may be express, or result from the incorporation of a set of supranational arbitration rules or – even if perhaps not too frequent – may be implied.

Selection of the Applicable Procedural Law/s

The selection of the applicable procedural law or laws is one of the elements which contribute to establish whether those proceedings are subject to international arbitration law. That may also be, in the absence of a selection by the parties, the result of a selection made by the arbitral tribunal.

Multilateral and Bilateral Conventions

Multilateral and bilateral conventions are so many that it seems preferable to refer to the various lists of conventions which have been published, limiting the reference here to:

- The General Protocol 1923[21];
- The Geneva Convention 1927[22];
- The New York 1958 Convention[23];
- The Inter Arab Convention[24] which has been followed by the Riyadh Convention[25];

21 Geneva Protocol on Arbitration Clauses, Geneva, September 24, 1923.

22 Geneva Convention on the Execution of Foreign Arbitral Awards, Geneva, September 26, 1927.

23 Convention on the Recognition and Enforcement of Foreign Arbitral Awards, New York, June 10, 1958.

24 The Inter Arab Convention on the Enforcement of Judgments and Awards, September 14, 1952.

25 Convention for Judicial Cooperation between States belonging to the Arab League, Riyadh April 4, 1983.

- the European Convention 1961[26];
- the Moscow 1972 Convention[27];
- the Panama Convention[28];
- the Port-Louis Convention.[29]

The United States has entered various bilateral treaties such as with France,[30] Japan,[31] Germany,[32] and the USSR.[33] In Asia, the Asian International Arbitration Centre (AIAC) has rules[34] which are being widely used in arbitration.

Supranational Arbitration Rules

The enormous – and probably excessive – number of arbitration rules includes various types of them. Many of them are for the domestic arbitrations. Some other ones are destined to a specific industry such as the Grain and Free Trade Association (Gafta), the Sugar Association, the Coffee Trade Association and the London Maritime Arbitrators' Association. Aside from these, there are supranational arbitrations which aim to regulate transnational and even national arbitrations. Among them, there are the UNCITRAL Arbitration Rules,[35] the UNCITRAL Model Law on International Commercial Arbitration,[36] the American Arbitration Association, the International Arbitration Court of the International Chamber of Commerce (ICC), the London Court of International Arbitration, the European Court of Arbitration, the German Arbitration Institute, the China International Economic and Trade Arbitration Commission and its Shanghai Branch, the Singapore International Arbitration Centre, the Australian Centre for International Commercial Arbitration, the Inter American Commercial Arbitration Commission, and the Hong Kong International Arbitration Centre.

---

26 The European Convention on International Commercial Arbitration, Geneva, 1961.

27 Convention on the Settlement by Arbitration of Civil Law Disputes arising out of economic scientific and technical cooperation relationships, Moscow, May 26, 1972.

28 *Inter-American Convention on international* commercial *law, Panama* City January 30, 1975

29 Which has *inter alia* created the "Organisation d'harmonisation en Afrique du droit des affaires" (the OHADA), Port-Louis, October 17, 1993.

30 11. U.S.T. 2398.

31 4. U.S.T. 2063.

32 7 U.S.T. 1839.

33 On October 18, 1972.

34 The last AIAC arbitration rules have been released in 2023.

35 2021.

36 1985.

The Selection of the Substantive Law

The so-called direct method is based, like in French law as it has been held in *Soerni*,[37] on a selection of the substantive law not based on national provisions but on general principles, which are based on the autonomy of arbitration, what leads, as held in *Dalico*[38] to establish the efficacy of arbitration based on the free selection by the parties, subject to any mandatory provision as well as to international public policy. French Courts has gone further by holding in *Hilmarton*[39] that an award made in Switzerland is an international award and does not fit into the legal order of that State. Supporters of such an approach aim to establish the existence of an "arbitration legal order."[40]

The Applicable Law

Rather than one applicable law, arbitration involves different applicable laws:

- the procedural law, which in case of conflict supersedes the incorporated arbitration rules;
- the substantive law;
- the law applicable to the capacity of the parties and to their authority to enter into the contract and the arbitration agreement;
- the law applicable to the form of the contract; and
- the *lex fori* and in particular his mandatory provisions.

Peculiarities of Arbitration

A comparison between arbitral and court proceedings evidences several peculiarities of arbitration.

The Selection of the Arbitrator

First as to the appointment of the arbitrator. While in Court proceedings the appointment of the judge depends on the assignment of a case to him based on the criteria of said Court, in arbitral proceedings the appointment of the arbitrator is made by the

---

37 *Soerni c. ASB*, Court of Cassation (France) July 8, 2009, Bulletin 2009, I, n. 165.

38 *Comité Popoulaire de la municipalité Khoms El Mergeb v. Dalico Contractors*, Court of Cassation (France) December 20, 1993, Clunet, 1964 at 437.

39 *Hilmarton Ltd v. Omnium de Traitement et de Valorisation S.A., Court of Cassation (France) March 23, 1994, Rev. arb. 1994 at 327.*

40 Pierre Mayer, *Droit intenational privé*, 10th ed., Montchrestien 2001.

parties or by the body designated for that by them, generally in their own arbitration rules or in those that have been incorporated in the arbitration agreement.

No *Lex Fori*

Contrary to court proceedings, arbitration has no *lex fori*, although it must comply with the mandatory provisions of the *lex fori*.

No Detailed Rules on Evidence

Contrary to court proceedings, generally arbitration rules do not contain the rules to govern the taking of oral and documentary evidence. The frequent reason provided for this is a deference to the decision by the arbitral tribunal. However, it is extremely rare that the members of the arbitral tribunal dedicate much time to a comparison between the rules of evidence in the various legal systems, in order to select the best ones.

This is not a minor weakness of such arbitrations. In fact, it is well-known that the rules of evidence of the various legal systems are not the same and frequently are extremely different. While in common law jurisdictions oral evidence is based on the right of the parties to call and examine witnesses including the parties, and orders for disclosure are issued in order to search documents which a party has preferred not to produce, in civil law jurisdictions, the rules are extremely different. In Court proceedings, Counsel for a party in some legal systems, such as in Italy,[41] may not examine witnesses. He must submit to the Court a proposal of the questions, one by one, which he would like to be put to the witness. The Court has discretion in allowing or rejecting such questions. Counsel must also list the witnesses which he would like to be called and here too the Court has discretion as to admitting them or rejecting the application as to some of them.

Eventually, the witnesses who have been admitted are examined only by the Court on the specific questions that have been admitted. Counsel for the parties have no right respectively to examine or cross examine the witnesses and may just ask the Court to put – if it agrees – some questions. The parties are not treated as witnesses and only any statement by them which is prejudicial to their case is taken into account.

In France, while witnesses may be heard by the Court,[42] this is not the legal practice, and each party produces just witness statements issued by the parties.[43]

41 Civil Procedure Code (Italy) *secs* 244 *et seq.*

42 Sec. 205 Civil Procedure Code (France).

43 Sec. 200 Civil Procedure Code (France).

The Court has the authority to hear a witness on his/her witness statement,[44] but this is most rare. Even in Spain Counsel for the parties may only attend to the taking by the Court of evidence from witnesses.[45] In Germany,[46] which in some respects is more liberal, it has been reported that a party may be authorized to examine a witness if the other party gives its consent. In this case, questions may be put by the parties and unless the Court does not allow them, are to be answered.

Even when the arbitrator is not bound to keep to the strict rules of evidence of court proceedings, and may be, and frequently is, definitely more liberal, the national procedural rules of an arbitrator tend to remain in his/her mind and on various occasions it results that the arbitrator has kept very close to them.

As to expert witnesses, while in common law jurisdiction each party calls her expert witness, who is examined and cross examined, in civil law jurisdictions generally the Court appoints its own expert who hears experts appointed by the parties, files a report, and is not subject to examination and cross examination but possibly to clarifications, frequently in writing.

It is suggested that the lack of clear rules of evidence forces the parties to enter into an arbitration, without knowing how they will be entitled to prove their case. This is like entering into a competition without knowing the rules of the game.

Arbitration Is Not Necessarily Subject to the Rules That Govern Court Proceedings at Its Seat

Another peculiarity related to the conduct of the arbitral proceedings is that it is not subject to the statutory provisions of the law of the seat unless the parties or the arbitrator so decide, and many rules of evidence in arbitration are quite different from court rules subject to compliance with the due process rule.

Duration

The duration of the arbitral proceedings should be and frequently is much shorter than in court proceedings.

Documents Only Arbitration and Pre-arbitral Referee

Arbitration may also avail itself – what is generally not provided for court proceedings – of documents only arbitration and of a prearbitral referee. However,

44 Sec. 203 Civil Procedure Code (France).

45 Francisco Ramos Mendez, *Derecho processual civil*, Bosch, 1990 at 554.

46 Matthias Schrader et al, *Introduction to German Civil Procedure, February 14,2023 German Law Series, Wilkie Farr and Gallagher.*

it must be noted that in some legal systems, proceedings similar to the pre-arbitral referee, such as an emergency arbitrator, are available, within the ambit of interlocutory injunctions, and that in principle, even in court proceedings, the judge may decide just based on the produced documents.

The New York Convention

A remarkable benefit of commercial arbitration vis-a-vis court proceedings derives – as it is well known – from the New York Convention 1958, which[47] allows "the recognition and enforcements of arbitral awards made in the territory of a State other than the State where the recognition and enforcement of such award are sought" and commits each Contracting State to recognize an arbitration agreement in writing. Their courts, at the request of one of the parties, when seized of a matter governed by an arbitration agreement, refer it to arbitration.

No Full De Novo Review of the Merits

The approach of a very large majority of the arbitration circles is against a review of the merits of the dispute decided by an arbitral tribunal. According to this approach, the award is off limits, and interventions as to breaches of due process and of public policy are just an inevitable exception and a sort of disease which must be tolerated. There is then a hostile approach to a review of the merits. This view has been confirmed in *Bridas*,[48] which has held that the raison d'être of arbitration consists in the merits of the dispute not being re-examined by State courts. This approach is not at all convincing.

Arbitrators are (normally!) human beings and as such they are exposed to making mistakes. It is a human and general reaction that such mistakes must be remedied. The raison d'être of arbitration is to provide a better service than Court proceedings or at least to be a useful alternative to them. Now, since in Court proceedings, a full review of the merits of the case is the practice, if in arbitration, it is excluded one is entitled to wonder whether arbitration is better or worse that Court proceedings.

A full de novo review of the merits should then be seen not as an obstacle to arbitration, but an important remedy to a serious weakness of the arbitral proceedings as to the merits. Those who are not totally against this remark might look for an alternative to intra-arbitration appeals, consisting in granting that authority

---

47 Art. I

48 *International Standard Electric Corp. v. Bridas Sociedad Anonima Petrolera* F. Supp. 72 (S.D.N.Y. 1990).

to State Courts. Exactly this alternative has given rise to a vivid debate between Ian Duncan Wallace, a top construction scholar[49] and Pierre Mayer, professor of conflicts at the University of Paris.[50]

Ian Duncan Wallace was in favor of a wider review of the award by State Court, the award and Court proceedings being, in his construction lawyer's solid language, "an essential building block." Pierre Mayer has objected that if arbitration becomes a prelude to a full review by State courts, this "would strike at the heart the very concept of arbitral process" and that the parties would be "betrayed." However, he has added that he was no objection to a review of the merits of the award by a second instance arbitral tribunal: "it would be perfectly possible not organize a system of arbitral appeals ... which would have the same additional assurance of quality of justice that is sought by creating appellate jurisdictions."

On the background of this learned discussion, a common concern seems to be shared by them, which is expressed by Wallace as follows:

> It is surely common sense to recognize that the further from outside control any person, whether arbitrator or not, believe himself to be, the greater, the likelihood of unjudicial or unfair attitude or conduct emerging.

Common sense, a good English quality, strengthens then the earlier remark that the introduction of a second instance arbitration is in line with the raison d'être of arbitration.

De Novo Review Around the World

The Washington Convention 1965

A review of the award by a new arbitral tribunal is provided for by the Washington Convention 1965,[51] which in a way has a role as to investment arbitration which reminds the New York Convention 1985 as to commercial arbitration. However, its second appellate arbitral tribunal, "the ad hoc Committee,"[52] has authority only to annul the award if the first instance arbitral tribunal has not been properly constituted or has manifestly exceeded its powers or there was corruption as to a member of that tribunal, or there has been a serious departure from a fundamental rule of procedure or no reason has been provided. Such specific

49 Ian Duncan Wallace, *Control by the Court: A Plea for More, Not Less*, 6 Arb. Int. 5, 253.

50 Pierre Mayer, *Seeking the Middle Ground of Court Control. A reply to I.N. Duncan Wallace*, 7, Arb. int. 4, 311.

51 Convention on the Settlement of International Disputes between States and Nationals of Other States, Washington March 18, 1965.

52 Art.52.

ground for the annulment of the first-instance award are consequently not a full de novo review of the merits. Furthermore, the second appellate arbitral tribunal must limits itself to set aside the first award.

Commodities Arbitrations

Commodities arbitration provide, under their standard provisions, a full review of the merits which generally focuses on decisions as to the quality of goods and on their delivery.

The European Court of Arbitration

This arbitration center[53] has introduced in 1997 intra-arbitration appellate proceedings, which hereafter will be dealt with more in detail.

The Arbitration Chamber of Paris

The rules of the Arbitration Chamber of Paris[54] have introduced in 2002 appellate arbitration proceedings.

AAA and CPR

The CPR (the International Institute for Conflicts Prevention and Resolution) has since some years also its Arbitration Appeal Procedure for CPR arbitration.

Even the American Arbitration Association has adopted[55] optional Appellate Arbitration Rules.

WTO

The World Trade Organization (WTO) also provides[56] for an appellate body which was established in 1995, which hears appeals from reports submitted by a WTO panel.

Documentary Evidence

Even as to documentary evidence, as to which in common law jurisdictions disclosure orders are known to be frequently granted and sometimes are invasive, in some civil law jurisdictions a party may only apply for an order to another party to produce a document. However, to obtain a similar order, it is generally needed to

---

53 Established in Strasbourg in 1959, as a legal entity under the patronage of many official entities among which the Council of Europe.

54 Art. 17 Arbitration Chamber of Paris Rules of Procedure (in force on January 5, 2002), Yearbook Commercial Arbitration 2003 at 305.

55 In 2023.

56 Understanding on Rules and Procedures Governing the Settlement of Disputes.

prove the existence of that document, that it is held by such other party and provide adequate information as to its content. All these requirements make it very difficult to achieve such an order and similar applications are in fact extremely rare in such jurisdictions. The result is that each party produces only the documents which support her case ad that the documents on which the Court will decide are only a small part of the documentary evidence while some main documents may remain concealed. This tendency remains in the background also of arbitral proceedings directed by an arbitrator formed by such national law.

How Appellate Arbitral Proceedings May Be Developed

A reference is made in this respect to the rules of the European Court of Arbitration (CEA).[57]

The Effects of a Challenge of the First-instance Award

If the first-instance award is challenged, the Rules provide[58] that the party that has been successful in the first instance undertakes not to enforce the first-instance award and agrees that the second-instance award will replace the first one.

Requirement to Appeal

Under these Rules,[59] the appeal, in the absence of a reasoned order by CEA, is admissible if at the same time a deposit is made to CEA of the principal amount, interest, and of the costs awarded by the first award or, if for any reason those figures are not available, of the amount determined by CEA. The purpose of the deposit is to ensure that the appellate award be enforced. The deposit may be replaced by an unconditional bond issued by a primary bank which operates in the same city of CEA's competent registrar. The contents of the unconditional bond must have been previously approved by CEA.

Deadline for the Appeal

The appeal may be filed within 40 days from the receipt of service of the first-instance award.

Full De Novo Review

The appeal involves a full review of the dispute by way of a rehearing, dealing in particular with the facts and the merits.

57 The last edition of which is dated 2022 and is effective from July 1, 2022.

58 Art. 28.2 CEA Arbitration Rules.

59 Art. 28.9 CEA Arbitration Rules.

Appointment of Three Arbitrators

All the three arbitrators that form the panel of the appellate proceedings are appointed by CEA.

Appellate Arbitration Rules

The appellate proceedings are governed by the arbitration rules which regulate the first instance proceedings and in addition – and if needed in replacement for them – by specific rules for appellate arbitration.

Term for the Appellate Award

The appellate award is to be filed within six months of the receipt of the case file by the arbitrators, or 9 months if evidence is taken at the appellate stage.

The Deposit Made for the Appeal

At the same time, the appellate award is filed by the arbitral tribunal CEA must[60]

> give instructions to the competent International Registrar and where appropriate to the guaranteeing bank, to return the fund deposited or to cancel the guarantee to it, or to cause the funds or guarantee to be returned, or to pay them immediately in part or in full to the party entitled to them under the appellate award and shall deliver the appellate award to the guaranteeing bank.

Arguments Raised Against Intra-arbitration Appellate Proceedings

Various arguments are used against intra-arbitration appellate proceedings, such as:

- the wish of many parties to have a one shot proceeding;
- a longer duration;
- higher costs; and
- the lack of certainty that the appellate tribunal will render a better decision than the first tribunal.

The winner's wish that the award be "one shot" is obvious but is strongly opposed by the expectations of the losing party to have a review of such award.

It is submitted that to deprive the loser of this right, in favor of the one shot approach to arbitration, is not only unfair but places – as earlier discussed – arbitration at a level lower rather than higher than court proceedings.

60 Art. 28 (9) CEA Arbitration Rules.

Even the objection that appellate arbitral proceedings would be longer is not convincing. In fact, by adding to the nine months provided for the first-instance award the six months for the second-instance award it makes a total of one year and there months. This time period is definitely shorter than the majority of commercial arbitrations. If one compares then this one year and three months' time period with the length of many arbitrations, the result is positive for a second degree of arbitration. The various steps of an investment arbitration have given rise in *Klöckner*[61] to nine years. Likewise in *Amco Asia*,[62] the various first-instance proceedings challenged before the Ad Hoc Committee were sent back from it to first instance proceedings, redecided from it, then back again to the Ad Hoc Committee and have lasted 10 years (due to proceedings being challenged before the Ad Hoc Committee, new proceedings being heard and then challenged again to the ad hoc Committee and being a second time set aside by the Ad Hoc Committee).

In order to consider the objection of higher costs, the calculation must start from the number of arbitrators under CEA arbitration rules (one arbitrator in the first instance, three in the second instance). The difference is just between three arbitrators in general in commercial arbitrations, and only one in CEA's first instance proceedings and three in the second one. A difference that is likely to further reduce since fees in the appellate proceedings are lower than for a full proceeding in the first instance.

This question applies to arbitration as well as to Court proceedings. If the higher number of members of the tribunal (3) in the second instance *vis-a-vis* 1 in the first instance is not sufficient, there is and there cannot be – in any proceedings – a certainty that the "higher" body decides in a better way.

Enforcement of an Award Set Aside in the State of Origin

A large debate has taken place as to granting in a State of the enforcement to an award which has been set aside in its State of origin. There is a tendency not to grant recognition and/or enforcement to an award that has been set aside in the State of origin on the grounds that its setting aside excludes the basic requirements for enforcing it in other States.

> The New York Convention provides[63] that recognition and enforcement of an award may be refused if .... "(e) the award has not yet become binding on the parties or has been set aside or suspended by a competent authority of the country in which, or under the law of which, that award was made."

---

61 *Klöckner Industrie-Anlagen GmbH et al v. United Republic of the Cameroon at al.* Icsid arbitration (from 1985 to 1994) ICSID Report 95, 122 [1994].

62 *Amco Asia Corp. V. Republic of Indonesia*, ICSID Arbitration, International Legal Materials, 365 [1985].

63 Art. V(1).

However, French Courts have taken an opposite view such as in *Putrabali*[64] holding that the issue is to be decided under French law and that an award set aside in another State would not prevent recognizing it, since international awards are governed by international public policy, and not by a national legal system. In *Hilmarton*[65] such basic attitude has been fortified by the remark that the application to recognize in France an award, which had been set aside in its country of origin (Switzerland), had not been made under the New York Convention and then just French law had to be applied.

If one goes back in general to the New York Convention, the term "*may be refused*" has been construed literally by various commentators who have drawn from this the conclusion that under the New York Convention the award had not to be refused cognition and simply might be refused. It is strongly objected to this that the New York Convention uses the term "may" as to all the grounds for the refusal of enforcement of an award (including the incapacity of the parties, the respondent having not been allowed to present his case and even when recognition would be contrary to the public policy of that State). If a literal construction of the term "may" would be proper, the refusal of recognition as to all its grounds would become merely optional, what seems to be exactly the opposite of the purpose of uniformity of the New York Convention.

The Proper Law of the Arbitration Agreement

As earlier discussed, the arbitration agreement is not only autonomous *vis-a-vis* the basic contract, but has occasionally been referred to as independent from the basic contract also as to its proper law. Two views oppose themselves on this aspect. According to one of them, the proper law of the arbitration agreement is the *lex contractus* (*i.e.*, the law which governs the basic contact). According to the opposite view, it is the law of the seat of the arbitration. The Court of Appeal of England and Wales has clearly described in *Sulamerica*[66] the various steps to be following to establish which law, "by introducing a three stages enquiry into (i) express choice (ii) implied choice and (iii) . . . closest and more real connection" in its more recent decision on *Kabab*.[67] The English Supreme Court has held that, in the absence of any choice of that law, the law that governs the contract "will generally apply to the arbitration agreement which forms part of the contract."

---

64 *P. T. Putrabali Adyamulia v. Société Rena Holding Company et Société Moguntia Est Epicos*, Court of Cassation (France) June 29, 2007 Bulletin 2007, I, N° 250.

65 *Hilmarton Ltd v. Omnium de Traitement et de Valorisation S.A., Court of Cassation (France) March 23, 1994, Yearbook Commercial Arbitration 1995, 663.*

66 *Sulamerica Cia Nacional de Seguros S.A. v. Enesa Engenharia S.A.*, Court of Appeal of England and Wales [2012] March 20, 2012 EWCA Civ 531.

67 *Kabab – Ji SAL (lebanono) v. KouT Food Group*, (Kuwait) [2012] UK SC48.

Conservatory measures, the grounds to set aside an award and an emergency arbitration, are among other common rules of commercial arbitrations.

Anti-suit and Anti-arbitration Injunctions

Anti-suit Injunctions in the United Kingdom

What Are Anti-suit Injunctions for?

Under English law, an anti-suit injunction is a court order that prohibits a party from starting or continuing legal proceedings in any jurisdiction or forum other than the one agreed upon in the contract. For instance, if a contract includes an arbitration clause but one party files a lawsuit in a national court in a jurisdiction outside of the seat of the arbitration the other party may apply for an anti-suit injunction from the courts of the seat of the arbitration.

An anti-suit injunction, if granted, typically targets the party attempting to initiate (or having initiated) proceedings in violation of an arbitration or jurisdiction agreement. Violating the order in the injunction may amount to contempt of court and may lead to serious legal consequences, including fines, asset seizure, or even imprisonment.

The granting of an anti-suit injunction is based on the binding effect of the contract. In other words, a breach of an exclusive jurisdiction clause constitutes a breach of contract, which must be addressed and sanctioned. In the *Angelic Grace* case,[68] Lord Millet explained:

> There is no good reason for diffidence in granting an injunction to restrain foreign proceedings on the clear and simple ground that the defendant has promised not to bring them.

The English courts have jurisdiction to grant an injunction in support of arbitration proceedings under Section 37(1) of the Senior Courts Act 1981, which provides that "the High Court may by order (whether interlocutory or final) grant an injunction or appoint a receiver in all cases in which it appears to the court to be just and convenient to do so." For instance, in *Axis Corporate Capital UK Ltd & Ors v. Absa Group Ltd & Ors*,[69] the court granted an anti-suit injunction against a defendant which initiated proceedings in South Africa in breach of an exclusive

68 Aggeliki Charis Compania Maritima SA v. Pagnan SpA (The Angelic Grace) [1995] 1 Lloyd's rep 87.

69 *Axis CorporateCapital UK Ltd & Ors v. Absa Group Ltd & Ors* [2021] EWHC 225 (Comm) (02 February 2021).

English jurisdiction agreement. Mr. J. Calver ruled that the jurisdiction agreement was established with "a high degree of probability" and therefore that "the court should grant an interim anti-suit injunction to restrain the South African proceedings or any other brought in breach of those exclusive jurisdiction agreements in the reinsurance contracts."

Standard of Proof

To grant an anti-suit injunction, the English court must be satisfied that an arbitration agreement exists and has been breached by the initiation of foreign court proceedings. This requires confirming that the party bringing the court proceedings is bound by the arbitration agreement, that the dispute and the claim in the foreign proceedings fall within the scope of the agreement, and that the claims made are arbitrable.

The standard of proof required to establish that an arbitration agreement exists and has been breached by the initiation of foreign court proceedings is a "high degree of probability." This ensures that anti-suit injunctions are not granted in cases where the existence of the arbitration agreement is too controversial, as confirmed by Mr. J. Christopher Clarke in *Transfield Shipping Inc v. Chiping Xinfa Huayu Alumina Co Ltd*[70]:

> The appropriate test is whether or not the applicant has shown on the material adduced at the interlocutory hearing a high degree of probability that there was such an [arbitration] agreement. It is one thing to enforce a clear agreement to arbitrate or one which on an interlocutory basis can be seen to be highly likely to be established. It is another to restrain a party from litigating in a foreign country where the position is less clear than that.

The "high degree of probability" threshold is also justified because an anti-suit injunction can have significant consequences for foreign jurisdictions. Although these injunctions target the parties rather than the foreign courts, they can still disrupt or even interfere with foreign court proceedings. In this regard, Mr. Simon Colton KC explained in *Marsh Ltd v. Greensill Bank AG & Anor* that "an interim ASI does not merely interfere with the private rights of the respondent, but may interfere, albeit indirectly, with the working or output of a foreign court, raising issues of comity."[71] As a result, in *Rochester Resources Ltd and others v. Lebedev*

70 *TransfieldShipping Inc v. Chiping Xinfa Huayu Alumina Co Ltd* [2009] EWHC 3629 (QB) (22 December 2009).

71 *Marsh Ltd v. Greensill Bank AG & Anor* [2024] EWHC 3068 (Comm) (03 December 2024).

and another,[72] the court refused to grant an anti-suit injunction as it was not satisfied that there was a high degree of probability that all the parties were bound by the relevant arbitration agreement.

In this case, a claimant initiated judicial proceedings in the Supreme Court of New York against two defendants based on an investment agreement. The defendants sought an anti-suit injunction on the basis of an arbitration agreement in a related acquisition agreement between two separate companies. Although neither the claimant nor the defendants were parties to the arbitration agreement, the defendants argued that, as the ultimate beneficiaries of the companies, they were bound by the arbitration agreement. However, the court concluded that it was not established with a sufficient degree of probability that the arbitration agreement was binding on all parties:

> The decision I have to make is whether it has been established to a high degree of probability that [the claimant] is bound by the arbitration clause [...] The parties have made it clear that the arbitration clause covers disputes "arising between the Parties." Had they not included these words, it might have been easier to conclude that the arbitration clause was intended to apply also to Affiliates [...] I conclude that [the claimant] is probably not bound by the arbitration clause, and that it has certainly not been established to a high degree of probability that he is bound.

Therefore, the application for an anti-suit injunction was rejected.

Once the applicant demonstrates a "high degree of probability" that a valid arbitration agreement exists and covers the disputes in question, the English Court will proceed to grant an anti-suit injunction, unless there are strong reasons not to do so.[73] A strong reason could be that the court is not convinced that the other party has either actually breached or is threatening to breach the arbitration agreement.[74]

When it is not established that the initiation of foreign court proceedings breaches the arbitration agreement, the English courts may nevertheless grant an anti-suit injunction on the basis that such proceedings are vexatious and oppressive. This type of remedy may be appropriate, for instance, when foreign proceedings are initiated by a third party, which is not a party to the arbitration agreement and which has colluded with the defendant to obstruct the arbitration.

---

72 *Rochester ResourcesLtd & Ors v. Lebedev & Anor* [2014] EWHC 2926 (Comm) (09 September 2014).

73 *Axis CorporateCapital UK Ltd & Ors v. Absa Group Ltd & Ors* [2021] EWHC 225 (Comm) (02 February 2021) at [30].

74 *Navigation Maritime Bulgare v. Rustal Trading* (The Ivan Zagubanski) [2002] 1 Lloyds Rep 106 at [112].

For instance, in *BNP Paribas SA v. Open Joint Stock Company Russian Machines & Anor*, Mr. Justice Blair stated as follows[75]:

> In order for the claimant to succeed on its claim on the merits it either must make out that the second defendant has acted in breach of and/or non compliance with the Arbitration Agreement, and/or that the second defendant has acted vexatiously, oppressively or unconscionably so that the ends of justice require that the claimant be granted the relief it seeks [...] I accept the claimant's contention that there is sufficient material to justify drawing the inference that the Russian proceedings are brought with a view to impeding the outcome of the arbitration.

Accordingly, the court granted an anti-suit injunction even though the party that initiated the foreign court proceedings was not a party to the arbitration agreement.

The Applicant Must Act Promptly

To prevent an anti-suit injunction from being used to exert illegitimate pressure on the opposing party, applicants must act promptly when seeking such injunctions, and, to the extent possible, before the foreign proceedings advance too far. In the Angelic Grace case, Lord Millet explained that the English Court should not hesitate to grant an anti-suit injunction "provided that it is sought promptly and before the foreign proceedings are too far advanced."[76]

Although there is no strict time limit for seeking an anti-suit injunction, the general position under English law is that a party should submit its application at the first available opportunity, unless there are valid reasons for delay. The longer a party waits before applying for such an injunction, the less likely the application is to succeed. In *Ecobank v. Tanoh*,[77] Lord Justice Christopher Clarke summarized the English Court's position as follows:

> Injunctive relief may be sought (a) before any foreign proceedings have begun; (b) once they have begun; (c) within a relatively short time afterwards; (d) when the pleadings are complete; (e) thereafter but before the trial starts; (f) in the course of the trial; (g) after judgment. The fact that at some stage the foreign court has ruled in favor of its own jurisdiction is not per se a bar to an anti-suit injunction: see AES. But, as each stage is reached, more will have

75 *BNP Paribas SA v. Open Joint Stock Company Russian Machines & Anor* [2011] EWHC 308 (Comm) (24 November 2011).

76 *Aggeliki Charis Compania Maritima SA v. Pagnan SpA (The Angelic Grace)* [1995] 1 Lloyd's rep 87.

77 *Ecobank Transnational Incorporated v. Tanoh* [2015] EWCA Civ 1309 (17 December 2015).

> been wasted by the abandonment of proceedings which compliance with an anti-suit injunction would bring about. That being so, the longer an action continues without any attempt to restrain it the less likely a court is to grant an injunction and considerations of comity have greater force.

In *Verity Shipping SA ("Owners") & Anor v. NVNorexa & Ors,*[78] Mr. Justice Teare explained that the requirement to act promptly in seeking an anti-suit injunction should be understood "in a common sense and straightforward manner." In this case, the Court denied the anti-suit injunction sought almost three years after the commencement of the proceedings before the foreign court.

On the other hand, in *Africa Finance Corporation & Ors v. Aiteo Eastern E & P Company Ltd,*[79] the High Court granted an anti-suit injunction sought thirteen months after the foreign court proceedings was commenced in Nigeria. In this case, the Court acknowledged that "[t]here was indeed a substantial delay of 13 months before relief was sought," but nonetheless granted the anti-suit injunction as "there was a reasonable explanation for that delay, namely, the existence of negotiations between the [parties] to restructure the Facility Agreements, assist the [defendant] to pay its debts, and to bring the Nigerian proceedings to an end."

In light of the above, applicants for anti-suit injunctions should bear in mind that, although an application for such relief may be made months (or even years) after foreign court proceedings were commenced in breach of the arbitration agreement, it should be made at the earliest opportunity to avoid the application being dismissed.

Anti-suit Injunctions and Arbitration

Many arbitration rules recognize the power of national courts to grant interim and conservatory measures, including anti-suit injunctions, in support of arbitration proceedings. For instance, Article 28 of the ICC Arbitration Rules 2021 provides that:

> Before the file is transmitted to the arbitral tribunal, and in appropriate circumstances even thereafter, the parties may apply to any competent judicial authority for interim or conservatory measures. The application of a party to a judicial authority for such measures or for the implementation of any such measures ordered by an arbitral tribunal shall not be deemed to

78 *Verity Shipping SA ("Owners") & Anor v. NVNorexa & Ors* [2008] EWHC 213 (Comm) (13 February 2008).

79 *Africa Finance Corporation & Ors v. Aiteo Eastern E & P Company Ltd* [2022] EWHC 768 (Comm) (01 April 2022).

> be an infringement or a waiver of the arbitration agreement and shall not affect the relevant powers reserved to the arbitral tribunal.

Similarly, Article 25.3 of the LCIA Arbitration Rules 2020 provides:

> A party may apply to a competent state court or other legal authority for interim or conservatory measures that the Arbitral Tribunal would have power to order under Article 25.1: (i) before the formation of the Arbitral Tribunal; and (ii) after the formation of the Arbitral Tribunal, in exceptional cases and with the Arbitral Tribunal's authorization, until the final award.

This allows parties in arbitration proceedings to approach national courts for anti-suit injunctions if national court proceedings are initiated in breach of arbitration agreements.

In the *Enka v. Chubb* case,[80] the UK Supreme Court affirmed the authority of the English courts to grant anti-suit injunctions in support of arbitrations seated in England and Wales, specifically to restrain a party from initiating court proceedings that breach an arbitration agreement specifying England and Wales as the seat of the arbitration. In this case, the defendant commenced judicial proceedings in Russia in breach of an arbitration clause providing that the seat of the arbitration was London, England. The Supreme Court confirmed the powers of English Courts to grant anti-suit injunctions to preserve England-seated arbitration as follows:

> By choosing a seat of arbitration the parties are choosing to submit themselves to the supervisory and supporting jurisdiction of the courts of that seat over the arbitration. A well established and well recognized feature of the supervisory and supporting jurisdiction of the English courts is the grant of injunctive relief to restrain a party from breaching its obligations under the arbitration agreement. A promise to arbitrate is also a promise not to litigate.

Additionally, English courts are not only empowered to grant anti-suit injunctions to preserve arbitration proceedings, but they also hold exclusive supervisory jurisdiction over arbitrations seated in England and Wales. Any attempt to seek a determination from another court constitutes a breach of contract and will be prevented with an injunction. This was for instance confirmed by Mr. J. Cooke in *C v. D*:[81]

80 *Enka Insaat Ve Sanayi AS v. OOO Insurance Company Chubb* [2020] UKSC 38.

81 *C v. D* [2007] EWHC1541 (Comm) (28 June 2007).

> The effect of them is that the agreement as to the seat of an arbitration is akin to agreement to an exclusive jurisdiction clause. Not only is there agreement to the arbitration itself but also to the courts of the seat having supervisory jurisdiction over that arbitration. By agreeing to the seat, the parties agree that any challenge to an interim or final award is to be made only in the courts of the place designated as the seat of the arbitration. This was confirmed by the England and Wales Court of Appeal, which indicated in Star Hydro Power Ltd v National Transmission and Despatch Company Ltd ([2025] EWCA Civ 928) that in relation to arbitration seated in England and Wales, pre-emptive challenges to arbitral awards cannot be initiated in foreign jurisdictions on the basis of the New York Convention 1958,. Challenges may be filed in England and Wales on the basis of the Arbitration Act 1996 (as recently amended by the Arbitration Act 2025).

In this regard, a choice of law for the arbitration agreement that differs from the seat of the arbitration is not, in itself, sufficient to negate the English courts' power to grant an anti-suit injunction in support of an arbitration seated in England and Wales. In *Enka v. Chubb*,[82] this was confirmed by Lord Hamblen and Lord Leggatt:

> The principles governing the grant of an anti-suit injunction in support of an arbitration agreement with an English seat do not differ according to whether the arbitration agreement is governed by English law or foreign law. Forum conveniens considerations are irrelevant and comity has little if any role to play. The court's concern will be to uphold the parties' bargain, absent strong reason to the contrary, and the court's readiness to do so is itself an important reason for choosing an English seat of arbitration.

Accordingly, for arbitrations seated in England and Wales, any attempt to involve the supervisory jurisdiction of another court constitutes a breach of contract and will be restrained by injunction.

The exclusive supervisory jurisdiction also extends to postarbitration proceedings, such as challenges to arbitral awards. Therefore, for arbitrations seated in England and Wales, English courts have exclusive jurisdiction to hear any challenge to the arbitral award. This was for instance confirmed in *Nigerian Agip Exploration Ltd v. GEC Petroleum Development Co Ltd.*[83] In this case, two Nigerian companies concluded various contracts containing arbitration clauses. The seat of the arbitration was London, England, and the governing law was Nigerian law.

82 *Enka Insaat Ve Sanayi AS v. OOO Insurance Company Chubb* (Rev1) [2020] UKSC 38 (09 October 2020).

83 *Nigerian Agip Exploration Ltd v. GEC Petroleum Development Co Ltd* [2023] EWHC 414 (Comm) (17 February 2023).

The arbitral tribunal issued an award, which was then challenged in the Nigerian courts by the losing party. The other party applied for an anti-suit injunction in England and Wales to discontinue the Nigerian proceedings. The High Court confirmed that these proceedings were initiated in breach of the arbitration agreements, as the English courts had exclusive jurisdiction to hear any challenge to the award:

> proceedings seeking to set aside an arbitration award resulting from arbitration proceedings with a London seat that were properly constituted is a matter exclusively within the jurisdiction of the English Court as the supervisory court of the arbitration, and it is as much a breach of the obligations arising out of the arbitration agreement to seek such relief from (in this case) the Nigerian Court as it is or would be to seek to place before the Nigerian Court (or any court) substantive claims, which, by the arbitration agreement, ought to have been referred to the arbitrators.

The question also arose as to whether English courts have jurisdiction to grant anti-suit injunctions in support of arbitrations not seated in England and Wales. In *UniCredit Bank GmbH v. RusChemAlliance LLC*,[84] an applicant sought an anti-suit injunction against a defendant that initiated court proceedings in Russia in breach of the applicable arbitration clause. The underlying contract was governed by English law and the seat of the arbitration was Paris, France. In this case, the Supreme Court, referring to the *Kabab* and *Enka* cases, confirmed that the arbitration agreement was governed by the law with the closest connection, which was English law as the law applicable to the contract, rather than French law as the law of the seat of the arbitration:

> The closest connection test applies when a question about the validity or scope of the arbitration agreement arises, as it did in Enka, before an award has been made [...] [The] argument that the arbitration agreements in the bonds are governed by French law is therefore the correct answer. It is as clear in this case as it was in Kabab Ji that, applying the rules of contractual interpretation of English law as the law of the forum, the parties have agreed that the arbitration agreements in the bonds are governed by English law.

The Supreme Court also assessed whether English courts had jurisdiction to issue an anti-suit injunction on the basis that the arbitration agreement was governed by English law, even though the seat of the arbitration was in France. In this case, the Supreme Court was satisfied that English Courts had jurisdiction to grant

84 *UniCredit Bank GmbH v. RusChemAlliance LLC* [2024] UKSC 30 (18 September 2024).

an anti-suit injunction on the ground that the contract – and therefore the arbitration clause – was governed by English law: "There is a substantial connection with England and Wales in the fact that the contractual rights which UniCredit is asking the court to enforce are rights governed by English law." It therefore appears that English courts can grant anti-suit injunctions in relation to contracts governed by English law, even if the seat of the arbitration is not in England.

However, the Arbitration Act 2025, which entered into force on 1 August 2025 amended the earlier Arbitration Act 1996 and overturned the position by introducing a new default rule whereby the law governing the arbitration agreement is the law of the seat of the arbitration (see section 6A of the Arbitration Act 2025). Therefore, it is no longer clear whether English courts would be willing to grant anti-suit injunctions in relation to arbitration proceedings where the seat of the arbitration is outside England and Wales, even though the contract is governed by English law.

Anti-suit Injunctions in France

France has traditionally been less inclined to grant anti-suit injunctions compared to many Common Law countries. French courts typically do not recognize or enforce anti-suit injunctions granted by foreign courts, reflecting a more restrictive approach to judicial intervention in cross-border disputes. For instance, in a 2004 decision, the French Court of Cassation ruled, in an *obiter dictum*, that anti-suit injunctions are contrary to French public policy in that they affect the jurisdiction of the State where the claim is brought.

However, in a 2009 decision,[85] the French Court of Cassation ruled that a foreign anti-suit injunction can be recognized in France. In this case, an American company concluded a contract with a French company including an exclusive jurisdiction clause referring to the State of Georgia, United States. As the French company commenced proceedings before the French Court, the American Company obtained an anti-suit injunction from the Georgian Courts. The French Court of Cassation confirmed that the Georgian anti-suit injunction was enforceable in France:

> Given that the jurisdiction clause was freely accepted by the parties, no fraud could result from the American company's referral to the court expressly designated as competent. Furthermore, based on its own reasoning and adopted grounds, there can be no denial of access to a judge, as the decision made by the Georgian judge specifically aims to rule on its own jurisdiction and ultimately to uphold the jurisdiction clause agreed upon by the

85 Cass. Civ. 1, 14 October 2009, 08-16.369 08-16.549.

> parties. Accordingly, an "anti-suit injunction" is not contrary to international public policy when, outside the scope of conventions or EU law, its sole purpose – as in this case – is to sanction the breach of a pre-existing contractual obligation.

In 2020, the Court of Appeal of Paris took a further step by confirming the first-instance judgment that granted an anti-anti-suit injunction (that is, an order against an anti-suit injunction), on the basis of Article 809(1) (now 835(1)) of the French Code of Civil Procedure.[86] In this case, a dispute arose between IPCom GmbH & Co. KG, a German company specializing in patent licensing, and the Lenovo Group, including its US and French subsidiaries. IPCom, which owns a portfolio of patents, accused Lenovo of infringing European Patent regulations. After failed negotiations over a FRAND (Fair, Reasonable, and Non-Discriminatory) license, Lenovo sought to pre-empt enforcement actions by filing an anti-suit injunction in California, aiming to prevent IPCom from pursuing parallel infringement proceedings in the UK and potentially in France.

IPCom reacted by filing emergency proceedings in France, seeking an order to prevent Lenovo from restricting its right to litigate. On 8 November 2019, the President of the Paris Court of First Instance ruled in favor of IPCom, ordering Lenovo to withdraw its anti-suit injunction and prohibiting it from filing similar applications in foreign jurisdictions. Lenovo appealed, arguing that French courts lacked jurisdiction and that its motion in California was justified to ensure global FRAND compliance. However, the Paris Court of Appeal upheld the anti-anti-suit injunction:

> It shall be held, regardless of the assessment whether the motion for anti-suit injunction filed by Lenovo companies before the Californian judge complies with the French concept of international public policy, that this measure characterizes a manifestly unlawful disturbance within the meaning of Article 835 of the Code of Civil Procedure, in view of the disturbance it generates by infringing a fundamental right, so that the first judge has, rightly and within the framework of the powers it derives from this article, ordered the Lenovo companies to withdraw under penalty the disputed motion for anti-trial injunction and that its order shall therefore be upheld.

It remains unclear whether this position should be interpreted as being limited to the granting of anti-anti-suit injunctions, or if, in the future, French courts will issue anti-suit injunctions to support French court proceedings and arbitration.

86 Court of Appeal of Paris, 3 March 2020, 19/21426.

The potential for such a shift in approach remains uncertain, leaving questions about the broader scope of French judicial intervention in international disputes.

Peculiarities of Construction Commercial Arbitration

In addition to the peculiarities of arbitration *vis-a-vis* court proceedings, construction commercial arbitration has peculiarities vis a vis commercial arbitration.

A Proliferation of Claims

The most important one is that disputes are frequently made by a large number of claims and counterclaims. Handling all of them at the same times makes the conduct of the proceedings heavy and may prevent to have a clear picture of each of them.

The ideal solution could be to deal with each claim separately from the other one, but this complicates the proceedings, if, for example, witnesses have to be heard separately as to each of sum claims or counterclaims.

Plurality of Involved Parties

The involvement of several parties, for example various contractors, possibly the Engineer and or the Designer, is a further peculiarity and difficulty. The role of the Engineer and of the Designer, even if they are not parties to a construction contract, creates a complicated background. While subcontractors are generally not parties to an arbitration between the Owner and the Contractor, they may be the object of a separate arbitration. The frequent flow down of provisions from the main contract to the subcontract creates a further difficulty in handling such proceedings. Eventually the technical aspect frequently dominates construction arbitration and requires either an official expert or hearing expert witnesses as to various areas with which frequently the arbitrator is not familiar. Wishing to fully understand the testimony of Engineer and or to the Designer, who frequently are not parties to the proceedings and are called as witnesses, is a further delicate aspect of construction arbitrations. All this tends to give rise to a very large production of documents, several of which are frequently not relevant but have to be read. The quantity of these peculiarities and the frequent coexistence of several of them frequently gives rise to more difficulties than in ordinary commercial arbitrations.

Continuous Search for New Horizons

As earlier suggested arbitration – as the majority of human things – is not perfect and one should feel the need to continuously search new horizons. It is suggested that the search must start from identifying the areas of arbitration which may be improved. They include:

- clear rule as to the taking of evidence;
- duration;
- costs;
- a full de novo review; and
- making certified arbitrators at the disposal of the parties through training and the Arbitrator's Pledge to the Parties.

Duration

As is well-known, one of the parties would generally wish to have the award in few days' time while the opposite party frequently tries to postpone the decision as much as possible. The arbitral rules of various arbitrations have provided for short periods, but frequently they allow many extensions of such deadline. It is suggested that, as a rule, arbitration should last no more than one year. This will require the arbitrator being firm and not allowing unnecessary extensions.

Costs

The costs of arbitration proceedings have become quite high. This may be to the detriment of a party with limited resources, which on some occasions has to waive to have access to arbitral justice and in some cases has even been excluded from the proceedings because it was unable to pay the requested advances or even just the "*cautio pro expensis.*"

No Full De Novo Review of the Case

It is suggested that a full de novo review of the merits of the award is a real must if – as discussed – arbitration wants to keep its "raison d'être".

A Sole Arbitrator

The scheme that each party is entitled to appoint his/her arbitrator is indeed appealing.

Unfortunately, many appointments are being challenged, sometimes rightly – because the arbitrator was not at all independent and impartial. On other occasions, a bias has been just invented in order to try to get rid of an arbitrator who looked not favorable to the applicant. Furthermore, the costs of three arbitrators is too high for small- and medium-sized arbitrations. On all these grounds, it is submitted that a sole arbitrator may in various cases be the proper solution.

Helping the Parties to Select the Right Horse for the Course

A review of the improvement, which arbitration needs in order to be able to perform its role at its best, concerns arbitrators. Rarely, a category of people like arbitrators is so heterogenous. In fact, it includes not only experienced counsel, academics, and retired judges but a great number of other persons with the most different backgrounds, which include colleagues of counsel for a party, relatives or friends of a party, people with no training in arbitration. Even an itinerant seller or a butcher may be appointed as arbitrator, in spite of a frequent total absence of training in arbitration.

Parties enjoy full freedom to choose their arbitrators and occasionally the appointment is made became the appointor relies on the friendship of the appointee. These appointers – it is suggested – have made their choice. The purpose is to help in their selection other people, those who earnestly look for a good arbitrator. Unfortunately not much information is available in order to make the best choice. Recommendations for other people are useful, but not sufficient. The examination of the CV of a candidate is also interesting but not decisive. It is suggested that more is needed and is available.

As it has been earlier suggested, it may be worth to inspire oneself to the important role played by the International Mediation Institute (IMI) as to mediators. Mediators certified by IMI are recognized as having a good standard.

Applying this to arbitration, one could envisage some steps. The first one might be to request arbitrators to issue an Arbitrator's Pledge panel to respond to standard queries listed in the attached questionnaire, which have the purpose to make known to everyone the position which an arbitrator has on very general legal issues. This allows to the parties to have a better notion of that arbitrator thanks to his/her views on basic issues, which are not related to any specific litigation. These very general issues include, for example, the arbitrator's recognition of the rights of the parties to prove their case, not limiting the number of witnesses, and their examination and cross-examination by Counsel (except as to matters which are not relevant or not appropriate), like requests to the parties to enter stipulations as to procedures, time, and costs, the commitment to devote to the proceedings all the time their require, as well as the commitment to be diligent and expeditions.

The Arbitrator's Pledge[87] and the possible Questionnaire by the Parties to the arbitrator are quoted below.[88]

---

87 Drafted by the European Court of Arbitration (Strasbourg).

88 Mauro Rubino-Sammartano, *A second (Quasi Perfect!) Storm also in Arbitration*, Journal of International Arbitration, Vol. 34, 2017, at 924.

European Court of Arbitration

Arbitrator's Pledge to the Parties

I ________________, ________________ [profession], born in ________________, resident at ________________, ________________ citizen, HEREBY UNDERTAKE toward any party which will appoint me as an arbitrator that, when acting as an arbitrator, I shall:

- disclose any past, present or potential future professional or personal link of which I am aware with any of the parties, its stake-holders, managers, staff, any of its associated company and counsel and with my co-arbitrators;
- remain fully independent, impartial and neutral during such proceedings and for two years after the award;
- bear in mind at any time that my fundamental task is to render justice, in compliance with the applicable law and/or rules;
- request the parties not to raise useless or totally ungrounded oppositions and motions, and not to behave in a delaying or obstructive way and I shall sanction such possible behavior, such as in the costs award;
- send to the parties the attached list of questions in order that they express their expectations as to the conduct of the proceedings;
- invite the parties to enter into a stipulation covering procedure, time and costs;
- conduct the proceedings very diligently and with all due expedition and deliver the award (except in very extraordinary situations, which I will state and certify), within nine months after my appointment;
- ensure that the right of the parties:
  - to present their case,
  - to call the relevant oral evidence,
  - to examine and cross-examine the witnesses,
  - as well as in respect of all relevant documentary evidence (the production of useless documents being subject to sanctions in terms of costs), be fully respected;
- devote to the proceedings all the time which is necessary;
- tax fees of counsel for the victorious party (or of each party, if each of them will be ordered to bear its costs), up to a maximum of ___ times (and, in very special circumstances to be stated and certified by myself, up to ___ times), the fees, which the parties have agreed

or are due as to such proceedings to myself as sole arbitrator or as a member of the Arbitral Tribunal;

- discharge personally my task as arbitrator;
- treat in strict confidence the proceedings and anything which is said or produced in such proceedings, as well as the award;
- respond to the Questionnaire as per attachment.

____

Date and place by____________________

Encl: Questionnaire
Questionnaire to the Arbitrator

1. **Priority of truth funding**
   Do you agree that truth funding is a priority in any proceedings?

   Yes ☐ No ☐

2. **Right of the parties to present and to prove their case**
   Do you agree that the parties are entitled to present and prove their case?

   Yes ☐ No ☐

3. **Case management**
   Do you agree that a calendar of the proceedings shall be prepared at the beginning of the proceedings and regularly reviewed and updated?

   Yes ☐ No ☐

4. **Duration of the proceedings**
   Do you agree that the proceedings must be dealt with expeditiously and that unless there are special reasons (to be stated) their ideal duration is 9 (maximum 15) months?

   Yes ☐ No ☐

5. **Raising all possible evidentiary material**
   Do you believe that truth funding requires that all possible evidentiary material be considered, even if this takes some more time?

   Yes ☐ No ☐

6. **Is there a priority of documents over oral testimony?**
   Do you believe that oral testimony is always less important than documentary evidence?

   Yes ☐ No ☐

7. **Short time period between one step of the proceedings and the following one**
   Do you believe that there should be a short time period between one step and the next in the proceedings?

   Yes ☐ No ☐

8. **Short time period between the closing of the proceedings and publishing the award**
   Do you agree that, in the absence of very important reasons (to be promptly stated), the award should be published not later than 60 days after the closing of the proceedings?

   Yes ☐ No ☐

9. **Early decision on preliminary issues or on a part of the claim which is manifestly well grounded**
   Do you agree that preliminary matters and issues which are manifestly well-grounded should be decided as early as possible?

   Yes ☐ No ☐

10. **Settlement facilitation**
    Do you believe that the arbitrators should go beyond inviting the parties to settle?

    Yes ☐ No ☐

11. **Witnesses**
    Do you agree that the selection and the examination of witnesses, and the cross-examination of the witnesses of the other parties, is a right of each party and that the role of the Arbitral Tribunal is to ensure that this takes place properly?

    Yes ☐ No ☐

12. **Witness preparation/coaching witness statements**
Do you believe that preparation/coaching of witnesses and witness statements should be authorized even if one of the parties object to it?

Yes ☐ No ☐

13. **Limiting witnesses' evidence**
Do you agree that the list of witnesses should not be reduced *a priori*, but rather by reserving the option to hear further witnesses on an issue if this has not been sufficiently dealt with by prior witnesses?

Yes ☐ No ☐

14. **Documents full disclosure**
Do you believe it is appropriate to order unlimited disclosure of documents or very wide disclosure of entire classes of documents?

Yes ☐ No ☐

15. **Experts nominations**
Do you believe that the Arbitral Tribunal should consult with the parties before appointing experts and putting questions to them?

Yes ☐ No ☐

16. **Adjournment of evidentiary hearing**
Do you agree that an evidentiary hearing should be adjourned (if the term for filing the award allows it) if a witness is not available (unless he/she is not necessary)?

Yes ☐ No ☐

17. **Parties' depositions**
Do you agree that the parties should be heard as witnesses (subject to taking account of their interest in the dispute)?

Yes ☐ No ☐

18. **Delegation**
Do you agree that no decision and no material research on a decisive issue and no drafting of a part of the award should

be delegated to the secretary to the Arbitral Tribunal or to any assistants?

Yes ☐ No ☐

19. Deliberations

Do you agree that the case must be decided after a full exchange of views among arbitrators, without any prior conclusion or drafting by any of them?

Yes ☐ No ☐

20. Dissenting opinion

Would you be disturbed by a dissenting opinion?

Yes ☐ No ☐

21. Costs

Do you believe it is appropriate for a party to recover all its (reasonable) costs, even if it has succeeded only on a part of its claim or only on some of its claims?

Yes ☐ No ☐

22. Costs

Do you believe that a party is entitled to receive, in case of success, the fees of various law firms which have been involved, even if their work was not absolutely necessary, or of a number of professionals beyond what was necessary for presenting and proving its case?

Yes ☐ No ☐

23. Costs

Do you believe that a party is entitled to be paid the contingent fees, which it has undertaken to pay in favor of its counsel, even when they exceed reasonable fees computed on another basis?

Yes ☐ No ☐

24. Questions to the parties

I enclose a questionnaire to the parties

________________ ________________

(date and place) (name)

Questionnaire

To the parties to arbitral proceedings administered by the European Court of Arbitration enclosed to the CEA Arbitrator's Pledge between __________________ Claimant and __________________ Respondent

1. Do you agree not to contact, and to cause your counsel not to contact, your nominated arbitrator in person or in writing, except officially during the proceedings with copy to the other arbitrators and to opposing counsel?

   Yes ☐ No ☐

2. Do you agree to limit the number of pleadings to the Statement of the Case, the Answer and to a simultaneous After Evidentiary Pleading followed by a simultaneous Rebuttal?

   Yes ☐ No ☐

3. Would you prefer to address the Arbitral Tribunal orally, rather than exchanging a Rebuttal subsequent to the After Evidentiary Pleading?

   Yes ☐ No ☐

4. Do you agree to limit the number of pages of your first pleading to 30 pages and subsequently to refer very succinctly to your previous pleadings without repeating your arguments?

   Yes ☐ No ☐

5. Do you agree that some hearings be held in a place and country different from the venue of the arbitral proceedings and more convenient?

   Yes ☐ No ☐

6. Do you agree to hear by videoconference those witnesses who cannot travel to attend the hearing?

   Yes ☐ No ☐

7. Do you accept that witnesses be contacted and prepared for their testimony?

   Yes ☐ No ☐

8. Do you agree to keep strictly confidential (and to cause your counsel, advisors, managers, employees and agents to do so) the file, the facts, the documents, the evidence of these arbitral proceedings and the reasons of the award?

Yes ☐ No ☐

9. Do you agree to conduct, and to cause your counsel to conduct, these proceedings in a manner which limits the duration of the proceedings to 9 months, maximum 15 months and to avoid the production of documents and the calling of witnesses who are unnecessary or irrelevant, restraining motion practices, delays, vexatious or repetitive conduct, and in general any overlawyering; and do you accept to pay to the other parties all legal costs caused by a breach of such commitment, even in the case of final success in the dispute?

Yes ☐ No ☐

10. Do you undertake to promptly reimburse any other party which should itself pay, in the absence of your own payment, your share of any advance requested by the European Court of Arbitration or its local competent chapter (if any) for the proceedings; and to recognize that the other party shall be entitled to an ex parte summary judgment, or other summary proceedings, for such repayment?

Yes ☐ No ☐

11. Do you agree that the arbitrator shall issue as soon as possible an award for the part of a claim or of a cross-claim which is undisputed or manifestly well grounded?

Yes ☐ No ☐

12. Do you agree that the award be signed by the last arbitrator in the place which is the venue of the proceedings? If not, where would you like the award to be made?

Yes ☐ No ☐

The second step should be to have a universally, or nearly universally, accepted certification like the IMI one for mediators. In this respect there is the big advantage of the existence of the Chartered Institute of Arbitrator's training which certifies arbitrators, granting them a various status, the top of which are Fellows of the Institute and Chartered Arbitrator. In order to make this status easier to be understood by users, the status of Fellow of the Institute might become "certified arbitrator." The Chartered Institute has been invited to formalize – so to say – such certification. This should allow people, who have no earlier contacts with the Institute, to easily understand that they are in front of a certified arbitrator. This might be decisive in making absolutely clear to those who are looking to select an arbitrator that they have a precise choice in front of them: whether to appoint a certified arbitrator or someone else who may have other qualities to his/her eyes but is not certified. Certified arbitrators would be requested to accept a Code of Ethics.

Many Code of Ethics

A unified Code of Ethics might better achieve the intended result, avoiding having to compare the various existing Codes of Ethics. The difference between arbitration rules, codes of ethics and other practices are not necessarily in favor of promoting arbitration. A possible, even if less obvious conclusion, might be to put on its way a procedure which in the end could give rise to a profession of arbitrator, a profession which inter alia will avoid the unsatisfactory view and effects of the practice of double hatting. An arbitration target not easy to be achieved but which might be worth being pursued.

Index

Note: Page numbers in *italics* refers to figures and **bold** refers to tables.

Major Issues in International Construction Law, First Edition. Mauro Rubino-Sammartano and Troy Harris.
Published 2026 by John Wiley & Sons, Inc.

d